THIN FILM MATERIALS TECHNOLOGY

Sputtering of Compound Materials

THIN FILM MATERIALS TECHNOLOGY

Sputtering of Compound Materials

by

Kiyotaka Wasa
Yokohama City University
Yokohama, Japan

Makoto Kitabatake
Matsushita Electric Industrial Co., Ltd.
Kyoto, Japan

Hideaki Adachi
Matsushita Electric Industrial Co., Ltd.
Kyoto, Japan

William Andrew publishing

Springer

Cover photo: Chamber for sputtering amorphous thin film on substrate cooled by liquid nitrogen.

Library of Congress Catalog Card Number: 2003018016
ISBN: 0-8155-1483-2
Printed in the United States

Co-published by:

William Andrew, Inc.
13 Eaton Avenue
Norwich, NY 13815
1-800-932-7045
www.williamandrew.com
www.knovel.com
(Orders from all locations in North and South America)

Springer-Verlag GmbH & Co. KG
Tiergartenstrasse 17
D-69129 Heidelberg
Germany
springeronline.com
ISBN: 3-540-21118-7
(Orders from all locations outside North and South America)

10 9 8 7 6 5 4 3 2 1

This book may be purchased in quantity at discounts for education, business, or sales promotional use by contacting the Publisher.

NOTICE

ISBN: 0-8155-1483-2 (William Andrew, Inc.)
ISBN: 3-540-21118-7 (Springer-Verlag GmbH & Co. KG)

Library of Congress Cataloging-in-Publication Data

Wasa, Kiyotaka
Thin film materials technology : sputtering of compound materials / by Kiyotaka Wasa, Makoto Kitabatake, Hideaki Adachi.
p. cm.
ISBN 0-8155-1483-2 (alk. paper)
1. Cathode sputtering (Planing process). 2. Thin films. I. Kitabatake, Makoto. II Adachi, Hideaki. III. Title.
TS695.W38 2003
621.3815'2--dc21
2003018016

Related Electronic Materials and Process Technology

COATING MATERIALS FOR ELECTRONIC APPLICATIONS: by James J. Licari

CHARACTERIZATION OF SEMICONDUCTOR MATERIALS, Volume 1: edited by Gary E. McGuire

CHEMICAL VAPOR DEPOSITION FOR MICROELECTRONICS: by Arthur Sherman

CHEMICAL VAPOR DEPOSITION OF TUNGSTEN AND TUNGSTEN SILICIDES: by John E. J. Schmitz

CHEMISTRY OF SUPERCONDUCTOR MATERIALS: edited by Terrell A. Vanderah

CONDUCTIVE POLYMERS AND PLASTICS: edited by Larry Rupprecht

CONTACTS TO SEMICONDUCTORS: edited by Leonard J. Brillson

DIAMOND CHEMICAL VAPOR DEPOSITION: by Huimin Liu and David S. Dandy

DIAMOND FILMS AND COATINGS: edited by Robert F. Davis

DIFFUSION PROCESSES IN ADVANCED TECHNOLOGICAL MATERIALS: edited by Devenda Gupta

ELECTROCHEMISTRY OF SEMICONDUCTORS AND ELECTRONICS: edited by John McHardy and Frank Ludwig

ELECTRODEPOSITION: by Jack W. Dini

FOUNDATIONS OF VACUUM COATING TECHNOLOGY: by Donald Mattox

HANDBOOK OF CARBON, GRAPHITE, DIAMONDS AND FULLERENES: by Hugh O. Pierson

HANDBOOK OF CHEMICAL VAPOR DEPOSITION, Second Edition: by Hugh O. Pierson

HANDBOOK OF COMPOUND SEMICONDUCTORS: edited by Paul H. Holloway and Gary E. McGuire

HANDBOOK OF CONTAMINATION CONTROL IN MICROELECTRONICS: edited by Donald L. Tolliver

HANDBOOK OF DEPOSITION TECHNOLOGIES FOR FILMS AND COATINGS, *Second Edition:* edited by Rointan F. Bunshah

HANDBOOK OF HARD COATINGS: edited by Rointan F. Bunshah

HANDBOOK OF ION BEAM PROCESSING TECHNOLOGY: edited by Jerome J. Cuomo, Stephen M. Rossnagel, and Harold R. Kaufman

HANDBOOK OF MAGNETO-OPTICAL DATA RECORDING: edited by Terry McDaniel and Randall H. Victora

HANDBOOK OF MULTILEVEL METALLIZATION FOR INTEGRATED CIRCUITS: edited by Syd R. Wilson, Clarence J. Tracy, and John L. Freeman, Jr.

HANDBOOK OF PHYSICAL VAPOR DEPOSITION PROCESSING: by Donald Mattox

HANDBOOK OF PLASMA PROCESSING TECHNOLOGY: edited by Stephen M. Rossnagel, Jerome J. Cuomo, and William D. Westwood

HANDBOOK OF POLYMER COATINGS FOR ELECTRONICS, *Second Edition:* by James Licari and Laura A. Hughes

HANDBOOK OF REFRACTORY CARBIDES AND NITRIDES: by Hugh O. Pierson

HANDBOOK OF SEMICONDUCTOR SILICON TECHNOLOGY: edited by William C. O'Mara, Robert B. Herring, and Lee P. Hunt

HANDBOOK OF SEMICONDUCTOR WAFER CLEANING TECHNOLOGY: edited by Werner Kern

HANDBOOK OF SPUTTER DEPOSITION TECHNOLOGY: by Kiyotaka Wasa and Shigeru Hayakawa

HANDBOOK OF THIN FILM DEPOSITION PROCESSES AND TECHNIQUES, *Second Edition:* edited by Krishna Seshan

HANDBOOK OF VACUUM ARC SCIENCE AND TECHNOLOGY: edited by Raymond L. Boxman, Philip J. Martin, and David M. Sanders

HANDBOOK OF VLSI MICROLITHOGRAPHY, *Second Edition:* edited by John N. Helbert

HIGH DENSITY PLASMA SOURCES: edited by Oleg A. Popov

HYBRID MICROCIRCUIT TECHNOLOGY HANDBOOK, *Second Edition:* by James J. Licari and Leonard R. Enlow

IONIZED-CLUSTER BEAM DEPOSITION AND EPITAXY: by Toshinori Takagi

MOLECULAR BEAM EPITAXY: edited by Robin F. C. Farrow

NANOSTRUCTURED MATERIALS: edited by Carl C. Koch

SEMICONDUCTOR MATERIALS AND PROCESS TECHNOLOGY HANDBOOK: edited by Gary E. McGuire

THIN FILM MATERIALS TECHNOLOGY: edited by Kiyotaka Wasa

ULTRA-FINE PARTICLES: edited by Chikara Hayashi, Ryozi Ueda and Akira Tasaki

WIDE BANDGAP SEMICONDUCTORS: edited by Stephen J. Pearton

Other Related Titles

CRYSTAL GROWTH TECHNOLOGY: by K. Byrappa

HANDBOOK OF ELLIPSOMETRY: edited by Harland G. Tompkins and Eugene A. Irene

HANDBOOK OF ENVIRONMENTAL DEGRADATION OF MATERIALS: edited by Myer Kutz

HANDBOOK OF FILLERS, *Second Edition:* edited by George Wypych

HANDBOOK OF HYDROTHERMAL TECHNOLOGY: edited by K. Byrappa and Masahiro Yoshimura

HANDBOOK OF INDUSTRIAL REFRACTORIES TECHNOLOGY: by Stephen C. Carniglia and Gordon L. Barna

HANDBOOK OF MATERIAL WEATHERING: edited by George Wypych

HANDBOOK OF PLASTICIZERS: edited by George Wypych

HANDBOOK OF SOLVENTS: edited by George Wypych

MECHANICAL ALLOYING FOR FABRICATION OF ADVANCED ENGINEERING MATERIALS: by M. Sherif El-Eskandarany

MEMS: A PRACTICAL GUIDE TO DESIGN, ANALYSIS, AND APPLICATIONS: edited by Oliver Paul and Jan Korvink

SOL-GEL TECHNOLOGY FOR THIN FILMS, FIBERS, PREFORMS, ELECTRONICS AND SPECIALTY SHAPES: edited by Lisa C. Klein

SOL-GEL SILICA: by Larry L. Hench

TRIBOLOGY OF ABRASIVE MACHINING PROCESSES: by Ioan Marinescu, Brian Rowe, Boris Dimitrov, Ichiro Inasaki

WEATHERING OF PLASTICS: edited by George Wypych

Preface

Bunsen and Grove first observed sputtering over 150 years ago in a discharge tube. Since then the basic level of understanding of the sputtering phenomena has been refined. The applications of sputtering, however, are still being developed on a daily basis. Sputtering deposition and sputtering etching have become common manufacturing processes for a wide variety of industries. First and foremost is the electronics industry, which uses sputtering technology to produce integrated circuits and magneto-optical recording media. This book describes many of the sputtering applications that are relevant to electronics.

Sputtering processes are also present in many other disparate areas. For example, sputter deposition is used to coat the mirrorlike reflective windows in many buildings. The hard coating of a machine tool is a well-known application of sputtering.

Sputtering is essential for the creation of new materials such as diamond thin films, high-T_c superconductors, and ferroelectric and magnetic materials like those used in random access memories. Nanometer materials are also provided by sputtering. It is important that the sputtering process is considered an environmentally benign production technology. The sputtering process is a key technology for material engineering in the twenty-first century.

In the last ten years, radical progress has been seen in sputtering technology. For production, an example is the high-rate sputtering technology using pulsed DC/MF dual-magnetron sputtering for coating large areas like window glass. Another production technology is the sputter-etching of deep trench structures using plasma-assisted long-throw magnetron sputtering systems. At the basic research level, epitaxial processing of complex oxides such as layered perovskite for high-T_c superconductors and ferroelectric superlattices of perovskites at the nanometer level were extensively studied, and commercial sputtering systems were developed.

The material in this book is based on the author's research works at Panasonic, Research Institute of Innovative Technology for the Earth, RITE (Japanese government institute for global environment issues), and Yokohama City University, for over forty years. This edition includes experimental sputtering data according to the author's recent experiments, and up-to-date references. The theoretical descriptions of the sputtering and film growth processes are geared to graduate students of materials engineering disciplines based on the author's lectures at Yokohama City University. Production level engineering data are included for engineers in industry.

Chapter 1 describes the special features of thin films from a materials science and engineering viewpoint. This chapter also reviews some of the devices and applications of sputtered thin films. Chapter 2 shows overviews of the thin film growth mechanisms, and basic deposition processes are discussed. Tables are included giving the properties of crystal substrates and summarizing the epitaxial relationships. The basic concepts of sputtering phenomena are described in Ch. 3 including discussions of ion energy, collisions, and sputtering yields. The systems used for sputtering deposition are characterized in Ch. 4. This chapter includes information on discharges, targets, and process monitoring. Chapter 5 shows a basic process design for the sputtering deposition of complex compound materials and a variety of experimental data for the application of thin films in electronics industry. Chapter 6 describes the basic sputtering process for a controlled microstructure showing examples of the deposition of perovskite ferroelectric thin films with a controlled microstructure, the application of the sputtering for nanometer thin film materials, and the interfacial control of high-k Si gate oxides by magnetron sputtering. Chapter 7 shows microfabrication methods using the sputtering etching process.

In the last ten years, several excellent tutorial texts on thin films have been published for young scholars. However, for the practical use of thin films, we should understand the physics and chemistry of thin film materials including crystal chemistry, vacuum engineering, and gas discharge and plasma physics. The study of thin film material engineering should include both the tutorial phenomena and practical engineering data, and few textbooks cover both. This text will act as a bridge between tutorial textbooks and practical application, and will be useful as a sub-textbook for graduated students and as an experimental guidebook for young scientists or engineers.

I owe my thanks to many senior material scientists including K. L. Chopra (Indian Institute of Technology), R. Roy, L. E. Cross, R. E. Newnham (Penn State University), and T. H. Geballe (Stanford University) for continuous discussion on ceramic thin films. I am also grateful to K. Uchino, S. Trolier-McKinstry, D. G. Schlom (Penn State University), and C. B. Eom (Wisconsin-Madison University), and S. Kisaka (Kanazawa Institute of Technology) for their helpful discussion and evaluation of the thin films. I also want to thank S. M. Rossnagel (IBM Watson Res. Center) whose kindly help has been invaluable. Thanks are due to many vacuum materials and equipment companies who supplied the practical data. I wish to thank students R. Ai, R. Suzuki, and K. Maeda at Wasa's Thin Film Lab., Yokohama City University, and my daughter Yasuko Hirai, for their help with the manuscript.

Finally, I pray for the repose of the souls of R. F. Bunshah (former professor, University of California, Los Angeles) and G. Narita (former Vice President, Executive Editor, Noyes Publications) who passed away before the publication of this edition.

November 2003

Kiyotaka Wasa
Yokohama, Japan

Table of Contents

1

Thin Film Materials and Devices

Thin films are fabricated by the deposition of individual atoms on a substrate. A *thin film* is defined as a low-dimensional material created by condensing, one-by-one, atomic/molecular/ionic species of matter. The thickness is typically less than several microns. Thin films differ from thick films. A *thick film* is defined as a low-dimensional material created by thinning a three-dimensional material or assembling large clusters/aggregates/grains of atomic/molecular/ionic species.

Historically, thin films have been used for more than a half century in making electronic devices, optical coatings, instrument hard coatings, and decorative parts. The thin film is a traditional well-established material technology. However, thin film technology is still being developed on a daily basis since it is a key in the twenty-first century development of new materials such as nanometer materials and/or a man-made superlattices.

Thin film materials and devices are also available for minimization of toxic materials since the quantity used is limited only to the surface and/or thin film layer. Thin film processing also saves on energy consumption in production and is considered an environmentally benign material technology for the next century.[1]

Thin film technology is both an *old and a current key material technology.* Thin film materials and deposition processes have been reviewed in several publications.[2] Among the earlier publications, the

Handbook of Thin Film Technology (Maissel and Glang) is still notable even though thirty years have passed since the book was published and many new and exciting developments occurred in the intervening years.

1.1 THIN FILM MATERIALS

Thin films are deposited on a substrate by thermal evaporation, chemical decomposition, and/or the evaporation of source materials by the irradiation of energetic species or photons. Thin-film growth exhibits the following features:

1. The birth of thin films of all materials created by any deposition technique starts with a random nucleation process followed by nucleation and growth stages.
2. Nucleation and growth stages are dependent upon various deposition conditions, such as growth temperature, growth rate, and substrate chemistry.
3. The nucleation stage can be modified significantly by external agencies, such as electron or ion bombardment.
4. Film microstructure, associated defect structure, and film stress depend on the deposition conditions at the nucleation stage.
5. The crystal phase and the orientation of the films are governed by the deposition conditions.

The basic properties of film, such as film composition, crystal phase and orientation, film thickness, and microstructure, are controlled by the deposition conditions. Thin films exhibit unique properties that cannot be observed in bulk materials:

1. Unique material properties resulting from the atomic growth process.
2. Size effects, including quantum size effects, characterized by the thickness, crystalline orientation, and multilayer aspects.

The properties of thin films are governed by the deposition method. The deposition process using the decomposition of source materials is known as *chemical vapor deposition* (CVD). The deposition process using the irradiation of energetic species is known as *sputtering.* Bunsen and Grove first observed sputtering in a gas discharge tube over 150 years ago. The cathode electrode material was disintegrated by the discharge. Since that time, the basic level of understanding of the sputtering process has become fairly well developed. It was known that the disintegration of the cathode material was caused by irradiation of the cathode surface by highly energetic ions. The removed particles, called sputtered species, were comprised of highly energetic atoms. Their energy ranges were 1 to 10 eV, which was higher than those of the other deposition processes such as thermal evaporation and chemical decomposition. The sputtering process achieves the deposition of a variety of materials without heating the source materials.

Now sputtering has become a common manufacturing process for a variety of industries. First and foremost is the semiconductor industry, where sputtering technology is used in the metallization process in the production of virtually every integrated circuit. The production technology for Si ICs has been established using an automatic sputtering deposition system.

Sputtering deposition is also present in many other disparate areas. For instance, sputter deposition is used to coat the mirrorlike windows and reflective layers in many tall buildings. For the stable production of reflective-coating glass windows, a special sputtering system was designed in the 1990s. The sputtering deposition process for material production, however, has a low level of efficiency. An optimum sputtering design for production is necessary for each material.

Bulk materials are usually sintered from powders of source materials. The particle size of these powders is of the order of 1 μm in diameter. Thin films are synthesized from atoms or a cluster of atoms. Sputtering deposition is unique, compared to other deposition processes, in that sputtering deposition is a quenched, or high-energy process. Films deposited by other processes, such as thermal evaporation and CVD, are formed under conditions of thermodynamic equilibrium. In the sputtering process, highly energetic sputtered species are quenched on the substrate surface. This dynamic quenching process allows the formation of novel thin-film materials. These ultrafine particles are quenched on substrates during film growth, and this non-equilibrium state can lead to the formation of exotic materials. A variety of abnormal crystal phases have been reported in thin films. Energetic sputtered particles lower the synthesis temperature of materials. A typical example is a diamond growth at room temperature.

Bulk diamonds are conventionally synthesized at high pressure (≈50,000 psi) and high temperature (2000°C). The deposition of diamonds from energetic carbon ions (≈10–100 eV) enables the growth of cubic diamond crystallites and/or diamond films at room temperature using a sputtering process.[3] A hexagonal diamond is also synthesized by sputtering. The natural diamond found on the earth is cubic diamond which is a stable phase. The hexagonal phase is not grown under thermodynamic equilibrium conditions; rather, it is grown under non-thermal equilibrium conditions.[4]

The high-T_c superconductors of layered perovskite discovered by Bednorz and Müller show different superconducting transition temperatures due to the numbers of copper oxide layers. The single phase was difficult for the bulk ceramics sintering process.[5] However, phase control of the high-T_c superconductors was successfully achieved by layer-by-layer deposition using the sputtering process.[6]

The sputtered, thin, two-dimensional structure fixed on the substrates modifies the material properties. It is reasonably considered that the thin films may show features that are different from the bulk materials in terms of mechanical strength, carrier transportation, superconductivity, ferroelectricity, magnetic properties, and optical properties. For instance, thin films may be characterized by a strong internal stress of 10^9–10^{10} dynes/cm^2 and a number of lattice defects. The density of the lattice defects can be more than 10^{11} dislocations/cm^2. These lattice defects have the effect of increasing the elastic strength. The strengths obtained in thin films can be up to two hundred times as large as those found in corresponding bulk material. The stress arises from the mismatch in the lattice parameter and the thermal expansion coefficient between the thin films and the substrates. The compressive stress elevates the Curie temperature of ferroelectric thin films of perovskite structure.[8][9] The superlattice of the ferroelectrics thin films shows a giant permittivity[12] and a pseudo-pyroelectric effect.[13] The stress affects the superconducting critical temperature at or below which electrical resistance vanishes. The tensile stress increases the critical temperature for metal superconducting films.[14] The compressive stress increases the critical temperature for high-T_c cuprate.[15]

The thin-film process is also essential for making nanometer materials. *Nanomaterials* are defined as follows: materials or components thereof in alloys, compounds, or composites having one or more dimensions of nanometer size (1 nm = 10^{-7} cm = 10 Å). Nanomaterials are classified into three types:

1. *Zero-dimensional nanomaterials* have all three dimensions of nanometer size (e.g., quantum dots).
2. *One-dimensional nanomaterials* have two dimensions of nanometer size (e.g., quantum wires).
3. *Two-dimensional nanomaterials* have one dimension of nanometer size (e.g., thin films, superlattices).

The phenomenological dimensionality of nanometer materials depends on the size relative to physical parameters such as quantum confinement regime (≤ 100 atoms), mean free path of conduction electron (< 10 nm), mean free path of hot electron (≤ 1 nm), Bohr excitation diameter (Si = 8.5 nm, CdS = 6 nm, GaAs = 196 nm), de Broglie wavelength (< 1 nm).[20]

The three types of nanometer materials have been successfully synthesized by thin-film processes such as codeposition, layer-by-layer deposition in an atomic scale, and nanolithography using a sputtering process.[21]

The current progress in thin-film research is much indebted to the atomic observation technology including the scanning tunneling microscope (STM) developed by Binnig and Rohrer.[22]

Figure 1.1 shows some photographs of thin-film materials. Table 1.1 summarizes the interesting phenomena expected in thin films

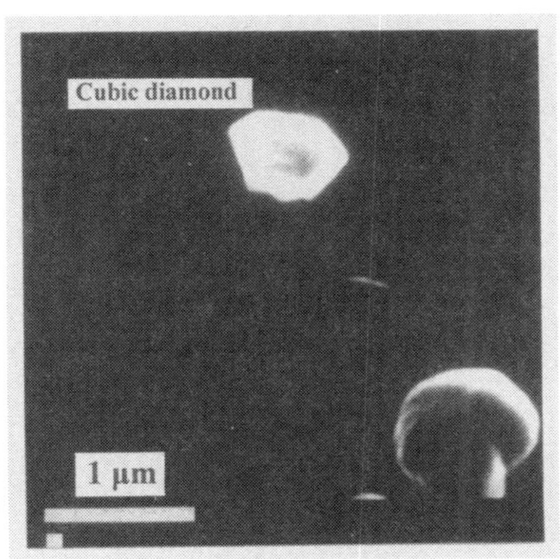

(a)

Figure 1.1. Photographs of some thin-film materials. (*a*) Diamond crystals prepared at room temperature by ion beam sputtering. (*b*) Cross-sectional TEM images of ferroelectric superlattice, $PbTiO_3/(Pb, La)TiO_3$ nanometer multilayers, prepared by magnetron sputtering.

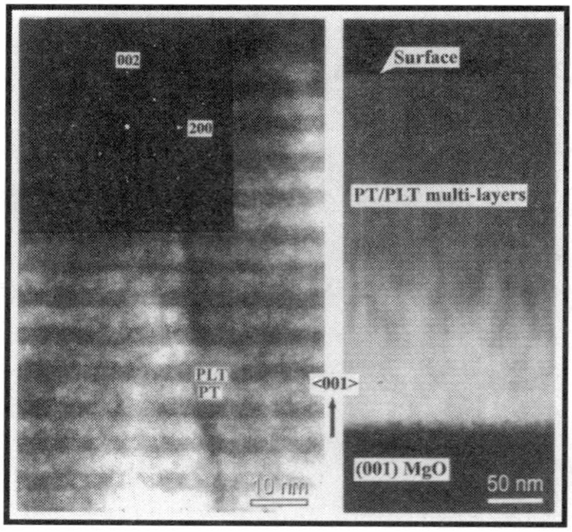

(b)

Figure 1.1. *(cont'd.)*

Table 1.1. Interesting Phenomena Expected in Thin Film Materials

Size Effect*
Increase of resistivity, ρ, in metal: $\rho_F/\rho_B \approx (4/3)[\gamma \ln(1/\gamma)]^{-1}$
Reduced TCR, α, in metal: $\alpha_F/\alpha_B \approx [\ln(1/\gamma)]^{-1}$
Reduced mobility, μ, in metal: $\mu_F/\mu_B \approx [\ln(1/\gamma)]^{-1}$
Anomalous skin effect at high frequencies in metal.
Reduced thermal conductivity, K, in metal: $K_F/K_B \approx (3/4)[\gamma \ln(1/\gamma)]$
Enhanced thermoelectric power, S, in metal: $S_F/S_B \approx 1 + (2/3)[(\ln\gamma - 1.42)/(\ln\gamma - 0.42)]$
Reduced mobility in semiconductor: $\mu_F/\mu_B \approx 1 + (1 + 1/\gamma)^{-1}$
Quantum size effects in semiconductors and semimetals, at $t < \lambda$, de Broglie wavelength: thickness-dependent oscillatory variation of resistivity, Hall coefficient, Hall mobility, and magnetoresistance. Galvanomagnetic surface effects on Hall effect and magnetoresistance due to surface scattering.
Note: $\gamma = t/l \ll 1$ where t is film thickness, l is mean free path of electrons, $\lambda = h/mv$ where h is the Planck constant, m is the mass of the particle, and v is the velocity. Electron transport phenomena (F: film, B: bulk).

(cont'd.)

Table 1.1. *(cont'd.)*

Field Effects
Conductance change in semiconductor surface by means of electric field, Insulated-gate thin transistor (TFT).
Space-Charge Limited Current (SCLC)
SCLC through insulator, J:$J = 10^{-13}\ \mu_d\, \varepsilon V^2/t^3$ (A/cm^2) (one-carrier trap-free SCLS). Note: μ_d, drift mobility of charge carriers, ε, dielectric constant, V, applied voltage.
Tunneling Effects
Tunnel current through thin insulating films, voltage-controlled negative resistance in tunnel diode.
Tunnel emission from metal, hot electron triode of metal-base transistor.
Electroluminescence, photoemission of electrons.
Tunnel spectroscopy.
Spin-dependent tunneling magnetoresistance (TMR) effects:[7] TMR=$2P_1P_2/(1- P_1P_2)$; P_1, P_2: spin polarization.
Tunnel current between island structure in ultra thin films.
Ferroelectricity
Increase of Curie temperature T_c by film stress.[8][9] $\Delta T_c = 2\varepsilon_0 C(Q_{11} + 2Q_{12})\sigma$ (cubic-tetragonal). Note: C: Curie constant; Q_{ij}: cubic electrostrictive constants; σ: hydrostatic stress.
Thickness dependence of dielectric constant.[10]
Crystalline size effects.[11]
Giant permittivity.[12]
Charge pumping, pseudopyroelectric effects.[13]

(cont'd.)

Table 1.1. *(cont'd.)*

Superconductivity
Superconductivity-enhancement: Increase of critical temperature, T_c, in metal with decreasing thickness, t: $\Delta T_c \approx A/t - B/t^2$ and/or crystallite size.
Stress effects: Tensile stress increases T_c. Compressive stress decreases T_c in metal.[14] Compressive stress along c axis decreases T_c. Compressive stress in a-b plane increases T_c in high-T_c cuprates.[15]
Proximity effects in superimposed films: Decrease of T_c in metal caused by contact of normal metal.
Reduced transition temperature, T_s: $(T_s/T_c)^2 = 1 - 1/(0.2 + 0.8t_s)$. Note: t_s is the ratio of thickness of superconducting films and a critical thickness below which no superconductivity is observed for a constant thickness of normal metal films.
Increase of critical magnetic field, H_c: At parallel field: $H_{CF}/H_{CB} \approx (\sqrt{24})(\lambda/t)$, where λ is the penetration depth due to Ginzburg-Landau theory. At transverse field, $H_{CF}/H_{CB} = (\sqrt{2})\, K$, where K is the Ginzburg-Landau parameter.
Reduced critical current, J_C: $J_{CF}/J_{CB} \approx \tanh(t/2\lambda)$, where J_{CB}, is the critical current of bulk and J_{CF} is the critical current of thin films)
Supercurrent tunneling through thin barrier, Josephson junction, and tunnel spectroscopy. Intrinsic Josephson junctions in high-T_c cuprates.[16]

(cont'd.)

Table 1.1. *(cont'd.)*

Magnetics
Increase in magnetic anisotropy: The anisotropies originate in a shape anisotropy, magnetocrystalline anisotropy, strain-magnetostriction anisotropy, uniaxial shape-anisotropy. Magnetic free energy (E) is expressed as: $E = K_u \sin^2\phi - \mathbf{M}\bullet\mathbf{H}$ where: K_u : magnetic anisotropy constant $\mathbf{M}$: magnetization $\mathbf{H}$: magnetic field ϕ : angle between **M** and easy axis.
Increase in magnetization and permeability in amorphous structure, and/or layered structure.
Giant magnetoresistance (GMR) effects in multilayers:[17][18] $\mathrm{MR} = (\rho_{AF} - \rho_F)/\rho_F$ where: ρ_{AF} : antiparallel resistivity ρ_F : parallel resistivity
GMR multilayer on V-groove substrate.[19] $\sigma_{\mathrm{CAP}} = \sigma_{\mathrm{CIP}} \cos^2\theta + \sigma_{\mathrm{CPP}} \sin^2\theta$ where σ_{CAP} : conductivity for current at an angle to plane σ_{CIP} : conductivity for current in plane σ_{CPP} : conductivity for current perpendicular to plane θ : angle of V-groove
Exchange coupling at the interface between ferromagnetic (FM) and antiferromagnetic (AF) layers. – Increase of coercive field (H_C) – Shift of *M-H* curve (exchange bias)

1.2 THIN FILM DEVICES

Since the latter part of the 1950s, thin films have been extensively studied in relation to their applications for making electronic devices. In the early 1960s, Weimer proposed thin film transistors (TFTs) composed of CdS semiconducting films. He succeeded in making a 256-stage, thin-film transistor decoder, driven by two 16-stage shift resistors, for television scanning, and associated photoconductors, capacitors, and resistors.[23] Although these thin-film devices were considered as the best development of both the science and technology of thin films for an integrated micro-electronic circuit, the poor stability observed in TFTs was an impediment to practical use. The bulk Si MOS (metal-oxide semiconductor) devices were successfully developed at the end of 1960s.[24] Thus, thin-film devices for practical use were limited to passive devices such as thin-film resistors and capacitors.

In the 1970s, several novel thin-film devices were proposed, including thin-film, surface acoustic wave (SAW) devices,[25] integrated thin-film bulk acoustic wave (BAW) devices,[26] and thin-film integrated optics.[27] A wide variety of thin-film devices were developed. Of these, one of the most interesting technologies was a thin-film amorphous silicon (a-Si) proposed by Spear using a CVD process.[28] This technology achieved a low-temperature doping of impurities into a-Si devices and suggested the possibility of making a-Si active devices such as a-Si TFT and a-Si solar cells.[29] In the 1980s, rapid progress was made in a-Si technology. Amorphous Si solar cells have been produced for electronic calculators, although the energy conversion efficiency is 5 to 7% and is lower than that of crystalline Si solar cells. In the middle of the 1980s, high quality a-Si technology led to the production of a liquid crystal television with a-Si TFT. Due to the improvement of a-Si thin film, the energy conversion efficiency of the a-Si solar cells has been improved and is now as high as 12%.[30] The a-Si/poly-Si stacked cell shows an efficiency of 21 to 23%,[31] which is in the same order of magnitude as the efficiency of single-crystal Si solar cells. The processing temperature is as low as 300°C for a-Si thin-film solar cells. Thin-film technology for making high-efficiency a-Si solar cells will be a key for the production of clean energy since a-Si solar cells consume much less energy to produce than single-crystal bulk Si solar cells, which also use the sputtering process.[32]

Other interesting thin-film devices are ZnO thin-film SAW devices and ZnO thin-film BAW devices for color televisions, mobile telephones, and communication systems.[33][34] ZnO is known as a piezoelectric material for making acoustic transducers. The sputtering process successfully deposits thin films of ZnO of transducer quality. The ZnO SAW and ZnO BAW devices act as solid-state resonators and/or band-pass filters in the frequency region of MHz to GHz band.

Silicon carbide (SiC) thin film, high-temperature sensors were another type of attractive thin-film device produced in the 1980s. Temperature sensors were developed using bulk SiC single crystals for satellite use because of their high radiation resistance. However, the difficulty of producing the precise equipment required to make them precluded their production for commercial use. Sputtering technology was able to achieve a low-temperature synthesis of high-temperature SiC materials to overcome the issues of producing SiC sensors with high accuracy.[35] These SiC devices are now developed as high-power, semiconducting integrated circuits and radiation-resistant semiconducting devices. The nanometer, multilayered structure provided by the δ-doping process made the high-mobility SiC MOS devices a reality. These SiC MOS devices have a high potential for saving energy in consumer electronics.[36]

The sputtering process produces a narrow magnetic gap for videotape recording systems and for computer disk applications. In the production of the magnetic gap, a nonmagnetic spacer is formed from glass material. Prior to the use of thin-film technology, the spacer manufacturing process was quite complex. For instance, magnetic-head core material was first immersed in a mixed solution of finely crushed glass, then taken out and subjected to centrifugation so that a homogeneous glass layer was deposited onto the surfaces of the core members of the opposing gap. After forming a glass film on the core surfaces by firing the deposited glass layer, the two opposing gap faces were butted against each other with the glass layer sandwiched between and then fused together by a heat treatment to form the desired operative gap. Since the width of the magnetic gap was around 0.3 μm, these traditional methods were difficult to use in production because of the difficulty in controlling the film thickness of the fired glass.

Thin-film deposition technology enabled the production of magnetic heads with narrow gap lengths of 0.3 μm.[37] The narrow-gap forming technology was based on the atomic scale achievable by thin-film deposition processes. Sputtering technology, with its precise, controlled deposition, is used to develop layered new materials including giant magnetoresistance (GMR) magnetic materials. The spin-dependent, tunneling-magnetoresistance (TMR) effects will provide a high-density memory disk of up to 200 $Gbit/inch^2$.[7]

Thin-film materials are used for the production of electronic devices such as high precision resistors, SAW devices, BAW devices, optical disks, magnetic tapes, magnetic disks, and sensors, and for active matrices for liquid-crystal TV. Thin films of high-T_c superconductors are used for the fabrication of superconducting planar filters with gigaband capability.[38] Additionally, integrated acousto-optic and magneto-optic devices for optical information processing have been developed by Tsai.[39]

The development of thin-film devices owes its development to the silicon large-scale integration (LSI) technology, including thin-film growth process, microfabrication, and analysis technology of both the surface and interfaces of thin films. It is noted that the ferroelectric dynamic random access memory (FEDRAM) has been developed and is now used in practice. Ferroelectric thin films were studied in the past for use in high-capacitive dielectric and pyroelectric sensors.[40] FEDRAM owes its development to the integration of Si LSI technology and ferroelectric thin-film technology. The superlattice of giant magnetoresistance also provides high-density magnetoresistance dynamic random access memory (MRDRAM).[41] FEDRAM and MRDRAM will be key next-generation LSI technologies.[42]

Figure 1.2 shows photographs of some thin-film devices.

(a)

Figure 1.2. Photographs of some thin-film devices. (*a*) A 5-GHz high-T_c superconducting planar disk resonator. (*b*) Submicron narrow-gap magnetic head for video system. (*c*) Optical disk, magneto-optical disk, and hard disk with Si wafers. *(Courtesy of Mitsubishi Materials Corporation.)*

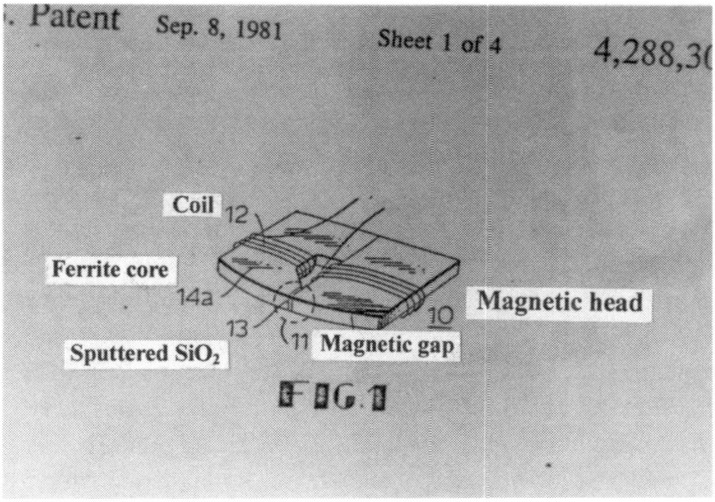

(b)

(c)

Figure 1.2. *(cont'd.)*

The pioneer researcher of ion-beam sputtering deposition, K. L. Chopra, said "The thin film was in past considered as the 5th state of matter next to plasma, since the reliable materials properties could not be obtained and thin films were considered to be different from bulk materials. At present the thin films are considered as the 1st state of matter. This is owed to establishment of scientific technology of the thin film growth kinetics."[43]

The sputtering deposition process is complicated. However, many of the problems associated with sputtering in the past are being eliminated. In this book, a number of experimental data on the sputtering deposition are presented for a variety of materials in relation to their structure and other electrical properties. These data are based on the author's experiments spanning the forty years since 1960, from basic research to production, and will be helpful for the research and production of new thin-film materials and devices.

Sputtering technology is becoming commonplace in many manufacturing disciplines, and the sputtering process is considered to be an environmentally benign thin-film process due to its small environmental load compared to the CVD process. New sputtering applications in technology are emerging.[1]

A pioneer researcher of sputtering physics, G. K. Wehner, believed that sputtering showed high potential for the deposition of high-quality semiconductors similar and/or superior to the MBE/CVD processes.[44] Further study on the growth kinetics for sputtering deposition will realize Wehner's concept.

REFERENCES

1. Wasa, K., *Bull. Mater. Res.,* 18:937, India (1995)
2. Maissel, L. I., and Glang, R., (eds.), *Handbook of Thin Film Technology,* McGraw-Hill, New York (1970); Chopra, K. L., *Thin Film Phenomena,* McGraw-Hill, New York (1969); Vossen, J. L., and Kern, W., (eds.), *Thin Film Processes*, Academic Press, New York (1978); Bunshah, R. F. (ed.), *Handbook of Deposition Technologies for Films and Coatings*, Noyes, NJ (1993); Elshabini Aicha, A. R., and Barlow, F. D, III, *Thin Film Technology Handbook,* McGraw-Hill, New York (1998)
3. Kitabatake, M., and Wasa, K., *J. Appl. Phys.,* 58:1693 (1985)

4. Silva, S. R. P., Amaratunga, G. A. J., Salje, E. K. H., and Knowles, K. M., *J. Mater. Sci.*, 29:4962 (1994)
5. Bednorz, J. G., and Müller, K. A., *Z. Phys. B,* 64:189 (1986)
6. Wasa, K., and Kitabatake, M., *Thin Film Processing and Characterization of High-Temperature Superconductors, Series 3* (J. M. E. Harper, R. J. Colton, and L. C. Feldman, eds.), American Vac. Society, New York (1988)
7. Julliere, M., *Physics Lett.,* 54A:225 (1975)
8. Yano, Y., Daitoh, Y., Terashima, T., Bando, Y., Watanabe, Y., Kasatani, H., and Terauchi, H., *J. Appl. Phys.,* 76:7833 (1994)
9. Rossetti, G. A., Jr., Cross, L. E., and Kushida, K., *Appl. Phys. Lett.,* 59:2524 (1991)
10. Feldman, C., *J. Appl. Phys.,* 65:872 (1989)
11. de Keijser, M., Dormans, G. J. M., van Veldhoven, P. J., and de Leeuw, D. M., *Appl. Phys. Lett.,* 59:3556 (1991)
12. Li, S., Eastman, J. A., Vetrone, J. M., Newnham, R. E., and Cross, L. E., *Philos. Mag. B*, 76:47 (1997)
13. Schubring, N. W., Mantese, J. V., Micheli, A. L., Catalan, A. B., and Lopez, R. J., *Phys. Rev. Lett.,* 68:1778 (1992)
14. Toxen, A. M., *Phys. Rev.,* 123:442; 124:1018 (1961)
15. Sato, H., and Naito, M., *Physica C,* 274:221 (1997)
16. Odagawa, A., Sakai, M., Adachi, H., and Setsune, K., *Jpn. J. Appl. Phys.,* 37:486 (1998)
17. Baibich, M. N., Broto, J. M., Fert, A., Nguyen van Dau, N., Petroff, F., Etienne, P., Cruezet, G., Friederich, A., and Chazelas, J., *Phys. Rev. Lett.,* 61:2472 (1988)
18. Itoh, H., Inoue, J., and Maekawa, S., *Phys. Rev. B,* 47:5809 (1993)
19. Shinjo, T., and Yamamoto, H., *J. Phys. Soc. Jpn.*, 59:3061 (1990)
20. Timp, G. (ed.), *Nanotechnology,* p. 7, Springer, New York (1998)
21. Schiffrin, D. J., *MRS Bull.,* 26:1015 (2001)
22. Binnig, G., Rohrer, H., Gerber, C., and Wiebel, E., *Phys. Rev. Lett.*, 50:120 (1983)
23. Weimer, P. K., *Proc. IRE,* 50:1462 (1962)
24. Kisaka, S., *History of Science for Electronics*, Nikkan Kogyou, Tokyo (2002)
25. Kino, G. S., and Wagers, R. S., *J. Appl. Phys.*, 44:1480 (1973)
26. Lakin, K. M., and Wang, J. S., *Proc. 1980 IEEE Ultrason. Symp.*, p. 829 (1980)

27. Tien, P. K., *Appl. Opt.,* 10:2395 (1971)
28. Spear, W. E., and LeComber, P. G., *J. Non-Cryst. Solids,* 11:219 (1972)
29. Spear, W. E., and LeComber, P. G., *Solid State Commun.*, 17:1193 (1975); Carlson, D. E., and Wronski, C. R., *RCA Review*, 38:211 (1977)
30. Hamakawa, Y., Okamoto, H., and Takakura, H., *18th IEEE Photovol. Spec. Conf.*, Las Vegas (1985); Hamakawa, Y., *Proc. of NESC 99*, p. 25, Osaka (1999)
31. Hamakawa, Y., Ma, W., and Okamoto, H., *MRS Bull.,* 18(10):38 (1993)
32. Jagannathan, B., Anderson, W. A., and Coleman, J., *Solar Energy Materials and Solar Cells,* 46:289 (1997)
33. Yamazaki, O., Mitsuyu, T., and Wasa, K., *IEEE Trans. Sonics and Ultrason.,* SU-27:369 (1980)
34. Nakamura, K., Kobayashi, H., and Kanbara, H., *Jpn. J. Appl. Phys.,* 39:3049 (2000)
35. Wasa, K., Tohda, T., Kasahara, Y., and Hayakawa, S., *Rev. Sci. Instr.,* 50:1086 (1979)
36. Kitabatake, M., private communication (Nov. 2000)
37. Wasa, K., U.S. Patent 4,288,307, Sept. 1981, assigned to Matsushita Electric Corp.
38. Enokihara, A., and Setsune, K., *J. Superconductivity,* 10:49 (1997)
39. Tsai, C. S., *Proc. IEEE,* 81:853 (1996)
40. Kusao, K., Wasa, K., and Hayakawa, S., *Jpn. J. Appl. Phys.,* 7:437 (1969); Okuyama, M., Matsui, Y., Seto, H., and Hamakawa, Y., *Jpn. J. Appl. Phys.,* Suppl. 20-1:315 (1981)
41. Auciello, O., Scott, J. F., and Ramesh, R., *Phys. Today,* 51:22 (Jul. 1998)
42. Parkin, S. S. P., Roche, K. P., Samnt, M. G., Rice, P. M., Beyers, R. B., Scheuerlein, R. E., O'Sullivan, E. J., Brown, S. L., Bucchigano, J., Abraham, D. W., Lu, Y., Rooks, M., Trouilloud, P. L., Wanner, R. A., and Gallagher, W. J., *J. Appl. Phys.*, 85:5828 (1999)
43. Chopra, K. L., *Thin Film Materials and Processing*, lecture at Yokohama City Univ., (Dec. 2001)
44. Wehner, W. K., private communication, San Diego (1989)

2

Thin Film Processes

Several publications have presented a detailed review of thin-film deposition processes;[1] thus only brief descriptions of the thin-film growth and deposition processes are presented in this chapter.

2.1 THIN FILM GROWTH PROCESS

Any thin-film deposition process involves three main steps:

1. Production of the appropriate atomic, molecular, or ionic species.
2. Transport of these species to the substrate through a medium.
3. Condensation on the substrate, either directly or via a chemical and/or electrochemical reaction, to form a solid deposit.

Formation of a thin film takes place via nucleation and growth processes. The general picture of the step-by-step growth process emerging from the various experimental and theoretical studies can be presented as follows:

1. The unit species, on impacting the substrate, lose their velocity component normal to the substrate (provided the incident energy is not too high) and are physically adsorbed on the substrate surface.
2. The adsorbed species are not in thermal equilibrium with the substrate initially and move over the substrate surface. In this process they interact among themselves, forming bigger clusters.
3. The clusters or the *nuclei*, as they are called, are thermodynamically unstable and may tend to desorb in time, depending on the deposition parameters. If the deposition parameters are such that a cluster collides with other adsorbed species before getting desorbed, it starts growing in size. After reaching a certain critical size, the cluster becomes thermodynamically stable and the *nucleation barrier* is said to have been overcome. This step involving the formation of stable, chemisorbed, critical-sized nuclei is called the *nucleation stage*.
4. The critical nuclei grow in number as well as in size until a saturation nucleation density is reached. The nucleation density and the average nucleus size depend on a number of parameters such as the energy of the impinging species, the rate of impingement, the activation energies of adsorption, desorption, thermal diffusion, and the temperature, topography, and chemical nature of the substrate. A nucleus can grow both parallel to the substrate by surface diffusion of the adsorbed species, and perpendicular to it by direct impingement of the incident species. In general, however, the rate of lateral growth at this stage is much higher than the perpendicular growth. The grown nuclei are called *islands*.
5. The next stage in the process of film formation is the *coalescence stage*, in which the small islands start coalescing with each other in an attempt to reduce the substrate surface area. This tendency to form bigger islands is termed *agglomeration* and is enhanced by increasing the surface mobility of the adsorbed species, by, for example, increasing the substrate temperature. In

some cases, formation of new nuclei may occur on areas freshly exposed as a consequence of coalescence.

6. Larger islands grow together, leaving channels and holes of uncovered substrate. The structure of the films at this stage changes from discontinuous island type to porous network type. Filling of the channels and holes forms a completely continuous film.

The growth process thus may be summarized as consisting of a statistical process of nucleation, surface-diffusion controlled growth of the three-dimensional nuclei, and formation of a network structure and its subsequent filling to give a continuous film. Depending on the thermodynamic parameters of the deposit and the substrate surface, the initial nucleation and growth stages may be described as *(a)* island type, called Volmer-Weber type, *(b)* layer type, called Frank-van der Merwe type, and *(c)* mixed type, called Stranski-Krastanov type. This is illustrated in Fig. 2.1. In almost all practical cases, the growth takes place by island formation. The subsequent growth stages for an Au film sputter-deposited on NaCl at 25°C, as observed in the electron microscope, are shown in Fig. 2.2.

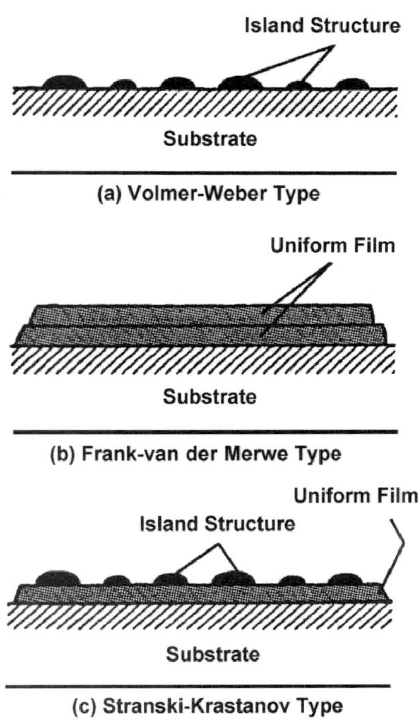

Figure 2.1. Three modes of thin film growth processes.

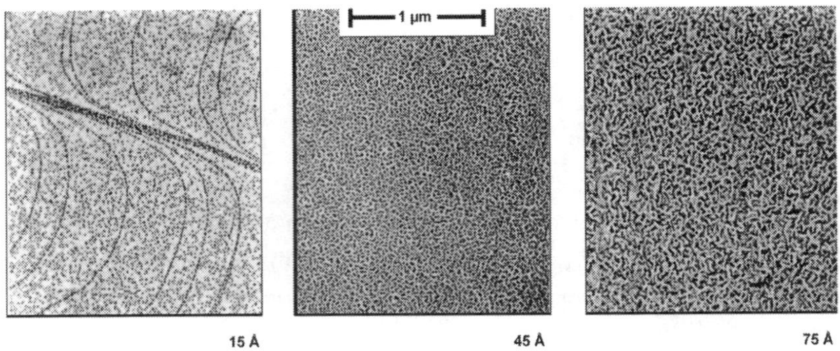

Figure 2.2. Transmission electron micrographs of 15, 45, and 75 Å thick argon-sputtered Au films deposited on NaCl at 25°C at a deposition rate of approximately 1 Å/sec.[2]

Except under special conditions, the crystallographic orientations and the topographical details of different islands are randomly distributed, so that when they touch each other during growth, grain boundaries and various point and line defects are incorporated into the film due to mismatch of geometrical configurations and crystallographic orientations, as shown in Fig. 2.3. If the grains are randomly oriented, the films show a ring-type diffraction pattern and are said to be polycrystalline. However, if the grain size is small (20 Å), the films show halo-type diffraction patterns similar to those exhibited by highly disordered or amorphous (noncrystalline) structures. Even if the orientation of different islands is the same throughout, as obtained under special deposition conditions described in Sec. 6.2.3 on suitable single-crystal substrates, a single-crystal film is not obtained. Instead, the film consists of single-crystal grains oriented parallel to each other and connected by low-angle grain boundaries. These films show diffraction patterns similar to those of single crystals and are called epitaxial single-crystal films.

Besides grain boundaries, epitaxial films may also contain other structural defects such as dislocation lines, stacking faults, microtwins, and twin boundaries, multiple-positioning boundaries, and minor defects arising from aggregation of point defects (for example, dislocation loops, stacking faults, and tetrahedra and small dotlike defects). Note that defects such as stacking faults and twin boundaries occur much less frequently in polycrystalline films. Dislocations with a density of 10^{10} to 10^{11} lines/cm^2 are the most frequently encountered defects in polycrystalline films and are largely incorporated during the network and hole stages due to displacement (or orientation) misfits between different islands. Some other mechanisms which may give rise to dislocations in thin films are (1) substrate film lattice misfit, (2) the presence of inherent large stresses in thin films, and (3) continuation into the film of the dislocations apparently ending on the substrate surface.

After a continuous film is formed, anisotropic growth takes place normal to the substrate in the form of cylindrical columns. The initial nucleation density determines the lateral grain size, or crystallite size.

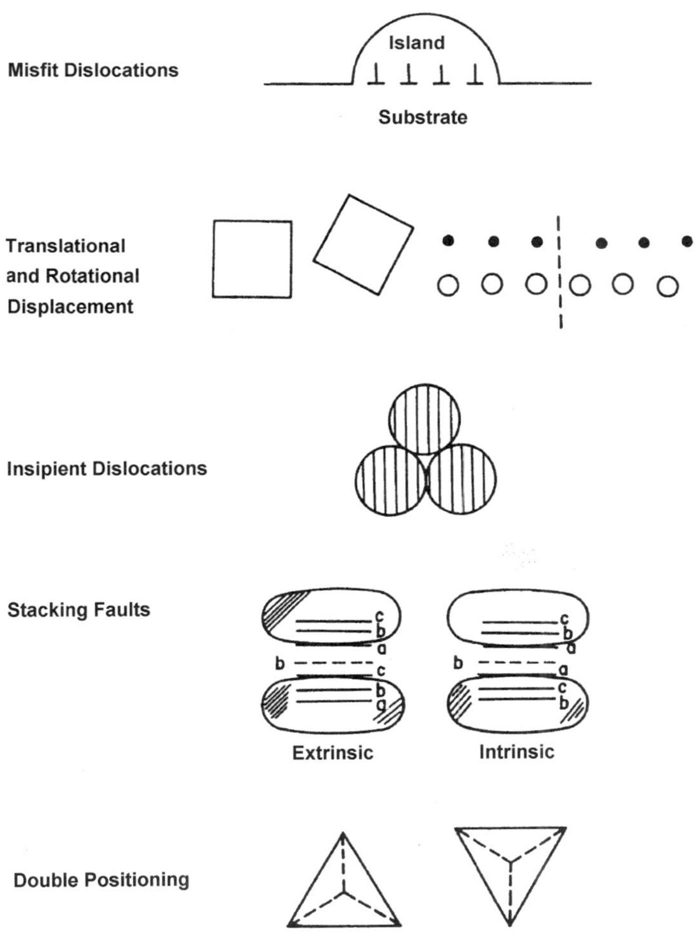

Figure 2.3. A schematic diagram showing the incorporation of defects in a thin film during growth.[2]

However, if recrystallization takes place during the coalescence stage, the lateral grain size is larger than the average separation of the initial nuclei, and the average number of grains per unit area of the film is less than the initial nucleation density. The grain size normal to the substrate is equal to the film thickness. For thicker films, renucleation takes place at the surface of previously grown grains, and each vertical column grows multigranularly with possible deviations from normal growth.

The film growth is initiated by the adatoms. The adatoms will be trapped at a nucleation center after a Brownian movement. The mean residence time of adatoms, τ_s, is estimated by

Eq. (2.1)
$$\tau_s = \tau_v \exp\left(\frac{E_{ad}}{kT}\right)$$

where τ_v is a period of vibration perpendicular to the surface assumed to be almost $1/\nu\,(=10^{-13}\,\text{s})$, where ν ($\approx 10^{13}$Hz) is a frequency of lattice thermal vibration, and E_{ad} is adsorption energy of adatoms on the substrates (= 0.1 to 1 eV). The thermal equilibrium time of the adatoms, τ_e, is expressed by

Eq. (2.2)
$$\tau_e = \tau_s \exp\left(\frac{-E_{ad}}{kT}\right)$$

If $E_{ad} \gg kT$, the adatoms will stay on the surface of substrates where $\tau_s \gg \tau_e$. If $E_{ad} \approx kT$, the adatoms will reevaporate from the substrates. The adatoms will diffuse on the surface showing Brownian movement, which is continuous random movement. The traveling time due to the diffusion on the surface, τ_d, is expressed by

Eq. (2.3)
$$\tau_d = \tau_p \exp\left(\frac{E_d}{kT}\right)$$

where τ_p is a period of vibration parallel to the surface assumed to be almost $1/\nu$ (= 10^{-13} s) and E_d is the surface diffusion energy for the adatoms against the potential barrier on the substrate surface.

The mean traveling distance of the adatoms, X, is expressed by

Eq. (2.4)
$$X = (D_s \tau_s)^{1/2}$$

Eq. (2.5)

$$D_s = \frac{a_0^2}{\tau_d}$$

where a_0 denotes a lattice spacing of surface atoms of the substrates. Since $\tau_p \approx \tau_v$, X is expressed by

Eq. (2.6)

$$X = a_0 \exp\left(\frac{E_{ad} - E_d}{2kT}\right)$$

Both E_{ad} and E_d are important for the growth of thin films. Assume $a_0 = 0.5$ nm, $E_{ad} = 0.2$ eV for adsorption, $E_d = 0.01$ eV, and $T = 300$ K, $X = 20$ nm ($= 39\ a_0$) and the mean residence time (physical) $\tau_s = 160$ ps. When $E_{ad} = 0.4$ eV, $X = 780$ nm ($= 1500\ a_0$) and $\tau_s = 0.24$ μs. This shows that the diffusion of adatoms strongly depends on the E_{ad} and E_d. In general, the diffusion length of the adatoms will be on the order of micrometers during the film growth. When $E_{ad} = 1$ eV for a chemical absorption, τ_s becomes 6×10^3 seconds at 300 K. The τ_s is reduced to 1 ms at 500 K. This estimation shows the potential utility of baking out high vacuum systems.

The growth stage of thin films is governed by the surface energy of thin films, γ_f, the surface energy of substrates, γ_s, and the interfacial energy between thin films and substrates, γ_{fs}. The island growth (Volmer-Weber mode) will be predominant at $(\gamma_s - \gamma_{fs}) < \gamma_f$, and the layer growth (Frank-van der Merwe mode) at $(\gamma_s - \gamma_{fs}) > \gamma_f$. In layer growth, the covering on the surface shows the minimum free energy. The binding energy of thin-film atoms at the coalescence stage is E_b where $E_b < E_{ad}$. The surface treatment before deposition also changes the E_{ad} and E_d. Neugebauer presented the critical review on nucleation and growth of the thin films.[3] Computer simulations are useful for understanding film growth.[4][5]

2.1.1 Structural Consequences of the Growth Process

The microstructure and topographical details of a thin film of a given material depend on the kinetics of growth and hence on the substrate temperature, the source and energy of impurity species, the chemical nature, the topography of the substrate, and gas ambients. These parameters influence the surface mobility of the adsorbed species: kinetic energy of the incident species, deposition rate, supersaturation (i.e., the value of the vapor pressure/solution concentration above that required for condensation into the solid phase under thermodynamical equilibrium conditions), the condensation or sticking coefficient (i.e., the fraction of the total impinging species adsorbed on the substrate), and the level of impurities. How the physical structure is affected by these parameters is described below.[6]

2.1.1.1 Microstructure

The lateral grain size is expected to increase with decreasing supersaturation and increasing surface mobility of the adsorbed species. As a result, deposits with well-defined large grains are formed at high substrate and source temperatures, both of which result in high surface mobility. Transmission electron micrographs of 100-Å-thick Au films deposited on NaCl at 100°, 200°, and 300°C by vacuum evaporation illustrate the effect of substrate temperature as shown in Fig. 2.4. Note that increasing the kinetic energy of the incident species (for example, by increasing the source temperature in the case of deposition by vacuum evaporation, or by increasing the sputtering voltage in the case of deposition by sputtering) also increases the surface mobility. However, at sufficiently high kinetic energies, the surface mobility is reduced due to the penetration of the incident species into the substrate, resulting in a smaller grain size. This effect of the kinetic energy of the impinging species on grain size is more pronounced at high substrate temperatures. Also, the effect of substrate temperature on grain size is more prominent for relatively thicker films.

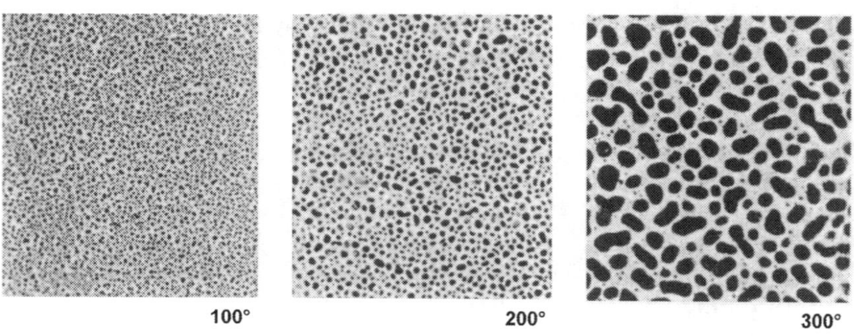

Figure 2.4. Transmission electron micrographs of 100-Å-thick Au films vacuum evaporated on NaCl at 100°, 200°, and 300°C.[2]

Giving the film a postdeposition annealing treatment at temperatures higher than the deposition temperature may also modify the grain size. The higher the annealing temperature, the larger the grain sizes obtained. The effect of heat treatment is again more pronounced for relatively thicker films. The grain growth obtained during postdeposition annealing is significantly reduced from that obtained by depositing the film at annealing temperatures; this is because of the involvement of the high activation-energy process of thermal diffusion of the condensate atoms in the former case compared to the process of condensation of mobile species in the latter.

For a given material-substrate combination and under a given set of deposition conditions, the grain size of the film increases as its thickness increases. However, beyond a certain thickness, the grain size remains constant, suggesting that coherent growth with the underlying grains does not go on forever, and fresh grains are nucleated on top of the old ones above this thickness. This effect of increasing grain size with thickness is more prominent at high substrate temperatures. The effect of various deposition parameters on the grain size is summarized qualitatively in Fig. 2.5. It is clear that the grain size cannot be increased indefinitely because of the limitation on the surface mobility of the adsorbed species.

The conditions favoring epitaxial growth are (1) high surface mobility as obtained at high substrate temperatures; (2) low supersaturation; (3) clean, smooth, and inert substrate surfaces; and (4) crystallographic compatibility between the substrate and the deposit material. Films in which only a particular crystallographic axis is oriented along a fixed direction (due to preferential growth rate) are called *oriented* films. In contrast to epitaxial films, which require a suitable single-crystal substrate, oriented films may also be formed on amorphous substrates. At the other extreme from thin-film microstructures, highly disordered, very fine-grained, noncrystalline deposits with a grain size of 20 Å that show halo-type diffraction patterns similar to those of amorphous structures (i.e., having no translational periodicity over several interatomic spacings) are obtained under conditions of high supersaturation and low surface mobility. The surface mobility of the adsorbed species may be inhibited, for example, by decreasing the substrate temperature, by introducing reactive impurities into the film during growth, or by codepositing materials of different atomic sizes and low surface mobilities. Under these conditions, the film is amorphous-like and grows layer-by-layer.

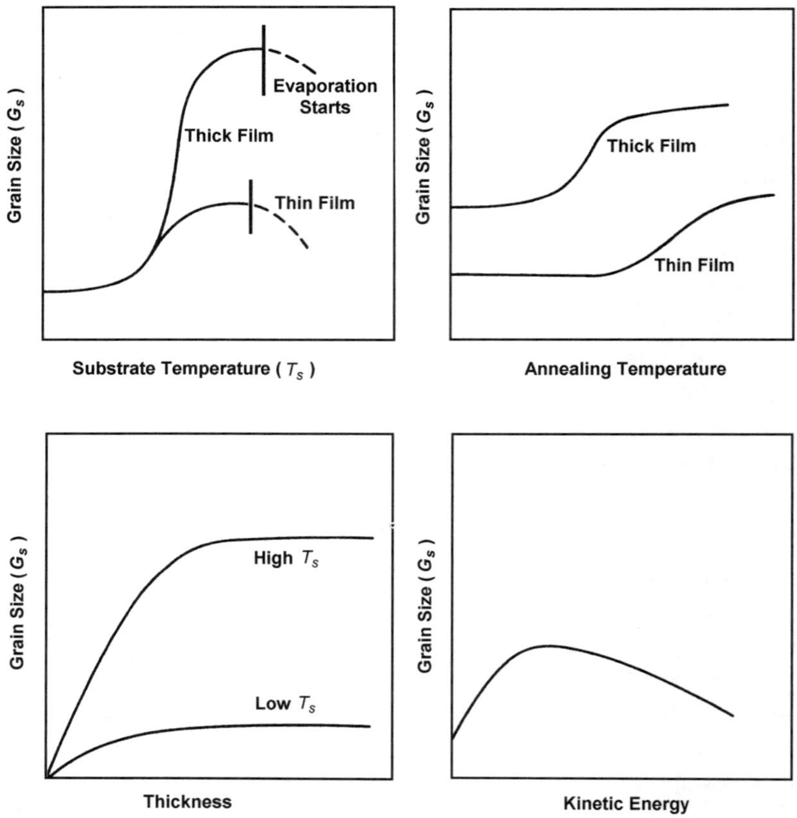

Figure 2.5. Qualitative representation of the influence of various deposition parameters on the grain size of thin films.[2]

2.1.1.2 Surface Roughness and Density

Under conditions of a low nucleation barrier and high supersaturation, the initial nucleation density is high and the size of the critical nucleus is small. This results in fine-grained, smooth deposits which become continuous at small thicknesses. On the other hand, when the nucleation barrier is large and the supersaturation is low, large but few nuclei are formed, resulting in coarse-grained rough films which become continuous at relatively large thicknesses. High surface mobility, in general, increases the surface smoothness of the films by filling in the concavities. One exception is the special case where the deposited material has a tendency to grow preferentially along certain crystal faces because of either large anisotropy in the surface energy or the presence of faceted roughness on the substrate.

A further enhancement in surface roughness occurs if the impinging species are incident at oblique angles instead of falling normally on the substrate. This occurs largely due to the shadowing effect of the neighboring columns oriented toward the direction of the incident species. Figure 2.6 shows the topography of two rough film surfaces, one (Fig. 2.6a) obtained by oblique deposition and the other (Fig. 2.6b) obtained by etching a columnar structure. Also shown are the topographies of rough (Fig. 2.6c) and smooth (Fig. 2.6d) CdS films prepared by controlled homogeneous precipitation under different conditions.[2]

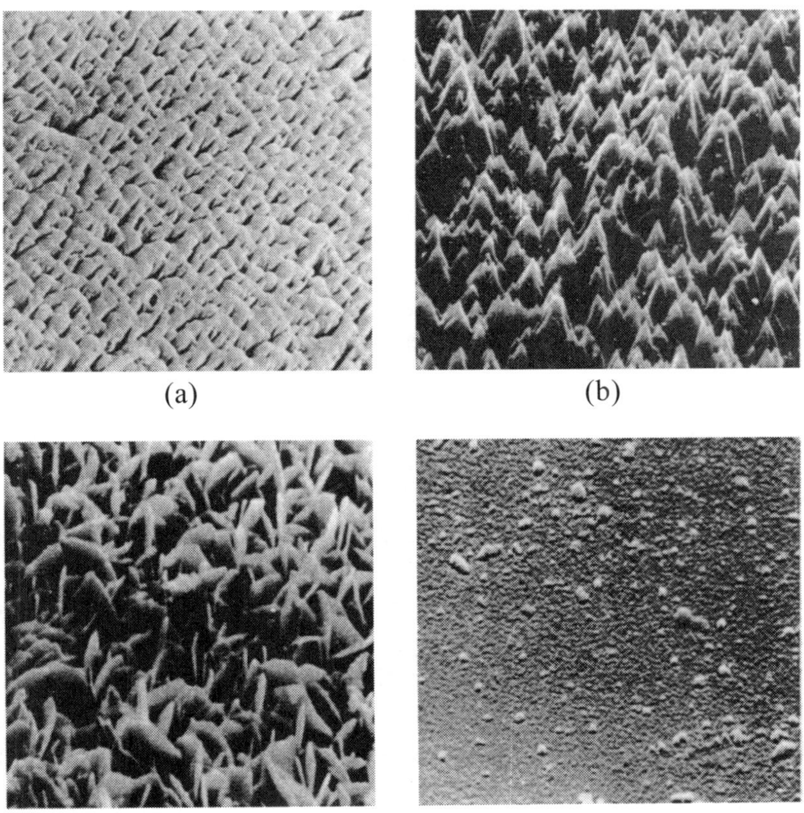

(a) (b) (c) (d)

Figure 2.6. Scanning electron micrographs showing topography of smooth and rough films; (*a*) obliquely deposited GeSe film; (*b*) etched CdS film (vacuum evaporated); (*c*) rough CdS film (solution grown); and (*d*) smooth CdS film (solution grown).[2]

A quantitative measure of roughness, the *roughness factor,* is the ratio of the real effective area to the geometrical area. The roughness factor, $\Delta\theta$, is given by

Eq. (2.7) $$\Delta\theta = \left[\left| \theta^2 - \left(\frac{1}{N} \sum_i h_i^2 \right) \right| \right]^{1/2}$$

where θ, the coverage or average film height, is defined by

Eq. (2.8) $$\theta = \frac{1}{N} \sum_i h_i$$

and N is the number of surface sites, and h_i is the film height of each site.

The variation of the roughness factor with thickness for a number of cases is qualitatively illustrated in Fig. 2.7a. In the case of porous films, the effective surface area can be hundreds of times the geometrical area. The deviation of the chemical composition also changes the roughness of compound thin films.

Density is also an important parameter of physical structure. It must be known for the determination of the film thickness by gravimetric methods. A general behavior observed in thin films is a decrease in the density with decreasing film thickness. This is qualitatively illustrated in Fig. 2.7b. Discrepancies observed in the value of the thickness at which the density of a given film approaches its bulk value are attributed to differences in the deposition conditions and measurement techniques employed by different observers.

Surface roughness is also essentially related to the modes of film growth described earlier in Sec. 2.1. The Frank-van der Merwe mode provides the smooth surface and Volmer-Weber mode provides the rough surface. The roughness is analyzed by computer modeling the film growth.[7] The strains due to the thermal expansion mismatch between a film and its substrate affect the surface microstructure and/or the roughness.[8] In the Ge/Si heterostructure, the compressive misfit strain in Ge thin films induces a transition of a planar film to a three-dimensional island morphology due to a reduction of strain.[9]

In a heteroepitaxial film, the density will change due to the deformation of the lattice structure. $PbTiO_3$ thin films epitaxially grown on

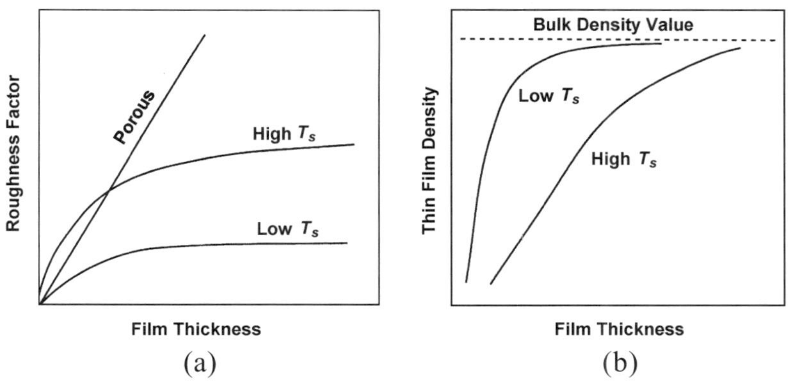

Figure 2.7. Qualitative variation of (*a*) the roughness factor and (*b*) the film density as a function of film thickness.[2]

$SrTiO_3$ substrates show an expansion of the c axis and the a/b axis. The density will be 90 to 95% of the bulk density. Figure 2.8 shows the surface AFM (atomic-force microscope) and SEM (scanning electron microscope) images of $PbTiO_3$ thin films epitaxially grown on vicinal $SrTiO_3$ substrates. The surface roughness was governed by the initial surface structure of the vicinal substrates as seen in Fig. 2.8a. The Pb-rich composition enhanced the growth of small islands on the surface probably due to the two-dimensional nucleation of Pb or PbO_x as seen in Fig. 2.8b. The film growth on the vicinal substrates is discussed in detail in Sec. 6.2.4.

2.1.1.3 Adhesion

The adhesion of a film to the substrate is strongly dependent on the chemical nature, cleanliness, and the microscopic topography of the substrate surface. The adhesion is better for higher values of (1) kinetic energy of the incident species, (2) adsorption energy of the deposit, and (3) initial nucleation density. The presence of contaminants on the substrate surface may increase or decrease the adhesion depending on whether the adsorption energy is increased or decreased, respectively. Also, the adhesion of a film can be improved by providing more nucleation centers on the substrate, for instance, by using a fine-grained substrate or a substrate precoated with suitable materials. Loose and porous deposits formed under conditions of high supersaturation and poor vacuum are less adherent than compact deposits.

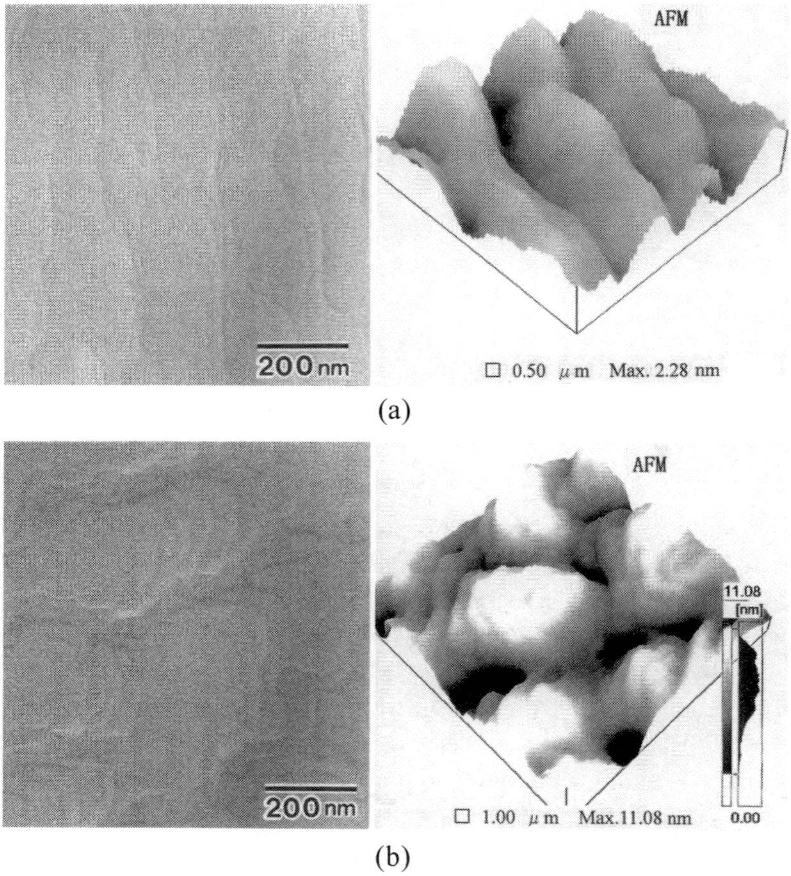

Figure 2.8. Surface SEM and AFM images of $PbTiO_3$ thin films; (*a*) stoichiometric films, (*b*) Pb-rich films grown on miscut (100) $SrTiO_3$.

It is known that the deposited films include internal stress during film growth. The elastic energy is stored in the films due to the stress. The elastic energy density, u_v, is expressed by

Eq. (2.9) $$u_v = \frac{\sigma^2}{2Y} \text{ (in J/m}^3\text{)}$$

where Y denotes Young's modulus of thin films. Thus, the stored energy at unit area of thin films, u_s, becomes

Eq. (2.10)
$$u_s = u_v d = \frac{\sigma^2 d}{2Y} \text{ (in J/m}^2\text{)}$$

where d shows the thickness of the deposited films. When u_s is larger than the adhesive energy, the deposited films will peel off. Under a given adhesive condition, there is a critical film thickness. Beyond the critical thickness, the deposited films will peel off. The internal stress originates mainly from adhesion to the substrates. Stress can be introduced in the films by differential thermal expansion between thin films and the substrates and/or by lattice misfit with crystalline substrates. The internal stress, σ, is estimated by measuring the strain, ε, using an XRD analysis. The stress, σ, is expressed by

Eq. (2.11)
$$\sigma = \frac{\varepsilon Y}{2\nu}$$

where ε denotes the strain and ν denotes the Poisson's ratio of the films. The ratio ν is a positive number for almost all materials and is less than one-half.[10]

2.1.1.4 Metastable Structure

In general, departures from bulk values of lattice constants are found only in ultrathin films. The lattice constants may increase or decrease, depending on whether the surface energy is negative or positive, respectively. As the thickness of the film increases, the lattice constants approach the corresponding bulk values.

A large number of materials, when prepared in thin-film form, exhibit new metastable structures not found in the corresponding bulk materials. These new structures may be purely due to deposition conditions or they may be impurity- or substrate-stabilized. Some general observations regarding these new structures in thin films are listed:

1. Most of the materials, in pure form or in combination with appropriate impurities, can be prepared in amorphous form.
2. The distorted NaCl-structure bulk materials tend to transform to the undistorted form in thin films.
3. The wurtzite compounds can be prepared in sphalerite form and vice versa.

4. Body-centered cubic (bcc) and hexagonal close packed (hcp) structures have a tendency to transform to face-centered cubic (fcc) structure.

Some common examples of such abnormal structures found in thin films are:

1. Amorphous Si, Ge, Se, Te, and As.
2. Face-centered cubic (fcc) Mo, Ta, W, Co, and b-Ta, etc. (all due to deposition conditions).
3. Face-centered cubic Cr/Ni, bcc Fe/Cu, and fcc Co/Cu (all due to the influence of the substrate).

Note that these abnormal metastable structures transform to stable normal structures upon annealing. This is illustrated in Fig. 2.9 for the case of ion-beam sputtered Zr films.

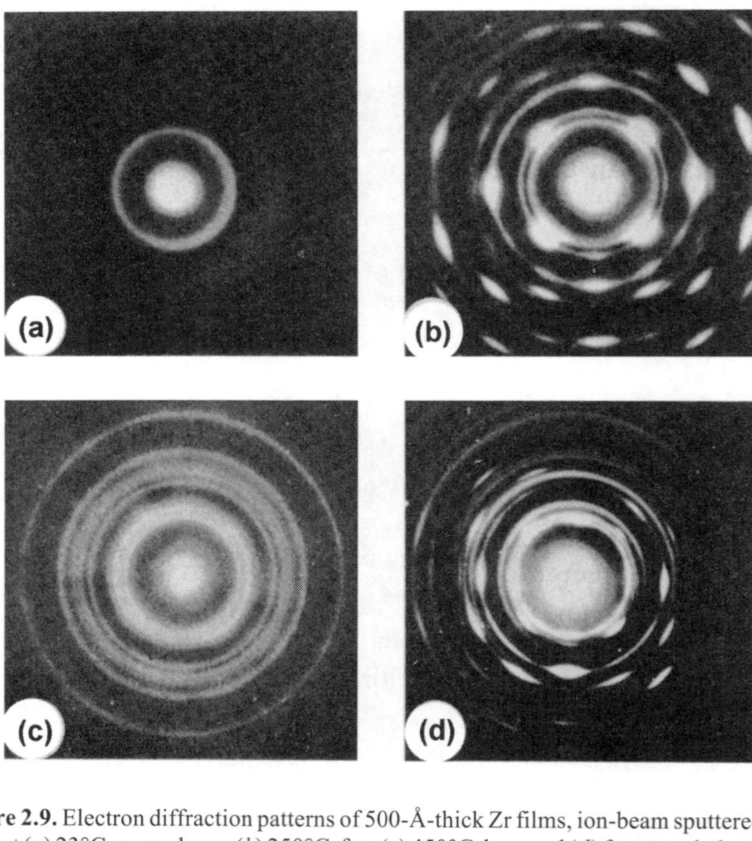

Figure 2.9. Electron diffraction patterns of 500-Å-thick Zr films, ion-beam sputtered onto NaCl at (*a*) 23°C, amorphous; (*b*) 250°C, fcc; (*c*) 450°C, hcp; and (*d*) fcc annealed at 675°C in vacuum, fcc + hcp.[2]

2.1.2 Solubility Relaxation

Another consequence of the thin-film growth process is the phenomenon of solubility relaxation. The atomistic process of growth during codeposition allows the doping and alloying of films. Since thin films are formed from individual atomic, molecular, or ionic species which have no solubility restrictions in the vapor phase, the solubility conditions between different materials upon codeposition are considerably relaxed. This allows the preparation of multicomponent materials, such as alloys and compounds, over an extended range of compositions compared to the corresponding bulk materials. It is thus possible to have materials tailor-made with desired properties adding a new and exciting dimension to materials technology. An important example of this technology of tailor-made materials is the formation of hydrogenated amorphous Si films for use in solar cells. Hydrogenation has made it possible to vary the optical bandgap of amorphous Si from 1 eV to about 2 eV and to decrease the density of dangling bond states in the bandgap so that n- and/or p-type doping is made possible.

2.2 THIN FILM DEPOSITION PROCESS

2.2.1 Classification of Deposition Processes

Typical deposition processes, physical and chemical, are shown in Fig. 2.10. The physical process is composed of the physical vapor deposition (PVD) processes, and the chemical processes are composed of the chemical vapor deposition (CVD) process and the chemical solvent deposition process. In this section, typical physical and/or chemical vapor processes are described.

2.2.1.1 PVD Processes

The PVD process is divided into two categories: (1) thermal evaporation and (2) sputtering. There are also miscellaneous processes discussed at the end of this section.

Thermal Evaporation. The thermal evaporation process comprises evaporating source materials in a vacuum chamber below 1×10^{-6} torr (1.3×10^{-4} Pa) and condensing the evaporated particles on a substrate. The thermal evaporation process is conventionally called *vacuum deposition*. Two types of thermal evaporation processes are shown in Fig. 2.11.

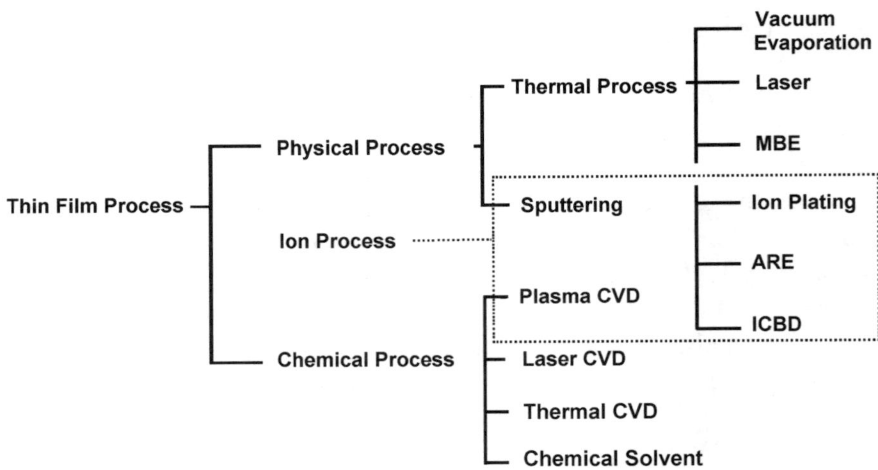

Figure 2.10. Thin-film deposition processes.[1]

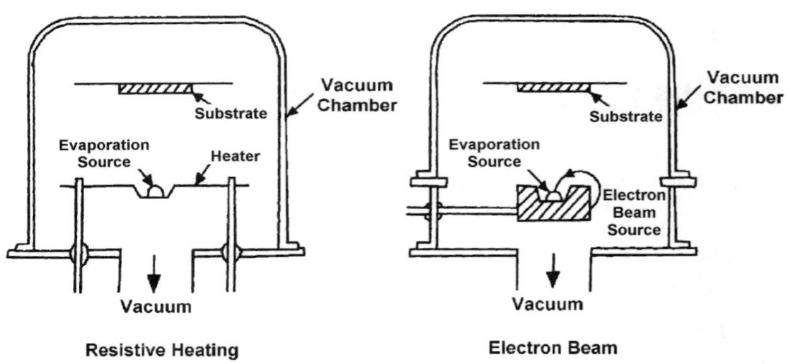

Figure 2.11. Thermal evaporation process: *ES*, evaporation source; *S*, substrate; *H*, Heater; *EB*, electron beam source.

Resistive heating is most commonly used for the deposition of thin films. The source materials are evaporated by a resistively heated filament or boat, generally made of refractory metals such as W, Mo, or Ta, with or without ceramic coatings. Crucibles of quartz, graphite, alumina, beryllia, boron-nitride, or zirconia are used with indirect heating. The refractory metals are evaporated by electron-beam deposition since simple resistive heating cannot evaporate high melting point materials.

Pulsed laser deposition (PLD) is an improved thermal process used for the deposition of alloys and/or compounds with a controlled chemical composition. In laser deposition, a high-power pulsed laser such as the KrF excimer laser (1 J/shot) is irradiated onto the target of source materials through a quartz window. A quartz lens is used to increase the energy density of the laser power on the target source. Atoms that are ablated or evaporated from the surface are collected on nearby sample surfaces to form thin films. A typical PLD system is shown in Fig. 2.12. The target material is locally heated to the melting point, melted, and vaporized in a vacuum. The laser pulse may also provide photoemitted electrons from the target to make a plasma plume and the evaporation mechanism may be complex since the process includes the thermal process and the plasma process. The PLD has the advantage of being simple in design, and the target has many forms such as a powder, sintered pellet, and single crystal. However, at present, the process has a limited area of uniform deposition, and microsized globules or particles are ejected from the target.[11] To avoid the deposition of the microsized ejections, the substrates are settled at an off-axis position.

The mechanism of evaporation is simply discussed by Mahan using the thermal evaporation model.[12] The irradiated laser power is absorbed on the surface of the target. The optical absorption depth, L_o, of an ultraviolet laser is on the order of 10 nm. The depth, L_o, is given by

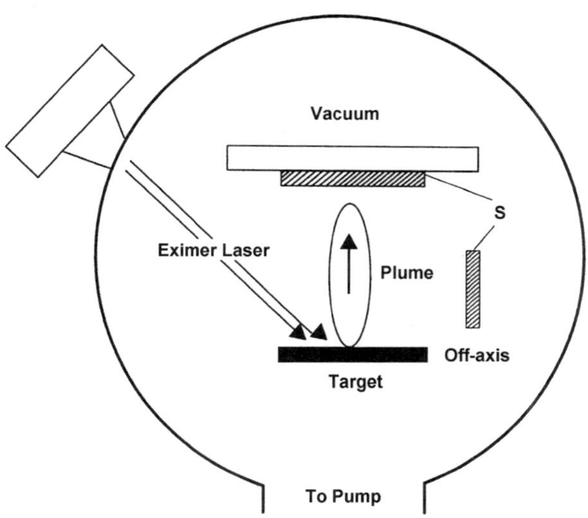

Figure 2.12. Pulsed laser deposition (PLD) uses a laser to ablate the source material from a target. The material is collected on substrates (*S*) in the form of thin films.

Eq. (2.12)

$$L_o = \frac{1}{\alpha}$$

where α is the optical absorption constant of the target. The thermal diffusion depth, L_t, will govern the heating of the target; this is given by

Eq. (2.13)

$$L_t = \left(2\delta t \frac{\kappa}{c n_{mol}}\right)^{1/2}$$

where δt is the pulse duration of the laser light, κ is the thermal conductivity of the target, c is the molar heat capacity of the target, and n_{mol} is the molar density of the target. The L_o and L_t characterize the mode of the heating as follows:

Eq. (2.14)

$$\begin{cases} L_o >> L_t \text{ (weakly absorbing)} \\ L_o << L_t \text{ (strongly absorbing)} \end{cases}$$

The values of L_t for copper and Si are typically in micrometers. The targets of the insulator and wide-bandgap semiconductors show weak absorption. Metal targets and narrow-bandgap semiconducting targets show strong absorption. The strong absorption effectively melts the target and thermally evaporates it. The laser irradiation may photoemit electrons from the target accompanied by photoionized atoms of target materials. The photoionized atoms create the plasma plume.

The *molecular beam epitaxy* (MBE) process is the most reliable deposition process in thermal evaporation. Figure 2.13 shows a typical MBE system. The system is a controlled process, where the evaporation rate of the source materials is controlled in situ by a computerized process control unit. The man-made superlattice structure composed of thin alternating layers of GaAs and GaAlAs can be successfully deposited as shown in Fig. 2.14.[13]

This kind of deposition process is now widely used for the controlled deposition of alloys and compounds. The system is generally composed of a growth chamber, the analysis chamber, and a sample chamber. A typical recent MBE system is shown in Fig. 2.15.[14] A vapor source was also available for the MBE system shown in Fig. 2.16, and metal-organic compounds were used for its source.[15]

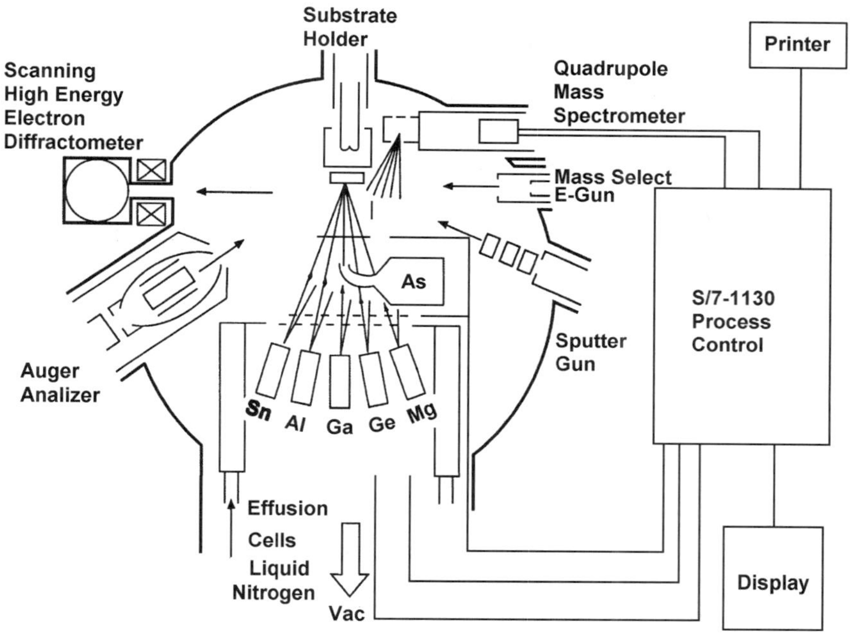

Figure 2.13. Molecular beam epitaxy (MBE) system.[13]

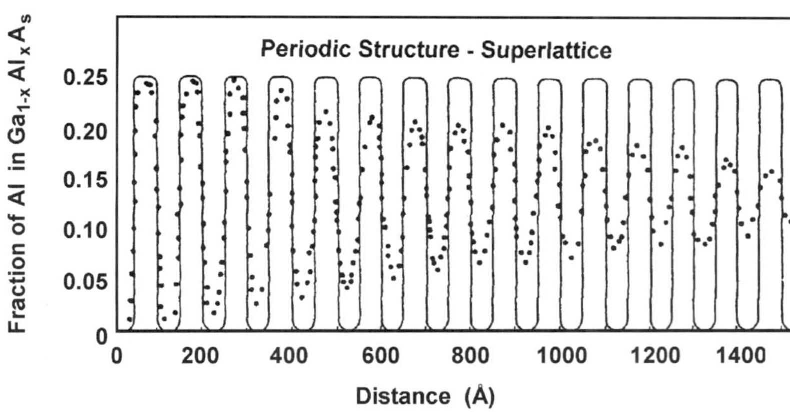

Figure 2.14. Man-made superlattice structure composed of thin alternating layers of GaAs and GaAlAs.[13]

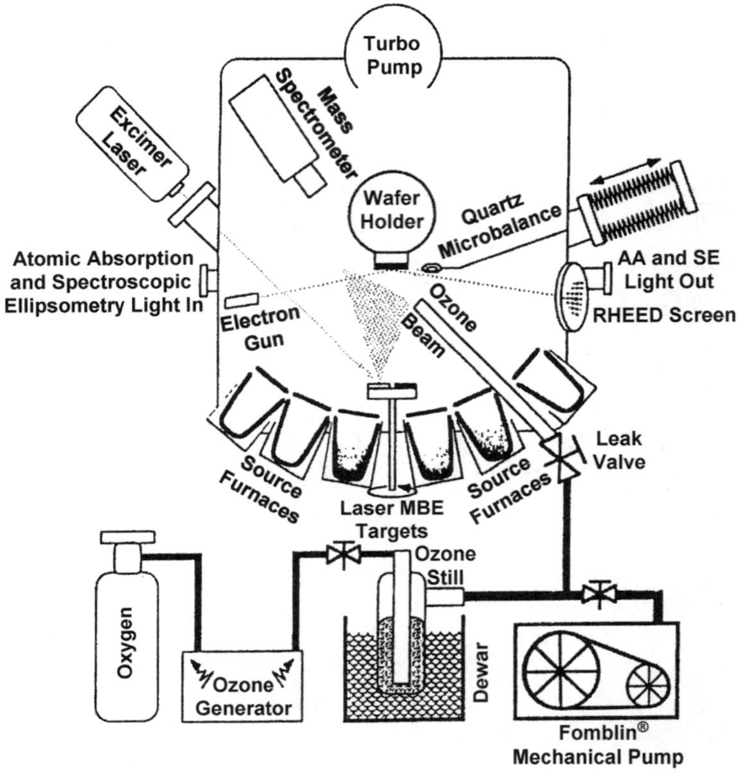

Figure 2.15. A schematic diagram of an MBE growth chamber for growing high T_c superconductors. The growth chamber combines MBE and laser-MBE, and includes in-situ spectroscopic elipsometry.[14]

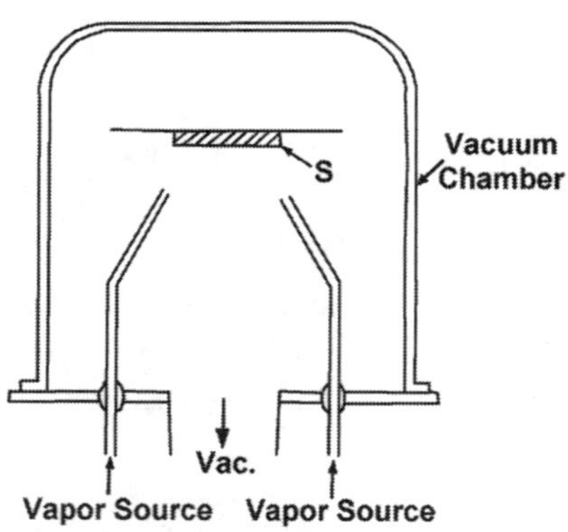

Figure 2.16. A schematic diagram of a vapor source MBE system.

Sputtering. When a solid surface is bombarded with energetic particles such as accelerated ions, surface atoms of the solid are scattered backward due to collisions between the surface atoms and the energetic particles, as shown in Fig. 2.17. This phenomenon is called *back-sputtering,* or simply *sputtering*. When a thin foil is bombarded with energetic particles, some of the scattered atoms transmit through the foil. The phenomenon is called *transmission sputtering.* The word "spluttering" is synonymous with "sputtering." *Cathode sputtering, cathode disintegration,* and *impact evaporation* are also used in the same sense.

Cathode sputtering is used for the deposition of thin films. Several sputtering systems are proposed for thin-film deposition including dc diode, rf diode, magnetron, and ion-beam sputtering. McClanahan and Laegreid presented an historical review of sputtering deposition.[16] Their basic constructions are shown in Fig. 2.18.

Among these sputtering systems, the simplest model is the dc diode sputtering system. The dc sputtering system is composed of a pair of planar electrodes. One of the electrodes is a cold cathode and the other is the anode. The front surface of the cathode is covered with target materials to be deposited. The substrates are placed on the anode. The sputtering chamber is filled with sputtering gas, typically argon gas at 5 Pa (4×10^{-2} torr). The glow discharge is maintained under the application of dc voltage between the electrodes. The Ar^+ ions generated in the glow discharge are accelerated at the cathode fall (sheath) and sputter the target, resulting in the deposition of the thin films on the substrates. In the dc sputtering system, the target is composed of metal since the glow discharge (current flow) is maintained between the metallic electrodes.

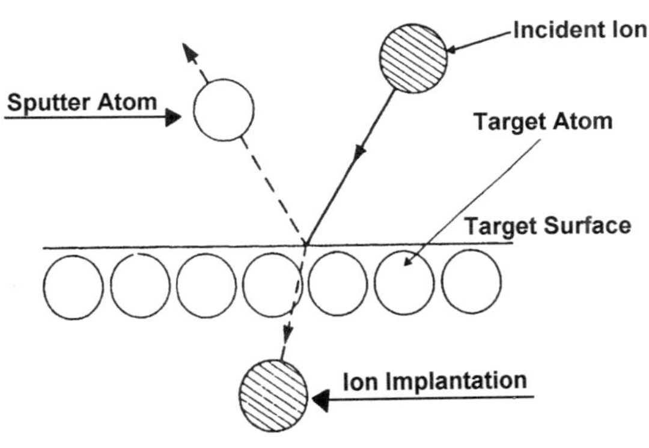

Figure 2.17. The physical sputtering processes.

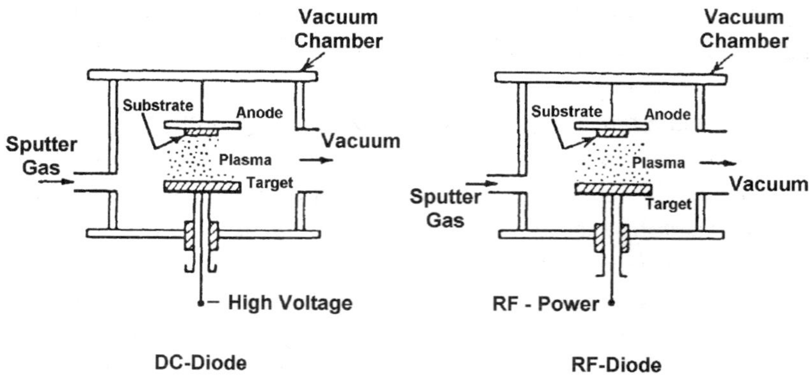

Figure 2.18. Sputter deposition systems.

By simple substitution of an insulator for the metal target in the dc sputtering discharge system, the sputtering discharge cannot be sustained because of the immediate buildup of a surface charge of positive ions on the front side of the insulator. To sustain the glow discharge with the insulator target, rf voltage is supplied to the target. This system is called rf-diode sputtering. In the rf-sputtering system, the thin films of the insulator are sputtered directly from the insulator target.

In magnetron sputtering, a magnetic field is superposed on the cathode and glow discharge, which is parallel to the cathode surface. The electrons in the glow discharge show cycloidal motion, and the center of the orbit drifts in the direction of $\mathbf{E}\times\mathbf{B}$ with the drift velocity of E/B, where **E** and **B** denote the electric field in the discharge and the superposed transverse magnetic field, respectively. The magnetic field is oriented such that these drift paths for electrons form a closed loop. This electron-trapping effect increases the collision rate between the electrons and the sputtering gas molecules. This enables one to lower the sputtering gas pressure as low as 10^{-5} torr, but more typically 10^{-2} torr. In the magnetron sputtering system, the magnetic field increases the plasma density which leads to increases in the current density at the cathode target, effectively increasing the sputtering rate at the target. Due to the gas's low working pressure, the sputtered particles traverse the discharge space without collisions, which results in a high deposition rate.

At present, the planar magnetron is indispensable for the fabrication of semiconductor devices. Historically, magnetron sputtering was first proposed by Penning in 1936.[17] A prototype of the planar magnetron was invented by Wasa in 1967,[18] and a practical planar magnetron system was devised by Chapin in 1974.[19] A typical construction is shown in Fig. 2.19.

When a reactive gas species such as oxygen or nitrogen is introduced into the chamber, thin films of compounds (i.e., oxides or nitrides) are deposited by the sputtering of the appropriate metal targets. This technique is known as *reactive sputtering* and may be used in either the dc or rf mode. Reactive sputtering is used in practice for the high-rate deposition of insulating metal oxide films. However, dc sputtering from a metal target is unstable in a standard large-scale dc sputtering operation due to the deposition of an insulating layer on the cathode leading to the buildup of a charge. To reduce the charge-up phenomenon, medium-frequency sputtering with twin targets is used in practice. A typical system is shown in Fig. 2.20. A medium-frequency power source is connected to the twin targets. The targets act as anode and cathode alternatively at the frequency of several tens of kilohertz, and deposition can be carried out with no appreciable decrease in the deposition rate.

In these glow-discharge systems, sputtering gas molecules, during thin-film growth, irradiate the sputtered films. This causes the inclusion of the gas molecules in the sputtered films. In ion beam sputtering systems, incident ions are generated at the ion source. The target is sputtered in a sputtering chamber separated from the ion source. The typical ion-beam current is 10 to 500 mA, with an ion energy from 0.5 to 2.5 kV. Since the ions are generated in the ion-source discharge chamber, the working pressure of the sputtering chamber can be reduced as low as 1×10^{-5} torr. This reduces the amount of gas molecules included in the sputtered films.

Chopra did pioneering work on the deposition of thin films by ion-beam sputtering in 1967.[20]

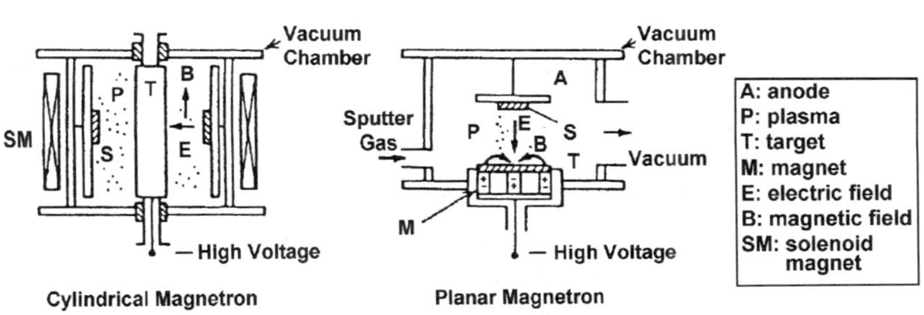

Figure 2.19. Sputter deposition systems.

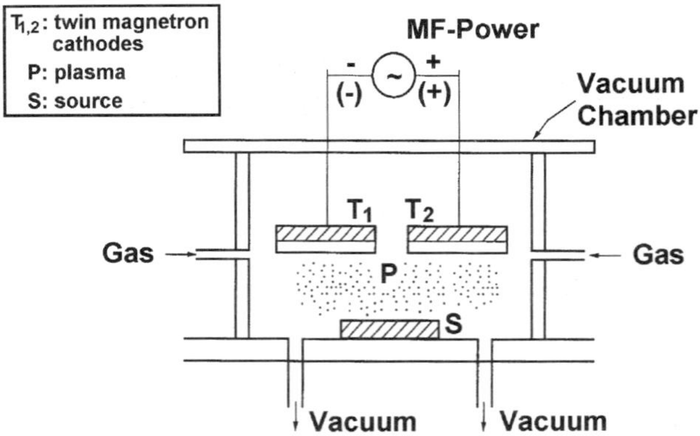

Figure 2.20. Medium frequency sputtering system. T_1, T_2: twin magnetron cathodes; P: plasma.

Although ion-beam sputtering is not widely used for thin-film deposition, this kind of system is widely used for the sputter-etching of semiconductor devices.[21] There has been recent interest in the synthesis of exotic thin films by ion-beam sputtering.[22] The basic sputtering data has been summarized by Behrisch.[23] The sputtering process is described in detail in Chs. 3 and 4.

Miscellaneous PVD processes. Ion plating was first proposed by Mattox in the 1960s.[24] The coating flux is usually provided by thermal evaporation. The evaporated atoms are ionized at the plasma region and accelerated by the electric field prior to deposition. A typical construction is shown in Fig. 2.21. The adhesion of thin films is improved by the acceleration of evaporated atoms.

Activated reactive evaporation (ARE), proposed by Bunshah, is commonly used for the deposition of metal oxides, carbides, and nitrides.[25] The configuration of the ARE system shown in Fig. 2.22 is similar to the ion plating system. Reactive gas is injected into the plasma region to achieve the reaction between evaporated atoms and the reactive gas atoms.

Ionized cluster beam deposition (ICBD), developed by Takagi in the 1970s, is a modification of ion plating.[26] Atoms are evaporated from a closed source through a nozzle. Cooling of the atoms upon expansion through the nozzle leads to cluster formation, which might have a few

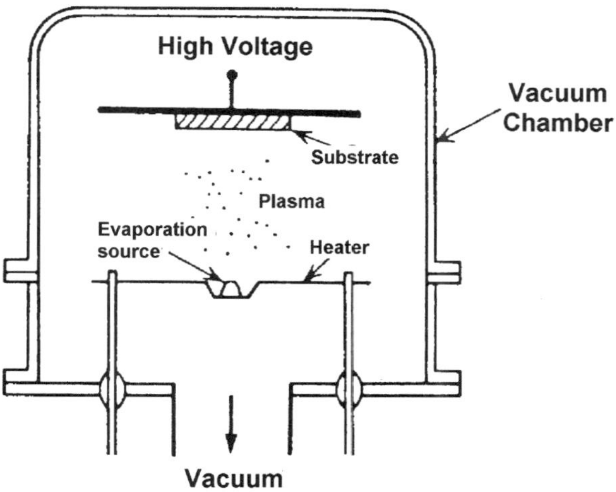

Figure 2.21. Ion plating (dc).

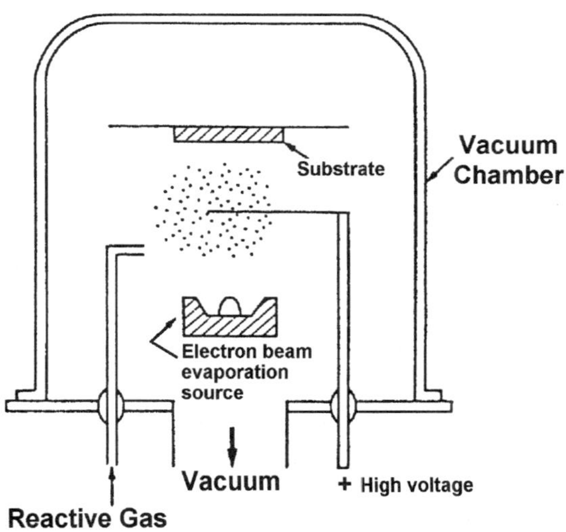

Figure 2.22. Activated reactive evaporation (ARE).

hundred to a thousand atoms per single cluster. The cluster is ionized through the plasma region and then is accelerated to the substrate. The average energy of the atoms in the accelerated cluster is in a range of 0.2 to several electron volts, even when the clusters themselves are accelerated to kilovolts. The relatively low energy of the adatoms reduces lattice damage to the substrate surface.

2.2.1.1 CVD Processes

When a volatile compound of the substance to be deposited is vaporized, and the vapor is thermally decomposed into atoms or molecules, and/or reacted with other gases, vapors, or liquids at the substrate surface to yield nonvolatile reaction products on the substrate, the process is called *thermal chemical vapor deposition* (CVD).

Most CVD processes operate in the range of a few torr to above the atmospheric pressure of the reactants. A relatively high temperature (near 1000°C) is required for CVD processes. Several CVD processes are proposed to increase the efficiency of the chemical reaction at lower substrate temperatures. A typical construction of a CVD deposition system is shown in Fig. 2.23.

Plasma-assisted chemical vapor deposition (PACVD) is one of the modifications of the conventional CVD process. The basic construction is shown in Fig. 2.24. In the PACVD system, electric power is supplied to the reactor to generate the plasma. The power is supplied by an induction coil from outside of the chamber, or directly by diode glow-discharge electrodes. Usually the working pressure is in the range of 10 to 100 Pa. In the plasma, the degree of ionization is typically only 10^{-4}, so the gas in the reactor consists mostly of neutrals. Ions and electrons travel through the neutrals and get energy from the electric field in the plasma. The average electron energy is 2 to 8 eV, which corresponds to electron temperatures of 23,000 to 92,800 K. In contrast, the heavy, much more immobile, ions cannot effectively get enough coupling energy from the electric field. The ions in the plasma show slightly higher energy than the neutral gas molecules at room temperature. Typically the temperature of the ions in plasma is around 500 K.

Since the electron temperature in the plasma is much higher than the gas temperature, thermal equilibrium is not maintained between electrons and neutral gas molecules. This suggests that the plasma in the glow discharge is a sort of "cold plasma" which comprises high temperature electrons (i.e., "hot electrons") and room temperature gas molecules.

The high-temperature electrons enhance the chemical reactions in the plasma as indicated in Table 2.1. This results in the lowering of the temperature of the reactions.

Several improved PACVD processes have been developed. In one major development, microwave-based plasmas have been used to reduce the working pressure. A magnetic field is superimposed on the microwave plasma at the appropriate field strength to cause a resonance between the electron cyclotron frequency and the applied electric field. This is known as an *electron cyclotron resonance* (ECR) condition. A typical construction is shown in Fig. 2.25.

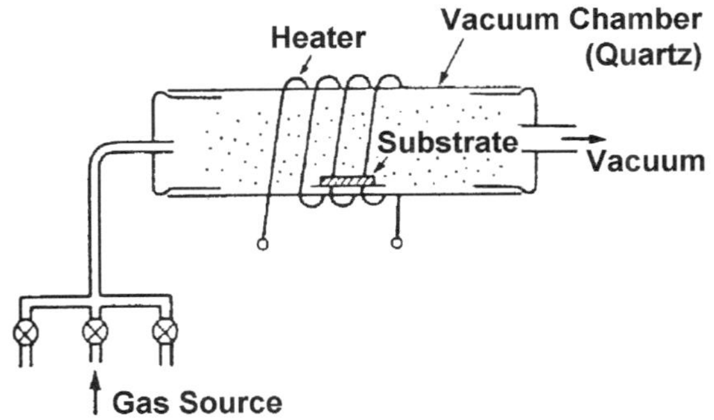

Figure 2.23. Chemical vapor deposition (CVD).

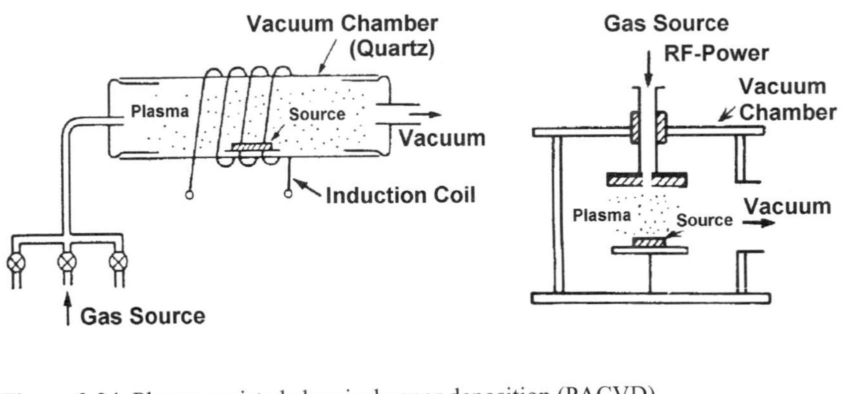

Figure 2.24. Plasma assisted chemical vapor deposition (PACVD).

Table 2.1. Plasma Enhanced Chemical Reactions

Electron neutrals	
Excitation	$e^- + A \rightarrow A^* + e^-$
Dissociation	$e^- + AB \rightarrow A + B + e^-$
Electron attachment	$e^- + A \rightarrow A^+ + 2e^-$
Dissociative attachment	$e^- + AB \rightarrow A^+ + B + 2e^-$
Electron ions	
Recombination	$e^- + A^+ \rightarrow A$
Dissociative recombination	$e^- + AB \rightarrow A^- + B$
Metastable collisions	$A^* + G \rightarrow G^* + A + e^-$
Ion neutrals	$A^+ + BC \rightarrow A^+B + C$
Surface reaction	$AB \rightarrow A\ (film) + B$ $A^+ + CB\ (film) \rightarrow A^+C\ (film) + B\ (film)$

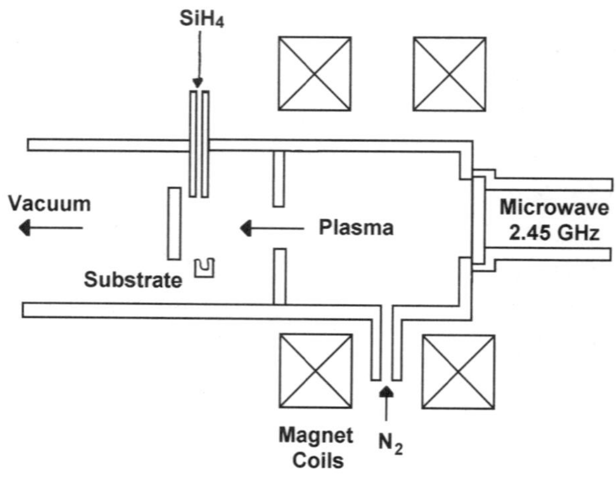

Figure 2.25. ECR plasma assisted chemical vapor deposition.

Laser-assisted chemical vapor deposition (laser CVD) has also been developed as a modification of CVD.[27] These CVD reactions are activated by the irradiation of ultraviolet laser light without the application of electric power.

Among these deposition processes, ion-assisted depositions such as sputtering, ion plating, and PACVD are called *ion processes* as shown in Fig. 2.10

2.2.2 Deposition Conditions

Most thin films are deposited by evaporation (vacuum deposition), sputtering, and/or chemical vapor deposition. The nature of deposited films is governed by the deposition parameters including deposition rate, substrate temperature, substrate materials, and deposition atmosphere.[28]

The chemical composition of deposited films is governed by the substrate temperature and/or the deposition atmosphere. Under low substrate temperatures, the chemical composition of deposited films coincides with that of the source materials. Under high substrate temperatures, the chemical composition of deposited films differs from the source materials due to the reevaporation of high vapor pressure materials from the films during the deposition.

The gas molecules of the deposition atmosphere are frequently included in the deposited films. The chemically active gas molecules react with the thin films during film growth and the resultant films become the compounds between the evaporated source and the active gas in the deposition atmosphere. Thin films of metal oxides, for instance, are prepared by reactive sputtering from a metal target in an oxygen atmosphere. A guide for the deposition of the elements is presented in Table 2.2. A suitable selection of deposition processes are required when thin films are used for the preparation of active electronic devices such as semiconducting devices, since the surface of semiconductive substrates is often damaged during deposition. The special features of the typical deposition processes are listed in Table 2.3.

To understand thin-film deposition, we should consider the growth rate of thin films. The kinetic theory of gas shows that the numbers, N_0, of evaporated gas molecules impinging on the substrate surface is expressed by

Table 2.2. Periodic Table of Elements with Deposition Condition

Element
Atomic weight
Density, g/cc
ρ, μΩcm
MP, °C
VP*
(S) structure
Deposition

VP*: Temperature for 1×10^{-4} Torr
S: Sputter yield at Ar^{+}600 eV

Deposition : examples for thin film deposition

EB : electron beam
R ( ) : resistive heating (coil materials)
CR ( ) : resistive heating (crucible materials)
BO ( ) : resistive heating (boat materials)
SP : sputtering

Crystal structure

cub	Cubic
fcc	Face-centered cubic
bcc	Body-centered cubic
hex	Hexagonal
dia	Diamond type
rmb	Rhombohedral
tet	Tetragonal
fct	Face-centered tetragonal
ort	Orthorhombic
mnc	Monoclinic

	Ia	IIa	IIIa	IVa	Va	VIa	VIIa	VIII	VIII	VIII	Ib	IIb	IIIb	IVb	Vb	VIb	VIIb	
1	METAL									H^{1} 1.00797 0.071 - −259.2 * -			METAL SEMICONDUCTOR		SEMI METAL NON-METAL			He^{2} 4.0026 - - - * -
2	Li^{3} 6.939 0.534 8.55 181 *404 bcc	Be^{4} 9.0122 1.85 2.8 1284 *997 hcp											B^{5} 10.811 2.34 1.8×10^{18} 2225 *1707	C^{6} 12.01115 2.26 (350~6300) 3827 *2137 dia hex	N^{7} 14.0067 0.81 - −210 * - cub	O^{8} 15.9994 1.14 - −218.8 * - ort	F^{9} 18.9984 1.5(−273°C) - −219.61 * -	Ne^{10} 20.183 1.20(−245°C) - −248.59 * - fcc
		EB											EB, SP	EB				
3	Na^{11} 22.9898 0.97 4.2 97.6 *93 bcc	Mg^{12} 24.312 1.74 3.9 650 *327 hcp											Al^{13} 26.9815 2.70 2.45 660.1 *972 (1.2) fcc	Si^{14} 28.086 2.33 3.5×10^{11} 1412 *1337 (0.2) dia	P^{15} 30.9738 1.82 - 44.0 *129 ort	S^{16} 32.064 2.07 - 119 *55 ort	Cl^{17} 35.453 2.2(−273°C) - −100.99 * - ort	Ar^{18} 39.948 1.65(−233°C) - −189.37 * - fcc
		EB											EB,R(W),SP	EB		CR(FQ)		
4	K^{19} 39.102 0.86 9.1 63.4 *123 bcc	Ca^{20} 40.08 1.55 3.43 839 *459 fcc (bcc hcp)	Sc^{21} 44.956 3.00 66 1539 *1107 hcp	Ti^{22} 47.90 4.51 50 1668 *1442 (0.6) hcp (bcc)	V^{23} 50.942 6.10 18.2 1905 *1547 bcc	Cr^{24} 51.996 7.19 12.7 1875 *1157 (1.3) bcc	Mn^{25} 54.9380 7.43 258 1244 *747 bcc (fcc cub)	Fe^{26} 55.847 7.86 8.9 1535 *1227 bcc fcc	Co^{27} 58.9332 8.90 5.6 1492 *1257 (1.4) hcp (fcc)	Ni^{28} 58.71 8.90 6.14 1453 *1262 fcc	Cu^{29} 63.546 8.96 1.56 1083 *1027 (2.3) fcc	Zn^{30} 65.37 7.14 5.5 419.505 *247 hcp	Ga^{31} 69.72 5.91 13.6 29.6 *907 ort	Ge^{32} 72.59 5.32 4.0×10^{7} 936 *1137 (1.2) dia	As^{33} 74.9216 5.72 32 817 *204 rmb	Se^{34} 78.96 4.79 8×10^{6} 217 *164 hex	Br^{35} 79.904 4.2(−273°C) - −7.2 * - ort	Kr^{36} 83.80 3.4(−273°C) - −157.3 * - fcc
		CR(FQ)	EB, SP	EB, SP	EB, SP	BO(W), SP	EB, BO(W)	EB, SP	EB, SP	EB, SP	EB, SP	EB, SP	CR(FQ)	EB	CR(Al_2O_3)	BO(W), EB		
5	Rb^{37} 85.47 1.532 11.0 38.6 *94 bcc	Sr^{38} 87.62 2.6 23 772 *404 fcc	Y^{39} 88.905 4.47 53 1502 *1332 hcp	Zr^{40} 91.22 6.49 40 1850 *1987 hcp (bcc)	Nb^{41} 92.906 8.4 13.9 2468 *2277 (0.65) bcc	Mo^{42} 95.94 10.22 5.2 2615 *2117 (0.9) bcc	Tc^{43} 99 11.5 - 2170 *2077	Ru^{44} 101.07 12.2 7.6 2280 *1987 hcp	Rh^{45} 102.905 12.4 4.3 1960 *1707 (1.5) fcc	Pd^{46} 106.4 12.02 10.0 1552 *1192 (2.4) fcc	Ag^{47} 107.868 10.5 1.51 960.8 *832 (3.8) fcc	Cd^{48} 112.40 8.65 6.8 321.03 *177 hcp	In^{49} 114.82 7.31 8.4 156.61 *742 fct	Sn^{50} 118.69 7.30 β11.5 231.91 *997 dia (tet)	Sb^{51} 121.75 6.62 39.0 630.5 *425 rmb	Te^{52} 127.60 6.24 4.2×10^{5} 449.6 *280 hex	I^{53} 126.9044 4.94 13×10^{21} 113.6 * - ort	Xe^{54} 131.30 - - −112.5 * - fcc
		EB, BO(W)	EB, SP	EB, SP	EB, SP	EB, SP	EB	EB, SP	EB, SP	R(W),EB,SP	EB,R(W),SP	CR(FQ)	EB, BO(W)	CR(Al_2O_3), SP	CR(Al_2O_3), SP	CR(FQ)		
6	Cs^{55} 132.905 1.90 18.8 28.6 *78 bcc	Ba^{56} 137.34 3.5 60 725 *462 bcc	57~71	Hf^{72} 178.49 13.1 29.6 2222 *1997 bcc	Ta^{73} 180.948 16.6 12.6 2998 *2587 (0.3) bcc	W^{74} 183.85 19.3 4.9 3380 *2757 (0.6) bcc	Re^{75} 186.2 21.02 18.6 3160 *2587 (0.9) hcp	Os^{76} 190.2 22.5 9.66 3027 *2487 (0.95) hcp	Ir^{77} 192.2 22.5 4.9 2443 *2107 fcc	Pt^{78} 195.09 21.45 9.81 1769 *1747 (1.6) fcc	Au^{79} 196.967 19.3 2.04 1063 *1132 (2.8) fcc	Hg^{80} 200.59 13.6 94.0766 −38.87 *−7 rmb	Tl^{81} 204.37 11.85 15 303 *463 hcp (bcc)	Pb^{82} 207.19 11.4 19.0 327.426 *547 fcc	Bi^{83} 208.980 9.8 107 271.375 *517 rmb	Po^{84} 210 9.2 - 246 *221 mnc	At^{85} 210 - - 302 *43	Rn^{86} 222 - - −71 * -
		EB	ACTINIDE	EB, SP	EB, SP	EB, SP	EB, SP	EB, SP	EB, SP	EB, SP	EB,R(W),SP		CR(FQ)	EB, BO(W)	R(W),CR(Al_2O_3)	CR(FQ)		
7	Fr^{87} 223 - - 24 *61	Ra^{88} 226 5.0 - 700 *417	Ac^{89} 227 10.07 - 1050 *1332	Th^{90} 232.038 11.7 13 1751 *1977 (0.7) fcc	Pa^{91} 231 15.4 - 1425 * -	U^{92} 238.03 19.07 32 1131 *1582 (1.0) ort (tet bcc)	Np^{93} 237 19.5 - 637 * - ort (tet bcc)	Pu^{94} 242 19~19.72 150(25°C) 640 *1207 mnc (ort bcc) (fcc, fct)	Am^{95} 243 11.7 - 995 *867	Cm^{96} 247 - - - * -	Bk^{97} 249 - - - * -	Cf^{98} 251 - - - * -	Es^{99} 254 - - - * -	Fm^{100} 253 - - - * -	Md^{101} 256 - - - * -	No^{102} - - - - * -	Lw^{103} - - - - * -	
						EB, SP												
LANTHANIDE		57~71	La^{57} 138.91 6.17 57 920 *1422 hcp (bcc fcc)	Ce^{58} 140.12 6.67 75 797 *1377 fcc, (hcp fcc bcc)	Pr^{59} 140.907 6.77 68 935 *1147 hcp (bcc)	Nd^{60} 144.24 7.004 64 1024 *1047 hcp (bcc)	Pm^{61} 147 - - 1035 * -	Sm^{62} 150.35 7.54 92 1072 *580 rmb	Eu^{63} 151.96 5.26 81 826 *466 bcc	Gd^{64} 157.25 7.895 134 1312 *1077 hcp (bcc)	Tb^{65} 158.924 8.272 116 1356 *1147 hcp	Dy^{66} 162.50 8.54 91 1407 *897 hcp	Ho^{67} 164.930 8.803 94 1461 *947 hcp	Er^{68} 167.26 9.051 86 1497 *947 hcp	Tm^{69} 168.934 9.332 90 1545 *680 hcp	Yb^{70} 173.04 6.96 28 824 *417 fcc (bcc)	Lu^{71} 174.97 9.842 68 1652 *1277 hcp	
			EB	EB	EB	EB		EB	EB	EB	EB, SP	EB	EB	EB	CR(Al_2O_3)			

Table 2.3. Special Features of Deposition Processes

Type of deposition	Evaporation		Sputtering		CVD		PLD*
Property	Resistive heating	Electron beam	Diode	Magnetron	Pyrolysis	Plasma	Excimer laser
Thin film material	Material of low melting point	Material of high melting point, refractory metals	Wide varieties of materials, compounds, refractory metals, alloys		Decomposition and/or chemical reaction of organometallic compounds or halides		Varieties of materials, compounds, refractory metals, alloys
Substrate temperature	Low		High (>300°C)	Low (≈ 100°C)	High (≈ 1000°C)	High (>300°C)	
Deposition rate	High Metal: ≈ 0.5–5 μm/min		Low Metal: ≈ 0.02–0.2 μm/min	High Same as evaporation method	High Same as evaporation method		
Gas pressure	Low $<10^{-5}$ torr		High ≈ 0.01–0.1 torr	Low ≈ 10^{-4}–10^{-3} torr	High 1 atm	High ≈ 1–10 torr	High ≈ 0.01–1 torr
Energy of evaporated atoms	≈ 0.–0.2 eV		≈ 10–200 eV	≈ 10–20 eV			≈ 0.1–100 eV
Energy of adatoms	≈ 0.1–0.2 eV		≈ 0.1–20 eV	≈ 0.2–10 eV			≈ 0.1–10 eV
Contamination: Residual gas atoms / adatoms	≈ 10^{-2}		50	1	High purity		

* Morimoto, S., private communication 2002

Eq. (2.15) $$N_0 = 3.5\times10^{22} \frac{p_0}{(MT)^{1/2}} \text{ (particles/cm}^2\text{ sec)}$$

or

Eq. (2.16) $$m_e = 5.8\times10^{-2} p_0 \left(\frac{M}{T}\right)^{1/2} \text{ (g/cm}^2\text{ sec)}$$

where p_0 is the vapor pressure (torr) of the evaporated gas molecules, M is the molar weight of the evaporated gas molecules, T(K) is the temperature of evaporated gas molecules, and m_e is the mass of impinging evaporated particles. When the substrates are in an atmosphere of evaporated source gas of pressure, p_0, the evaporated gas molecules will arrive at the substrates, and thin films will be grown on the substrates. If we assume the whole molecules are deposited on the substrates without reevaporation from the substrates, the growth rate of thin films, D, is expressed by

Eq. (2.17) $$D = 5.8\times10^{5} \frac{p_0 (M/T)^{1/2}}{\rho} \text{ (nm/sec)}$$

where ρ is the density of thin films.

It is known that a source vapor pressure, p_0, of 10^{-2} torr gives a useful deposition rate of thin films. As an example, for the deposition of copper thin films, taking $p_0 = 10^{-2}$ torr, $M = 63.5$ g, $T = 300$K, and $\rho = 8.95$ g/cm^3, D becomes 300 nm/sec. In a thermal evaporation process, the practical evaporation temperature for metallic elements (e.g., vapor pressure: 10^{-2} torr) is almost equal to the melting point. Table 2.2 shows the characterizing temperatures of elements giving the vapor pressures in units of 10^{-4} torr. Source vapor pressures of 10^{-2} torr will be obtained at 200° to 300°C higher than the characterizing temperatures. In a vacuum system, to suppress outgassing from the wall materials, the temperature of the wall should be 200° to 300°C lower than the characterizing temperatures of the elements of the wall materials. These lower temperatures of 200° to 300°C reduce the outgas pressure to below 10^{-6} torr.

The crystalline properties of deposited films are controlled by the selection of the substrate materials and the substrate temperatures.

Amorphous thin films are prepared on glass and/or ceramic substrates at substrate temperatures below the crystallization temperature of the thin films. Polycrystalline films are prepared on glass and/or ceramic substrates at substrate temperatures above the crystallization temperature. Single-crystalline films are, in general, grown on single-crystal substrates due to the epitaxial growth process at substrate temperatures above the epitaxial temperature. In the epitaxial growth process, the epitaxial temperature is governed by the relationship

Eq. (2.18)
$$R = a\exp\left(-\frac{Q}{kT_e}\right)$$

where R denotes the deposition rate, T_e is the minimum temperature for the epitaxial growth (epitaxial temperature), and Q is the activation energy for the epitaxial growth. The epitaxial temperature is about 400°C for vacuum-evaporated Ge thin films, and about 1,100° to 1,200°C for Si thin films. A low deposition rate results in a lower epitaxial temperature. These epitaxial processes are called *vapor-phase epitaxy*. In contrast, polycrystalline films deposited on single-crystalline substrates typically become single-crystalline films due to a postannealing process. This case is called *solid-phase epitaxy*. The relation between the deposition conditions and the crystalline properties of deposited films are summarized in Table 2.4.

Typical substrate materials of glass, ceramics, and single crystals are shown in Tables 2.5, 2.6, and 2.7, respectively. The epitaxial relationships of single-crystal substrates are summarized for various thin films in Table 2.8.

Table 2.4. Deposition Conditions and Crystalline Properties

Amorphous phase	$T_s < T_c$
Polycrystalline	$T_s > T_c$; $T_s < T_c$, post-annealing
Single crystals	$T_s > T_{epi}$
(single crystal substrate)	$T_s < T_c$, post-annealing (solid-phase epitaxy)

Table 2.5. Typical Glass Substrates for Growth of Polycrystalline Thin Films

Material		Microsheet	Pyrex	Alumino-borosilicate	Fused silica
Corning Code No.		0211	7740	7059	7940
Density (g/cm^3)		2.57	2.23	2.76	2.2
Thermal expansion ($\times 10^{-7}$/°C) (≈ 0–300°C)		73.8	32.5	46	5.5
Strain point (°C)		508	510	593	956
Anneal point (°C)		550	560	639	1084
Softening point (°C)		720	821	844	1580
Thermal conductivity (cal/cm/sec/°C) (25°C)		-	0.0027	-	0.0034
Hardness (KHN_{100})		608	418	-	489
Young's modulus ($\times 10^3$ kg/cm^2)		7.59	6.4	6.89	7.4
Poisson's ratio		0.22	0.20	0.28	0.16
Resistivity	25°C	-	15	-	17
(log ρ, Ωcm)	250°C	8.3	8.1	13.1	11.8
Dielectric constant (1 MHz, 20°C)		6.7	4.6	5.84	3.8
Tan δ (%)		3.1	2.6	0.58	0.0038
Refractive index (5893, 5876 Å)		1.523	1.474	1.530	1.459
Optical transparency		>90% ≈ 0.36–2.5 μm (1 mm thick)	>90% ≈ 0.36–2.2 μm (2 mm thick)	—	>90% ≈ 0.36–2.2 μm (2 mm thick)
Composition		SiO_2 64.4% B_2O_3 10.3% Na_2O 6.2% ZnO 5.4% K_2O 6.9% Al_2O_3 4.1% TiO_2 3.1%	SiO_2 80.5% B_2O_3 12.8% Na_2O 4.1% Al_2O_3 2.5% K_2O 0.5%	SiO_2 49.9% BaO 25.1% B_2O_3 10.5% Al_2O_3 10.3% CaO 4.3%	SiO_2 99.5%

Table 2.6. Typical Ceramic Substrates for the Growth of Polycrystalline Thin Films

Material	Steatite	Forsterite	Alumina		Beryllia	Spinel
	$MgO{\cdot}SiO_2$	$2MgO{\cdot}SiO_2$	Al_2O_3 ≈ 96–97%	Al_2O_3 99.50%	BeO	$MgO{\cdot}Al_2O_3$
Density (g/cm^3)	≈ 2.6–2.7	≈ 2.7–2.8	3.8	≈ 3.8–3.9	2.9	3.3
Thermal expansion ($\times 10^6/°C$) (≈ 40–400°C)	≈ 7–8	10	6.7	6.8	6	8.1
Max. safe temp. (°C)	1000	1000	1600	1750	1600	1200
Thermal conductivity (cal/cm/sec/°C) (20°C)	0.006	0.008	0.05	≈ 0.06–0.07	≈ 0.5–0.55	0.04
Specific heat (cal/g·°C)	-	-	0.19	0.19	-	-
Hardness, Hv (kg/mm^2)	650 (500 g)	900 (500 g)	1600 (500 g)	1600 (500 g)	-	-
Compressive strength (kg/cm^2)	9000	9000	17,500	25,000		
Flexural strength (kg/cm^2)	1650	1650	≈ 2800–3500	≈ 2800–4000	-	1500
Young's modulus ($\times 10^{-6}$ kg/cm^2)	-	-	≈ 3.4–3.5	≈ 3.7	-	-
Dielectric strength (kV/mm)	9	9~10	10	10	9	10
Resistivity (log ρ, Ωcm) 25°C	>14	>14	>14	>14	>14	>14
Resistivity (log ρ, Ωcm) 500°C	8	10	11	11		11
Dielectric constant (1 MHz) (25°C)	6.3	≈ 6–6.5	9.4	9.7	≈ 6.3–6. 4	8
tan δ (%) (1 MHz) (25°C)	0.06	≈ 0.01–0.03	0.02	0.02		0.01
Miscellaneous property			surface roughness Rz = 1.3 μm Ra = 0.3 μm	Rz = 0.5 μm Ra = 0.075 μm		

Table 2.7. Typical Crystal Substrates for the Epitaxial Growth of Thin Films

Crystal Material	α-Al_2O_3 (sapphire)	Si	Ge	GaAs
Crystal system	trigonal	cubic	cubic	cubic
Lattice constant (Å)	a = 4.763 c = 13.003	a = 5.431	a = 5.657	a = 5.654
Density (g/cm^3)	3.97	2.33	5.32	5.32
Thermal expansion ($\times 10^{-7}$/°C) (25°C)	53 ($C_{\parallel}$) 45 ($C_{\perp}$)	26	58	57
Melting point (°C)	2053	1430	937	1238
Thermal conductivity (cal/cm/sec/°C) (25°C)	0.1	0.34	0.15	0.11
Specific heat (cal/g·°C)	0.18	0.17	0.077	0.038
Hardness (mhos)	9 (Hv = 2300 kg/cm^2)	7	6.5	
Compressive strength (kg/cm^2)	30000			
Flexural strength (kg/cm^2)	7000			
Young's modulus ($\times 10^6$ kg/cm^2)	4.8	≈ 1–1.6	≈ 1–1.6	≈ 0.6–1.2
Dielectric strength (kV/mm)	48 (25°C)			
Resistivity (log ρ, Ωcm)	16 (25°C) 11 (500°C)	5.8 (25°C)	1.66 (25°C)	
Dielectric constant (1 MHz) (25°C)	11.5 ($C_{\parallel}$) 9.3 ($C_{\perp}$)	11.7	15.8	12.9
Tan δ (%) (1 MHz) (25°C)	0.01			
Refractive index	N_o = 1.768 N_e = 1.760 (λ = 0.63 μm)	N_o = 3.47 (λ = 1.66 μm)	N_o = 4.10 (λ =2.06 μm)	N_o = 3.36 (λ = 1 μm)
Optical transparency	>80% 0.25 ≈ 4 μm (1 mm thick)			
Miscellaneous property	(0001), c plane $(\bar{1}102)$, r plane $(1\bar{1}20)$, a plane			

(cont'd.)

Table 2.7. *(cont'd.)*

Crystal Material	MgO	$MgO \cdot Al_2O_3$ (spinel)	NaCl
Crystal system	cubic	cubic	cubic
Lattice constant (Å)	a = 4.203		a = 5.628
Density (g/cm^3)	3.65	≈ 3.58–3.61	2.16
Thermal expansion ($\times 10^{-7}/°C$) (25°C)	138	76	
Melting point (°C)	2800	2135	801
Thermal conductivity (cal/cm/sec/°C) (25°C)	0.06	0.06	0.0155
Specific heat (cal/g· °C)	0.209	0.2	0.204
Hardness (mhos)	5.5	8	2.5
Compressive strength (kg/cm^2)			
Flexural strength (kg/cm^2)		≈ 2800–3500	
Young's modulus ($\times 10^6$ kg/cm^2)	≈ 2.5–3.5		
Dielectric strength (kV/mm)			
Resistivity (log ρ, Ωcm)		> 14 (25°C)	
Dielectric constant (1 MHz) (25°C)	9.65	8.4	5.62
Tan δ (%) (1 MHz) (25°C)		0.008	
Refractive index	$N_o = 1.74$ ($\lambda = 0.633$ μm)	$N_o = 1.718$ ($\lambda = 0.5893$ μm)	
Optical transparency		>85% ≈ 0.3–5 μm	
Miscellaneous property			

(cont'd.)

Table 2.7. *(cont'd.)*

Crystal Material	Mica	SiO_2	$LiNbO_3$
Crystal system	monoclinic	hexagonal	trigonal
Lattice constant (Å)			a = 5.148 c = 13.163
Density (g/cm^3)	2.76 ~ 3.0	2.65	4.64
Thermal expansion ($\times10^{-7}$/°C) (25°C)	110 ($C_{//}$) 200 ($C_{\perp}$)	79.7 ($C_{//}$) 133.7 ($C_{\perp}$)	
Melting point (°C)		1425	1253
Thermal conductivity (cal/cm/sec/°C) (25°C)	0.0016	0.03	
Specific heat (cal/g· °C)	0.21	0.191	
Hardness (mhos)	2.5 ~ 3.0	7	5
Compressive strength (kg/cm^2)			
Flexural strength (kg/cm^2)			
Young's modulus ($\times10^6$ kg/cm^2)			
Dielectric strength (kV/mm)			
Resistivity (log ρ, Ωcm)		16 (20°C)	
Dielectric constant (1 MHz) (25°C)	6.5 ~ 9.0	4.6 ($C_{//}$) 4.5 ($C_?$)	
Tan δ (%) (1 MHz) (25°C)			
Refractive index		N_o = 1.544 (Na D)	N_o = 2.2967 (λ = 0.633μm) N_e = 2.2082
Optical transparency			
Miscellaneous property	chemical formula $KH_2Al_3(SiO_4)_3$		

Table 2.8. Summaries of Epitaxial Relationships of Single Crystal Substrates for Various Thin Film Materials

Thin film	Substrate	Epitaxial relation	Ref.
Ag	Si	Ag(111) // Si(111)	
Au	NaCl	(100)Au // (100)NaCl , [110]Au // [110]NaCl	
Cu	Ti	(111)Cu // (0001)Ti , $[1\bar{1}0]$ Cu // $[11\bar{2}0]$ Ti	
	W	(111)Cu // (110)W , $[\bar{1}\bar{1}2]$ Cu // $[\bar{1}10]$ W	
Co	MgO	(001)Co // (001)MgO , [100]Co // [100]MgO	
Fe	MgO	(001)Fe // (001)MgO , [100]Fe // [110]MgO	
Pt	MgO	(001)Pt // (001)MgO , [100]Pt // [100]MgO	[29]
		(110)Pt // (110)MgO , [001]Pt // [001]MgO	
		(111)Pt // (111)MgO , [110]Pt // [110]MgO	
Si	α-Al_2O_3	(100)Si // $(1\bar{1}02)$ Al_2O_3 , [110]Si // $[\bar{2}201]$ Al_2O_3	[30]
		(111)Si // (0001)Al_2O_3 , $[1\bar{1}0]$ Si // $[\bar{1}230]$ Al_2O_3	
		(111)Si // $(11\bar{2}0)$ Al_2O_3 , $[1\bar{1}0]$ Si // $[\bar{2}201]$ Al_2O_3	
		(111)Si // $(11\bar{2}4)$ Al_2O_3 , $[1\bar{1}0]$ Si // $[\bar{1}100]$ Al_2O_3	
	MgO·Al_2O_3 (spinel)	(100)Si // (100)spinel	
		(110)Si // (110)spinel	
		(111)Si // (111)spinel	
SiC	Si	(001)SiC // (001)Si , [100]SiC // [100]Si	[31]
		(111)SiC // (111)Si , [110]SiC // [110]Si	
Ge	CaF_2	(111)Ge // (111)CaF_2	[32]
CdS	NaCl	(0001)CdS // (001)NaCl , $[11\bar{2}0]$ CdS // [110]NaCl	[33]
	Mica	(0001)CdS // (0001)mica , $[10\bar{1}0]$ CdS // $[10\bar{1}0]$ mica	
ZnO	α-Al_2O_3	(0001)ZnO // (0001)Al_2O_3 , $[11\bar{2}0]$ ZnO // $[10\bar{1}0]$ Al_2O_3	[34]
		$(11\bar{2}0)$ ZnO // $(01\bar{1}2)$ Al_2O_3 , [0001]ZnO // $[0\bar{1}11]$ Al_2O_3	

(cont'd.)

Table 2.8. *(cont'd.)*

Thin film	Substrate	Epitaxial relation	Ref.
AlN	α-Al_2O_3	(0001)AlN // (0001)Al_2O_3 , [$\bar{1}$210] AlN // [$\bar{1}$100] Al_2O_3	[35]
		(11$\bar{2}$0) AlN // (10$\bar{1}$2) Al_2O_3 , [0001]AlN // [01$\bar{1}$1] Al_2O_3	[36]
	Si	(0001)AlN // (111)Si , [11$\bar{2}$0] AlN // [1$\bar{1}$0] Si	[37]
		(0001)AlN // (110)Si , [1$\bar{1}$20] AlN // [1$\bar{1}$0] Si	
		(0001)AlN // (100)Si , [11$\bar{2}$0] AlN // [011]Si	
	MgO	($\bar{2}$110) AlN // (110)MgO , [0111]AlN // [001]MgO	[38]
GaN	α-Al_2O_3	(0001)GaN // (0001)Al_2O_3 , [$\bar{1}$210] GaN // [$\bar{1}$100] Al_2O_3	[35]
		(11$\bar{2}$0) GaN // (01$\bar{1}$2) Al_2O_3 , [0001]GaN // [01$\bar{1}$1] Al_2O_3	[39]
GaAs	α-Al_2O_3	(111)GaAs // (0001)Al_2O_3 , [0$\bar{1}$1] GaAs // [11$\bar{2}$10] Al_2O_3	[35] [40]
	MgO·Al_2O_3 (spinel)	(111)GaAs // (111)spinel , [01$\bar{1}$] GaAs // [01$\bar{1}$] spinel	[35]
		(100)GaAs // (110)spinel , [011]GaAs // [$\bar{1}$10] spinel	[41]
		(111)GaAs // (100)spinel , [11$\bar{1}$] GaAs // [$\bar{1}$10] spinel	
	BeO	(100)GaAs // (10$\bar{1}$1) BeO	[35]
		(111)GaAs // (0001)BeO	[42]
		(111)GaAs // (10$\bar{1}$1) BeO	
GaP	α-Al_2O_3	(111)GaP // (0001)Al_2O_3 , [1$\bar{1}$0] GaP // [11$\bar{2}$0] Al_2O_3	[35]
	MgO·Al_2O_3 (spinel)	(111)GaP // (111)spinel , [01$\bar{1}$] GaP // [01$\bar{1}$] spinel	[43]
	Si/α-Al_2O_3	(100)GaP / (100)Si / (01$\bar{1}$2) Al_2O_3	
InSb	NaCl	(0001)InSb // (001)NaCl , [10$\bar{1}$0] InSb // [110]NaCl	[44]

(cont'd.)

Table 2.8. *(cont'd.)*

Thin film	Substrate	Epitaxial relation	Ref.
TiN	MgO	(100)TiN // (100)MgO, [001]TiN // [001]MgO	[45]
	Si	(111)TiN // (111) Si, [110]TiN // [110] Si	
$SrRuO_3$	$SrTiO_3$	(110)$SrRuO_3$ // (100)$SrTiO_3$	[46]]
ZrO_2:Y (YSZ)	Si	(100)YSZ // (100) Si, [001]YSZ // [001] Si	[47]
$LaMnO_3$	$SrTiO_3$	(001)$LaMnO_3$ // (001)$SrTiO_3$	[48]
	MgO	(001)$LaMnO_3$ // (001)MgO	
	YSZ	(110)$LaMnO_3$ // (001)YSZ	
$PbTiO_3$ (PT)	α-Al_2O_3	(111)PT // (0001)Al_2O_3	
PLZT	MgO	(100)PLZT // (100)MgO	[49]
	α-Al_2O_3	(111)PLZT // (0001)Al_2O_3	[50]
	$SrTiO_3$	(100)PLZT // (100)$SrTiO_3$	
Fe_3O_4 (magnetite)	MgO	(100)Fe_3O_4 // (100)MgO, [001]Fe_3O_4 // [001]MgO	[51]
		(110)Fe_3O_4 // (110)MgO, [001]Fe_3O_4 // [001]MgO	
$LiNbO_3$ (LN)	α-Al_2O_3	(0001)LN // (0001)Al_2O_3	[52]
	$LiTaO_3$ (LT)	(0001)LN // (0001)LT	[53]
$Bi_{12}TiO_{20}$ (BTO)	$Bi_{12}GeO_{20}$ (BGO)	(111)BTO // (111)BGO	[54]

2.3 CHARACTERIZATION

A measurement of thin-film properties is indispensable for the study of thin-film materials and devices. The chemical composition, crystalline structure, and optical, electrical, and mechanical properties must be considered in evaluating thin films.

Several methods are proposed for the evaluation of the thin films, summarized in Table 2.9. Among these processes, rapid progress has been made in evaluating the surface and thin-film composition for semiconductor materials. Several methods have been proposed for the evaluation of the surface or thin films. Typical methods are listed in Table 2.10. There are a number of major considerations that determine the choice of an instrumental method to solve a specific problem in the surface or thin films, including area and depth to be sampled, sensitivity and reproducibility, and the number of detectable elements.

In studying thin-film materials, evaluate the thin-film properties listed in Table 2.9 and notice the correlation between growth conditions and the properties of the resultant thin films shown in Fig. 2.26. In-situ evaluations of the surface or thin-film properties are needed for determining the relationship between the growth condition and the film properties. Auger electron spectrometry (AES) is widely used to measure the chemical composition of the thin film during film growth. The condensation process and surface-crystal properties of thin films are evaluated by reflection high energy electron diffraction (RHEED).

Aside from instrumental analyses, evaluating electrical properties and physical properties are also important. For example, the impurity contents in a metal film are evaluated by measuring an electrical resistance and its temperature coefficient.[55]

Table 2.9. Summaries of Thin Film Evaluation Methods

Film properties	Evaluation methods	Remarks
Thickness	Optical	
	color comparison	Transparent films on substrate. Simple, range: ≈ 500–15000 Å, accuracy: ≈100–200 Å.
	interferometer	Step and reflective coating, range: ≈ 10–20000 Å, accuracy: ≈ 2–30 Å
	ellipsometry	Transparent films, range: few Å to micron
	cross-sectional SEM & TEM	Accuracy: up to atomic level
	Mechanical stylus	Step required, simple, range 20 Å to no limit, accuracy: ≈ 10 Å
Surface roughness	Mechanical stylus and/or optical microscope	Simple, mechanical stylus, resolution: ≈ 10 Å
	Scanning electron microscope	Conductive coating needed for dielectric films, resolution: ≈ 10 Å
	Scanning tunneling microscope	Conductive films, high resolution: ≈ 1 Å. Atomic scale
	Atomic force microscope	Simple, high resolution: ≈ 1 Å. Atomic scale
Chemical composition	Inductively coupled plasma optical emission spectroscopy (ICP)	
	Rutherford backscattering spectroscopy (RBS)	High Sensitivity obtained by ICP and SIMS; detection limit: ≈ 0.1 ppm
	Auger electron spectroscopy (AES)	Non-destructive, quantitative analysis; by RBS and/or EPMA
	Electron probe microanalysis (EPMA)	Depth profile; by RBS, SIMS, and AES
	X-ray photoelectron spectroscopy (XPS), and/or secondary ion mass spectroscopy (SIMS)	Simple analysis; by ICP and/or EPMA

(cont'd.)

Table 2.9. *(cont'd.)*

Film properties	Evaluation methods	Remarks
Structure	Electron and/or X-ray diffraction analysis X-ray photoelectron spectroscopy (XPS) and electron energy loss spectroscopy (EELS) Optical absorption	Microstructural analysis, by TEM. Amorphous films, by IR (infrared) absorption and/or EXAFS (extended X-ray absorption fine structure). Electronic states and valence states, by EELS and XPS, respectively. Refractive index, by ellipsometry.
Adhesion	Peeling method, scratching method, and pulling method	For weak adhesion, film is peeled off using a backing of adhesive tape. For strong adhesion (>1 kg/cm^2), scratching and/or pulling methods.
Stress	Disk method, bending-beam method, and x-ray diffraction method.	In the disk method, the film stress is measured by observing the deflection of the center of a circular plate; in the bending-beam method, by observing the deflection of a beam. 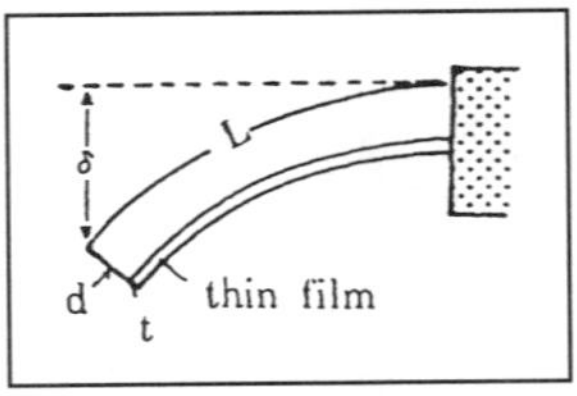 Bending-beam method: Relation for stress σ, $\sigma = Ed\delta/3L^2(1-\nu)t$ E, ν d, L, denote Young's modules, Poisson's ratio, thickness, and length of substrate beam, respectively, t, film thickness.
Hardness	Micro-Vickers hardness measurement and nano identation measurement.	Measured at light load of indentor. Hardness of substrate affects the hardness of the films. Extraporated values at zero indentor load gives the true values of film hardness. Ultra-thin films, surface are evaluated from identation load-displacement data using a nano-identation tip*
Wear & Friction	Wear test between film coated ball and iron plate. Sand blast method.	Strong adhesion onto the substrate is necessary for the wear test.

* Oliver, W. C., and Pharr, G. M., *J. Mater. Res.,* 7:1564 (1992).

(cont'd.)

Table 2.9. *(cont'd.)*

<table>
<tr><th>Film properties</th><th>Evaluation methods</th><th>Remarks</th></tr>
<tr><td>Electrical resistivity, ρ</td><td>Standard four terminals resistive measurements</td><td>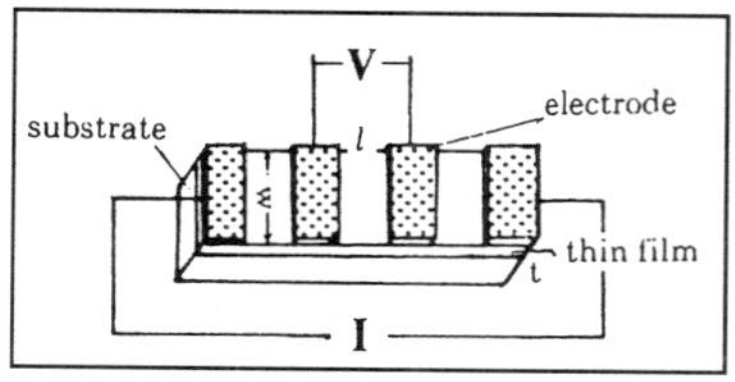

Four terminals measurements. $\rho = V/I \cdot Wt/l$</td></tr>
<tr><td>Dielectric constant, ε^*f</td><td>Dielectric measurements: at sandwich structure; evaporated electrode / dielectric film / evaporated electrode on substrate, or at interdigital electrodes (IDE) on dielectric films.</td><td>Sandwich structure: $\varepsilon^*f = Ct/e_0 S$, $\varepsilon^*f, \varepsilon_0$; dielectric constant of thin films, free space, respectivively; C, capacitance; t, film thickness; S, electrode area. Interdigital electrodes: $C = Knl\{[(\varepsilon^*_s + 1) + (\varepsilon^*_f - \varepsilon^*_s)[1 - \exp(-4.6\, t/L)]\}$, $K \equiv [6.5 (D/L)^2 + 1.08\, D/L + 2.37] \times 10^{-12}$ F/m; ε^*_s, dielectric constant of substrates; n, l, D, L; numbers of pair, length, width, pitch of the IDE, respectively.</td></tr>
<tr><td>Piezo-electricity</td><td>Electromechnical coupling k_t: admittance measurements at sandwich structure; evaporated electrode / piezoelectric film / evaporated electrode on fuzed quartz substrate. d_{33}, at sandwich structure; piezoelectric film / conductive substrate, Si, La-doped $SrTiO_3$. d_{31}, at micro-cantilevers</td><td>Electromechanical coupling, k_t $k_t^2 \approx G_A X_C \pi Z_M / 4Z_T$ where G_A, X_C, are the conductance and capacitive reactance at antiresonant frequency, Z_M, Z_T; acoustic impedance of substrate and piezoelectric films, respectively. d_{33}: measurement, Newton meter;[1] $d_{31} = (dL/L)(t/v)$,[2] L, length of cantilever beam, t, piezoelectric film thickness, dL, deflection of beam at voltage v.</td></tr>
<tr><td></td><td colspan="2">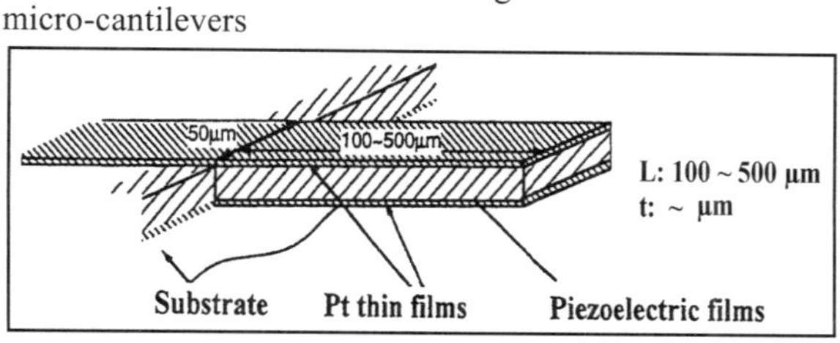
</td></tr>
</table>

[1] Uno, T., Noge, S., *Int. Conf. New Piezo. Mater. & High Performance AW Devices,* Tokyo (Jan. 2002).

[2] Kanno, I., Fujii, S., Kamada, T., and Takayama, R., *Appl. Phys. Lett.,* 70:1378 (1997).

Table 2.10. Summaries of Surface and Thin Film Analysis Methods

	Techniques	Incident beam (particle)	Emitted beam (particle)	Spatial resolution (μm) dia	Spatial resolution (μm) thick	Detection limits (at. ppm)	Accuracy (%)	Elements	Other features
XRF	X-ray fluoresence spectroscopy	X-rays	X-rays	10000	30	≈ 1–100	±1	$Z \geq 9$	Quantitative Non-destructive
EPMA	Electron microprobe analysis	Electrons	X-rays	1	1	≈ 0.01–0.1%	±2	$Z \geq 4$	Quantitative Non-destructive
PIXE	Particle induced x-ray emission	Ions	X-rays	1	1	1	--	$Z \geq 4$	Elemental analysis
RBS	Rutherford back-scattering spectroscopy	Ions	Ions	3	0.03	≈ 0.01–0.1%	±3	$Z \geq 5$	Quantitative Non-destructive Depth profile
ISS	Ion scattering spectroscopy	Ions	Ions	1000	0.0003	≈ 0.1–1%	±20	$Z \geq 6$	Semi-quantitative Depth profile
SIMS	Secondary ion mass spectrometry	Ions	Ions	1	0.003	≈ 0.1–100	±20		Semi-quantitative Depth profile Conductor
AES	Auger electron spectroscopy	Electrons	Electrons	0.05	0.003	≈ 0.1–1%	±20	$Z \geq 3$	Elemental analysis Electronic States

(cont'd.)

Table 2.10. *(cont'd.)*

	Techniques	Incident beam (particle)	Emitted beam (particle)	Spatial Resolution (μm) dia	thick	Detection limits (at. ppm)	Accuracy (%)	Elements	Other features
EELS	Electron energy loss spectroscopy	Electrons	Electrons	0.01	0.05		--		Semiquantitative Valance states Non-destructive
XPS	X-ray photoelectron spectroscopy	X-rays	Electrons	150	0.003	≈ 0.1–1%	±20	$Z \geq 3$	Surface structure
RHEED	Reflection high energy electron diffraction	Electrons	Electrons	100 Å	100 Å				Surface structure
LEED	Low energy electron diffraction	Electrons	Electrons	300 Å	3 Å				Surface/interface structure
TEM	Transmission electron microscopy	Electrons	Electrons	100 Å	2 Å				Surface topography Conductor
SEM	Scanning electron microscopy	Electrons	Electrons	20 Å	30 Å				Surface topography Conductor
STM	Scanning tunneling microscopy	Distance (≈ 1–2 Å)	Tunneling (≈ 1–10 nA)	3 Å	0.01 Å				Bias (≈ 1 mV to ≈ 1 V)

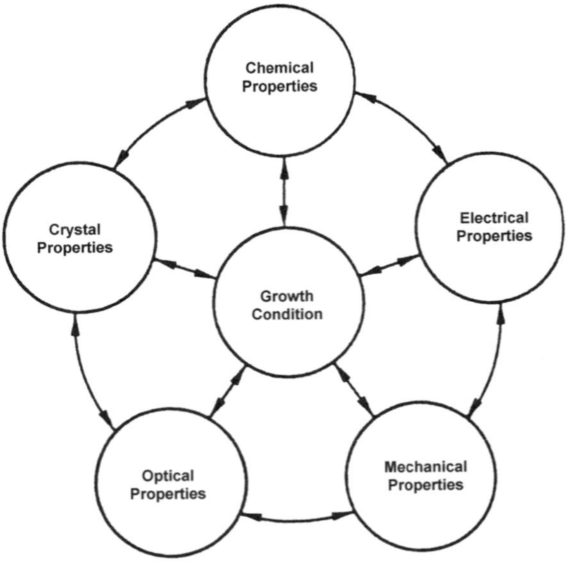

Fig. 2.26. Correlations between growth conditions and the properties of the resultant thin films

REFERENCES

1. Bunshah, R. F., (ed.), *Deposition Technologies for Films and Coatings*, Noyes Publications, NJ, (1982); Mattox, D. M., *Handbook of Physical Vapor Deposition Processing,* Noyes Publications, NJ (1998); Mahan, J. E., *Physical Vapor Deposition of Thin Films*, John Wiley & Sons, New York (2000); Elshabini-Riad, A., and Barlow, F. D., III, (ed.), *Thin Film Technology Handbook*, McGraw-Hill, New York (1997)
2. Chopra, K. L., and Kaur, I., *Thin Film Device Applications*, Plenum Press, New York (1983)
3. Neugebauer, C. A., Condensation, Nucleation, and Growth of Thin Films, in: *Handbook of Thin Film Technology,* (Maissel and Glang, eds.), p. 8-3, McGraw-Hill (1970)
4. Venables, J. A., Doust, T., and Kariotis, R., *Mat. Res. Symp. Proc.,* 94:3 (1987); Grabow, M. H., and Gilmer, G. H., *Mat. Res. Symp. Proc.*, 94:15 (1987)

5. Nishinaga, T., Nucleation and Surface Diffusion in Molecular Beam Epitaxy, in: *Handbook of Crystal Growth, Vol. 3* (D. T. J. Hurle, ed.), p. 164, Elsevier (1994)
6. Kaur, I., Pandya, D. K., and Chopra, K. L., *J. Electrochem. Soc.,* 127:943 (1980)
7. Burmester, C. P., Wille, L. T., and Gronsky, R., *Mat. Res. Symp. Proc.,* 389:59 (1995)
8. Genin, F. Y., *Mat. Res. Symp. Proc.,* 389:83 (1995)
9. Jesson, D. E., Chen, K. M., and Pennycook, S. J., *MRS Bulletine,* 21:31 (1996)
10. Tu, K. N., Mayer, J. W., and Feldman, L. C., *Electronic Thin Film Science,* p. 77, Macmillan Publ. Co., New York (1992)
11. Koga, T., *Off-axis Pulsed Laser Deposition of YBaCuO Superconducting Thin Films,* MS Thesis, Royal Institute of Technology, Stockholm, Sweden (1994)
12. Singh, R. K., and Narayan, J., *Phys. Rev.,* B 41:8843 (1990); Mahan, J. E., *Physical Vapor Deposition of Thin Films,* p. 133, John Wiley & Sons, New York (2000)
13. Esaki, L., Proc. 6th Int. Vacuum Congr., Kyoto, *Jpn. J. Appl. Phys.,* 13:Suppl. 2(Pt. 1):821–828 (1974)
14. Schlom, D. G., and Harris, J. S., Jr., MBE Growth of High Tc Superconductors, in: *Molecular Beam Epitaxy,* (R. F. C. Farrow, ed.), p. 505, Noyes Publications, NJ (1995)
15. Tokumitsu, E., Kudou, Y., Konagai, M., and Takahashi, K., *J. Appl. Phys.,* 55:3163 (1984)
16. McClanahan, D., and Laegreid, N., Production of Thin Films by Controlled Deposition of Sputtered Material, in: *Sputtering by Particle Bombardment III, Topics in Applied Physics*, (R. Behrisch and K. Wittmaack, eds.), 64:339, Springer Verlag, Berlin (1991)
17. Penning, F. M., U.S. Patent 2,146,025 (Feb. 1935)
18. Wasa, K., and Hayakawa, S., *Rev. Sci. Instrum.,* 40:693 (1969)
19. Chapin, J. S., *Res./Dev,* 25:37 (1974)
20. Chopra, K. L., and Randlett, M. R., *Rev. Sci. Instr.,* 38:1147 (1967)
21. Vossen, J. L., and Kern, W., (eds.), *Thin Film Processes*, Academic Press, New York (1978)
22. Kitabatake, M., and Wasa, K., *J. Appl. Phys.,* 58:1693 (1987)
23. Behrisch, R., (ed.), *Sputtering by Particle Bombardment I. II, Topics in Applied Physics,* 47:52, Springer Verlag, Berlin (1981), (1983)
24. Mattox, D. M., *J. Vac. Sci. Technol.,* 10:47 (1973)

25. Bunshah, R. F., and Raghuram, A. C., *J. Vac. Sci. Technol.,* 9:1385 (1972)

26. Takagi, T., Yamada, I., and Sasaki, A., *J. Vac. Sci. Technol.,* 12:1128 (1975); Takagi, T., Ionized Cluster Beam Deposition and Epitaxy, Noyes Publ., NJ (1988)

27. Nishizama, J., Kurabayashi, T., Abe, H., and Sakurai, N., *J. Vac. Sci. Technol.,* A5:1572 (1987)

28. Khan, I. H., in: *Handbook of Thin Film Technology,* (L. Maissel and R. Clang, eds.) 10-1, McGraw-Hill, New York (1970)

29. Lairson, B. M., Visokay, M. R., Sinclair, R., Hagstrom, S., and Clemens, B. M., *Appl. Phys. Lett.,* 61:1390 (1992)

30. Maissel, L. I., and Glang, R., (eds.) *Handbook of Thin Film Technology,* pp. 10–12, McGraw Hill, New York (1970)

31. Nishino, S., Powell, J. A., and Will, H. A., *Appl. Phys. Lett.,* 42:460 (1983)

32. Cho, C. C., Liu, H. Y, Magel, L. K., and Anthon, J. M., *Appl. Phys. Lett.,* 63:3291 (1993)

33. Shimaoka, G., *J. Cryst. Growth,* 45:313 (1978)

34. Mitsuyu, T., Ono, S., and Wasa, K., *J. Appl. Phys.,* 51:2464 (1980)

35. Wang, C. C., McFarlane, S. H., III, *Thin Solid Films,* 31:3 (1976)

36. Morita, M., Uesugi, N., Isogai, S., Tsubouchi, K., and Mikoshiba, N., *Jpn. J. Appl. Phys.,* 20:17 (1981)

37. Morita, M., Isogai, S., Shimizu, N., Tsubouchi, K., and Mikoshiba, N., *Jpn. J. Appl. Phys.*, 20:L173 (1981)

38. Heffelfinger, J. R., Medlin, D. L., and McCarty, K. F., *J. Mater. Res.,* 13:1414 (1998)

39. Thorsen, A. C., Manasevit, H. M., *J. Appl. Phys.,* 42:2519 (1971)

40. Chin, A., Bhattacharya, P., Chang, K. H., and Biswas, D., *J. Vac. Sci. Technol.,* 10:7 (1989)

41. Wang, C. C., Dougherty, F. C., Zanzucchi, P. J., McFarlane, S. H., III, *J. Electrochem. Soc.,* 121:571(1974)

42. Thorsen, A. C., Manasevit, H. M., Harada, R. H., *Solid-State Electron.,* 17:855 (1974)

43. Wang, C. C., McFarlane, S. H., III, *J. Cryst. Growth.,* 13-14:262 (1972)

44. Greene, J. E., and Wickersham, C. E., *J. Appl. Phys.,* 47:3630 (1976)

45. Sano, K., Oose, M., and Kawakubo, T., *Jpn. J. Appl. Phys.,* 34:3266 (1995)

46. Eom, C. B., Cava, R. J., Fleming, R. M., Phillips, J. M., van Dover, R. B., Marshall, J. H., Hsu, J. W. P., Krajewski, J. J., and Peck, W. F., Jr., *Science,* 258:1766 (1992)

47. Hata, T., Nakano, S., Masuda, Y., Sasaki, K., Haneda, Y., and Wasa, K., *Vacuum,* 51:583 (1998)
48. Gorbenko, O. Y., Graboy, I. E., Kaul, A. R., and Zandbergen, H. W., *J. Magn. Mater.,* 211:97 (2000)
49. Okuyama, K., Usui, T., and Hamakawa, Y., *Appl. Phys.,* 21:339 (1980)
50. Ishida, M., Tsuji, S., Kimura, K., Matsunami, H., and Tanaka, T., *J. Cryst. Growth,* 45:393 (1978)
51. Matsuda, H., Sakakima, H., Adachi, H., Odagawa, A., and Setsune, K., *J. Mater. Res.*, 17:1985 (2002)
52. Takada, S., Ohnishi, M., Hayakawa, H., and Mikoshiba, N., *Appl. Phys. Lett.*, 24:490 (1974)
53. Miyazawa, S., Fushimi, S., and Kondo, S, *Appl. Phys. Lett.,* 26:8 (1978)
54. Mitsuyu, T., Wasa, K., and Hayakawa, S., *J. Cryst. Growth*, 41:151 (1977)
55. Huttemann, R. D., Morabito, J. M., Stieidel, C. A., and Gerstenberg, D., *J. Appl. Phys.,* 13:Suppl. 2(Pt-1):513 (1974)

3

Sputtering Phenomena

Grove first observed sputtering in a dc gas discharge tube in 1852. He discovered that the cathode surface of the discharge tube was sputtered by energetic ions in the gas discharge, and cathode materials were deposited on the inner wall of the discharge tube.

At that time, sputtering was regarded as an undesired phenomenon since the cathode and grid in the gas discharge tube were destroyed. Today, however, sputtering is widely used for surface cleaning and etching, thin-film deposition, surface and surface layer analysis, and sputter ion sources.

In this chapter, the fundamental concepts of the various sputtering technologies are described. The energetic particles in sputtering may be ions, neutral atoms, neutrons, electrons, or photons. Since most relevant sputtering applications are performed under bombardment with ions, this text deals with that particular process.

3.1 SPUTTER YIELD

The sputter yield, S, which is the removal rate of surface atoms due to ion bombardment, is defined as the mean number of atoms removed from the surface of a solid per incident ion and is given by

Eq. (3.1) $$S = \frac{\text{atoms removed}}{\text{incident ions}}$$

Sputtering is caused by the interactions of incident particles with target surface atoms. The sputter yield will be influenced by the following factors:

1. Energy of incident particles
2. Target materials
3. Incident angles of particles
4. Crystal structure of the target surface

The sputter yield, *S*, can be measured by the following methods:

1. Weight loss of target
2. Decrease of target thickness
3. Collection of the sputtered materials
4. Detection of sputtered particles in flight

The sputter yield is commonly measured by weight loss experiments using a quartz crystal oscillator microbalance (QCOM) technique. Surface analysis techniques, including Rutherford back scattering spectroscopy (RBS), are available for measuring the change in thickness or composition of targets on an atomic scale during sputtering. RBS is essentially nondestructive and the dynamic sputter yield is determined with a priori accuracy of some 10%. Scanning electron microscope (SEM) and stylus techniques are used for measuring minute changes in target thickness. These techniques need an ion erosion depth in excess of around 0.1 μm. The QCOM technique is a sensitive probing method with sub-monolayer resolution.[1]

Both electron- and proton-probe beam techniques are successfully used dynamically in situ and provide absolute yield determinations. Auger electron spectroscopy (AES) could also be used for the determination of monolayer thickness. Proton-induced x-ray emission (PIXE) with proton energies of 100 to 200 keV[2] and electron-induced x-ray emission with electron energies of around 10 keV are also used for sputter yield measurements.[3] The PIXE technique can quantify both initial surface impurities as well as the pure sputter yield of the target.

3.1.1 Ion Energy

Figure 3.1 shows a typical variation of the sputtering yield with incident ion energy. The figure suggests the following:

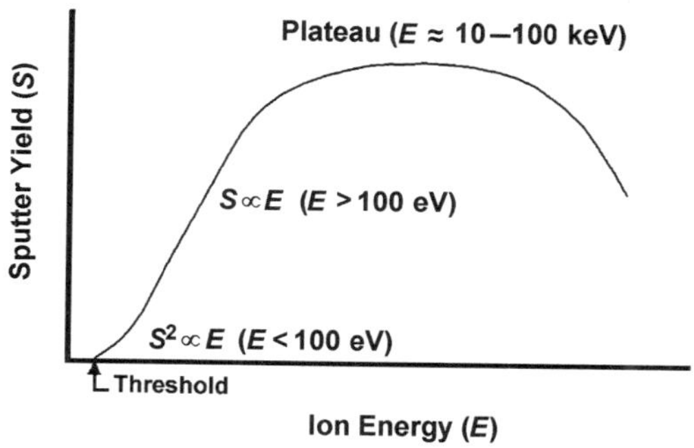

Figure 3.1. Variations of sputter yield with incident ion energy.

1. In a low energy region, a threshold energy exists for sputtering.
2. The sputter yield shows maximum value in a high-energy region.

Hull first observed the existence of the sputtering threshold in 1923. He found that the Th-W (thorium-tungsten) thermonic cathode in gas rectifier tubes was damaged by bombardment with ions when the bombarding ion energy exceeded a critical value, which was in the order of 20 to 30 eV.[4] Many workers have studied the sputtering threshold because it is probably related to the mechanism of sputtering. Threshold values obtained by these workers ranged from 50 to 300 eV.[5][6] Their results were somewhat doubtful because the threshold energy was mainly determined by measurements of small weight loss from the cathode, in the range of 10^{-4} atoms/ion. The threshold energy is very sensitive to contamination of the cathode surface. In addition, the incident angle of ions and the crystal orientation of cathode materials also change the threshold values.

In 1962, Stuart and Wehner skillfully measured reliable threshold values by the spectroscopic method.[7] They observed that threshold values are in the order of 15 to 30 eV, similar to the observations of Hull, and are roughly four times the heat of sublimation of cathode materials.

A gas discharge tube used by Stuart and Wehner is shown in Fig. 3.2.[7] The target is immersed in plasma generated by a low-pressure mercury discharge. In a noble gas discharge, the mercury background pressure is about 10^{-5} torr (1.33×10^{-3}Pa) or less. The temperature of the

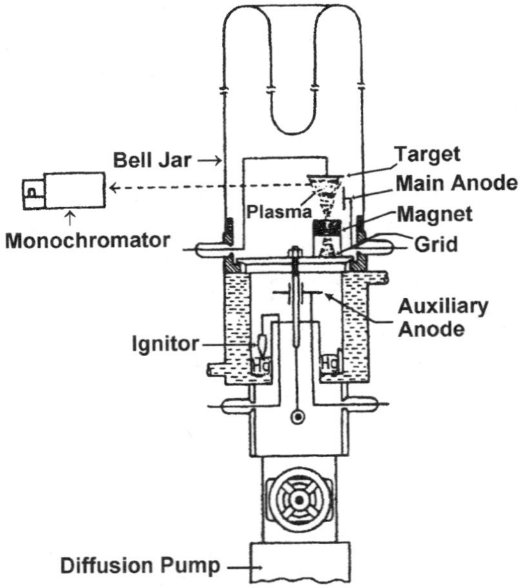

Figure 3.2. Experimental gas discharge tube for the measurement of threshold energy.[7]

target is kept at about 300°C to reduce condensation of the mercury vapor. The sputtered target atoms are excited in the plasma and emit the specific spectrum. The sputter yields at low energy are determined by the intensity of the spectral line. This technique eliminates the need for very sensitive weight measurements.

Typical sputter yields in a low-energy region measured by Stuart and Wehner are shown in Fig. 3.3. The threshold values determined in the sputter yields are on the order of 10^{-4} to 10^{-5} atoms/ion. Table 3.1 summarizes the sputtering threshold energy measured by the spectroscopic method for various target materials.[7][8]

The table suggests that there is not much difference in threshold values. The lowest value, which is nearly equal to four times the heat of sublimation, is observed for the best mass fit between target atom and incident ion. The higher threshold energy is observed for poor mass fits.

The threshold energy also strongly depends on the particular sputtering collision sequence involved. High threshold energy (i.e., $E_{th}/U_0 > 10$ for $Ar^+ \rightarrow Cu$ where E_{th} denotes the threshold energy; U_0, the heat of sublimation) will be expected in the collision sequence where primary recoil produced in the first collision is ejected directly. Lower values will be observed for the multiple sputtering collisions. An incident angle of around 40° to 60° offers the minimum threshold energy (i.e., $E_{th}/U_0 \approx 2$ for $Ar^+ \rightarrow Cu$) under the multiple sputtering collisions.[9]

The sputter yield, S, varies with the incident ion energy, E. In the low-energy region near the threshold, S obeys the relation $S \propto E^2$, as seen in Fig. 3.1. At the energy region in the order of 100 eV, S is expressed by $S \propto E$.[6][10] In this energy region, the incident ions collide with the surface atoms of the target, and the number of displaced atoms due to the collision will be proportional to the incident energy.

At higher ion energies of 10 to 100 keV, the incident ions travel beneath the surface and the sputter yields are not governed by surface scattering, but by the scattering inside the target. Above 10 keV, the sputter yields will decrease due to energy dissipation of the incident ions deep in the target.

Maximum sputter yields are seen in the ion energy region of about 10 keV. Figure 3.4 summarizes the energy dependence of the yield, i.e., the sputter rate, as reported by Sigmund.[11]

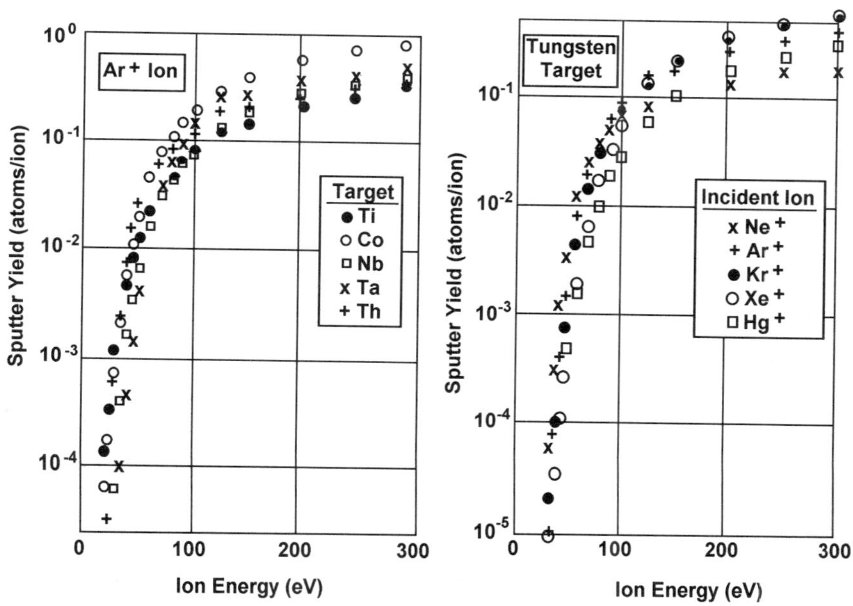

Figure 3.3. Sputter yield in a low-energy region.[7]

Table 3.1. Sputtering Threshold Data for Various Target Materials[7][8]

Target Material	Bombarding Ions					Heat of sublimation (eV)
	Ne	Ar	Kr	Xe	Hg	
Be	12	15	15	15	-	-
Al	13	13	15	18	18	-
Ti	22	20	17	18	25	4.40
V	21	23	25	28	25	5.28
Cr	22	22	18	20	23	4.03
Fe	22	20	25	23	25	4.12
Co	20	25	23	22	-	4.40
Ni	23	21	25	20	-	4.41
Cu	17	17	16	15	20	3.53
Ge	23	25	22	18	25	4.07
Zr	23	22	18	25	30	6.14
Nb	27	25	26	32	-	7.71
Mo	24	24	28	27	32	6.15
Rh	25	24	25	25	-	5.98
Pd	20	20	20	15	20	4.08
Ag	12	15	15	17	-	3.35
Ta	25	26	30	30	30	8.02
W	35	33	30	30	30	8.80
Re	35	35	25	30	35	-
Pt	27	25	22	22	25	5.60
Au	20	20	20	18	-	3.90
Th	20	24	25	25	-	7.07
U	20	23	25	22	27	9.57
Ir		8				5.22

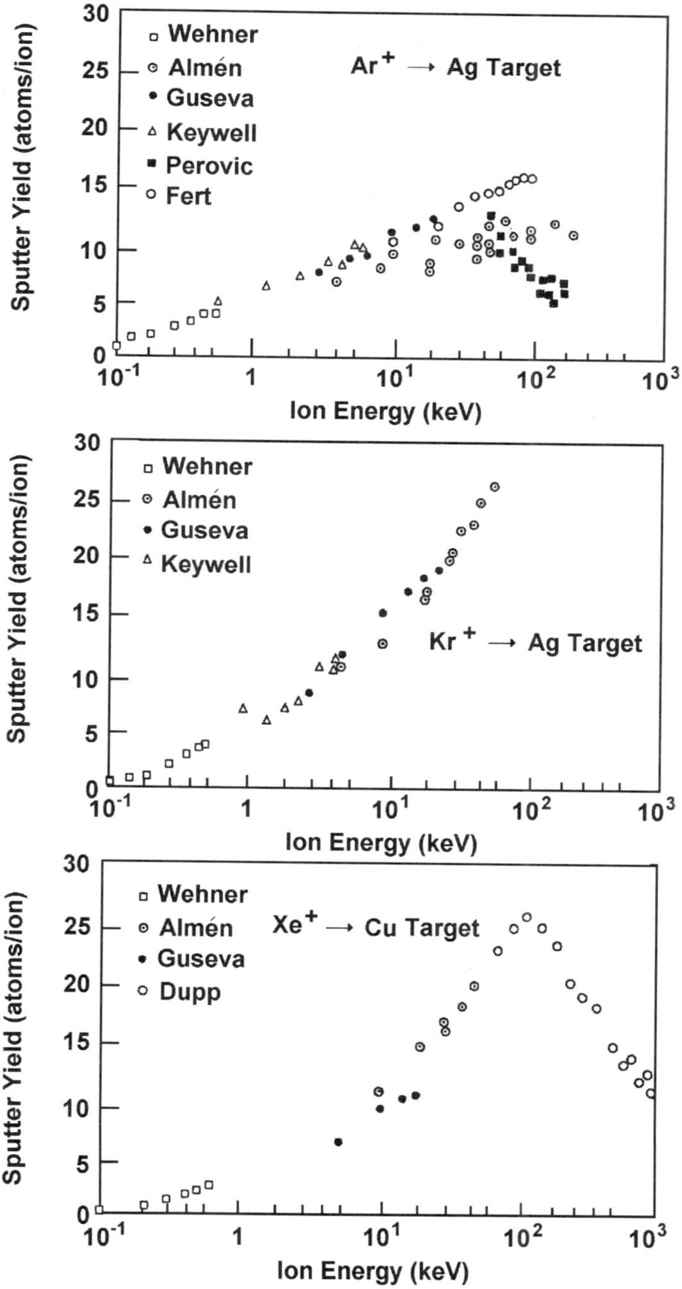

Figure 3.4. Sputter yield as a function of incident ion energy.[11]

3.1.2 Incident Ions, Target Materials

Sputter yield data have been extensively accumulated in relation to gas discharges, sputter deposition, etching, surface analysis, and radiation damage. At first, sputter yields were measured in the cold-cathode discharge tube,[12][13] but these measurements did not offer reliable data because the incident ions and sputtered atoms frequently collide with discharge gas molecules in the cold-cathode discharge tube.

Laegreid, Wehner, and Meckel, in 1959, accumulated the first reliable data of sputter yields in a low gas-pressure discharge tube.[14] At present, this yield data is still widely used for sputtering applications. Figure 3.5 shows Wehner's sputtering stand.[6][15] The system is based on a mercury discharge tube. The discharge is maintained at low gas-pressure by a thermionic mercury cathode. The discharge gas is 1×10^{-3} torr for Hg, $\approx (2–5) \times 10^{-3}$ torr for Ar, and 4×10^{-2} torr for Ne. The sputter yield, *S*, is determined by

Eq. (3.2) $$S = 10^5 \frac{W}{AIt}$$

where *W* denotes the weight loss of target during the sputtering time, *t*, with the ion current, *I*, to the target, and *A* denotes the atomic number of the target materials.

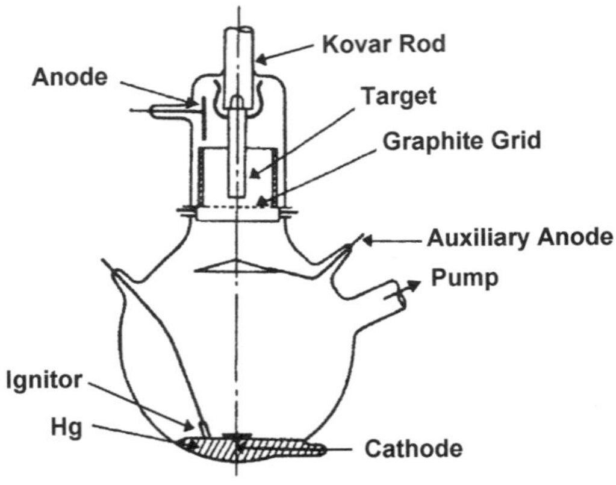

Figure 3.5. Wehner's experimental apparatus for the measurement of sputter yield.[6]

Typical results are shown in Fig. 3.6 and Table 3.2. Polycrystalline targets were used in the measurements so the effects of the crystal orientation could be neglected. It is noted that this yield data correspond to $S/(1+\gamma)$, where γ is the secondary electron emission coefficient of the target materials. The values of γ for various combinations of incident ions and target materials are shown in Table 3.3.[16] For sputter deposition, the range of the incident ion energy is below 1,000 eV where the $\gamma = 0.1$ for Ar^+. The error of Wehner's data will be typically less than 10% even if the effects of the secondary electron emission are not taken into consideration.

Recent PIXE techniques suggest that the sputter yield of Cr, S(Cr), measured by Wehner is larger than S (Cr) measured by PIXE. Sartwell's PIXE measurements indicate that S (Cr) = 0.93 atoms/ion at 1.0 keV Ar^+ bombardment.[2] Wehner has reported a value of S (Cr) = 1.18 atoms/ion at 500 eV. Wehner's measurements of S (Cr) indicated in Fig. 3.6 and Table 3.2 are 1.5 times larger that the PIXE measurements.[2]

As indicated in Fig. 3.6, sputter yields vary periodically with the element's atomic number. Comparing various materials, the yields increase consistently as the electronic d shells of target materials are filled, with Cu, Ag, and Au having the highest yields. Conversely, elements with the most open electronic structure result in the least sputtering yield. The sputtering yields are proportional to the energy transfer factor, and are a function of the hardness of the atoms since hard collisions increase the yield. Periodicity is also observed in sputtering thresholds.[17] Sigmund detailed the theoretical considerations.[11]

3.1.3 Effects of Incidence Angle

Sputter yields vary with the angle of incident ions. Fetz studied the influence of incident ions in 1942 and, later, Wehner studied the topic in detail.[18][19] Metals such as Au, Ag, Cu, and Pt, which have high sputtering yields, show very slight "angle effects." Iron, Ta, and Mo, that have low sputtering yields, show very pronounced angle effects. The yield increases with the incident angle and shows a maximum at angles between 60° and 80°, while it decreases rapidly for larger angles. The influence of the angle is also governed by the surface structure of the target. Theoretical studies have been done by several workers.[20] Detailed results obtained by Bay and Bohdansky are shown in Fig. 3.7.[21]

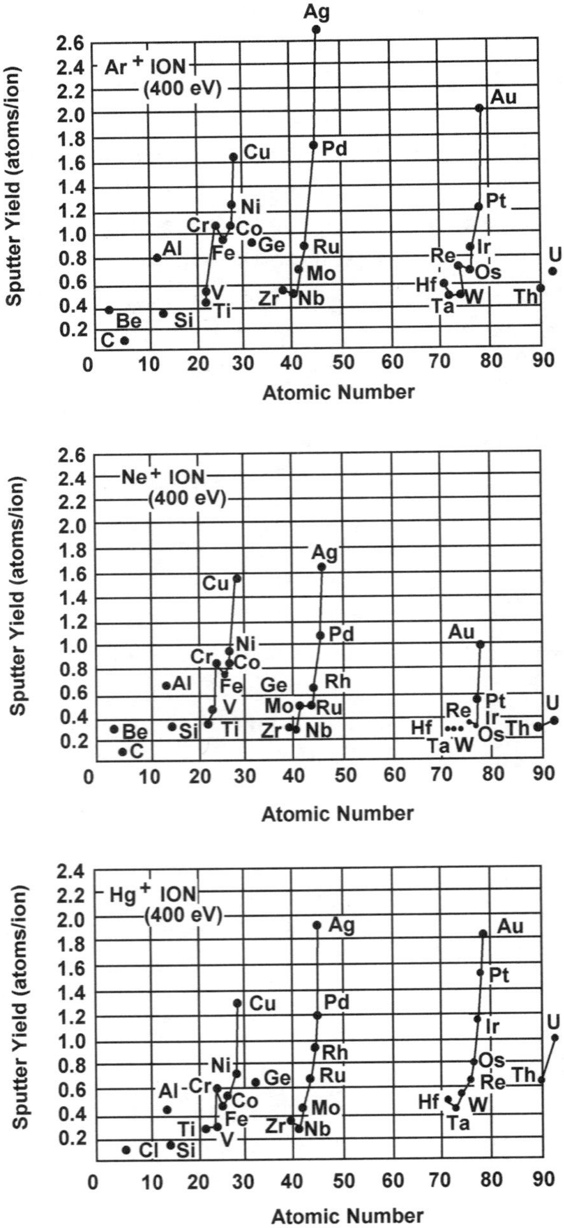

Figure 3.6. Sputter yield vs atomic number for the impingement of Ar^+, Ne^+, and Hg^+.[7]

Table 3.2. Sputter Yield[15]

Target	Ne+				Ar+			
	100 (eV)	200 (eV)	300 (eV)	600 (eV)	100 (eV)	200 (eV)	300 (eV)	600 (eV)
Be	0.012	0.10	0.26	0.56	0.074	0.18	0.29	0.80
Al	0.031	0.24	0.43	0.83	0.11	0.35	0.65	1.24
Si	0.034	0.13	0.25	0.54	0.07	0.18	0.31	0.53
Ti	0.08	0.22	0.30	0.45	0.081	0.22	0.33	0.58
V	0.06	0.17	0.36	0.55	0.11	0.31	0.41	0.70
Cr	0.18	0.49	0.73	1.05	0.30	0.67	0.87	1.30
Fe	0.18	0.38	0.62	0.97	0.20	0.53	0.76	1.26
Co	0.084	0.41	0.64	0.99	0.15	0.57	0.81	1.36
Ni	0.22	0.46	0.65	1.34	0.28	0.66	0.95	1.52
Cu	0.26	0.84	1.20	2.00	0.48	1.10	1.59	2.30
Ge	0.12	0.32	0.48	0.82	0.22	0.50	0.74	1.22
Zr	0.054	0.17	0.27	0.42	0.12	0.28	0.41	0.75
Nb	0.051	0.16	0.23	0.42	0.068	0.25	0.40	0.65
Mo	0.10	0.24	0.34	0.54	0.13	0.40	0.58	0.93
Ru	0.078	0.26	0.38	0.67	0.14	0.41	0.68	1.30
Rh	0.081	0.36	0.52	0.77	0.19	0.55	0.86	1.46
Pd	0.14	0.59	0.82	1.32	0.42	1.00	1.41	2.39
Ag	0.27	1.00	1.30	1.98	0.63	1.58	2.20	3.40
Hf	0.057	0.15	0.22	0.39	0.16	0.35	0.48	0.83
Ta	0.056	0.13	0.18	0.30	0.10	0.28	0.41	0.62
W	0.038	0.13	0.18	0.32	0.068	0.29	0.40	0.62
Re	0.04	0.15	0.24	0.42	0.10	0.37	0.56	0.91
Os	0.032	0.16	0.24	0.41	0.057	0.36	0.56	0.95
Ir	0.069	0.21	0.30	0.46	0.12	0.43	0.70	1.17
Pt	0.12	0.31	0.44	0.70	0.20	0.63	0.95	1.56
Au	0.20	0.56	0.84	1.18	0.32	1.07	1.65	2.43 (500)
Th	0.028	0.11	0.17	0.36	0.097	0.27	0.42	0.66
U	0.063	0.20	0.30	0.52	0.14	0.35	0.59	0.97

Table 3.3. Secondary Electron Coefficients, γ[16]

Target Material	Incident Ion	Ion Energy (eV)		
		200	600	1000
W	He^+	0.524	0.24	0.258
	Ne^+	0.258	0.25	0.25
	Ar^+	0.1	0.104	0.108
	Kr^+	0.05	0.054	0.058
	Xe^+	0.016	0.016	0.016
Mo	He^+	0.215	0.225	0.245
	He^{++}	0.715	0.77	0.78
Ni	He^+		0.6	0.84
	Ne^+			0.53
	Ar^+		0.09	0.156

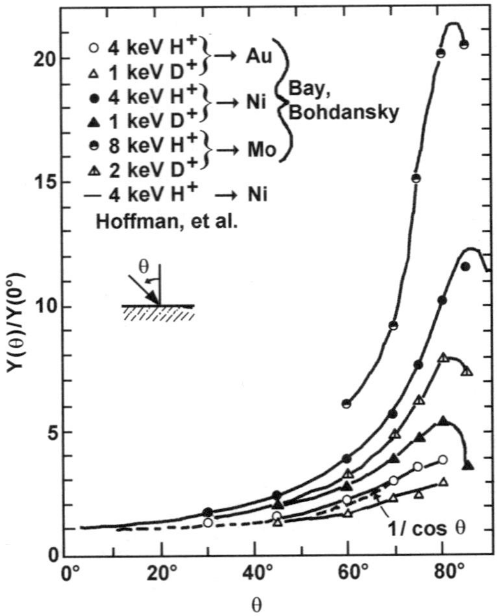

Figure 3.7. Sputter yield vs incident angle of bombarding ions.[21]

Figure 3.8 shows the angular distribution of sputtered atoms for the oblique incidence of bombarding ions measured by Okutani, et al.[22] The atoms are ejected preferentially in a forward direction.

The angular distributions of sputtered atoms for normal-incidence ion bombardment were studied in various ranges of incident ion energy. Seeliger and Sommermeyer measured them in a high-energy region of 10 keV. These experiments suggested that the angular distribution was governed by Knudsen's cosine law, which was observed in a thermal evaporation process.[23] Wehner and Rosenberg have measured the angular distribution in the lower energy region of 100 to 1,000 eV in a low-pressure mercury-discharge tube, which is shown in Fig. 3.9. The target strip is mounted in the center of the glass cylinder, which holds the collecting glass ribbon on its inside wall. The density of material sputtered over the entire 180° of the ribbon gives the angular distribution. Typical results are shown in Fig. 3.10.[19] It suggests that the angular distribution is "under cosine," i.e., much more material is ejected to the sides than in the direction normal to the target surface. The distribution approaches a cosine distribution at higher ion energies. Molybdenum and Fe show more pronounced tendencies to eject to the sides than Ni or Pt. At energies higher than 10 keV, the distribution shows "over cosine."

Angular distributions are related to the sputtering mechanism and are also considered in several applications including secondary ion mass spectrometry (SIMS), sputter deposition, and sputter etching. Angular distributions have been studied in detail using ion-beam sputtering systems.[24]

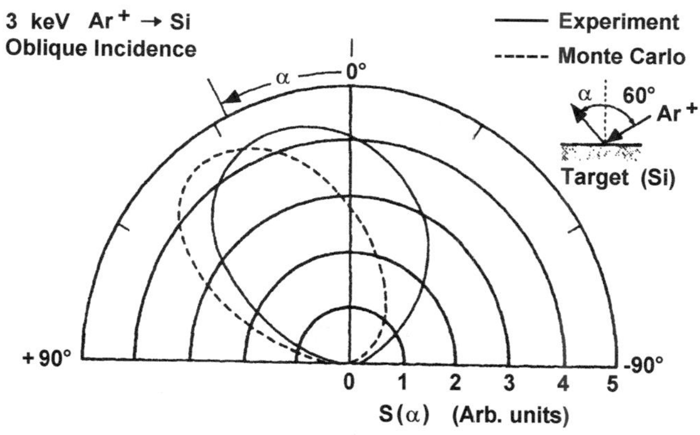

Figure 3.8. Angular distributions of sputtered Si atoms for 3 keV Ar^+ ion bombardment at an incident angle of 60°.[22]

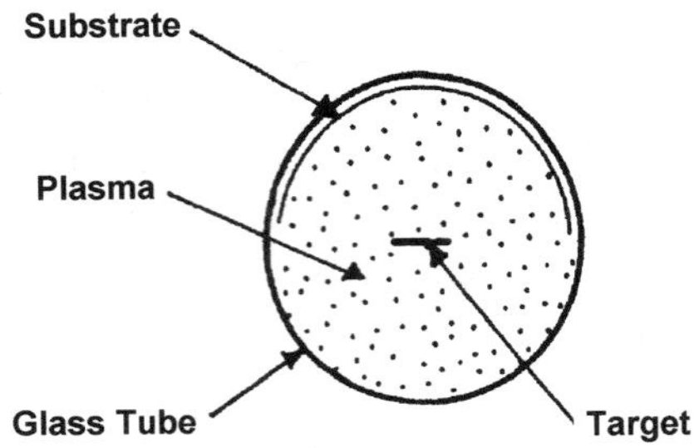

Figure 3.9. Experimental sputtering apparatus for the measurements of angular distributions.[19]

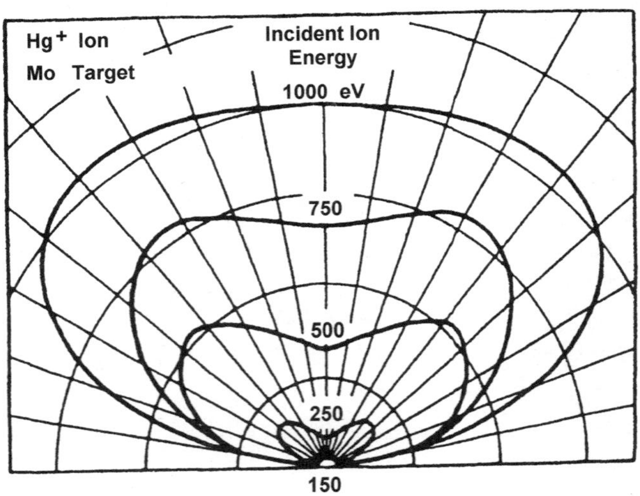

Figure 3.10. Angular distributions of sputtered particles from a polycrystalline target.[19]

3.1.4 Crystal Structure of Target

It is well known that the sputtering yield and angular distribution of the sputtered particles are affected by the crystal structure of the target surface. As described in the previous section, the angular distribution may be either under cosine law or over cosine law when the target is composed of polycrystalline materials. Nonuniform angular distribution is often observed from the single-crystal target.

Wehner studied, in detail, the nonuniform angular distribution of single crystals and found that deposited patterns appeared.[6] He suggests that, near the threshold, the sputtered atoms are ejected in the direction of close-packed atoms. For instance, in fcc Ag, the close-packed direction corresponds to ⟨110⟩. When an fcc (111) Ag target is sputtered, a threefold symmetrical pattern appears since there are three close-packed directions in the (111) plane.

Table 3.4 shows a summary of the sputtered patterns for single-crystal targets.

At higher ion energy, additional atoms are freed from the more numerous positions where neighbor atoms may interfere with the direction of close-packed rows. This causes deviations from these directions.

Measurements of Ar^+ incident onto fcc crystals such as Cu, Ag, Au, and Al confirm the general features; the sputter yields show that $S_{111} > S_{poly} > S_{110}$ for incident energies of a few keV.[25] At low incident energies, < 100 eV, the ordering of the sputter yields becomes $S_{110} > S_{100} > S_{111}$.[6] The theoretical approaches suggest that the ordering of the sputter yields relates to that of the binding energy $U_{111} > U_{100} > U_{110}$.[26]

The sputter yields of hcp crystals for a few keV Ar^+ bombardment exhibit $S_{0001} > S_{1010} > S_{1120}$ for Mg, $S_{1010} > S_{0001} > S_{1120}$ for Zr, and $S_{0001} > S_{1010} > S_{1120}$ for Zn and Cd. The ordering of these sputtering yields relates the ordering of the interatomic distance, t, for [0001] and [1010] direction with the value c/a: $t_{0001} > t_{1010} > t_{1120}$ for Cd and Zn, $t_{1010} > t_{0001}$ for Mg and Zr.[25] Typical experimental results are shown in Fig. 3.11.[27]

Several models have been considered to explain the angular dependence of the sputter yield for the monocrystalline target including the transparency model by Fluit[28] and the channeling model by Onderdelinden.[29] Figure 3.12 shows the angular dependence of the sputter yield measured at 27-keV Ar^+ on a (111) Cu crystal turned around the (112) axis with theoretical values due to the channeling model.[30]

Table 3.4. Sputtered Pattern for Single Crystal Targets

Target	Crystal Structure	Direction of Sputtered Pattern
Ag	fcc	<110>
W, Mo, α-Fe	bcc	<111>
Ge	diamond	<111>
Zn, Ti, Re	hexagonal	$<11\bar{2}0>$, $<20\bar{2}3>$

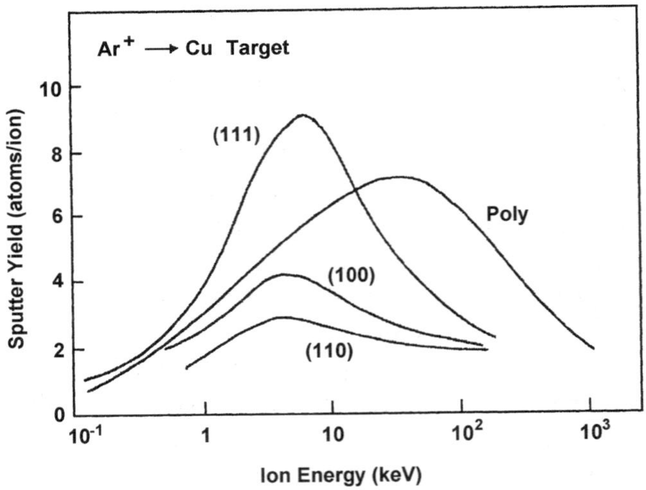

Figure 3.11. Energy dependence of the sputter yields of Ar^+ on the (110), (100), and (111) planes of Cu.[27]

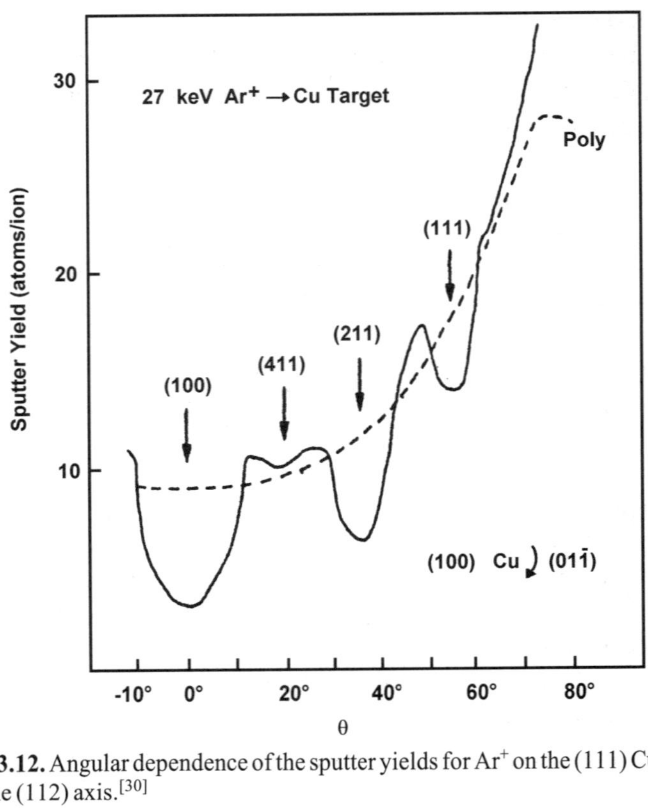

Figure 3.12. Angular dependence of the sputter yields for Ar^+ on the (111) Cu plane rotated about the (112) axis.[30]

3.1.5 Sputter Yields of Alloys

Experience has shown that the chemical composition of the deposited film sputtered from an alloy target is very close to that of the target. This suggests that sputtering is not governed by thermal processes, but by a momentum transfer process. When the temperature of the target is so high that the composition of the alloy target changes due to the thermal diffusion, the resultant films show different chemical compositions. Under even higher substrate temperature, the resultant films also show different chemical compositions because of the reevaporation of the deposited films.

Several alloys are routinely sputtered including Permalloy (81Ni-19Fe) and Ni-Cr alloys. Flur and Riseman found that films sputtered from a Permalloy target had the same composition as the target.[31]

Patterson and Shirn suggested that the composition of the Ni-Cr target was preserved in the sputtered films.[32] Sputtering conditions will alter the composition of the sputtered films. The correlation between the compositions of the targets and the deposited films of alloys and compound materials need to be studied.

The nature of changes in the alloy target is significant. Patterson showed that variation of the surface composition during sputtering deposition of a binary alloy target (A, B) was expressed by the following relations:[32]

Eq. (3.3) $$\mathrm{A} = \left(A_0 - \frac{S_\mathrm{B} N_0 A_0}{A_0 S_\mathrm{B} + B_0 S_\mathrm{A}} \right) \exp - \left(\frac{A_0 S_\mathrm{B} + B_0 S_\mathrm{A}}{N_0^2} Ft \right) + \frac{S_\mathrm{B} N_0 A_0}{A_0 S_\mathrm{B} + B_0 S_\mathrm{A}}$$

Eq. (3.4) $$\mathrm{B} = \left(B_0 - \frac{S_\mathrm{A} N_0 A_0}{A_0 S_\mathrm{B} + B_0 S_\mathrm{A}} \right) \exp - \left(\frac{A_0 S_\mathrm{B} + B_0 S_\mathrm{A}}{N_0^2} Ft \right) + \frac{S_\mathrm{A} N_0 A_0}{A_0 S_\mathrm{B} + B_0 S_\mathrm{A}}$$

where A_0 and B_0 denote the surface density of the composition A, B at the initial stage of the sputtering, respectively. The sputter yields of A, B are S_A and S_B, respectively; N_0 is the atomic density of the alloy target surface, F is the ion current density at the target surface, and t is the sputtering time.

These relations are obtained under the assumption that values S_A and S_B for the alloy are the same as those of each element, A and B, respectively. We know these relations are useful when each element, A and B, shows small differences in its atomic weight. A typical example is Ni-Cr alloys.

Equations (3.3) and (3.4) suggest that the surface composition of the target will change during the sputtering time. The time constant for this change is

Eq. (3.5)
$$\left[\frac{A_0 S_B + B_0 S_A}{N_0^2} F\right]^{-1}$$

As $t \rightarrow \infty$

Eq. (3.6)
$$\begin{cases} A \rightarrow \dfrac{A_0 S_B N_0}{A_0 S_B + B_0 S_A} \\ \\ B \rightarrow \dfrac{B_0 S_A N_0}{A_0 S_B + B_0 S_A} \end{cases}$$

and both A and B approach a constant value.

Consider, for example, the sputtering of Ni-20Cr ($A_0 = 0.8N_0$, $B_0 = 0.2N_0$) by 600 eV Ar ions; the sputter yield for Ni (S_A) is 1.5 atoms per ion and Cr (S_B) = 1.3. The surface density, N_0, is about 2×10^{15} at.cm^{-2}. If $F = 6 \times 10^{15}$ cm^{-2} sec^{-1} (1 mA/cm^2), the time constants become 250 msec. The calculated time constant is fairly shorter than the sputtering time for the deposition. It can be shown from Eqs. (3.5) and (3.6) that the surface of the target will change from 80Ni-20Cr to 78Ni-22Cr during one time constant.

The instantaneous film compositions for A and B are expressed by $\int_0^t S_A A\,dt$ and $\int_0^t S_B B\,dt$, respectively. Then, the ratio of the each composition of the sputtered film at $t \rightarrow \infty$, (A_s/B_s) becomes

Eq. (3.7)
$$\frac{A_s}{B_s} = \frac{A_0}{B_0}$$

Equation (3.7) suggests that the sputtered film should have the same composition as the target in the steady state when the solid-state diffusion in the target is neglected. At ordinary temperatures (a few hundred degrees centigrade), such diffusion will be unimportant.[32]

Liau has found that the surface change of the target composition for a Pt and Si alloy is not governed by Eqs. (3.5) and (3.6).[33] He studied the surface composition by Rutherford backscattering techniques and showed that the Si, which has a small elemental sputter yield, had been preferentially sputtered off. After sputtering, the Pt, which has a high elemental yield, increased in the target surface.

Table 3.5 shows the steady-state composition of the surface layers for 40-keV Ar ion sputtering for various binary alloys allied to Pt/Si. It was suggested that these phenomena were chiefly observed in the binary alloys which were composed of light components and heavy components. The heavy components were generally enriched in the surface layer.

Haff has studied the phenomena of surface-layer enrichment and suggests that it is related to the collision cascade between two species of binary alloys in the surface layer.[34] Tarng and Wehner showed that the surface change of a Cu-Ni alloy is governed by Eqs. (3.5) and (3.6), although Cu is much heavier than Ni.[35]

As described above, the compositional change of the target surface is complicated. The composition of the sputtered alloy films, however, is generally equal to the target composition when the target is cooled during deposition.[36] Table 3.6 shows some examples of the composition of alloy films sputtered by conventional magnetron sputtering.[37]

Table 3.5. Variations of Surface Compositions Under Ion Bombardment with 40-keV Ar^+ Ions[33]

Bulk	Surface
$Au_{0.19}Ag$	$Au_{0.23}Ag$
Cu_3Al	$Cu_3Al_{1.1}$
Au_2Al	$Au_{3.3}Al$
$AuAl_2$	$Au_{1.3}Al_2$
Pt_2Si	$Pt_{3.3}Si$
PtSi	$Pt_{2.1}Si$
NiSi	$Ni_{1.6}Si$
InP	same as bulk
GaP	same as bulk
GeSi	same as bulk
Ta_2O_5	$Ta_{4.5}O_5$

Table 3.6. Compositions of Al Alloy Thin Films Deposited by Magnetron Sputtering[36]

Materials	Target Composition	Film Composition
Cu	3.9–5%	3.81%
Si	0.5–1%	0.86%
Mn	0.4–1.2%	0.67%
2014 Al alloy Mg	0.2–0.8%	0.24%
Fe	1.0%	0.21%
Cr	0.10%	0.02%
Zn	0.25%	0.24%
(Al+Cu+Si)Cu	4%	3.4%
Si	2%	2.8%
(Al+Si)Si	2%	2% ±0.1%

3.2 SPUTTERED ATOMS

3.2.1 Features of Sputtered Atoms

In a conventional sputtering system, sputtered atoms are generally composed of neutral single atoms of the target material when the target is sputtered by bombardment with ions having a few hundred electron volts. These sputtered atoms are partially ionized (i.e., a few percent of the sputtered atoms are ionized) in the discharge region of the sputtering system.

Woodyard and Cooper have studied the features of sputtered Cu atoms under bombardment with 100 eV Ar ions, using a mass spectrometer. They found that 95% of the sputtered atoms are composed of single Cu atoms, and the remaining 5%, Cu_2 molecules.[38]

Under higher incident-ion energy, clusters of atoms are included in the sputtered atoms. Herzog showed that the sputtered atoms were comprised of clusters of Al_7 when the Al target was sputtered by 12-keV Ar ions. Under bombardment with Xe ions, clusters of Al_{13} were detected.

For an alloy target, the features of the sputtered atoms are similar to those of the single-element target. Under low incident-ion energy, most of the sputtered atoms are composed of the single element of the alloy. Clusters are predominant when the ion energy is higher than 10 keV.

3.2.2 Velocity and Mean Free Path

3.2.2.1 Velocity of Sputtered Atoms

The average energy of sputtered neutral atoms is much higher than that of thermally evaporated atoms in a vacuum. Guenthershulze,[39] Mayer,[40] and Spron[41] first observed the high energy of the sputtered atoms.

Wehner has studied the velocity of sputtered atoms in detail. He measured the velocity by a quartz balance shown in Fig. 3.13.[42] A low-pressure Hg plasma is maintained between an anode and an Hg pool cathode. A quartz helix balance with a little quartz pan is suspended in the upper part of the tube. When the sputtered atoms deposit on the underside of the pan, the atoms exert a force, $(dM/dt)\overline{v}_z$, which displaces the pan upward by a certain distance, where dM/dt denotes the mass-per-second arriving, and $\overline{v}_z$ is the average velocity component normal to the pan surface. The continuous deposition increases the weight of the pan with time, and the pan will return to its original position. When one measures the time interval, t, required for the pan to return to its original position, $\overline{v}_z$ is estimated from the relation $(dM/dt)\overline{v}_z = (dM/dt)tg$, where g is the acceleration of gravitational force. Then, $\overline{v}_z = tg$ and $g = 981$ cm/sec^2 without knowledge of dM/dt.

The quartz-balance method is also used for determining the average velocity of vacuum-evaporated particles. Wehner has estimated the average velocity by this method and suggested that the average velocity of Pt, Au, Ni, and W, sputtered by Hg ions of a few hundred eV, is of the order of 3×10^5 to 7×10^5 cm/sec, corresponding to the average kinetic energy of 10 to 30 eV. The kinetic energy is more than 100 times higher than thermal evaporation energy.

Both the calorimetric method and time-of-flight method are also used for determining the average energy of sputtered atoms. In the calorimetric method, the temperature rise of the substrate due to the impact of sputtered atoms determines the average energy. However, this method must take into account the heat of condensation and the various energetic processes occurring in the plasma. Detailed data, however, have been obtained by the time-of-flight method.

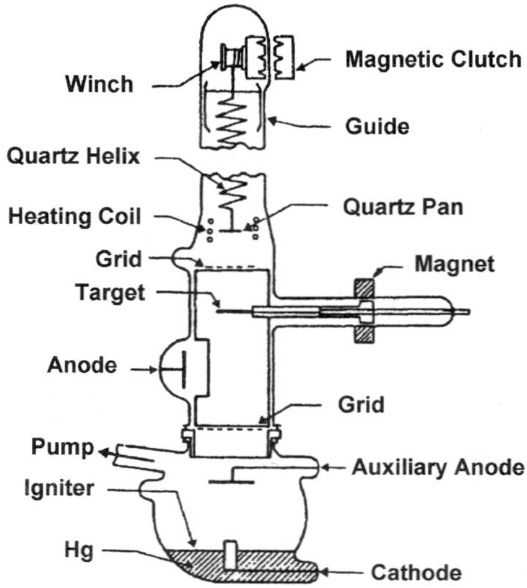

Figure 3.13. A quartz balance for measuring the velocity of sputtered atoms.[42]

Figure 3.14 shows the experiment used by Stuart and Wehner for the time-of-flight measurements.[43] The target in low pressure, dense, dc plasma is pulsed to a fixed negative voltage so that atoms are sputtered from the target as a group. When the group of atoms travels in the plasma, the atoms are excited and emit their characteristic spectrum. The energy distribution is determined by the measurement of the time shift of the emitted spectrum.

Typical results for a Cu target, sputtered by Kr ions, are shown in Fig. 3.15.[43] It shows that the kinetic energy of sputtered atoms is in the range of 0 to 40 eV for Cu (110) normally bombarded by 80 to 1,200 eV Kr ions. The energy distribution peaks at a few electron volts and more than 90% of the sputtered atoms have energies greater than 1 eV.

The energy of the sputtered atoms is dependent on both the incident ion species and incident bombardment angles. Typical results are shown in Figs. 3.16 and 3.17. Oblique incident ions shift the energy distribution to a higher region. Figure 3.18 shows the variation of the average energy of the sputtered atoms with the incident-ion energy. Note that the average energy of the sputtered atoms shows a slight dependence on the incident-ion energy. The value will saturate near 10 eV.

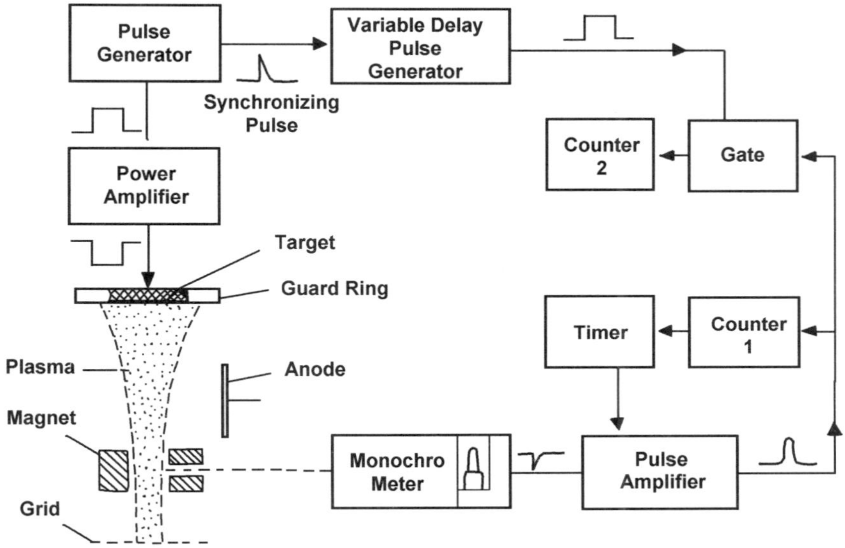

Figure 3.14. Time-of-flight method.[43]

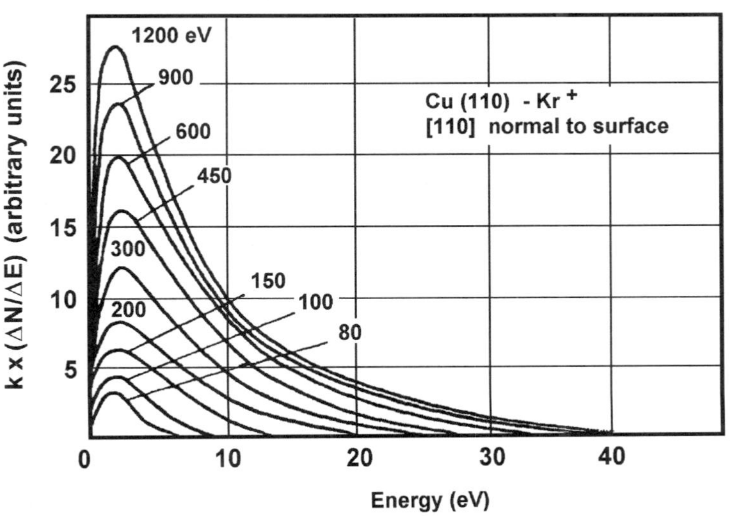

Figure 3.15. Energy distributions of sputtered atoms for various incident-ion energies.[43]

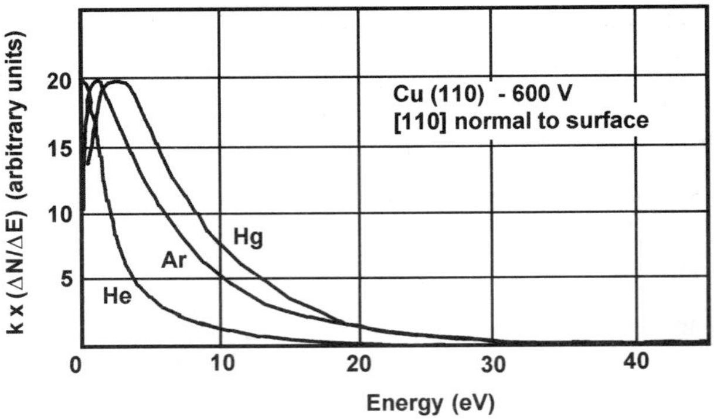

Figure 3.16. Energy distributions of sputtered atoms for various incident ions.[43]

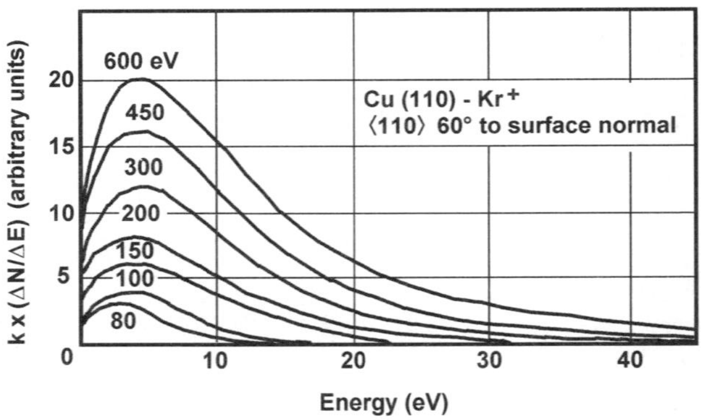

Figure 3.17. Energy distributions of sputtered atoms for various ion energies.[43]

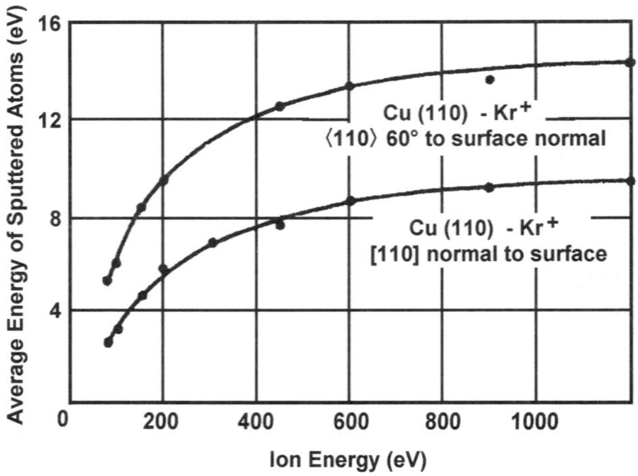

Figure 3.18. Average energy of sputtered atoms.[43]

These results suggest that when the incident-ion energy increases, energy losses in the target also increase such that the energy of incident ions is not effectively transferred to sputtered atoms. For light ion bombardment, ions will penetrate beneath the target surface and their energy loss will increase. This will shift the energy distribution to a lower energy level.

In the case of sputtered ions, their average energy is higher than that of sputtered neutral atoms described above. This experiment is complicated by the presence of a strong electric field at the target surface in a plasma-based sputtering measurement. This strong field will tend to return emitted positive ions to the target surface. Sputtered ions which escape from the target are limited to very highly energetic sputtered ions. In an ion-beam experiment which has no strong field at the cathode surface, the amount of sputtered ions is on the order of 1% of the sputtered atoms.

The knowledge of the composition and the kinetic properties of sputtered particles is important for understanding the sputtering mechanism, thin-film growth, and surface science. Extensive studies have continued on sputtered particles[44] including the postionization by an electron cyclotron resonance (ECR) plasma,[45] multiphoton resonance ionization (MPRI),[46] and laser-induced fluorescence spectroscopy (LFS).[47]

Since the vast majority of sputtered particles are emitted as neutrals, most experiments rely on some sort of ionization technique and subsequent detection of the ions. The simplest technique relies on postionization by an electron beam. However, the postionization probabilities are estimated to approach, at best, values around 10^{-4}. Postionization by electron impact in a low-pressure, noble gas plasma excited by an ECR plasma is much more efficient in ionizing the sputtered atoms, as is the MPRI technique. LFS is also an attractive method for understanding the energy distribution of sputtered particles. The velocity of the sputtered particles can be measured by the Doppler-shift of the emitted light. These experimental arrangements are shown in Fig. 3.19.[45][48]

Typical energy distributions measured by the ECR plasma ionization system are shown in Fig. 3.20.[45] Experimentally, the E^{-2} dependence at the high-energy slope of $N(E)$ has been observed, which denotes the validity of the theory, developed by M. W. Thompson, of isotropic sputtering cascades in solid targets.[49]

The LFS study suggests that under anisotropic conditions such as a low-energy region and oblique incidence, the isotropic theory is insufficient. The velocity distributions have a "hot" tail compared to the

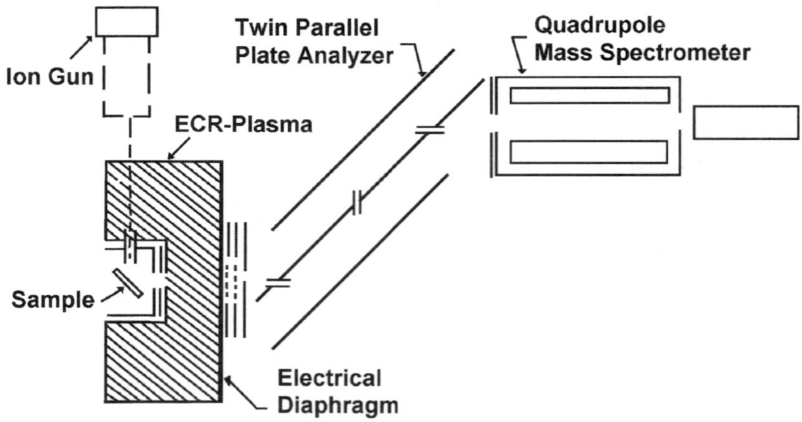

Figure 3.19. Experimental arrangements for the study of sputtered particles post-ionized by an ECR plasma source.[48]

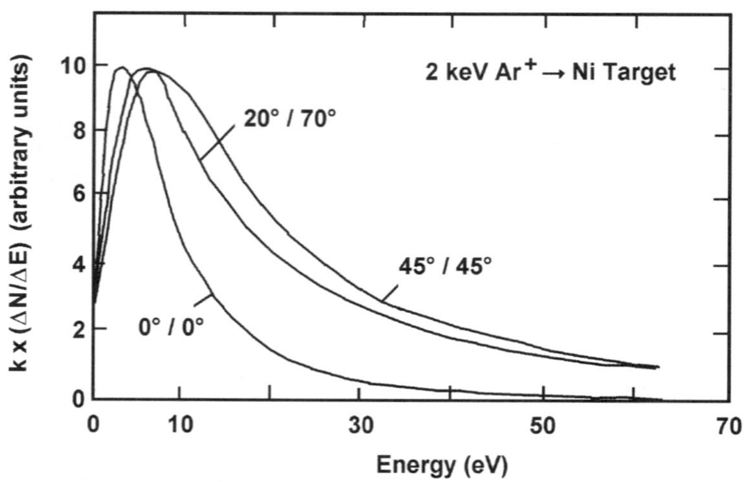

Figure 3.20. Energy distributions of sputtered atoms for various incident/escape angles for Ar^+ bombardment of Ni.[45]

isotropic model.[49] The energy and angular distribution of the sputtered particles are completely reviewed by Hofer.[50]

3.2.2.2 Mean Free Path

Most of the sputtered atoms pass through the discharge space in the form of neutral atoms. Their mean free path before they collide with discharge gas molecules is given by

Eq. (3.8) $$\lambda_1 \approx c_1/v_{12}$$

where c_1 is the mean velocity of sputtered atoms and v_{12} is the mean collision frequency between sputtered atoms and discharge gas molecules. Since the velocity of sputtered particles is much larger than the gas molecules, v_{12} is given by

Eq. (3.9) $$v_{12} \approx \pi\left(r_1 + r_2\right)^2 \bar{c}_1 n_2$$

where r_1 and r_2 are the atomic radii of sputtered atoms and discharge gas molecules, respectively, and n_2 is the density of discharge gas. Then, the mean free path is simply given by

Eq. (3.10) $$\lambda_1 \approx 1/\pi\left(r_1 + r_2\right)^2 n_2$$

In the case of Ar ion bombardment of a copper target, taking $r_1 = 0.96 \times 10^{-8}$ cm, $r_2 = 1.82 \times 10^{-8}$ cm, $n_2 = 3.5 \times 10^{16}$ cm^3 (at 0°C, 1 torr) λ_1 becomes 11.7×10^{-3} cm. These estimated values are slightly longer than the mean free path between neutral gas molecules at room temperature.

3.3 MECHANISMS OF SPUTTERING

Two theoretical models were originally proposed for sputtering:

1. The thermal-vaporization theory: the surface of the target is heated enough to be vaporized due to the bombardment of energetic ions.

2. The momentum-transfer theory: surface atoms of the target are emitted when kinetic moments of incident particles are transferred to target surface atoms.

The thermal-vaporization theory was supported by Hippel in 1926, Sommermeyer in 1935, and Townes in 1944 due to their experimental observations of Knudsen's cosine emission distribution. At that time, thermal vaporization was considered the most important mechanism.

The momentum-transfer theory was first proposed by Stark in 1908 and Compton in 1934. Detailed studies by Wehner in 1956, including the observation of spot patterns in single-crystal sputtering, suggested that the most important mechanism is not thermal vaporization, but momentum transfer.

At present, sputtering is believed to be caused by a collision cascade in the surface layers of a solid.

3.3.1 Sputtering Collisions

Electron clouds screen the nuclei of the target atoms. The type of collision between an incident particle and the target is determined by incident-ion energy and the degree of electron screening.[51] The effects of electron screening are considered Coulomb collisions. The interaction between two atoms is given by

Eq. (3.11) $$\pm\frac{z_1 z_2 e^2}{r}\exp\left(-\frac{r}{a}\right)$$

where r is the distance between incident ions and target surface atoms, z_1e and z_2e are the nuclear charge of incident ions and target atoms respectively, and a is the screening radius. The a varies with the degree of electron screening. For the Thomas-Fermi potential, $a = ca_h/z_3^{1/3}$, where a_h is the Bohr radius ($= \hbar^2/me^2$) and $c \approx 1$.

The degree of electron screening expressed by Eq. (3.11) and the distance of closest approach determine the type of collisions between the incident ions and the target atoms. In the case of head-on collisions between these atoms, the distance of closest approach b is given by the relation

Eq. (3.12)
$$\frac{z_1 z_2 e^2}{b} = \frac{M_1 M_2 v_1^2}{2(M_1 + M_2)}$$

where M_1 and M_2 are the mass of the incident ion and the target atom, respectively, and v_1 is the velocity of the incident ions. Combining Eqs. (3.11) and (3.12) leads to

Eq. (3.13)
$$\frac{b}{a} = \frac{2 z_1 z_2 e^2}{M_1 v_1^2} \frac{z_2^{1/3}}{a_h} = z_1 z_2^{4/3} \frac{2R_h}{E}$$

where R_h is the Rydberg energy, $e^2/2a_h = 13.54$ eV, and E is the the incident-ion energy.

The state of collisions varies with the ratio of b/a. At high energies (i.e., $b/a \ll 1$), the incident ions are scarcely screened by the electron clouds of the target atoms. The incident ions are scattered by target atoms. This is similar to Rutherford scattering. At low energies (i.e., $b/a \gg 1$), the incident ions are screened by the electron clouds of target atoms. The classical hard-sphere model describes the collisions.

At moderate energies, the collision cascade in the surface layers of the target is believed to cause the sputtering phenomenon. For higher-order collisions near the target surface, the energy of the collision atoms will be much less than the incident recoil energy from the first collision (several hundred eV). According to Eq. (3.13), this suggests that the approximation $b/a \ll 1$ is valid. The approximation $b/a \gg 1$, however, will not fit for the primary collision, so the whole sputtering process can not be understood by the simple hard-sphere model, where $b/a \gg 1$. The simple hard-sphere model can only roughly explain sputtering.

For $b/a \ll 1$ and $b/\lambda\!\!\!^{-} \gg 1$, the collision is governed by the classical Rutherford scattering where $\lambda\!\!\!^{-}$ denotes the wavelength of incident atoms ($= \hbar/\mu v$, $\mu = M_1 M_2/(M_1 + M_2)$, and v is the velocity of incident atoms). A differential cross section $R(\theta)dw$ (dw denotes an elements of solid angle) is expressed by

Eq. (3.14)
$$R(\theta) = \frac{b^2}{8} \operatorname{cosec}^4 \frac{\theta}{2}, \quad \tan\frac{\theta}{2} = \frac{b}{2p}$$

For $b/a \ll 1$ and $b/ƛ \ll 1$, the differential cross section $B(\theta)dw$ is given by

Eq. (3.15)
$$B(\theta) = \frac{R(\theta)}{\left\{1 + \left[\frac{\left(\frac{ƛ}{2a}\right)}{\sin\frac{\theta}{2}}\right]^2\right\}^2}$$

under the assumption that the collision is screened Coulomb scattering event with Born's approximation.

For $b/ƛ \gg b/a \gg 1$, the collision is described by the elastic-collision theory. The collision is isotropic and the total scattering cross section is expressed by πR^2. From Eq. (3.11), we have $R = a\log(z_1 z_2 e^2/RE)$ for an incident-ion energy, E. For a small E, the R is approximately the atomic radius of the incident particles.

For $b/ƛ < b/a$, the cross section is not calculated simply by Born's approximation, except in the case $b/ƛ \ll (b/a)^{1/2}$ where the collision is isotropic.

3.3.2 Sputtering

A sputtering event is initiated by the first collision between incident ions and target surface atoms, followed by the second and third collisions between the target surface atoms. The displacement of target surface atoms will eventually be more isotropic due to successive collisions, and atoms may finally escape from the surface. Figure 3.21 shows the features of sputtering collision in the target surface.

The sputtering process is considered in detail for the following three different energy regions of the incident ions.[52]

1. Threshold region (< 100 eV)
2. Low-energy region (>100 eV)
3. High-energy region (10–60 keV)

Detailed studies have been performed with the aid of computer simulation.[53][54]

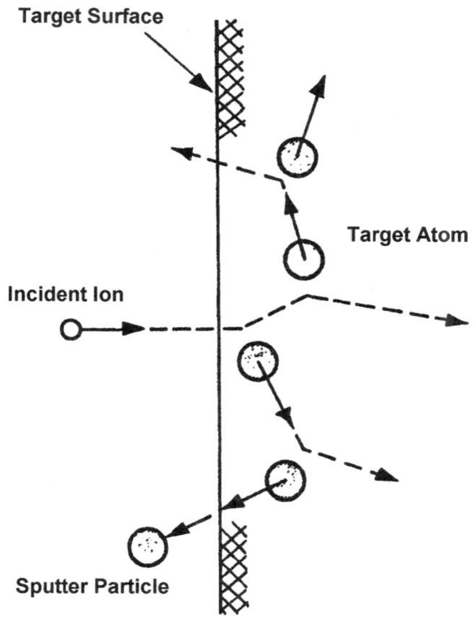

Figure 3.21. Incident ions and the sputtered particles.

3.3.2.1 Classical Empirical Formula of Sputtering Yield

According to the elastic-collision theory, the maximum possible energy transferred in the first collision, T_m, is given by

Eq. (3.16)
$$T_m = \frac{4M_1M_2}{(M_1+M_2)^2}E$$

where M_1 and M_2 are the masses of the incident ions and target atoms respectively, and E is the energy of incident ions. In the first order of approximation, the sputter yield, S, is proportional to the T_m, and the sputter yield of a given target material bombarded with a different element is given by

Eq. (3.17)
$$S = k\frac{1}{\lambda(E)\cos\theta}\frac{M_1M_2}{(M_1+M_2)^2}E$$

where k is a constant that includes different target material constants, λ is the mean free path for elastic collisions near the target surface, and θ is the angle between the normal on the target surface and the direction-of-incidence ions. The mean free path is given by

Eq. (3.18)
$$\lambda = \frac{1}{\pi R^2 n_0}$$

where n_0 is the number of lattice atoms per unit volume, and R is the collision radius.

The collision radius, R, for the rigid-sphere model can be calculated for a screened potential as

Eq. (3.19)
$$R = C \frac{a_h}{\left(z_1^{2/3} + z_2^{2/3}\right)^{1/2}} \ln \frac{z_1 z_2 e^2}{\varepsilon_0 R E'}$$

where $E' = M_1 E/(M_1 + M_2)$, C is a constant, a_h is the radius of the hydrogen atom ($=0.57 \times 10^{-8}$ cm), e the elementary charge, ε_0 is the dielectric constant in the vacuum, and $z_1 e$ and $z_2 e$ are the nuclear charges for M_1 and M_2, respectively. The relation in Eq. (3.17) gives qualitative information about the sputter yield.

Rol, et al., have suggested that by setting $k = 1.67 \times 10^{-11}$ m/eV and $C = 1$, the sputter yield measured for the ion bombardment of a copper target with 5 to 20 keV Ar^+ or N^+ ions fits the theoretical relationship Eq. (3.17).[55] Almén, et al., have shown that the constant, k, is expressed by the experimental relationship

$$k = A \exp\left(\frac{-b\sqrt{M_1}}{M_1 + M_2} E_B \right)$$

and found the sputter yield as

Eq. (3.20)
$$S = 4.24 \times 10^{-8} n_0 R^2 E \frac{M_1 M_2}{(M_1 + M_2)^2} \exp\left(-10.4 \frac{\sqrt{M_1}}{M_1 + M_2} E_B \right)$$

where E_B is the binding energy of the target materials.[56]

Goldman and Simon have shown a theoretical model of the sputtering of copper with 500 keV deuterons.[57] Rutherford scattering will take place in the collisions between the deuterons and target copper atoms. In these collisions, the mean free path of the deuteron is of the order 10^{-4} cm. The energy for the displacement of target atoms from their normal lattice is about 25 eV. The average energy of the target recoil atoms after the deuteron bombardment at the first collision is estimated to be 200 eV.

When incident atoms with energy, E, and mass, M_1, elastically collide with the target atom with mass M_2, the mean free path of the incident atom, λ_i, is expressed by

Eq. (3.21) $$\lambda_i = (n_0\sigma_d)^{-1},\ \sigma_d = \pi\frac{M_1 z_1{}^2 z_2{}^2 e^4}{M_2 E E_d}$$

where n_0 is the atomic density of the target, and E_d is the energy for the displacement from the normal lattice. The average energy of the recoil atoms, $T_{d,av}$, is expressed by

Eq. (3.22) $$\overline{T}_{d,av} = E_d \ln\frac{4M_1M_2E}{(M_1+M_2)^2 E_d}$$

The simple hard-sphere model considers the collisions after the first collision. The mean free path after the first collision is of the order of 10^{-7} cm. The diffusion of these knock-on atoms is treated by simple theory, and the sputter yield is given by the relationship

Eq. (3.23) $$S \propto \frac{M_1}{M_2}\frac{\ln E}{E}\frac{1}{\cos\theta}$$

where θ denotes the incident angle of ions. Almén, et al., have experimentally shown the validity of the relationship in Eq. (3.23).

3.3.2.2 Linear Cascade Collision Theory

Modern theoretical and experimental works on linear cascade collision theory started in the 1960s. Sigmund, Thomson, and Wehner did the seminal works. Sigmund has studied the theory of sputtering in detail. He assumed that sputtering of the target by energetic ions or recoil atoms results from cascades of atomic collisions. The sputtering yield is calculated under the assumption of random slowing in an infinite medium. The theoretical formula was compared with the experimental results given by Wehner.[58] Sigmund's cascade collision theory is the most acceptable for understanding sputtering phenomena in the kiloelectron-volt energy range.

In the theory, the collision is considered to be the hard-sphere model. The sputter yield, $S(E)$ is expressed by the following relationships:

Eq. (3.24) $$S(E) = \frac{3}{4}\pi^2 \frac{\alpha T_{\mathrm{m}}}{U_0}, \qquad <1\ \mathrm{keV}$$

Eq. (3.25) $$S(E) = 0.420 \frac{\alpha S_{\mathrm{n}}(E)}{U_0}, \qquad \approx 1\text{–}10\ \mathrm{keV}$$

where: $$T_{\mathrm{m}} = \frac{4M_1M_2}{(M_1+M_2)^2}E$$

U_0 = heat of sublimation

α = function of M_2/M_1

$$S_{\mathrm{n}}(E) = 4\pi z_1 z_2 e^2 a_{12} \frac{M_1}{(M_1+M_2)} s_n(\varepsilon)$$

$$\varepsilon = \frac{M_2 E/(M_1+M_2)}{z_1 z_2 e^2 a_{12}}$$

$$a_{12} = 0.8853 a_{\mathrm{h}} \left(z_1^{2/3} + z_2^{2/3}\right)^{-1/2}$$

$s_n(\varepsilon)$ = reduced nuclear stopping cross section for Thomas-Fermi interaction.

The a_{h} is the Bohr radius ($a_{\mathrm{h}} = 0.53 \times 10^{-8}$ cm). The values α and $S_n(\varepsilon)$ are shown in Fig. 3.22 and Table 3.7. A Born-Mayer potential [$V(r) = Ae^{-r/a}$, $a = 0.219$ Å, $A = 52\,(Z_1Z_2)^{3/4}$ eV] is applied in the low-energy region, Eq. (3.24), while the Thomas-Fermi potential is applied in the high-energy region, Eq. (3.25).

Figure 3.23 shows the comparison of measured and theoretical yield values at energies below 1 keV. The agreement is quite good. The periodicity relates to the U_0. The U_0 is equal to the surface binding energy, U_{sb}. The values of U_0 for elements tabulated by Gschneidner are shown in Table 3.8. Figure 3.24 shows sputter yields of polycrystalline copper for Kr ions. At ion energies below 10 keV, Eqs. (3.24) and (3.25) agree well with the experiments.

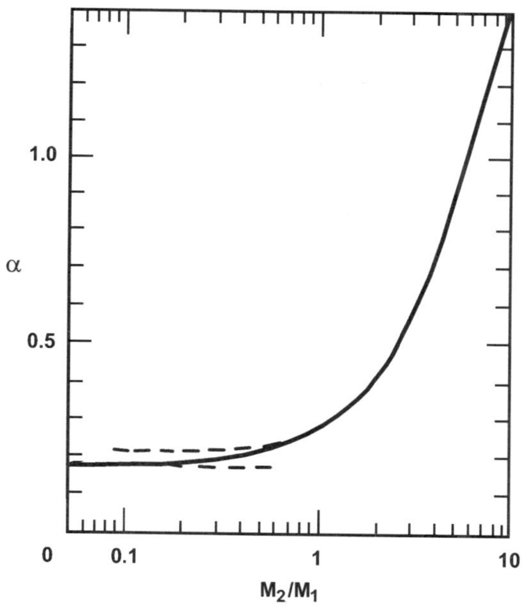

Figure 3.22. The value of α (where $\alpha = M_2/M_1$) according to Sigmund.[11]

Table 3.7. Values of Reduced Nuclear Stopping Cross Sections for Thomas-Fermi Interactions, $S_n(\varepsilon)$[11]

ε	$S_n(\varepsilon)$	ε	$S_n(\varepsilon)$
0.002	0.120	0.4	0.405
0.004	0.154	1.0	0.356
0.01	0.211	2.0	0.291
0.02	0.261	4.0	0.214
0.04	0.311	10	0.128
0.1	0.373	20	0.0813
0.2	0.403	40	0.0493

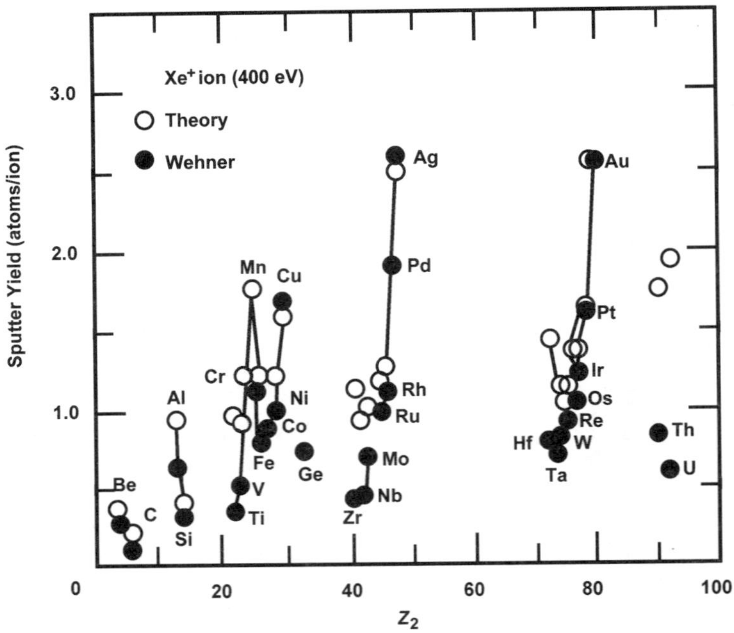

Figure 3.23. Sputter yields for various materials compared to Sigmund's theory.[11]

Table 3.8. Surface Binding Energy, U_0[59]

Target U_0 (eV/atom)	Li 1.68	B 5.73	C 7.42	Mg 1.55	Al 3.36	Si 4.69
Target U_0 (eV/atom)	Ti 4.90	Cr 4.12	Mn 2.92	Fe 4.34	Co 4.43	Ni 4.47
Target U_0 (eV/atom)	Zn 1.35	Ga 2.82	Ge 3.88	Se 1.74	Sr 1.70	Y 4.25
Target U_0 (eV/atom)	Zr 6.34	Nb 7.60	Mo 6.86	Ru 6.68	Rh 5.77	Ag 2.97
Target U_0 (eV/atom)	Cd 1.16	In 2.49	Sn 3.12	Hf 6.34	Ta 8.12	W 8.68
Target U_0 (eV/atom)	Ir 6.90	Pt 5.86	Au 3.80	Hg 0.64	Pb 2.03	Bi 2.17

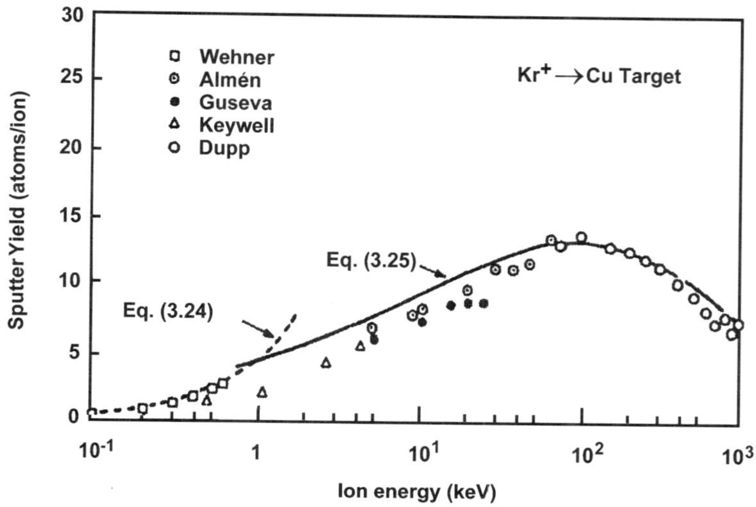

Figure 3.24. Sputter yields calculated by Sigmund.[11]

The linear cascade theory developed by Sigmund can satisfactorily account for the sputtering behavior of amorphous or polycrystalline elemental targets. A possible exception to this theory is the case of anisotropic collisions including the single target, the low-energy region, and the oblique incidence.

The nature of atomic collision cascades has been continuously studied since they are important not only as a fundamental science, but also for their technological applications in solids. They initiate the process of radiation damage, sputtering, and ion implantation.

Surface analysis methods, such as AES, ESCA, and SIMS, are useful for examining the sputtering features of an alloy. Altered layers are formed on the alloy target, and enhanced-surface and near-surface diffusions are observed during sputtering. This sometimes causes difficulty in the controlled deposition of alloy films by direct sputtering of the alloy target.[60]

Hofer reviewed modern sputtering theory using a linear cascade model and showed the angular, energy, and mass distribution.[50] Thomson also reviewed the sputtering model;[61] he evaluated the velocity of sputtered gold atoms using time-of-flight techniques and compared this with his linear cascade theory. The experiments with polycrystalline gold targets show that the energy spectra at different energies, 10, 20, 40 keV, have asymmetric peaks at about 2 eV and tails that behave functionally as

$1/E^2$. Thomson's linear cascade theory shows that the energy distribution of sputtered particles, dY/dE_0, is expressed by

Eq. (3.26)
$$\frac{dY}{dE_0} \propto \frac{E_0 U_b}{(E_0 + U_b)^3}$$

where E_0 denotes the incident energy and U_{sb} denotes the surface binding energy of target material. The experimental observations are well-understood by the formula in Eq. (3.26): a maximum at $U_{sb}/2$ (U_{sb}= 4 eV for gold), and $1/E^2$ slope at higher energies. Sigmund's theory is useful for understanding linear cascade collisions. However, there are still experimental results that cannot be explained by the cascade collision theory. Thomson pointed out the experiments on the energy spectra shown in Fig.3.25 which indicate the presence of sputtering from thermal spikes when the target was kept at elevated temperature. This is probably due to the evaporation at local patches of the target surface heated by the subsiding

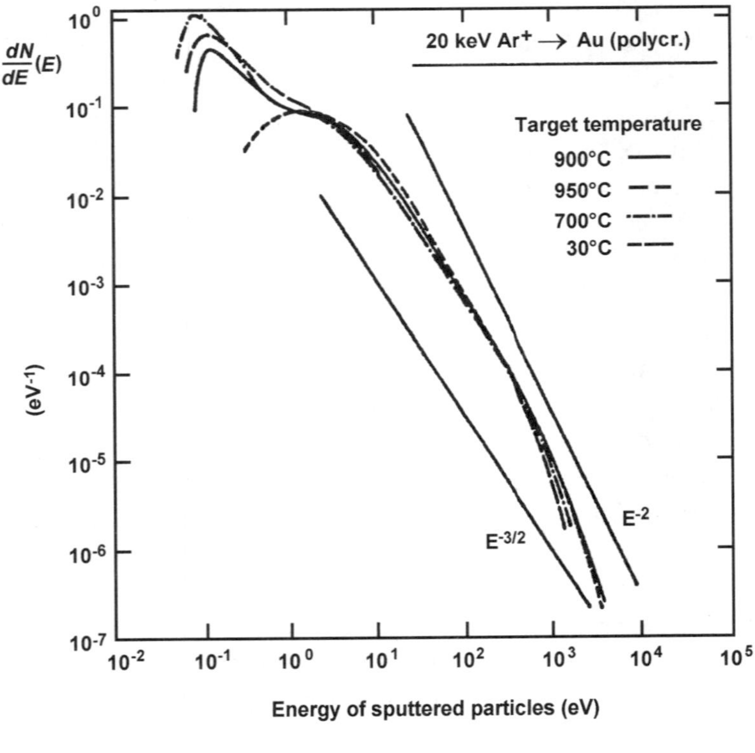

Figure 3.25. Change in energy distribution at higher temperatures.[61]

cascade similar to thermal evaporation. This process is considered a nonlinear cascade collision and/or thermal sputtering, in constrast to the linear cascade collision model.

3.3.2.3 Simplified Model and Modern Yield Formula

Mahan has reviewed empirical and theoretical formulas related to linear cascade collisions with experimental data and discusses several useful formulas for yields of sputtering deposition.[62] An empirical formula by Bohdansky for the 1-keV energy range is given by Eq. (3.27).[63]

Eq. (3.27) $$Y = 6.4\times10^{-3} m_r \gamma^{5/3} E^{0.25}\left(\frac{1-E_{th}}{E}\right)^{3.5}$$

where γ is the energy transfer mass factor expressed by $\gamma = 4m_p m_r/(m_p + m_r)^2$, and m_p and m_r are the projectile mass and the recoil ion mass, respectively. The projectile energy and sputtering threshold energy in electron volts are represented by E and E_{th}, respectively. The empirical expression for E_{th} includes the surface binding energy, U_{sb}, and is given by

Eq. (3.28) $$\left.\begin{aligned} E_{th} &= \frac{U_{sb}}{\gamma(1-\gamma)} && \text{for } \left(\frac{m_p}{m_r}\right) < 0.3 \\ E_{th} &= 8U_{sb}\left(\frac{m_p}{m_r}\right)^{2/5} && \text{for } \left(\frac{m_p}{m_r}\right) > 0.3 \end{aligned}\right\}$$

Mahan shows an example of the calculation using the sputtering of gold by Ar ions at 500 eV. Taking $m_p = 39.9$ amu, $m_r = 197$ amu, and $U_{sb} = 3.8$ eV, we have $\gamma = 0.56$, $E_{th} = 15.4$ eV, and $Y = 2.03$ (Ref 62, p. 211). These values are close to Wehner's classical experimental values of $E_{th} = 20$ eV and $Y = 2.43$.

Mahan proposed a simplified collision model of sputtering yield for the linear cascade regime. Mahan assumed that the recoils were uniformly distributed in a cylindrical volume and made a simple assumption about the distribution function.[64] Mahan's yield expression for normal incidence is given by

Eq. (3.29) $$Y = \left(\frac{1}{4}\right) f N, \quad f = \left(\frac{R^{\mathrm{p}}_{\mathrm{r,eff}}}{R^{\mathrm{p}}_{\mathrm{p}}(E)}\right), \quad N = \frac{E}{E_{\mathrm{i,av}}}$$

where f is the fraction of the total number of recoils, $R^{\mathrm{p}}_{\mathrm{r,eff}}$ (escape depth) $= \pi(R^{\mathrm{p}}\mathrm{r} + R^{\mathrm{m}}_{\mathrm{p}})^2 R^{\mathrm{p}}_{\mathrm{r,eff}}$, and $R^{\mathrm{p}}\mathrm{p}$ (uniform depth distribution of recoils) $= \pi(R^{\mathrm{p}}_{\mathrm{r}} + R^{\mathrm{m}}_{\mathrm{p}})^2 R^{\mathrm{p}}_{\mathrm{p}}$, N is the effective number of recoils at energy incident E, $E_{\mathrm{i,av}}$ (average energy of recoils) $= U_{\mathrm{sb}}$, $\ln(\gamma E/U_{\mathrm{sb}})$ for surface binding energy U_{sb}, and the mean value of the escape probability for all the recoils is 25%. The simplified equation (Eq. 3.29) shows a qualitative model. Near the threshold, it greatly overestimates the yield values. However, at the energy range of 0.1 to 1 keV, the calculated yields are very close to empirical yields measured by Matsunami.[65]*

---

*The other expression for yield is Matsunami's empirical formula:[65]

$$Y(E) = 0.42[\alpha^* Q s_{\mathrm{n}}(\varepsilon)]/U_{\mathrm{s}}[1 + 0.35 U_{\mathrm{s}} s_{\mathrm{e}}(\varepsilon)]$$
$$\times\ 8.478 Z_{\mathrm{p}} Z_{\mathrm{r}} m_{\mathrm{p}}/(Z^{2/3}{}_{\mathrm{p}} + Z^{2/3}{}_{\mathrm{r}})^{1/2}(m_{\mathrm{p}} + m_{\mathrm{r}})$$
$$\times\ [1 - (E_{\mathrm{th}}/E)^{1/2}]^{2.8}$$

where U_{s} is the sublimation energy and two empirical parameters α^* and E_{th} are expressed by the following empirical formulas,

$$\alpha^* = 0.08 + 0.164(m_{\mathrm{r}}/m_{\mathrm{p}})^{0.4} + 0.0145(m_{\mathrm{r}}/m_{\mathrm{p}})^{1.29}$$
$$E_{\mathrm{th}} = U_{\mathrm{s}} \times [1.9 + 3.8(m_{\mathrm{r}}/m_{\mathrm{p}})^{-1} + 0.134(m_{\mathrm{r}}/m_{\mathrm{p}})^{1.24}]$$

The $s_{\mathrm{n}}(\varepsilon)$ and $s_{\mathrm{e}}(\varepsilon)$ are expressed by following approximated formulas:

$$s_{\mathrm{n}}(\varepsilon) = 3.441\sqrt{\ln(\varepsilon + 2.718)}/[1 + 6.355\sqrt{\varepsilon} + \varepsilon(-1.708 + 6.882\sqrt{\varepsilon})]$$
$$s_{\mathrm{e}}(\varepsilon) = 0.079[(m_{\mathrm{p}} + m_{\mathrm{r}})^{2/3}/m_{\mathrm{p}}{}^{3/2} m_{\mathrm{r}}{}^{1/2}] \times [Z_{\mathrm{p}}{}^{2/3} Z_{\mathrm{r}}{}^{1/2}(Z_{\mathrm{p}}{}^{2/3} + Z_{\mathrm{r}}{}^{2/3})^{3/4}]$$

The values of Q is numerically given for several elements as follows:

(Be) 2.17 ±0.82	(Zr) 0.70 ±0.16	(Hf) 0.75 ±0.08	(Ta) 0.78 ±0.19
(B) 4.6 ±1.5	(Cr) 1.23 ±0.21	(Nb) 1.02 ±0.09	(W) 1.10 ±0.18
(C) 3.1 ±0.9	(Mo) 0.84 ±0.24	(Ru) 1.52 ±0.20	(Re) 1.27 ±0.22
(Mg)1.09 ±0.14	(Fe) 1.06 ±0.18	(Rh) 1.26 ±0.18	(Os) 1.47 ±0.19
(Al) 0.78 ±0.17	(Co) 1.0 ±0.32	(Pd) 1.10 ±0.25	(Ir) 1.37 ±0.22
(Ti) 0.58 ±0.10	(Cu) 1.30 ±0.22	(Ag) 1.21 ±0.19	(Pt) 1.13 ±0.17
(V) 0.9 ±0.3	(Ni) 1.06 ±0.26	(Sn) 0.47 ±0.14	(Au) 1.04 ±0.23

It was surprising that Wehner's classical yield data of Cu at 400 eV Ar^+ was almost the same as the recent yield data measured by Matsunami.

Theoretical studies of sputtering using computer simulation are useful for understanding sputtering phenomena. Thomson says that computer simulation of the behavior of irradiated solids can be performed at relatively low cost. These include molecular dynamic simulations, binary collision approximation simulations, and Monte Carlo simulation. TRIM (transport of ions in matter) is the most widely used computer model.[66] TRIM simulates the average ion trajectory and lattice damage with sputtering yield.

The detailed analysis should be studied by molecular dynamic simulation. Harrison, et al., have done seminal work on the computer simulation of sputtering using the molecular dynamic method.[54] They suggest that the emission of clusters of two and three atoms are high-probability events in sputtering of (100) copper by Ar ions in the energy range of 0.5 to 5.0 keV. Atomic motion in crystals after ion bombardment is clearly demonstrated by the molecular dynamic simulation.[67]

REFERENCES

1. Zalm, P. C., in *Handbook of Ion Beam Processing Technology* (J. J. Cuomo, S. M. Rossnagel, and H. R. Kaufman, eds.), p. 78, Noyes Publications, NJ (1989)
2. Sartwell, B. D., *J. Appl. Phys.*, 50:78 (1979)
3. Kirschner, J., and Etzkorn, H. W., *Appl. Phys.* A 29: 133 (1982)
4. Hull, A. W., and Winter, H. F., *Phys. Rev.*, 21:211 (1923)
5. Holland, L., *Vacuum Deposition of Thin Films*, p. 408, Chapman & Hall Ltd., London (1961)
6. Wehner, G. K., *Phys. Rev.*, 102:690 (1956), ibid., 108:35 (1957), ibid., 112:1120 (1958)
7. Stuart, R. V., and Wehner, G. K., *J. Appl. Phys.*, 33:2345 (1962)
8. Harrison, D. E., and Magnuson, G. D., *Phys. Rev.*, 122:1421 (1961)
9. Yamamura, Y., and Bohdansky, J., *Vacuum*, 35:561 (1985)
10. Wehner, G. K., *Phys. Rev.*, 108:35 (1957)
11. Sigmund. P., *Phys. Rev.*, 184:383 (1969)
12. Wehner, G. K., *Advances in Electronics and Electron Physics, Vol. VII.*, Academic Press, Inc., New York (1955)

13. Massey, H. S. W., and Burhop, E. H. A., *Electronics and Ionic Impact Phenomena*, Oxford Univ. Press, New York (1952)
14. Laegreid, N., Wehner, G. K., and Meckel, B., *J. Appl. Phys.*, 30:347 (1959)
15. Laegreid, N., and Wehner, G. K., *J. Appl. Phys.*, 32:365 (1961)
16. Brown, S. C., *Basic Data of Plasma Physics*, p. 230, MIT Press, Cambridge, MA, (1959)
17. Henschke, E. B., *Phys. Rev.*, 106:737 (1957)
18. Wehner, G. K., *J. Appl. Phys.*, 30:1762 (1959)
19. Wehner, G. K., and Rosenberg, D. L., *J. Appl. Phys.*, 31:177 (1960)
20. Hoffmann, T., Dodds, H. L, Robinson, M. T., and Holmes, D. K., *Nucl. Sci. Eng.,* 68:204 (1978)
21. Anderson, H. H., Bay, H. L., in *Sputtering by Particle Bombardment I* (R. Behrisch, ed.), p. 202, Springer Verlag, Berlin (1981)
22. Okutani, T., Shikata, M., Ichimura, S., and Shimizu, R., *J. Appl. Phys.,* 51:2884 (1980)
23. Seeliger, R., and Sommermeyer, K., *Z. Physik*, 93:692 (1935)
24. Maissel, L. I., and Glang, R., (eds.), *Handbook of Thin Film Technology*, p. 3–25, McGraw Hill Inc., New York (1970)
25. E. Roosendaal, H. E., in *Sputtering by Particle Bombardment, Vol. I.* (R. Behrish, ed.) p. 217 Springer Verlag, Berlin (1983)
26. Jackson, D. P., *Radiat. Elf.*, 18:185 (1973)
27. Roosendaal, H. E., in *Sputtering by Particle Bombardment, Vol. I* (R. Behrish, ed.), p. 224, Springer Verlag, Berlin (1983)
28. Fluit, J. M., Rol. P. K, and Kristemaker, J., *J. Appl. Phys.*, 34:3267 (1963)
29. Onderlinden, D., *Can. J. Phys.*, 46:739 (1968)
30. Elich, J. J. Ph., Roosendaal, H. E., and Onderlinden, D., *Radiat. Elf.*, 14:93 (1972)
31. Flur, B. L., and Riseman, J., *J. Appl. Phys.*, 35:344 (1964)
32. Patterson, W. L., and Shirn, G. A., *J. Vac. Sci. Technol.*, 4:343 (1967)
33. Liau, Z. L., Brown, W. L., Homer, R., and Poate, J. M., *Appl. Phys. Lett.*, 30:626 (1977)
34. Haff, P. K., and Switkowski, Z. E., *Appl. Phys. Lett.*, 29:549 (1976)
35. Tarng, M. L., and Wehner, G. K., *J. Appl. Phys.*, 42:2449 (1971)
36. Wilson, R. L., and Terry, L. E., *J. Vac. Sci. Technol.*, 13:157 (1976)
37. Anderson, G. S., *J. Appl. Phys.*, 40:2884 (1969)
38. Maissel, L. I., and Glang, R. (eds.), *Handbook of Thin Film Technol.*, p. 3–23, McGraw Hill, New York (1970)

39. Guenthershulze, A., *Z. Physik.*, 119:79 (1942)
40. Mayer, H., *Phil. Mag.*, 16:594 (1933)
41. Sporn, H., *Z. Physik.*, 112:278 (1939)
42. Wehner, G. K., *Phys. Rev.*, 114:1270 (1959)
43. Stuart, R. V., and Wehner, G. K., *J. Appl. Phys.,* 35:1819 (1964)
44. Anderson, H. H., and Bay, H. L., in *Sputtering. by Particle Bombardment, Vol. I* (R. Behrisch, ed.), p. 157, Springer Verlag, Berlin (1983)
45. Oechsner, H., in *Handbook of Ion Beam Processing Technology* (J. J. Cuomo, S. M. Rossnagel, and H. R. Kaufman, eds.), p. 145, Noyes Publications, NJ (1989)
46. Pappas, D. L., Winograd, N., and Kimok, F. M., in *Handbook of Ion Beam Processing Technology* (J. J. Cuomo, S. M. Rossnagel, and H. R. Kaufman, eds.), p. 128, Noyes Publications, NJ (1989)
47. Calaway, W. F., Young, G. E., Pellin, M. J., and Gruen, D. M., in *Handbook of Ion Beam Processing* (J. J. Cuomo, S. M. Rossnagel, and H. R. Kaufman, eds.), p. 112, Noyes Publications, NJ (1989)
48. Wucher, A., and Oechsner, H., *Surface Sci.*, 199:567 (1988)
49. Thompson, M. W., *Phil. Mag.*, 18:337 (1968)
50. Hofer, W. O., in *Sputtering by Particle Bombardment, Vol. III* (R. Berisch, and K. Wittmaack, eds.), p. 15, Springer-Verlag, Berlin (1991)
51. Kotani, M., Banno, Y., and Fukada, E., *Radiation Physics*, p. 18, Iwanami, Tokyo (1959)
52. Wilson, R. G., and Brown, G. R, *Ion Beams*, p. 322, John Wiley & Sons, New York (1973)
53. Harrison, D. E., Levy, N. S., Johnson, J. P., and Effron, H. M., *J. Appl. Phys.*, 39:3742 (1968)
54. Harrison, D. E., and Delaplain, C. B., *J. Appl. Phys.*, 47:2252 (1976)
55. Rol, P. K., Fluit, J. M., and Kistemaker, J., *Physica.*, 26:1009 (1960)
56. Almén, O., and Bruce, G., *Nucl. Instr. Methods*, 11:257, 279 (1961)
57. Goldman, D. T., and Simon, A., *Phys. Rev.*, 111:383 (1958)
58. Rosenberg, D. L., and Wehner, G. K., *J. Appl. Phys.*, 33:1842 (1962)
59. Gschneidner, K. A., *Solid State Phys.*, 16:344 (1964)
60. Saeki, N., and Shimizu, R., *Jpn. J. Appl. Phys.,* 17:59 (1978)
61. Thomson, M. W., *Vacuum*, 66:99 (2002)
62. Mahan, J. E., *Physical Vapor Deposition of Thin Films*, p. 253, John Wiley & Sons, NY (2000)
63. Bohdansky, J., *Nucl. Instr. Methods*, B2:587 (1987)

64. Mahan, J. E., and Vantomme, *J. Vac. Sci. Technol.*, A15(4):1976 (1997)

65. Matsunami, N., Yamamura, Y., Itikawa, Y., Itoh, N., Kazumata, Y., Miyazawa, S., Morita, K., Shimizu, R., and Tawara, H., Atom. Data Nucl. Data Tables, 31:1 (1984)

66. Ziegler, J. F., *The Transport of Ions in Matter* (1992)

67. Kitabatake, M., and Greene, J. E., *Thin Solid Films*, 272:271 (1996)

4

Sputtering Systems

The understanding of glow discharges is important in order to master the sputter deposition system since virtually all of the energetic incident particles originate in the plasma.

In this chapter, the basic concepts of the glow discharge are given, after which the construction and operation of the sputtering deposition system are described.

4.1 DISCHARGE IN A GAS

4.1.1 Cold Cathode Discharge

In a diode gas-discharge tube, the minimum voltage, which initiates the discharge, i.e., the spark or breakdown voltage, V_s, is given by

Eq. (4.1) $$V_s = a\frac{pl}{\log pl + b}$$

where p is the gas pressure, l is the electrode spacing, and a and b are constant. The relationship between spark voltage, V_s, and the gas pressure, p, is called Paschen's Law. A typical experimental result is shown in Fig. 4.1, which shows the existence of minimum spark voltage at gas pressure p_m.[1]

In a conventional sputtering system, the gas pressure is kept below the pressure p_m. To initiate the discharge, the gas pressure, p_s, under a given electrode separation, l, is expressed by

Eq. (4.2) $$p_s \geq \frac{\lambda_0}{l}$$

where λ_0 is the mean free path of electrons in the discharge gas, given by $\lambda_0 = 1/p_c$, where p_c denotes the elastic collision cross section between electrons and gas atoms. The p_c values for Ne, Ar, Kr, and Xe are shown in Fig. 4.2.[2] Taking $p_c = 20\ \mathrm{cm}^{-1}\cdot\mathrm{torr}^{-1}$ in Ar at 100 eV, λ_0 becomes 0.05 cm. By putting $l = 10$ cm, p_s becomes ≥ 5 millitorr. This suggests the gas pressure should be higher than 5 millitorr (6.5×10^{-1} Pa) for initiating the breakdown and forming the discharge.

In a diode discharge tube, when the discharge is initiated in a low-pressure gas with a high-impedance dc power supply, the mode of discharge varies with the discharge current. Figure 4.3 shows the discharge modes.[3]

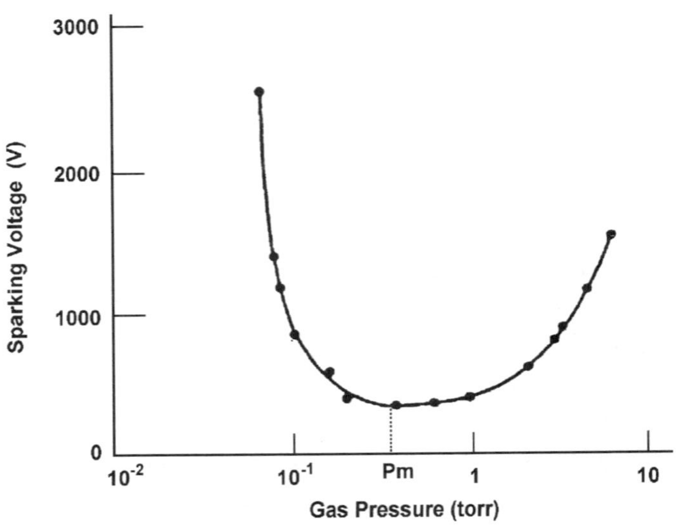

Figure 4.1. Spark voltage vs gas pressure measured for Cu electrodes in air (electrode spacing, 5 mm).

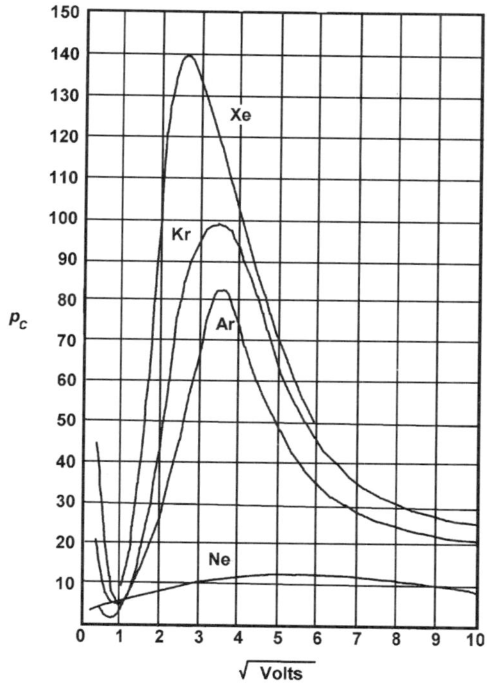

Figure 4.2. Probability of elastic collision of electron p_c ($cm^{-1} \cdot torr^{-1}$) in Ne, Ar, Kr, and Xe.[2]

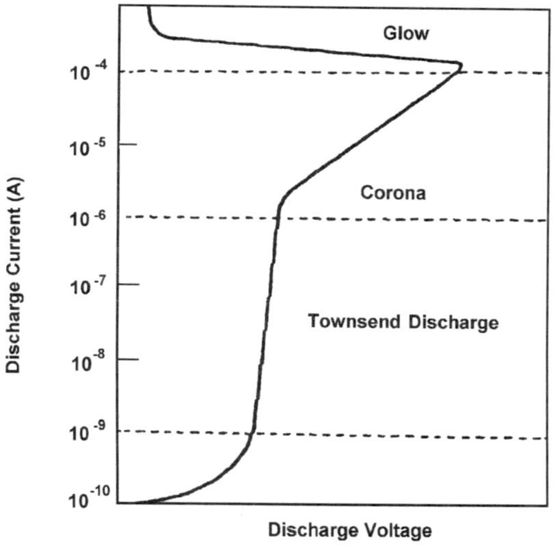

Figure 4.3. A classification of gas discharge modes for various discharge currents.[2]

At a discharge current below 0.1 mA/cm^2, the Townsend discharge appears. A small, non-self-sustained, continuous current is maintained in the Townsend discharge. Positive ions are created close to the anode due to a multiple avalanche, traverse the whole gap, and produce a uniform space charge. However, the effects of the space charge are not dominant.

At a discharge current above 0.1 mA/cm^2, a luminous glow is observed near the cathode. This region is called a *glow discharge* and the effects of the space charge are dominant.

At the low-current end of the glow-discharge region, the glow partially covers the cathode surface. This mode is called a *normal* glow discharge. When the current density increases, the glow covers the whole cathode. This mode is called an *abnormal* glow discharge. In the abnormal glow discharge, if the cathode is not cooled and the current increases above 0.1 A/cm^2, thermionic electrons are emitted. This causes a transition into an arc discharge.

The basic ionization process in the gas discharge is as follows. When the electrons collide with gas molecules, the latter are ionized and positive ions appear. The energy of the electrons should be higher than the ionization energy of the gas molecules. At the beginning of the discharge, the primary electrons from the cathode are accelerated by the electric field near the cathode. The energy is in excess of the ionization energy of the gas molecules. These energetic electrons collide with the gas molecules and generate positive ions before they travel to the anode. The positive ions bombard the cathode surface, which results in the generation of secondary electrons from the cathode surface. The secondary electrons increase the ionization of the gas molecules and generate a self-sustained discharge.

When the discharge current is below 10^{-9} A, the secondary electrons are insufficient in number to cause enough ionization to produce a self-sustained discharge as shown in Fig. 4.3. The self-sustained discharge appears when the discharge current is above 10^{-9} A. This discharge is characterized by positive space charge effects in the edge regions of the plasma.

When a single primary electron collides with *m* gas molecules and creates *m* numbers of electrons and ions, the self-sustained discharge will appear if

Eq. (4.3) $$m\gamma = 1$$

where γ denotes the number of secondary electrons per incident ion impact on the cathode surface (also known as the secondary electron coefficient).

The rate at which electrons are created in the discharge due to collisions with the secondary is given as $dn_e/dx = \alpha n_e$, where n denotes the number of electrons and α is the primary ionization coefficient. Thus the single primary electron creates $\exp(\alpha l)$ secondary electrons during the travel across the space between the electrodes. The multiplication coefficient of the primary electron m is given by $m = [\exp(\alpha l) - 1]$. Since $m\gamma = 1$,

Eq. (4.4)
$$\alpha l = \ln\left(1 + \frac{1}{\gamma}\right)$$

The ionization coefficient, α, is empirically expressed by the following relationship:

Eq. (4.5)
$$\frac{\alpha}{p} = A\exp\left[\frac{-B}{(E/p)}\right]$$

where E denotes the electric field in the discharge region, p is the pressure of the discharge gas, and A and B are constants. Assuming $E = V_s/l$, Paschen's Law, described in Fig. 4.1, is given by

Eq. (4.6)
$$V_s = \frac{B(pl)}{\ln(pl) + \ln\left[\frac{A}{[\ln(1+1/\gamma)]}\right]}$$

Table 4.1 shows the values A and B with ionization energy.

In a glow discharge, the potential distribution between electrodes is not uniform due to the presence of the charge as shown in Fig. 4.4. There is a voltage drop, V_c, near the cathode, which is known as the *cathode fall*. The cathode fall region corresponds to a so-called *cathode dark space* or *Crookes' dark space*. The spacing, d, corresponds to the region through which the electron gains the ionization energy of gas molecules. The cathode fall satisfies the relationship

$$\exp\int_o^d \alpha\, dx = 1 + \frac{1}{\gamma}$$

Table 4.1. Ionization Energy, V_i, and Constants A and B for Various Gases[2]

Gas	A (Ion pairs)/(cm·torr)	B V/(cm·torr)	E/p V/(cm·torr)	V_i (V)
H_2	5	130	150~600	15.4
N_2	12	342	100~600	15.5
O_2	-	-	-	12.2
CO_2	20	466	500~1,000	13.7
air	15	365	100~800	-
H_2O	13	290	150~1,000	12.6
HCl	25	380	200~1,000	-
He	3	34 (25)	20~150 (3~10)	24.5
Ne	4	100	100~400	21.5
Ar	14	180	100~600	15.7
Kr	17	240	100~1,000	14
Xe	26	350	200~800	12.1
Hg	20	370	200~600	10.4

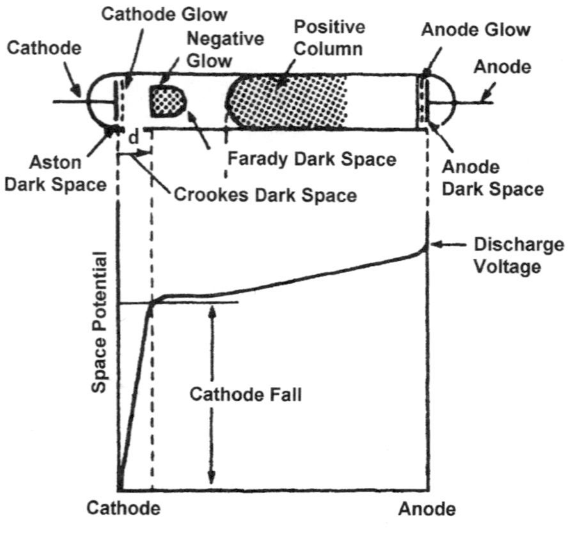

Figure 4.4. Features of a glow discharge.

The electrical field, $E(x)$, in Crookes' dark space is empirically shown by the relationship $E = k(d - x)$. The spatial distribution of the potential $V(x)$ is given by $V(x) = \int_o^x E dx = \int_o^x k(d-x)dx$. Since $V(x)$ becomes the cathode fall, V_c, at $d = x$, the $V(x)$ and $E(x)$ are expressed by

Eq. (4.7) $$V(x) = V_c x(2d - x)/d^2$$

and

Eq. (4.8) $$E(x) = 2V_c(d - x)/d^2$$

The space charge of positive ions is accumulated in Crookes' dark space and the space charge is estimated by Poisson's relationship $d^2V/dx^2 = -\rho/\varepsilon_0$. The space charge, ρ, is given by the relationship

Eq. (4.9) $$\rho = 2\varepsilon_0 V_c / d^2$$

In the cathode fall region, the current flow is carried by ions and is equal to ion current j_+. The ion current is governed by the space charge limited current (SCLC), since the space charge given by the relationship in Eq. (4.9) appears. The ions are accelerated by the potential $V(x)$ in the cathode fall region. The velocity of the ions j_+ is given by $v_+ = \sqrt{2eV/M}$, where e and M are the charge and mass of the ions, respectively. Since $j = \rho v_+$, $d^2V/dx^2 = (j_+/\varepsilon_0)\sqrt{M/2eV}$. Then, the current density j_+ is given by

Eq. (4.10) $$j_+ = \frac{4}{9}\varepsilon_0 \left(\frac{2e}{M}\right)^{1/2} \frac{V_c^{3/2}}{d^2}$$

under the condition $V = 0$ at $x = 0$.

The glow discharge is maintained by secondary electrons produced at the cathode by positive ion bombardment. The current density at cathode j_s is given by

Eq. (4.11) $$j_s = j_+(1 + \gamma)$$

In glow-discharge sputtering, the energy of the incident ions is close to the cathode fall potential. Since the anode fall, which is 10 to 20 V, and the potential drop across the positive column are much smaller than the cathode fall, the incident ion energy is roughly equal to the discharge potential.

Gas atoms are excited by inelastic collisions between the energetic electrons and gas molecules, resulting in a negative glow. The abnormal glow discharge is used for sputtering systems and most processing plasmas. In this case, the energy of incident ions is nearly equal to the discharge voltage, since $V_s = V_c$. Note that V_s and j_s cannot be changed independently. The V_c also varies with j_s/p^2 as indicated in Fig. 4.5.[3] The minimum V_c is obtained at normal glow discharge. In the sputtering system, the V_c is 500 to 1,000 V, which is characteristic of the abnormal discharge. It is understood that lowering the gas pressure causes a decrease of the current density under constant V_c. In order to keep the current density constant, the V_c should be increased.

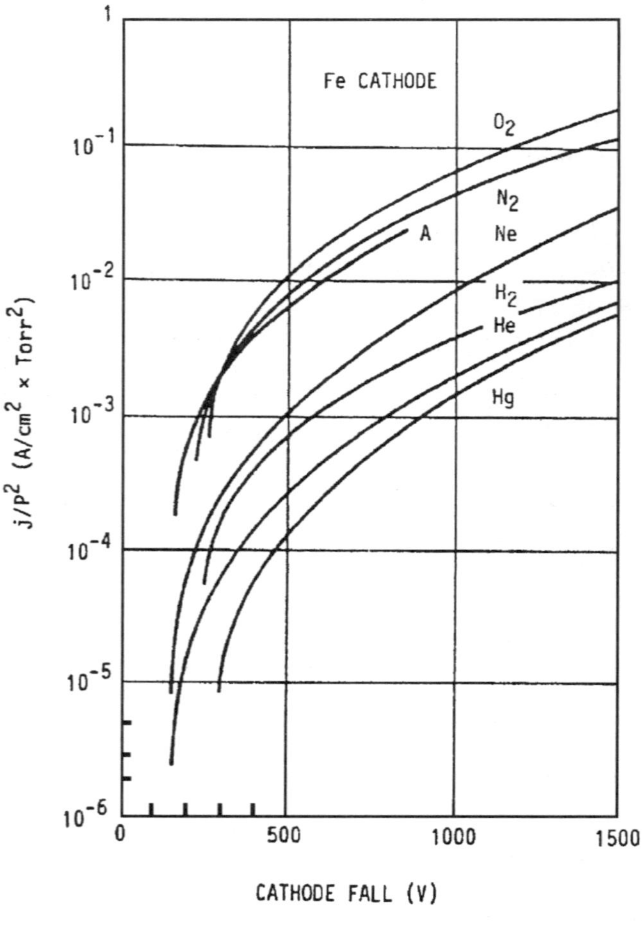

Figure 4.5. Cathode fall for an abnormal glow discharge.[3]

In glow discharge, the values (*pd*) are essentially constant, where *p* is the discharge gas pressure, and *d* is the width of the cathode dark space. However, experimental values vary with discharge parameters. Table 4.2 shows typical experimental values of (*pd*). The values of (*pd*) in normal glow discharges are slightly larger than values at abnormal glow discharge.

A superposition of magnetic field or high-frequency electrical field somewhat extends the controllable range of discharge parameters. In high-frequency discharges, the electrons in the discharge region will have cyclic motions and the number of the collisions between the electrons and the gas molecules will increase. This increases the efficiency of the gas discharge. The modified Paschen's Law expresses the effects of the alternating electric field of the discharge, or spark, voltage:

Eq. (4.12)
$$V_s = f\left(pl, \frac{p}{\omega}\right)$$

where ω is the frequency of the alternating electric field.

Table 4.2. Observed (*pd*) Values (torr × cm) For an Iron Cathode[3]

Gas	(*pd*) in (torr × cm)			
	Normal grow	Abnormal glow discharge		
		V_e = 500 V	V_e = 1000 V	V_e = 2000 V
H_2	0.9	0.46	0.25	0.2
He	1.3	0.8	0.6	0.5
N_2	0.4	0.17	0.1	0.1
Hg	0.3	0.2	0.12	-
O_2	0.3	0.1	0.07	0.07
Ar	0.25	0.07	-	-

Note: (*pd*) is the discharge gas pressure, *p*, times the width of the cathode dark space, *d*.

4.1.2 Discharge in a Magnetic Field

4.1.2.1 Spark Voltage in a Magnetic Field

In the presence of a magnetic field, electrons in a gas-discharge tube show orbital motion around the magnetic lines of force as shown in Fig. 4.6. In a magnetic field of strength B, the radius of the orbital motion is $r = mv/eB$, where e, m, and v are the electron charge, mass, and velocity, respectively. When the magnetic field is longitudinally superposed on the gas discharge, the electrons in the discharge are twined around the magnetic line of force. This reduces the loss of electrons in the discharge region and increases the discharge current density. However, strong additional effects of the magnetic field are observed in the presence of a transverse magnetic field.

When the transverse magnetic field, B, is superposed on the electric field, E, the electron shows cycloidal motion with an angular velocity $\omega = eB/m$ and center of orbit drift in the direction of $\mathbf{E}\times\mathbf{B}$ with the velocity of E/B as shown in Fig. 4.7. These electron motions increase the collision probability between electrons and molecules enabling the use of lower gas pressure, as low as 10^{-5} torr. Figure 4.8 shows the effects of the transverse magnetic field on the spark voltage measured in coaxial cylindrical electrodes, in which the magnetic field is applied in an axial direction.[1] It shows the effects of the magnetic field observed at a critical value B_c, which is the so-called "cutoff field." In the magnetic field below B_c, the primary electrons from the vicinity of the cathode will reach the anode without performing the cycloidal motion between the electrodes since the radius of the cyclotron motion is greater than the spacing of the electrodes.

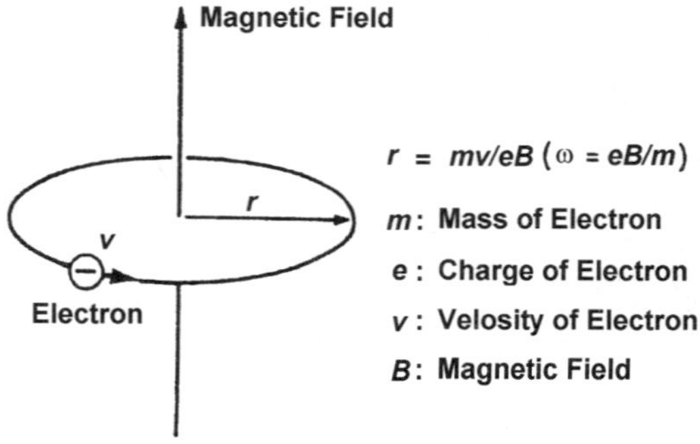

Figure 4.6. Cycloidal motion of electrons in a magnetic field.

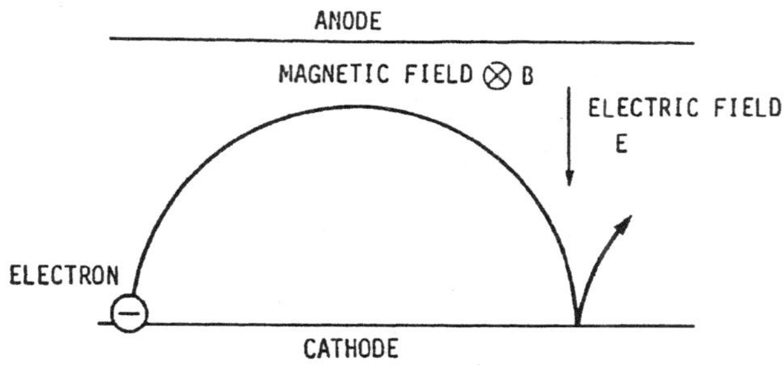

Figure 4.7. Electron trajectory in a crossed electric and magnetic field.

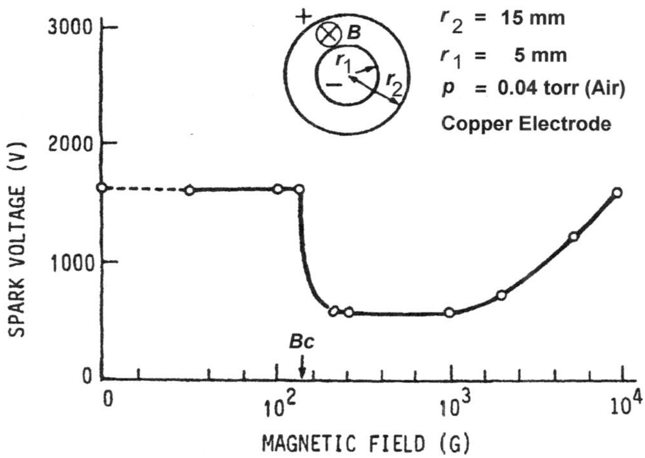

Figure 4.8. Spark voltage vs magnetic field strength measured between coaxial cylindrical electrodes with an axial magnetic field.

The cutoff field, B_c, is expressed by the following relationships:

Eq. (4.13)

$$B_c = \left(\frac{2mV_c}{e}\right)^{1/2}\left(\frac{2r_2}{r_2^2 - r_1^2}\right) \quad \text{(cylindrical electrode)}$$

$$B_c = \left(\frac{2mV_c}{e}\right)^{1/2}\left(\frac{1}{l}\right) \quad \text{(planar electrode)}$$

where V_c is the applied voltage between the electrodes, r_2 and r_1 are the radii of the outer and inner electrodes, respectively, and l is the electrode spacing.

The abrupt decrease of the discharge voltage in the magnetic field above B_c results from the increase in the number of collisions between the primary electrons and neutral gas molecules. In a significantly higher magnetic field, the spark (discharge) voltage increases with the magnetic field. The increase of voltage in the strong magnetic field is induced by the increase of electron energy losses through successive collisions.

The effects of the transverse magnetic field are qualitatively considered to be an effective pressure, p_e, given by

Eq. (4.14)
$$\frac{p_e}{p} \cong \left[1 + (\omega\tau)^2\right]^{1/2}$$

where ω is the cyclotron frequency of the electron and τ is the mean free time of the electron.[4] Since

$$\omega = \frac{eB}{m} \quad \text{and} \quad \tau = \frac{\lambda_0}{p\left[2(e/m)V_0^{1/2}\right]}$$

Eq. (4.15)
$$\omega\tau \cong \frac{\lambda_0 B (e/m)^{1/2}}{\sqrt{2}\, p V_0^{1/2}}$$

where λ_0 is the mean free path of the electron at 1 torr, B is the strength of the magnetic field, (e/m) is the specific charge of the electron, and V_0 is the acceleration voltage for the electron. Taking $B = 100$ G (0.01 T), $p = 1 \times 10^{-5}$ torr (Ar), $V_0 = 100$ V, and $\lambda_0 = 0.05$ cm, we obtain $\omega\tau = 5 \times 10^3$, and $p_e = 0.05$ torr.

Theoretically, the spark voltage in the transverse magnetic field is estimated by the change of ionization coefficients α and γ with the superposition of the crossed field. Table 4.3 shows the theoretical formulas of the ionization coefficient α in the crossed electric and magnetic field. The effect of the crossed field on γ is not well understood. In general, the secondary electrons released from the cathode surface will be partially recaptured at the cathode surface since the crossed field bends their trajectory. This may reduce the effective secondary electron coefficient.

Table 4.3. Ionization Coefficient, α, in a Crossed Electric and Magnetic Field (Wasa, Ph.D. Thesis, Osaka Univ., 1967)

References	Relations
Kugler, et al.	$\alpha = \frac{3eB^2}{4mE} \;:V_e = \frac{16}{3\pi}\frac{mE^2}{eB^3\lambda}$
Haefer, Somerville	$\alpha = \frac{3eB^2}{4mE}\left(1 - \frac{B^2V_i^{\,2}}{4E^2}\right)^{1/2}$
Redhead	$\alpha = \frac{1}{g}\frac{3eB^2}{4mE}\left(1 + \frac{B^2V_i^2}{8E^2}\right)^{-1} \quad : \; g \sim 3$
Somerville	$\alpha = \frac{3}{4}AL\frac{eB^2}{mE}\exp\left(-\frac{3}{4}bL\frac{eB^2}{mE}\right)$
Blevin and Haydon	$\alpha = Ap\left(1 + \frac{B^2e^2L^2}{p^2m^2u^2}\right)^{1/2}\exp\left[-\frac{bp}{E}\left(1 + \frac{B^2e^2L^2}{p^2m^2u^2}\right)^{1/2}\right]$

λ: electron mean free path; *L*: λ at 1 torr; *u*: electron velocity.

4.1.2.2 Glow Discharge in a Magnetic Field

Under the superposition of a magnetic field crossed on an electric field, the width of the Crookes' dark space decreases. This suggests that the superposition of the magnetic field equivalently increases the gas pressure in the discharge region, as shown in the relationship described in Eq. (4.14).

Consider now the glow discharge with parallel electrodes shown in Fig. 4.9. The motion of a primary electron is determined by the equations

Eq. (4.16) $$m\frac{d^2x}{dt^2} = eE - Be\frac{dy}{dt};\; m\frac{d^2y}{dt^2} = Be\frac{dx}{dt}$$

where $E = 2V_c(d-x)/d^2$. From Eq. (4.16), the equation of electron motions becomes

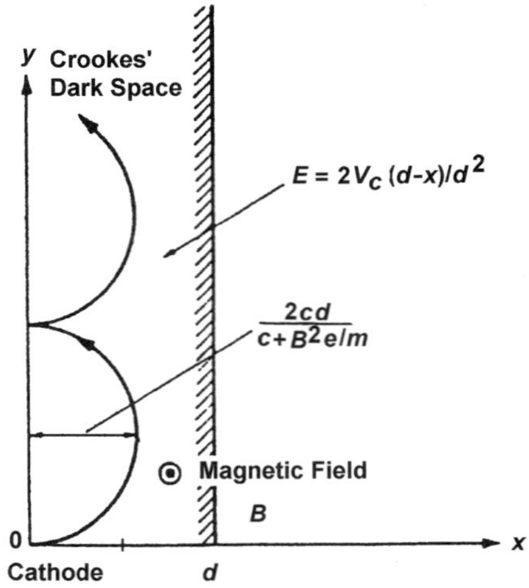

Figure 4.9. Electron motion in a Crookes' dark space.

Eq. (4.17) $$m\frac{d^2x}{dt^2}+\left(ce+\frac{B^2e^2}{m}\right)x=cde$$

where $c=2V_c/d^2$. On the assumption that the electron starts from the cathode with zero initial velocity, it moves in a cycloidal path. The electron motion in the x-direction is given by

Eq. (4.18) $$x=\frac{cd}{c+B^2\dfrac{e}{m}}(1-\cos\omega\tau)$$

where

$$\omega=\frac{\left(ce+B^2\dfrac{e^2}{m}\right)}{m}$$

The maximum displacement in the direction of the electric field, D, is given by

Eq. (4.19) $$D=\frac{2cd}{c+B^2\dfrac{e}{m}}$$

The magnetic field affects the discharge when

Eq. (4.20) $$D = \frac{2cd}{c + B^2 \dfrac{e}{m}} < d \quad \text{or} \quad B^2 \frac{e}{m} > \frac{2V_c}{d^2}$$

The rf discharge with a transverse magnetic field is often used for the thin-film processing. From Eq. (4.16), the electron motions in the rf discharge are given by

Eq. (4.21) $$m\frac{d^2x}{dt^2} = eE_0 \sin \omega t - Be\frac{dy}{dt}; \quad m\frac{d^2y}{dt^2} = Be\frac{dx}{dt}$$

when we substitute E with $E_0 \sin \omega t$. From Eq. (4.21), the electron velocity in the direction of electric field dx/dt is given by

Eq. (4.22) $$\frac{dx}{dt} = v_0 \sin \omega_H t + u_0 \cos \omega_H t - \frac{eE_0}{m}\frac{\omega}{\omega_H^2 - \omega^2}\cos \omega_H t + \frac{eE_0}{m}\frac{\omega}{\omega_H^2 - \omega^2}\cos \omega t$$

where u_0 and v_0 denote the initial velocity of electrons in x and y directions, respectively, and $\omega_H = eB/m$ denotes the electron cyclotron frequency. The energy, ε, transferred to the electron during the cycloidal motion, is given by

Eq. (4.23) $$\varepsilon = \frac{eE_0^2}{8m}\left[\frac{1}{(\omega - \omega_H)^2} + \frac{1}{(\omega + \omega_H)^2} + \frac{2\cos 2\omega t}{\omega_H^2 - \omega^2}\right]$$

Equations (4.22) and (4.23) show that electrons will effectively receive energy at the cyclotron frequency $\omega = \omega_H$. The discharge voltage exhibits a minimum point at $\omega = \omega_H$.

4.1.2.3 Glow Discharge Modes in a Transverse Magnetic Field

In the presence of a transverse magnetic field, there appears to be two different modes of glow discharge in a dc cold cathode discharge tube.

One is a positive space-charge–dominated mode (PSC mode), which appears in a weak magnetic field. The other is a negative space-charge mode (NSC mode), which appears in a strong magnetic field. The appearance of the different modes is interpreted in the following manner: In the presence of a magnetic field, the path of the electrons is deflected. The electrons perform a cycloidal motion in the plane perpendicular to the magnetic field between the electrodes. As a result, the electron radial velocity is reduced, and decreases with the increase of the strength of the magnetic field. The ion radial velocity is only marginally changed by the magnetic field. When the magnetic field is so strong that the electron velocity is smaller than the ion velocity, the NSC mode appears. In this mode, the cathode fall is very small and, inversely, the anode fall is very large. The electron gains sufficient energy to cause ionization of neutral molecules in the anode fall. When the magnetic field is so weak that the electron velocity is larger than the ion velocity, the PSC mode appears. In this mode, the anode fall is very small and, inversely, there is a large cathode fall by which the electrons gain energy to cause ionization. Typical distributions are shown in Fig. 4.10.[5]

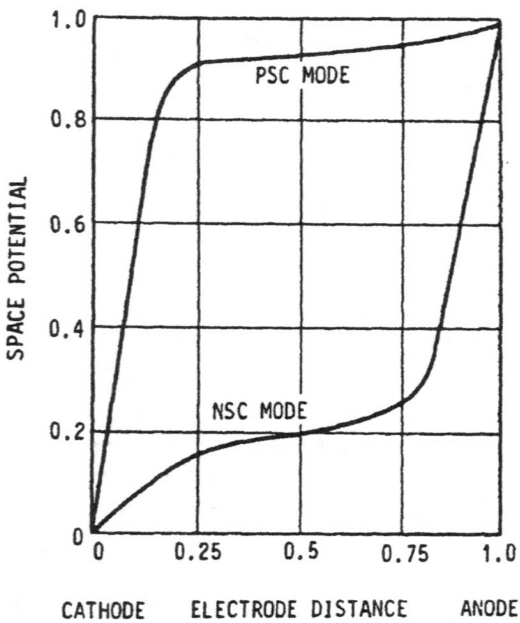

Figure 4.10. Special potential distributions for a glow discharge in the presence of a crossed electric and magnetic field.

The magnetic field prolongs the electron path between electrodes, and thus increases the probability of electron-neutral collisions through which the neutral molecules can be ionized. This enables the discharge of both the PSC and NSC modes to be sustained at a low gas pressure where the mean free path of the electrons is longer than the electrode spacing, i.e., $\lambda > l$ or $p < \lambda_0/l$. Here, λ denotes the electron mean free path at a given pressure p, λ_0 the electron mean free path at 1 torr, and l the electrode spacing. At such low pressure, the discharge cannot be sustained without the magnetic field. Thus, both modes may be used for low-pressure sputtering. Kay suggests that, at low gas pressure, the PSC mode is much more applicable for sputtering than the NSC mode, since the former has a higher sputtering rate than the latter due to the large cathode fall.[6] He suggests that in order to have the large cathode fall, the magnetic field should be as small as possible. The electron cutoff field, below which the glow discharge cannot be sustained, then determines the optimum magnetic field.

The properties of the NSC mode are somewhat modified at low gas pressure under the influence of a very strong magnetic field. Figure 4.11

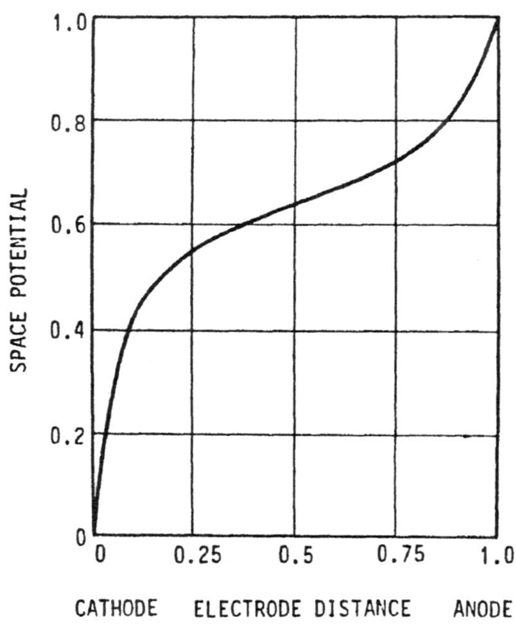

Figure 4.11. A spatial potential distribution in a strong magnetic field.

shows the potential distribution across the electrodes. Significant voltages at both the cathode fall and the anode fall are observed. In this modified mode, the electron can gain the ionization energy in both falls. The cathode fall results from the loss of electrons due to their recapture at the cathode surface. Since the large cathode fall results in significant levels of sputtering, this mode is useful for many practical sputtering applications.[7] In this mode, the strong magnetic field can increase the ionization efficiency remarkably. Thus the operating pressure can be as low as 10^{-5} torr or less for an electrode spacing of about 1 cm. This modified mode can be adapted to a practical low-pressure sputtering system. A transition magnetic field where the PSC mode changes into the NSC mode is roughly estimated from the following relationship: $v_i/v_e \sim 1$, where v_i denotes the radial ion velocity and v_e the radial electron velocity. In low pressure such as 1 millitorr, the ion radial velocity depends on the square root of the discharge voltage since the ion mean free path is longer than the electrode spacing and the ion mass is so great that the magnetic field hardly causes a deflection of of the ion. Then the ion velocity v_i is given by $v_i = (2eV_s/m_i)^{1/2}$, where e denotes the electronic charge, V_s the discharge voltage, and m_i the mass of the ions. The electron drifts in the radial direction due to collisions with the gas molecules, and v_e is given by $v_e = 16m_eE^2/\pi^2 e\lambda B^3$, where m_e denotes the mass of the electrons, E the radial electric field, and B the strength of a given magnetic field. The v_i/v_e becomes

Eq. (4.24) $$\frac{v_i}{v_e} = \left(\frac{2eV_s}{m_i}\right)^{1/2} \frac{\pi e\lambda B^3}{16m_eE^2}$$

At the transition condition of discharge mode, $v_i/v_e = 1$, taking $e = 1.6 \times 10^{-19}$ C, $\lambda = 0.2$ m at 1 millitorr, $m_e = 9.1 \times 10^{-31}$ kg, $V_s \sim 1500$ V, $m_i = 6.6 \times 10^{-26}$ kg (Ar), and $E \sim 7.5 \times 10^4$ V/m, we found that the transition magnetic field is estimated to be about 140 G. The modified NSC mode may appear in a much higher magnetic field of more than 1000 G, regardless of the electrode size.

Computer modeling of the discharge is also studied to understanding the energy, angle, and spatial distribution of ions in magnetron plasma.[8] Most of the computer modeling is based on basic experimental data of the discharge, i.e., space potential distribution and/or ion velocity. Computer modeling may be a useful tool for designing sputtering systems.

4.1.2.4 Plasma in a Glow Discharge

In the positive column of the glow discharge there is plasma composed of the same number of electrons and ions. The energy of electrons and ions in the plasma is estimated as follows: the electrons (mass m) and ions (mass M) in the plasma, whose initial velocity is zero, are accelerated by the electric field E with an acceleration rate of Ee/m and Ee/M, respectively. The energy given for the electron and ion from the electric field E in time t is expressed by

Eq. (4.25) $$\frac{(Eet)^2}{2m} \text{ (electron)}, \quad \frac{(Eet)^2}{2M} \text{ (ion)}$$

Since $m << M$, most of the energy is transferred to the electron from the electric field in the initial stage. In low-pressure gas discharge, the collision rate between electrons and gas molecules is not frequent enough for a nonthermal equilibrium to exist between the energy of the electrons and the gas molecules, so the highly energetic particles are mostly composed of electrons while the energy of the gas molecules is around room temperature.

We have $T_e \gg T_i \gg T_g$, where T_e, T_i, and T_g are the temperatures of the electron, ion, and gas molecules, respectively. This type of plasma is called "cold plasma." In a high-pressure gas discharge, the collisions between electrons and gas molecules occur frequently. This causes thermal equilibrium between the electrons and the gas molecules. We have $T_e \sim T_g$. We call this type of plasma "hot plasma."

In cold plasma, the degree of ionization is below 10^{-4}. The electrons receive energy from the electrical field and collide with the neutral atoms and/or molecules. This results in excitation and/or ionization of the atoms and gas molecules.[9]

In plasma-assisted deposition systems, thin films are grown on the negatively biased electrode. The negative bias is induced even in rf discharge, since electron mobility is much higher than that of ions. The negative bias is induced at the rf electrode with the blocking capacitor, as shown in Fig. 4.12.

The induced negatively biased V is estimated as follows: The positive ions of mass M come from the plasma and traverse the dark spaces without making any collisions, and form a space charge region near the electrode with a current density j:

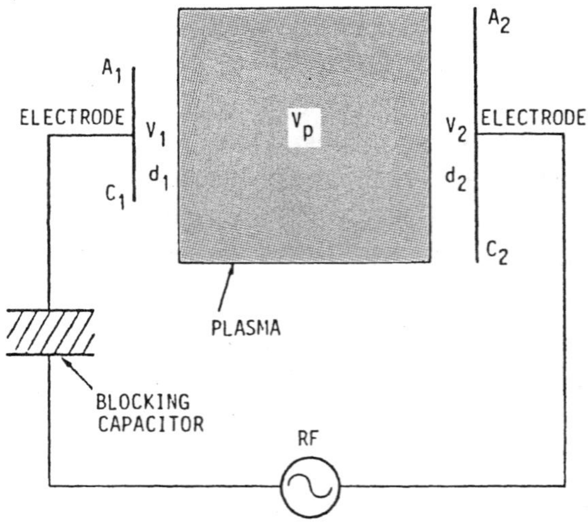

Figure 4.12. Potential induced on electrodes which are immersed in a plasma.

Eq. (4.26)
$$j = \frac{KV^{3/2}}{M^{1/2}d^2}$$

where d is the thickness of the space charge zone and K is a constant.

The total current flow is equal at both electrodes,

Eq. (4.27)
$$j_1 A_1 = j_2 A_2$$

Combining this with Eq. (4.25) gives

Eq. (4.28)
$$\frac{A_1 V_1^{3/2}}{d_1^2} = \frac{A_2 V_2^{3/2}}{d_2^2}$$

The capacitance across the dark space is proportional to the electrode area and inversely proportional to the dark space thickness. The rf voltage is capacitively divided between the two sheaths near the electrodes: We have

Eq. (4.29)
$$\frac{V_1}{V_2} = \frac{c_2}{c_1} = \frac{A_2}{d_2}\frac{d_1}{A_1}$$

where c_1 and c_2 are the capacitances across the dark space for each electrode. Combining Eqs. (4.28) and (4.29), we have

Eq. (4.30)
$$\frac{V_1}{V_2} = \left(\frac{A_2}{A_1}\right)^2$$

This suggests that the self-bias voltage ratio V_1/V_2 is proportional to the square of the inverse area ratio A_2/A_1. It is noted that under an assumption that the current density of the positive ions is equal at both electrodes, the V_1/V_2 is proportional to $(A_2/A_1)^4$. However, the relationship in Eq. (4.30) shows the actual power dependence.

4.2 SPUTTERING SYSTEMS

There are several sputtering systems for the purpose of thin-film deposition. Their designs are shown in Fig. 4.13. Among these sputtering systems, the basic model is the dc diode sputtering system. The other sputtering systems are improvements on the dc diode sputtering system.[9]

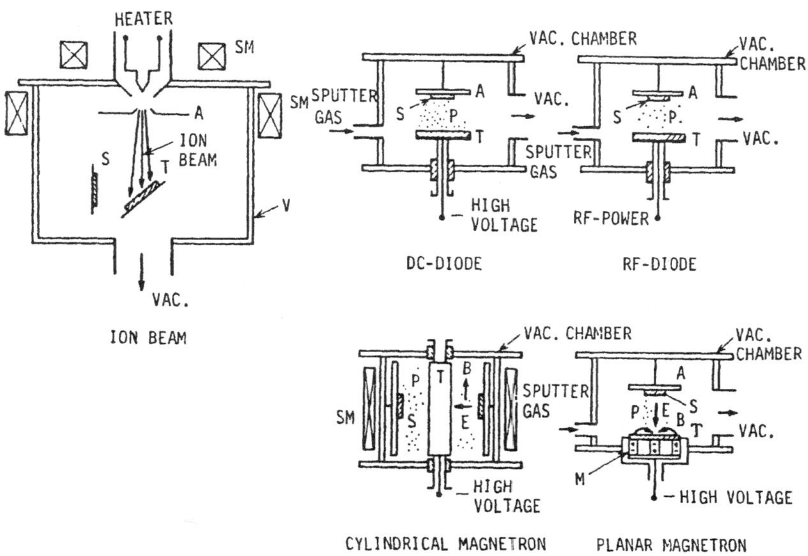

Figure 4.13. Basic configurations of sputtering systems.

4.2.1 DC Diode Sputtering

The dc diode sputtering system is composed of a pair of planar electrodes. One of the electrodes is a cold cathode and the other is an anode. The top plasma-facing surface of the cathode is covered with a target material and the reverse side is water-cooled. The substrates are placed on the anode. When the sputtering chamber is kept in argon gas at 0.1 torr and several kilovolts of dc voltage with a series resistance of 1 to 10 kΩ are applied between the electrodes, the glow discharge is initiated. The Ar ions in the glow discharge are accelerated at the cathode fall and sputter the target, resulting in the deposition of thin film on the substrates.

In the dc diode system, sputtered particles collide with gas molecules and then eventually diffuse to the substrate since the gas pressure is so high and the mean free path of the sputtered particles is less than the electrode spacing. The amount of sputtered material deposited on a unit substrate area W is then given by

Eq. (4.31)
$$W = \frac{k_1 W_0}{pl}$$

and the deposition rate R is given by

Eq. (4.32)
$$R = \frac{W}{t}$$

where k_1 is a constant, W_0 is the amount of sputtered particles from the unit cathode area, p is the discharge gas pressure, l is the electrode spacing, and t is the sputtering time.

The amount of sputtered particles from the unit cathode area W_0 is given by

Eq. (4.33)
$$W_0 = \left(\frac{j_+}{e}\right) St \left(\frac{A}{N}\right)$$

where j_+ is the ion current density at the cathode, e is the electron charge, S is the sputter yield, A is the atomic weight of the sputtered materials, and N is Avogadro's number.

With the assumption that the ion current is nearly equal to the discharge current I_s, and the sputter yield is proportional to the discharge voltage V_s, the total amount of sputtered particles becomes $V_s I_s t/pl$. Thus, the sputtered deposit is proportional to $V_s I_s t$.

4.2.2 RF Diode Sputtering

By simple substitution of an insulator target for the metal target in a dc diode sputtering system, the sputtering glow discharge cannot be sustained because of the immediate buildup of a surface charge of positive ions on the front side of the insulator.

To sustain the glow discharge with an insulator target, the dc voltage power supply is replaced by an rf power supply. This system is called an rf sputtering system. Robertson and Clapp observed sputtering in the rf discharge in 1933.[10] They found that the glass surface of the discharge tube was sputtered during the rf discharge.

In the 1960s, sputtering in the rf discharge was used for the deposition of dielectric thin films as a practical rf sputtering system.[11] The rf sputtering system holds an important position in the deposition of thin films. A typical sputtering system is shown in Fig. 4.14.

The rf diode sputtering system requires an impedance-matching network between the power supply and discharge chamber. A typical network for impedance matching is shown in Fig. 4.15.[12] The impedance of the rf power supply is almost always 50 Ω. The impedance of the glow discharge is on the order of 1 to 10 kΩ. In rf diode sputtering, the cathode current density is given by

Eq. (4.34)
$$i_s \cong C\frac{dV}{dt}$$

where C is the capacitance between the discharge plasma and the target, and dV/dt denotes the time variations of the target surface potential. This indicates that the increase of the frequency increases the cathode ion currents. In practical systems, the frequency used is 13.56 MHz.

Note that in the rf discharge system the operating pressure is lowered to as low as 1 millitorr, since the rf electrical field in the discharge chamber increases the collision probability between secondary electrons and gas molecules. In the rf sputtering system, a blocking capacitor is connected between the matching network and the target. The target area is

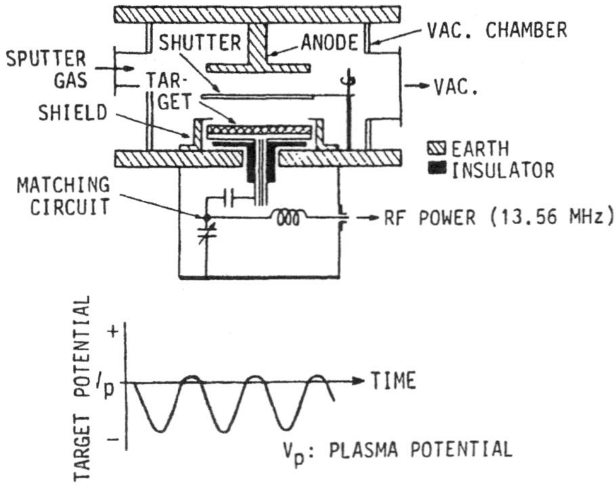

Figure 4.14. Radio-frequency diode sputtering system.

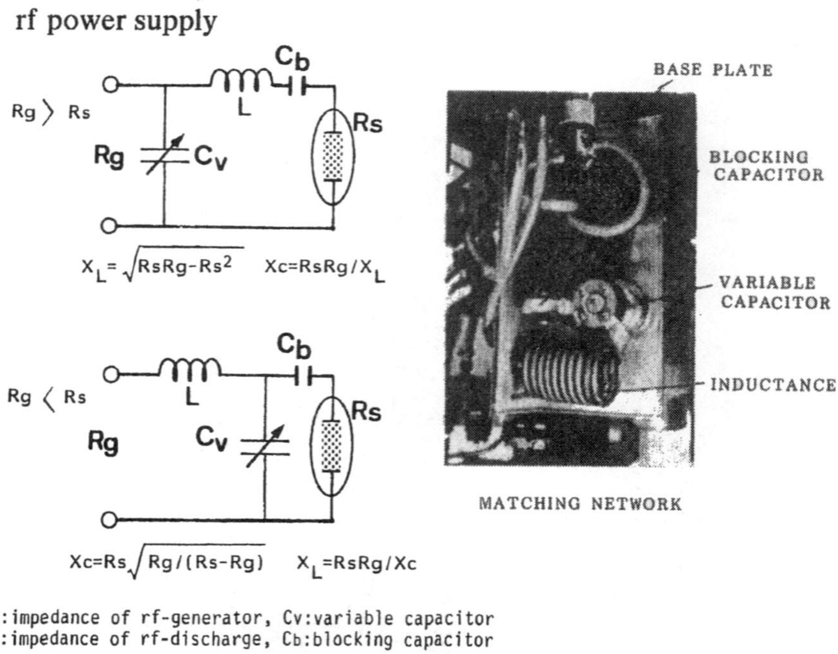

Figure 4.15. Impedance matching networks for rf sputtering system.

much smaller than the grounded anode and the chamber wall. This asymmetric electrode configuration induces negative dc bias on the target, and this causes sputtering in the rf system. The dc bias is on the order of one half of the peak-to-peak voltage of the rf power supply.

Figure 4.16 is a photograph of the sputtering target during deposition. In rf sputtering systems, the target and inductance in the matching network are always cooled by water. The electrical resistivity of the cooling water should be high enough to serve as electrical insulation.

4.2.3 Magnetron Sputtering

Low-pressure sputtering is one of the most promising techniques for the production of thin-film devices. A wide variety of thin films can be made with little film contamination and at a high deposition rate by the low-pressure sputtering technique.

In 1935, Penning first studied low-pressure sputtering in which a transverse magnetic field was superposed on a dc glow-discharge tube as shown in Fig. 4.17.[13]

The experimental system was composed of coaxial cylindrical electrodes with an axial magnetic field, similar to a cold cathode magnetron. He found that superimposition of the magnetic field of 300 G lowered the sputtering gas pressure by a factor of ten, and increased the deposition rate of sputtered films. However, this kind of system was not used in practice.

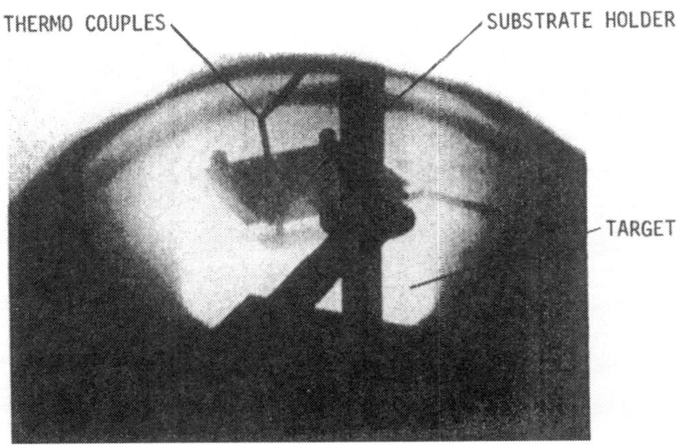

Figure 4.16. Photograph showing the rf sputtering system during a deposition.

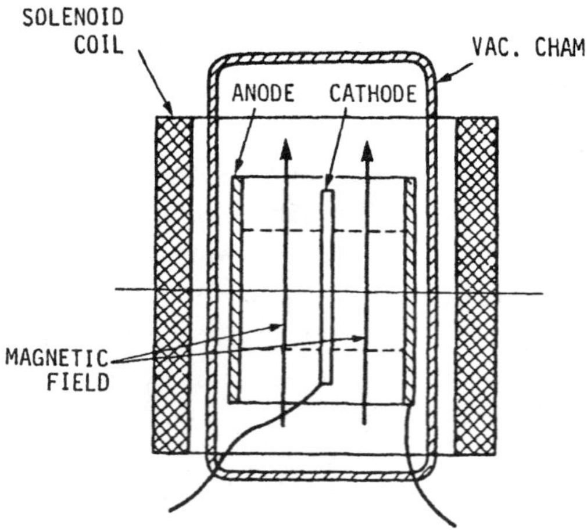

Figure 4.17. Penning's sputtering system.[13]

In the early 1960s, magnetron sputtering was reconsidered an attractive process for thin-film deposition. Kay studied the glow discharge in the presence of a magnetic field in relation to thin-film deposition. He found that a quadrupole magnetic field increases ion current density at cathode by more than one order of magnitude accompanied by an increase in deposition rate.[14] The electrode arrangement with quadrupole magnetic field is shown in Fig. 4.18. Gill and Kay proposed an inverted magnetron sputtering system and demonstrated that the sputtering gas pressure was as low as 10^{-5} torr, which was two orders of magnitude lower than conventional sputtering systems. They applied the inverted magnetron for the deposition of superconducting thin films.[15] The strength of the magnetic field was several hundred gauss and the PSC mode was dominant in the sputtering discharge.

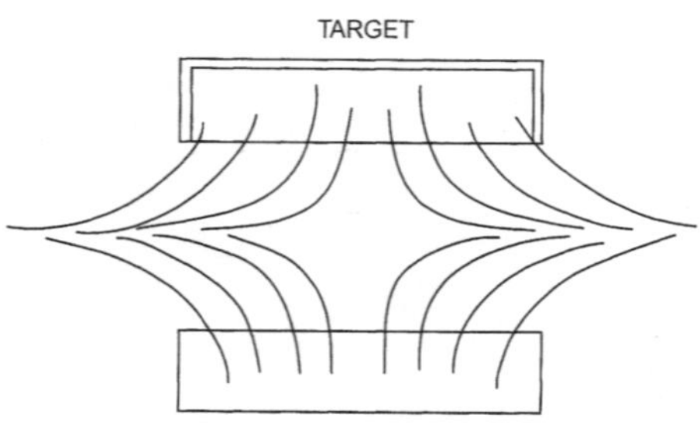

Figure 4.18. A quadrupole magnetic field for sputtering system.[14]

Hayakawa and Wasa also studied this type of discharge in relation to the creation of high-current rotating plasma for nuclear fusion.[16] They also studied an effect of a transverse magnetic field on a glow-discharge mode,[17] plasma instability,[18] and cathode sputtering.[19]–[21] It was found that the presence of a strong magnetic field enhanced the cathode sputtering. In a glow discharge, the mode of discharge changed from conventional PSC mode to NSC mode with a large anode fall. They first proposed the planar magnetron sputtering system with a solenoid coil. Figure 4.19 shows the construction of the system.[22] The planar magnetron arrangement was very convenient to use for the deposition of thin films, compared to the cylindrical magnetron. The original planar magnetron was based on a modification of a nuclear fusion system of current loop apparatus shown in Fig. 4.20.[23] However, in the 1960s, this type of magnetron sputtering system was not used in practice.

In the 1970s, Clarke developed the practical magnetron sputter system.[24] The system, called an S-gun, was a modification of the cylindrical magnetron, and is shown in Fig. 4.21. Also, Chapin developed the planar magnetron system, shown in Fig. 4.22.[25]

The seminal work on planar magnetron sputtering was done by Thornton.[26] Magnetron sputtering became more widely used for its potential applications for Si integrated circuits as well as for various other thin-film coating processes. Today, it is in broad-scale practical use[27] in industries ranging from microelectronics, to automobiles, to architectural glass, to hard coatings.

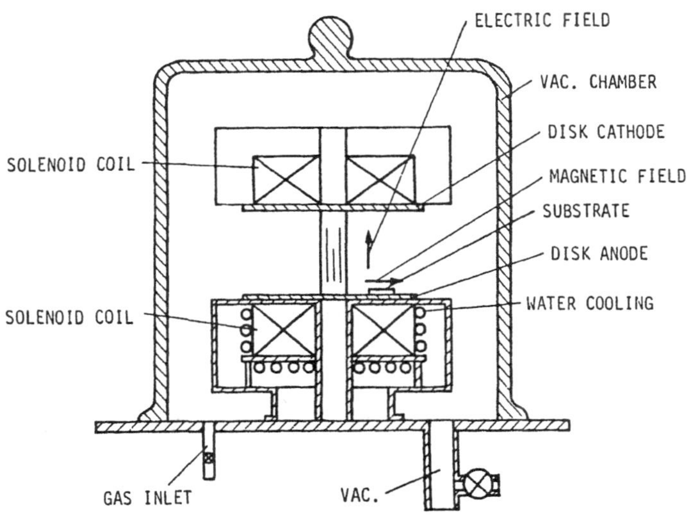

Figure 4.19. Construction of a planar magnetron sputtering system with solenoid coil.[22]

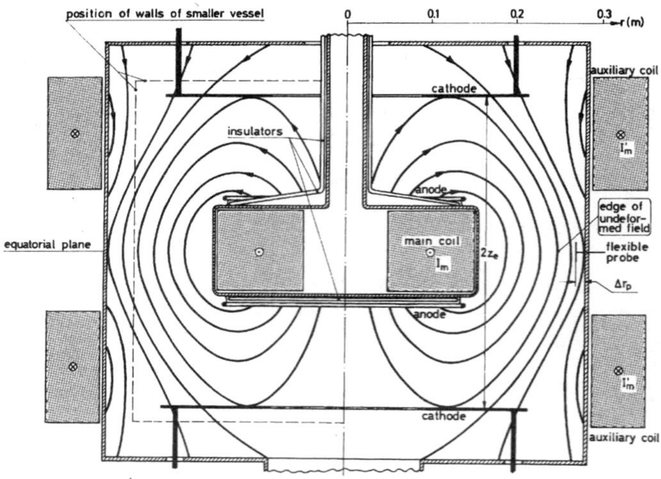

Figure 4.20. Construction of a high current discharge system for a nuclear fusion using a planar magnetron configuration.[23]

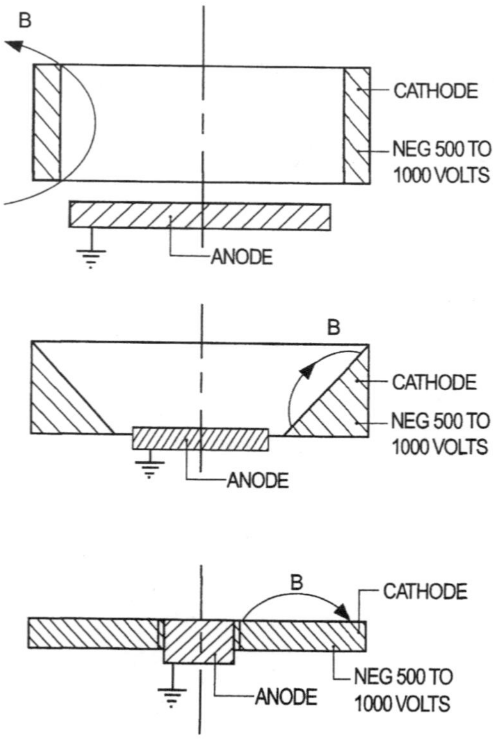

Figure 4.21. Construction of a magnetron sputter gun.[24]

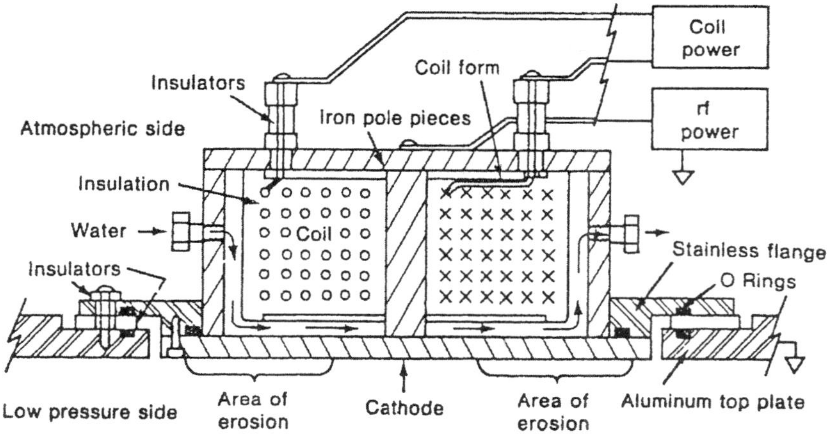

Figure 4.22. Construction of a planar magnetron cathode.[25]

Two types of magnetron sputtering systems are widely used for thin-film deposition. One is a cylindrical type; the other is a planar type. Typical cathode configurations are shown in Fig. 4.23. Within the cathode target, permanent magnets are embedded such that the resultant magnetic field is several hundred gauss. The glow discharge is concentrated in the high-magnetic-field region, thus a circular cathode glow is observed, as shown in Fig. 4.24. The surface of the cathode is nonuniform due to the circular cathode glow. This shortens the actual life of the cathode target. Several types of improved magnetron targets have been proposed, including a magnetron target with moving magnets or multi-magnets, in order to have a uniform erosion area and also to extend the actual life of the target, as shown in Fig. 4.25.

At one time magnetron sputtering sources were considered to be of limited use when magnetic target materials were used for the deposition. Chopra and Vankar have designed and fabricated a sputtering cathode, which offers both of these advantages (uniform erosion area and extended life) and, in addition, yields excellent uniformity of the deposited films without employing a cumbersome substrate rotation. This geometry can also be used with magnetic materials with equal ease. Figure 4.26 shows the construction of the planar magnetron sputtering source designed by Chopra and co-workers.[28] It consists of a water-cooled cathode (A) made of copper on which planar targets (B) of any material bounded to a copper backing plate can easily be screw-mounted. The cathode is insulated from a water-cooled aluminum shield (C) with a Teflon® spacer (D) and is kept in position using a stainless-steel nut (I). The magnets (J) are mounted outside the shield. The whole assembly is affixed to a stainless-steel base plate (G), which is placed in a bell jar in which sputtering is carried out. The discharge voltage is 300 to 800 V where the maximum sputtering yield per unit energy is obtained.

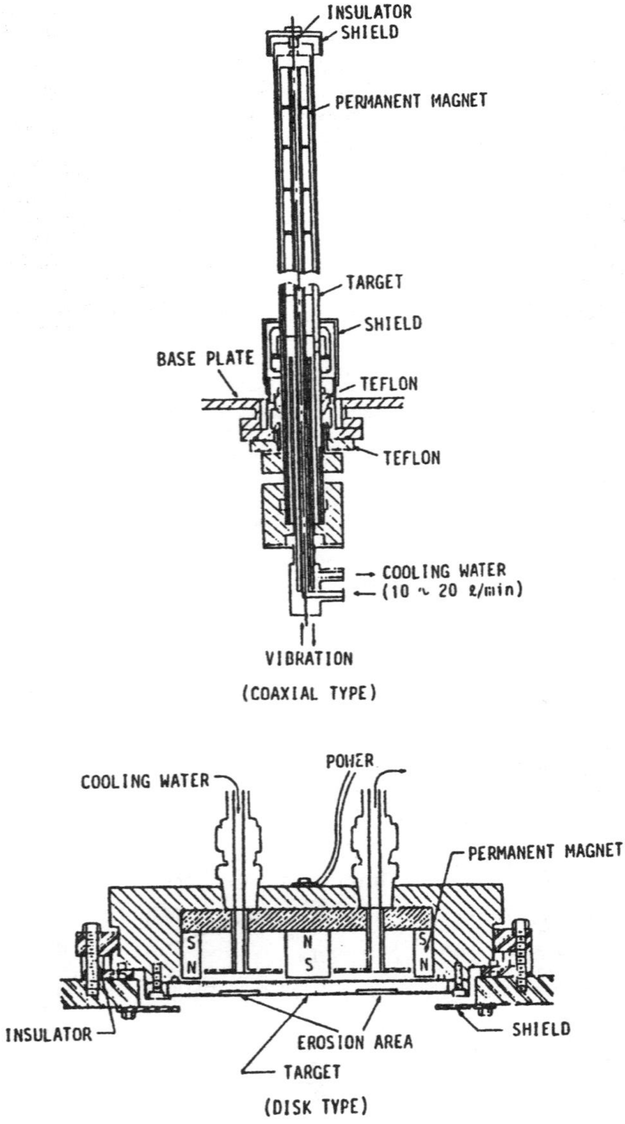

Figure 4.23. Construction of commercialized magnetron cathodes.

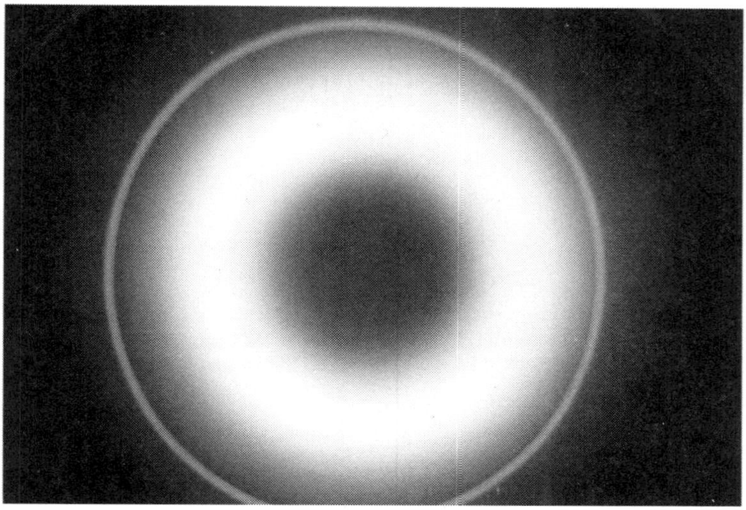

Figure 4.24. Photograph showing a top view of a planar disk target in a magnetron discharge; target 100 mm diameter.

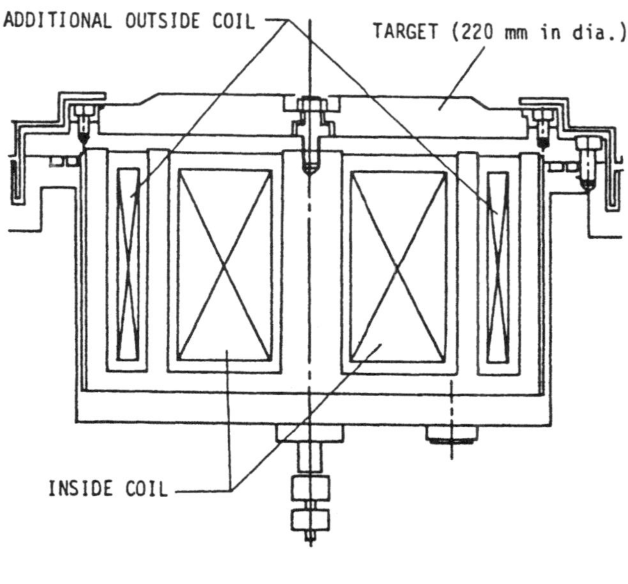

Figure 4.25. Construction of a magnetron target with an additional solenoid coil.

(a)

(b)

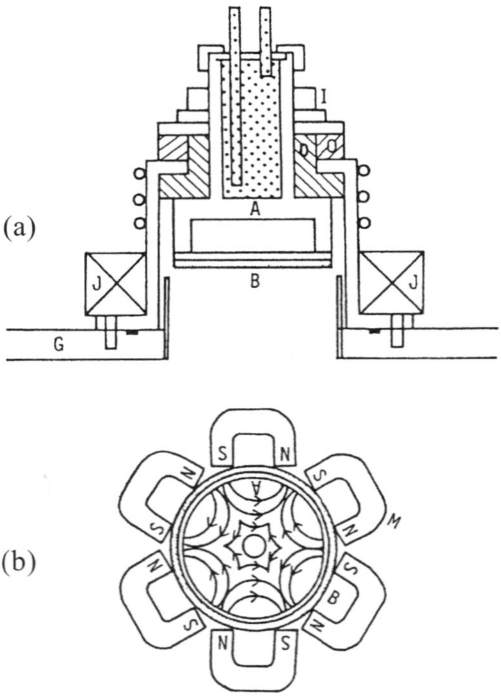

Figure 4.26. Cross-sectional view (*a*) and top view (*b*) of planar magnetron sputtering source. (A) cathode, (B) planar target, (C) aluminum shield, (D) Teflon® spacer, (I) stainless-steel nut, (J) magnets, and (G) stainless-steel base plate.[28]

At present the sputtering process is widely used for the deposition of magnetic materials, enabled by the improved design of the magnetic flux distribution with a strong rare-earth permanent magnet.

In the magnetron sputtering system, the working pressure is 10^{-5} to 10^{-3} torr, and the sputtered particles traverse the discharge space without collisions. Thus the deposition rate R is simply given by

Eq. (4.35)
$$R = \frac{kW_0}{t}$$

where $k = 1$ for the planar system, $k = r_c / r_a$ for the cylindrical system, r_c is the cathode radius and r_a is the anode radius, and W_0 is the amount of sputtered particles given by Eq. (4.30).[20] The typical experimental results for the magnetron sputtering system are shown in Fig. 4.27.

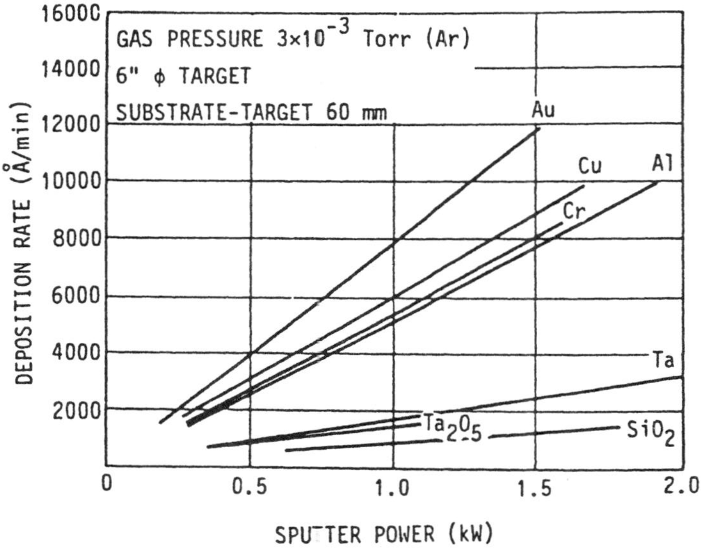

(a)

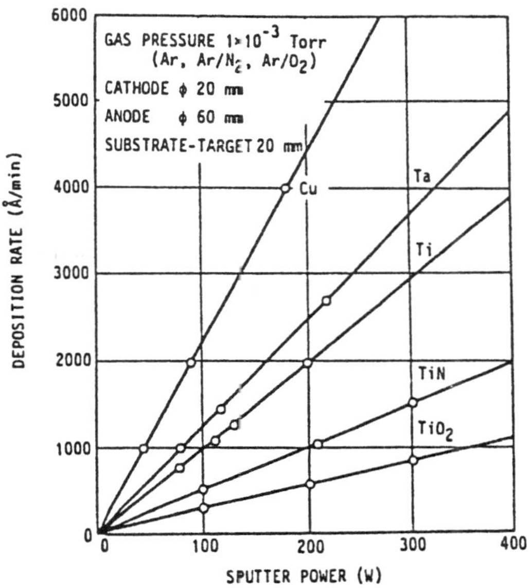

(b)

Figure 4.27. (*a*) Deposition rate as a function of sputtering power for a planar magnetron sputtering system, dc magnetron for a metal target, rf magnetron for a dielectric target. (*b*) Deposition rate as a function of sputtering power for a dc coaxial magnetron.

In a traditional planar magnetron, the magnetic flux on the cathode surface is terminated to the magnetic core as shown in Fig. 4.28a. The magnetron is called a *balanced* magnetron. When an additional magnetic flux is superposed to the balanced magnetron shown in Fig. 4.28b, the magnetron is called an *unbalanced* magnetron. Several types of modified magnetron sputtering systems are proposed based on the unbalanced magnetron system. In the balanced magnetron, energetic electrons escape from the primary magnetic trap near the cathode surface. These electrons go to the anode. In the unbalanced magnetron, the excess magnetic field lines trap the escaping energetic electrons, and the electrons make ionizing collisions. Secondary plasma is generated near the substrates. The incident flux of ionized particles increases in the unbalanced magnetron. The incident ionized flux modifies the film-growing process similar to ion-assisted deposition.[29] The unbalanced magnetron is useful for the reactive deposition of oxides or nitrides on metal at metallic mode under substrate bias conditions, since the ion currents to the substrates increase due to secondary plasma ionization. High ion current density will enhance the chemical reaction at the substrate during the film growth.[30]

Several types of compact sputtering cathodes of the unbalanced magnetron have been commercially developed. Typical magnetron cathodes are shown in Fig. 4.29. These compact cathodes look like a sputter gun. The sputtering deposition system is made just by fastening the sputter cathode via a vacuum con-flat flange and/or fittings to the vacuum chamber.

The other modified magnetron cathode is the sputtering system with inductively coupled plasma shown in Fig. 4.30.[31] The rf coil generates plasma and enhances the ionization at the magnetron cathode as seen in Fig. 4.30a. The working pressure of the modified magnetron sputtering system is as low as 5×10^{-4} torr. The working pressure is one order of a magnitude lower than that of balanced magnetron sputtering. The construction and the sputtering discharge of the modified magnetron cathode are shown in Fig. 4.30b. The in-situ monitoring of the crystal growth of the thin films by RHEED is available under the low-pressure operation. This kind of system shows uniform thickness distribution and is useful for the deposition of ultrathin magnetic materials and/or layered structure. The ionized particles will modify the physical and/or chemical properties of sputtered films similar to the unbalanced magnetron.

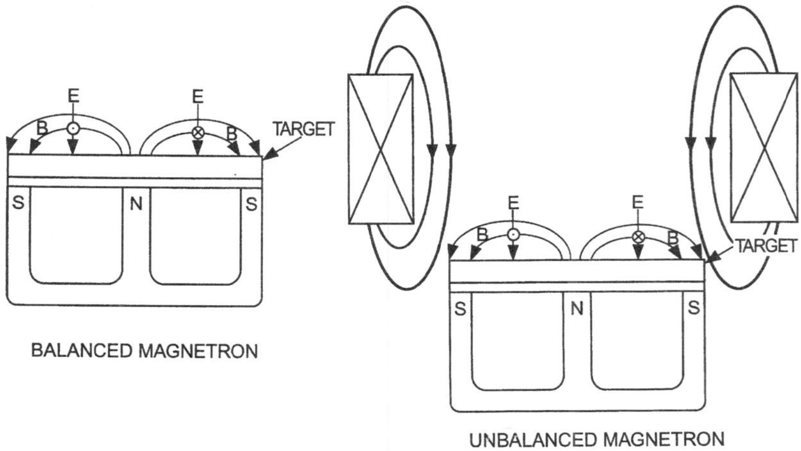

Figure 4.28. Balanced and unbalanced magnetrons.

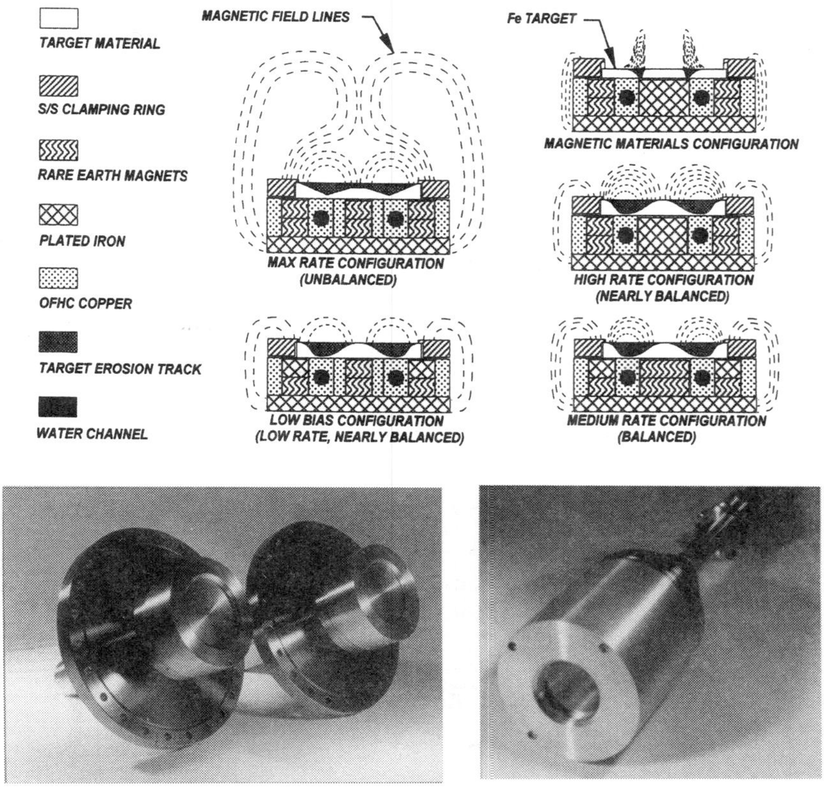

Figure 4.29. Commercial sputtering cathode *(Gencoa catalog).*

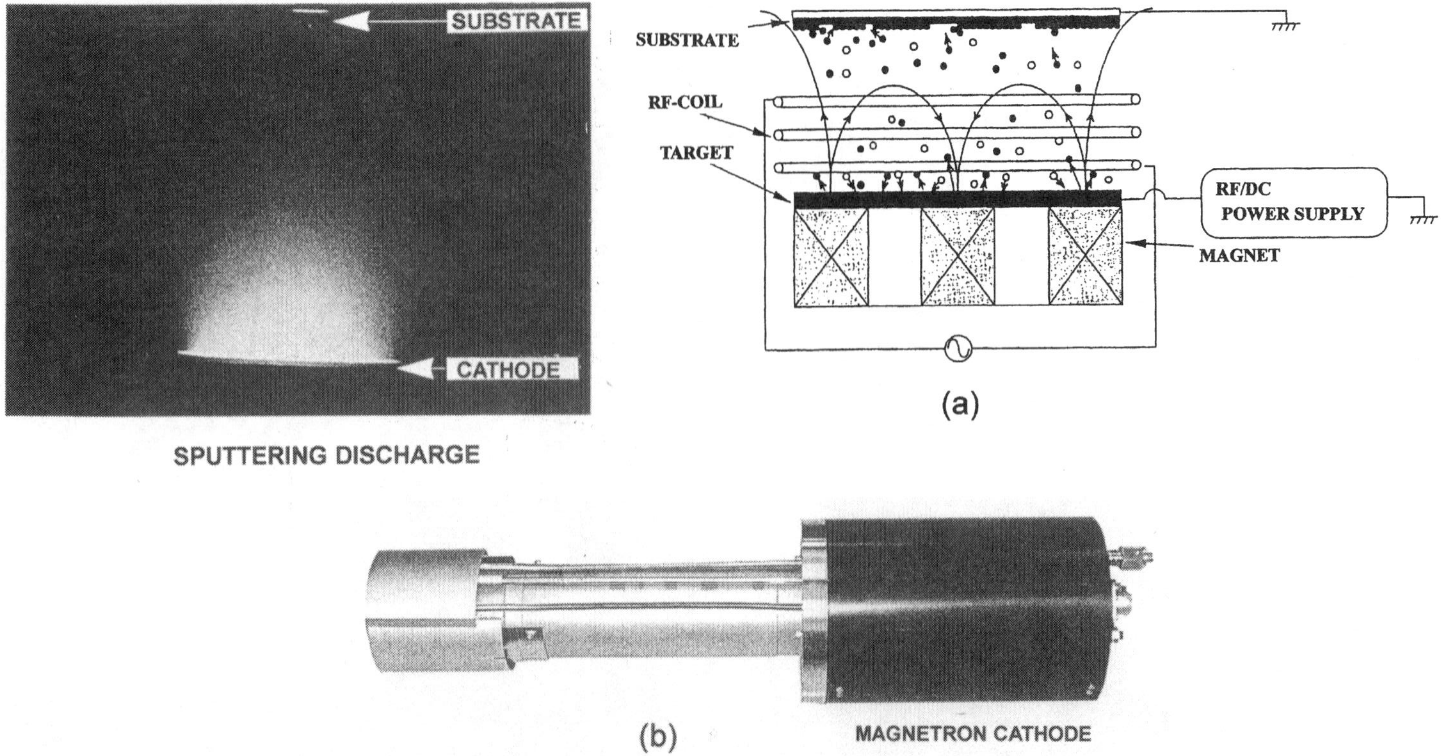

Figure 4.30. (*a*) Schematic diagram of the magnetron sputtering source with an inductively coupled rf plasma.[28] (*b*) Photograph of magnetron cathode (*ULVAC catalog, MHS-HC series, N4515/97043000PK 1997*).

4.2.4 Ion Beam Sputtering

In the glow-discharge system, the gas pressure is so high that gas molecules irradiate the sputtered films during thin-film growth. This causes the inclusion of gas molecules in the sputtered films. In the ion-beam sputtering system, the incident ions are generated at the ion source. The target is bombarded by the ion beam in a sputtering chamber separated from the ion source as shown in Chopra and Randlett's sputtering system, Fig. 4.31. Chopra and Randlett did pioneering work on the deposition of thin films by ion-beam sputtering.[32] They constructed the ion-beam sputtering system and showed how it could be used for deposition of thin films of metals and insulators. The ion source is composed of the von Ardene type of duoplasmatron in which an arc discharge is maintained to create the ions for the sputtering. A well defined ion beam with a current of up to 500 μA over an area of 1 cm^2 can be extracted into the vacuum chamber at a pressure of 10^{-5} torr with an accelerating voltage of 50 V to 2 kV.

Kaufman-type, broad-beam ion sources have been developed for use in thin-film applications. This type of source typically has a plasma chamber within the ion source, with a hot filament cathode. Ions created

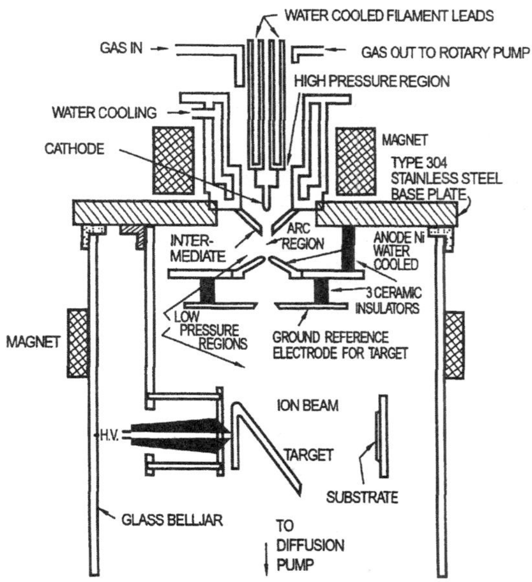

Figure 4.31. Construction of the duoplasmatron argon ion source for sputtering deposition.[32]

in the source can be accelerated by means of multi-aperture grids to form a large-area, intense beam. The ion-beam current ranges from 10 mA to several amperes, depending on the dimensions of the source. The ion energy can vary from 0.5 to 2.5 kV. The gas pressure of the sputtering chamber is typically in the high 10^{-5} to low 10^{-4} torr range. This kind of system is widely used for sputter etching of semiconducting devices. Increasing interest has been recently paid to ion-beam sputtering not only for semiconducting processes but also for developing exotic materials. Figure 4.32 shows a conventional ion-beam sputtering system with a Kaufman ion source. The typical operating properties of the commercial ion-beam source are shown in Table 4.4. The hollow-cathode, ion-beam source and/or rf ion-beam source are used for the oxygen ion source.

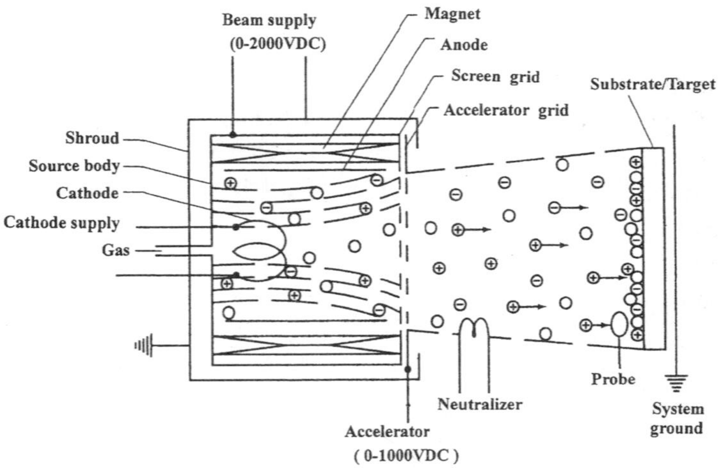

Figure 4.32. Construction of conventional ion beam sputtering system. *(Ion Tech, Inc.)*

Table 4.4. Ion Source *(Ion Tech, Inc.)*

	Beam Diameter				
Beam Size (at grids)	1.2" (3 cm)	2" (5 cm)	4.3" (11 cm)	6" (15 cm)	8.3" (21 cm)
Maximum Beam Current (mA)	100	200	350	500	900
Beam Energy (ev)	50–1200	50–2000	50–2000	50–2000	50–2000
Gas Flow, Argon (sccm)	1–4	1–5	2–8	2–8	3–10

4.2.5 ECR Plasma

Electron cyclotron resonance (ECR) microwave-based plasma has been developed primarily for applications with reactive etching or deposition.[33] However, a variation of the source can be configured for sputter deposition or reactive sputter deposition. The ECR discharge is sustained in an rf electric field with static magnetic field. The ECR conditions are given by

Eq. (4.36) $$f = 1/2\pi \frac{eB}{m}$$

where f denotes the frequency of the rf electric field, B, magnetic field strength, and e and m, electron charge and mass, respectively. For the conventional ECR system, $f = 2.45$ GHz and $B = 874$ gauss. A typical ECR sputtering system is shown in Fig. 4.33. The system comprises a negatively biased ring target settled at the outlet of the discharge chamber. The target bias is 0.4 to 1.0 keV. Since the system uses a chemically stable cold cathode, reactive gases could also be used for the sputtering. The operating pressure is as low as 10^{-5} torr so the sputtering deposits atoms onto samples in a line-of-sight mode. Table 4.5 summarizes the operative properties of these sputtering systems.

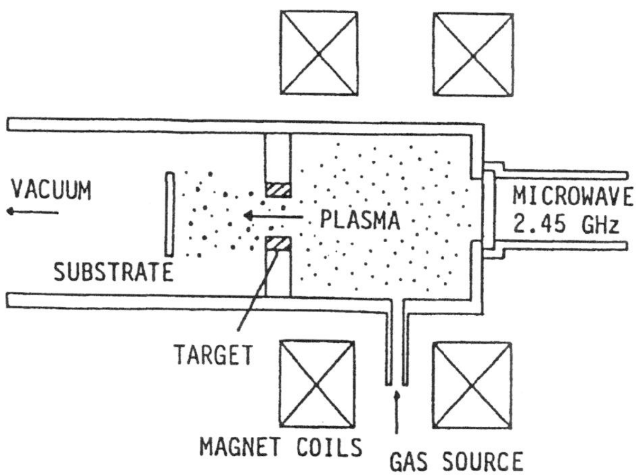

Figure 4.33. Electron cyclotron resonance (ECR) plasma sputtering system with ring target.

Table 4.5. Operating Properties of Sputtering Systems

Operating Properties		DC, RF-Diode Sputter	Planar Magnetron	Ion Beam Sputter	ECR Sputter
Operating pressure (Pa)		0.1~100	0.01~10	0.001~0.1	0.001~0.1
Ionization degree		10^{-5}~10^{-4}	10^{-5}~10^{-2}	--	10^{-3}~10^{-1}
Electron temperature (K)		10^{4}~10^{5}	10^{4}~10^{5}	--	5×10^{4}~10^{6}
Ion temperature (K)		~10^{3}	~10^{3}	--	10^{3}~10^{4}
Gas temperature (K)		~10^{3}	~10^{3}	--	~10^{3}
Plasma density (cm^{-3})		10^{8}~10^{10}	10^{9}~10^{12}	--	10^{9}~10^{12}
Particle energy	Adatoms (eV)	1~10	1~10	1~10	1~10
	Secondary electron bombardment	large	small	--	small
Ionization degree of adatoms		10^{-3}~10^{-2}	10^{-3}~10^{-1}	10^{-3}~10^{-2}	10^{-2}~10^{-1}

4.2.6 Medium-Frequency Sputtering

Sputtering of dielectrics is done using rf magnetron sputtering. In general, the deposition rate from the dielectric target is slower than the reactive sputtering from the metal target. For high-rate deposition of dielectrics, dc reactive sputtering from the metal target under a metallic mode, sputtering from metal cathode, and oxidation at the substrate are promising. However, arcing hampers reactive sputtering, which appears mainly as a breakthrough of dielectric layers grown on the target surface and charged up by energetic positive ions. The insulating film also builds up on the surface of the chamber and/or anode. When the insulating layer on the anode becomes thick, the sputtering discharge becomes unstable. This phenomena is called "disappearing anode." The anode problem is solved by the introduction of a self-cleaning anode.[34] One way to achieve a self-cleaning anode is to use dual/twin magnetron cathodes.

Figure 4.34 shows the twin magnetron cathode, which is available for the fast and stable deposition of metal oxide and/or metal nitride layers in production. The magnetron cathode comprises a pair of metal targets. The cathode is connected to an ac medium frequency (MF) power supply or dc pulsed power supply in a frequency range of 10 to 100 kHz, typically 40 kHz.[35] Bipolar pulse units periodically reverse the voltage of a single magnetron in order to allow the neutralization of surface charges. The reactive sputtering produces the oxide and/or nitride layers. At any time, one of the magnetron cathodes has a negative potential and acts as a sputter cathode, while the other acts as an anode. The alternating polarity keeps the anode clean, even during lengthy coating runs, which prevents the generation of the arc discharge.

The dual magnetron is a successful solution to the problem of the "disappearing anode." It requires two targets. Figure 4.35 shows a single magnetron cathode with self-cleaning anodes.[36] Two anodes are provided, a dual-anode system, each connected to one side of a center-tapped ac power source. Each of the anode elements acts alternatively as true anodes (electron collectors) and sputtered cathodes (ion collectors), depending on the polarity of the ac power supply. The continuous reversal of voltage and current keep any insulating regions of an anode from charging. The dual-anode system can provide sustainable anodes and periodically discharge both anode and cathode target surfaces, avoiding arcing.

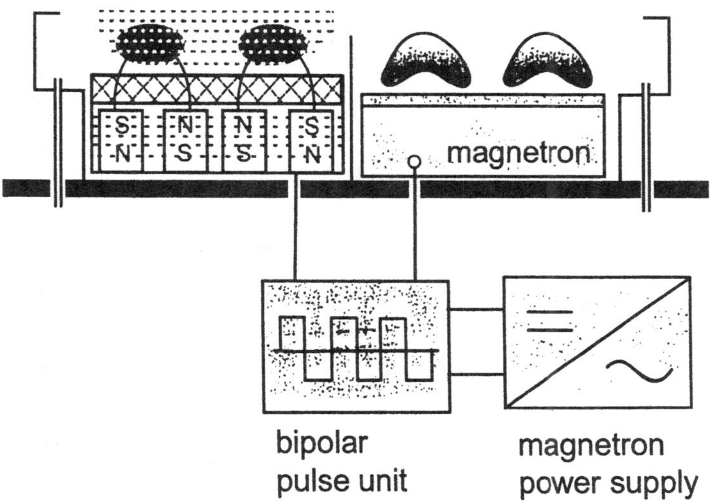

Figure 4.34. Schematic construction of dual magnetron sputtering system.

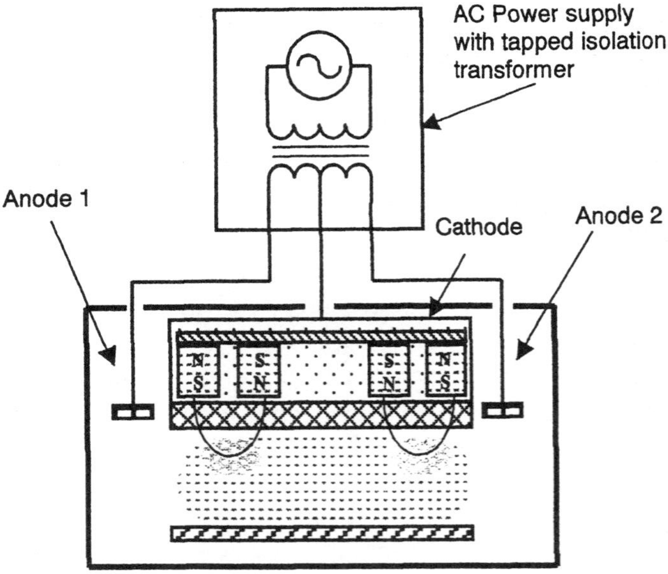

Figure 4.35. Schematic construction of dual anode sputtering system.[36]

In the pulsed magnetron, the medium-frequency range is suitable for construction of a large-area coating system and realizes the long-term stable operation of high-rate sputter processes in the reactive deposition of dielectric materials like SiO_2 or Al_2O_3.

The MF pulsed sputter technology took place after the 1990s. The technology is also used for substrate pre- and post-treatment, pulsed plasma activation, and pulsed bias sputtering.[37]

4.3 PRACTICAL ASPECTS OF SPUTTERING SYSTEMS

For the operation of the sputtering systems, several kinds of equipment are prepared. These include sputtering targets, sputtering gas, substrate holders, and monitoring systems.

4.3.1 Targets for Sputtering

The target is generally made of a metal or alloy disk. Hot-pressed sintered disks are used for depositing compound thin films. The diameter is 2.5 to 10 cm when used for research, and 15 to 30 cm for production. The large size of the rectangular sputter cathode, 300 mm x 550 mm, is also provided for production of a high-rate glass and web coating. Typical sputtering targets are listed in Table. 4.6. Figure 4.36 shows a photograph of the targets.

Sputtering is an inefficient process, and most of the power input to the system appears ultimately as target heating; thus these targets are usually mounted on a water-cooled backing plate. The target is fixed on the backing plate by mounting clips or other mechanical support. A dark space shield, known as a ground shield, surrounds the target so that only target material is sputtered. The construction of the target with the shield is shown in Fig. 4.37. The spacing between the target and the ground shield must be less than the thickness of the dark space, λ_0/p. If we design the maximum operating pressure to be 20 Pa (0.15 torr), taking $\lambda_0 = 0.05$ cm, and $p = 0.15$ torr, λ_0/p becomes 3 mm. The condition of the target strongly affects the properties of the resultant thin films. Note that the resistivity of the cooling water should be high enough so as to keep an electrical insulation between the target and the grounded chamber.

When compound materials such as oxides and/or nitrides are deposited by sputtering, a sintered, ceramic sputter target is used. The sintered target provides stable sputtering deposition, although the deposition rates are smaller than reactively sputtered metal targets. Manufacturers supply sintered targets; however, when just starting to study thin-film deposition, sputter targets for trial experiments can be made by methods described below.

4.3.1.1 Compound Targets

Thin films of alloys for lab scale can be provided from a composite target as shown in Fig. 4.38. The composition of the thin films is controlled by the area ratio of each element. Taking the sputtering yield and area ratio of each element, (s_1, a_1), (s_2, a_2), (s_3, a_3), ..., (s_n, a_n), partial composition of each element becomes $s_1a_1(A_1/N)$, $s_2a_2(A_2/N)$, $s_3A_3(A_3/N)$, ..., $s_nA_n(A_n/N)$, where $A_1, A_2, A_3, \ldots, A_n$ denote the atomic weights of each element, and N is Avogadro's number. When the number of partitions is

Table 4.6. Sputtering Targets and Purities

Target	Purity (%)				Target	Purity (%)			
Ag			99.99		Ni		99.9		
Al			99.99	99.999	Pd			99.99	99.999
Au			99.99		Pt			99.99	
Bi		99.9	99.99	99.999	Sb				99.999
C				99.999	Si				99.999
Co		99.9			Sn				99.999
Cr		99.9	99.99		Ta			99.99	
Cu			99.99		Ti		99.9		
Fe		99.9			V		99.9		
Ge				99.999	W			99.99	
Hf		99.9			Y		99.9		
In			99.99		Zn				99.999
Mo			99.99		Zr		99.9		
Nb			99.99						
AlN	99				Mo-Si_2	99			
Al_2O_3		99.9	99.99		NbN	99			
B_4C	99				Nb_2O_5		99.9	99.99	
BN	99				PbO				99.999
$BaTiO_3$	99				PbS		99.9		
BiSrCaCuO					$PbTiO_3$		99.9		
Bi_2O_3		99.9			SiC	99			
Bi_2Te_3				99.999	Si_3N_4	99	99.9		
CdS				99.999	SiO_2			99.99	
CdTe				99.999	SnO_2		99.9		
$CrSi_2$	99				TaS_2		99.9		
Cu_2S			99.99		TiC	99			
Fe_2O_3		99.9	99.99	99.999	TiN	99			
Fe_3O_4		99.9			WO_3			99.99	
HfO_2		99.9			Y_2O_3			99.99	
In_2O_3			99.99		YBaCuO		99.9		
ITO		$(In_2O_3)_{0.8}$	$(SnO_2)_{0.2}$		ZnO				99.999
LaSrCuO					ZnS			99.99	99.999
$LiNbO_3$		99.9	99.99		ZnSe				99.999
$LiTaO_3$		99.9	99.99						
MgO		99.9	99.99						

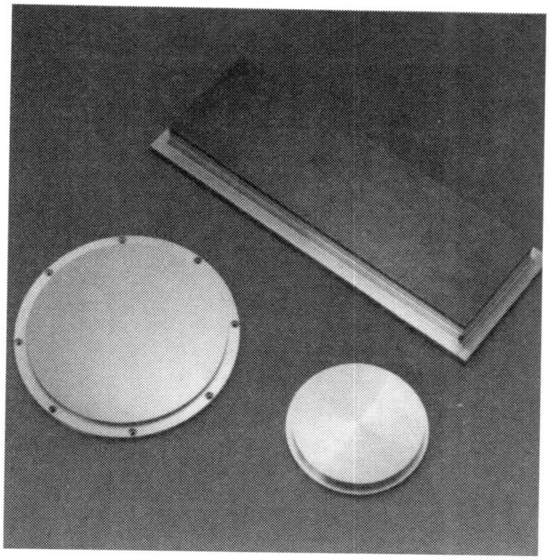

Figure 4.36. Photographs of sputtering targets. *(Courtesy of Mitsubishi Materials Corporation.)*

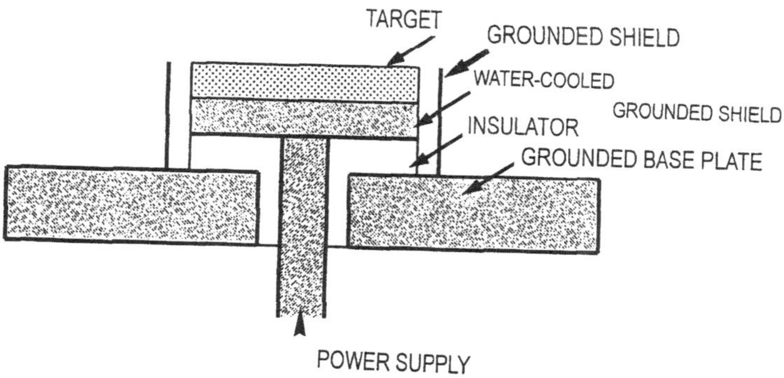

Figure 4.37. Construction of target with ground shield.

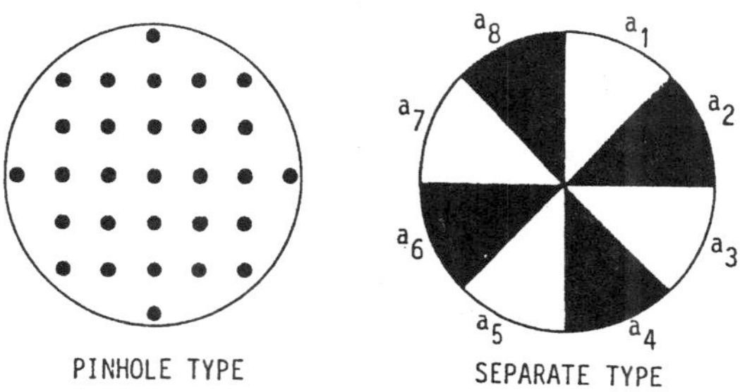

Figure 4.38. Construction of composite target for the deposition of alloy, compound thin films.

small, the composition of the sputtered films will distribute nonuniformly over the substrate. A rotating substrate holder is often used to obtain uniform composition over the entire substrate. Hanak suggested that a binary compound target achieved deposition of thin films with various composition ratios.[38] Figure 4.39 shows the binary compound target. It suggests thin films of different composition are obtained at different substrate positions.

4.3.1.2 Powder Targets

Thin films of compounds, including metal oxides, nitrides, and carbides, can be deposited by direct sputtering from the sintered powder of these compounds. A stainless dish, which is mounted on the backing plate, is filled with the sintered powder. The construction of the powder

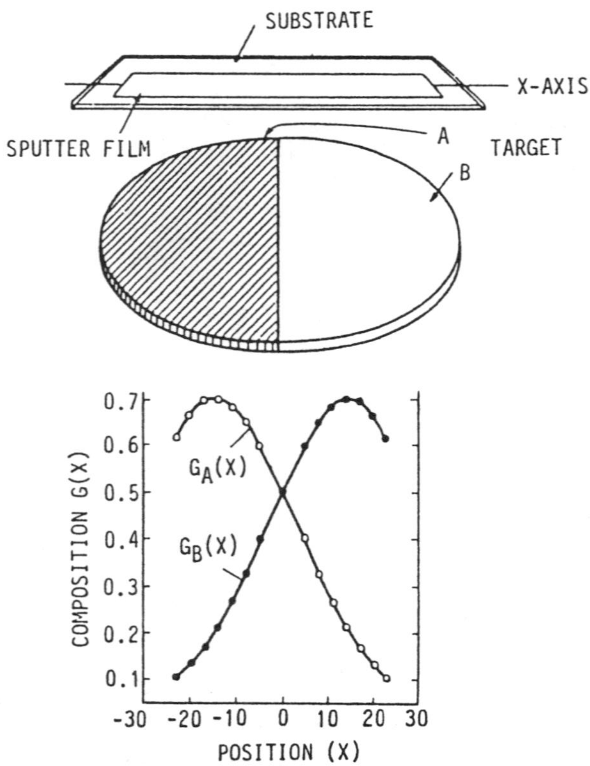

Figure 4.39. A composite target for depositing a binary alloy.[38]

target with the stainless dish is shown in Fig. 4.40. A wide variety of compound thin films can be sputtered from a powder target. The powder target is also useful for doping a foreign element. When using a powder target, pre-sputtering for several hours sinters the surface of the target slightly; then the target shows a stable operation. The excess power input to the powder target makes cracks. The drawback to powder targets is lower sputtering rates than those of sintered targets. Their advantage is that they make it easy to get a variety of compositions, since a mixture of their compositional elements and/or compositional oxides or nitrides achieves a composition of complex compounds without sintering the target materials.

4.3.1.3 Auxiliary Cathode

Small amounts of foreign metals can be added to thin films by co-sputtering an auxiliary cathode made of foreign metals. Varying the sputtering current to the auxiliary cathode changes the amount of foreign metal in the resultant films.[39]

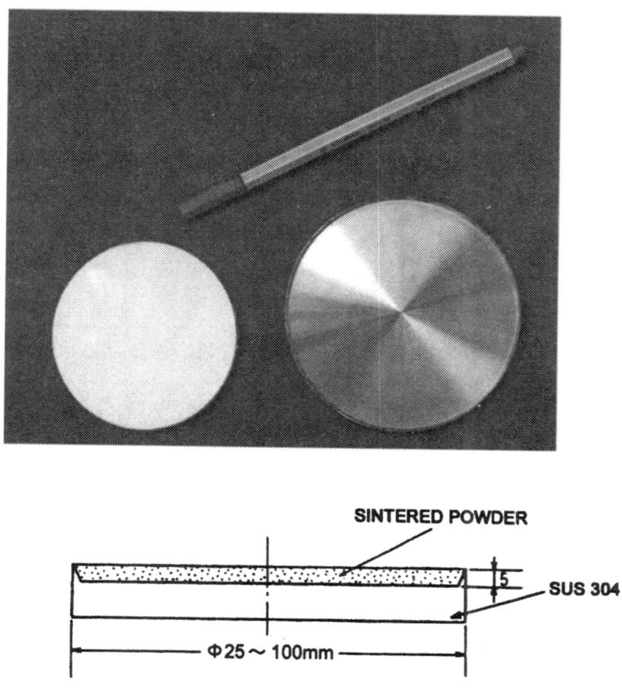

Figure 4.40. Powder target and a construction of the stainless dish for the powder target.

4.3.2 Sputtering Gas

For the deposition of metals, pure Ar gas (purity 4N to 6N) is introduced through a variable leak valve. An automatic gas-flow controller is also usually used for the sputtering system. A typical gas-flow system is shown in Fig. 4.41. The vacuum system should be water- and/or oil-vapor free. A liquid-nitrogen trap should be used for the oil diffusion pump system.

When using a reactive gas such as oxygen, thin films of metal oxides can be deposited from a metal target, and is known as *reactive sputtering*. In reactive sputtering, the reaction takes place both at the cathode surface and the substrate during deposition. An impingement of the reactive gas of the cathode surface forms the compounds such as metal oxides, and the resultant compounds will be sputtered. This leads to the deposition of compounds. The optimum concentration of the reactive gas is determined by the reactions at cathode and substrate:

Eq. (4.37)

$$kN_g / N_c > 1 \text{ (cathode)}$$

$$kN_g / N_s > 1 \text{ (substrate)}$$

where N_g denotes the number of reactive gas molecules which strike a unit area of the cathode surface or the substrate per unit time, N_c is the number of sputtered atoms from the unit area of cathode per unit time, N_s is the number of deposited atoms per unit area of substrate per unit time, and k is the absorption coefficient of reactive gas at the surface of cathode and sputtered films.

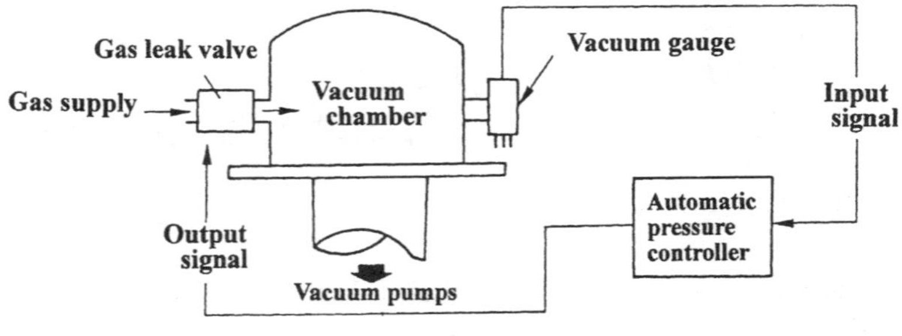

Figure 4.41. Typical gas flow system for the sputtering system.

In the case of sputtering a Ti cathode in the reactive gas O_2, N_g is 3.5×10^{20} P_{O2} molecules/sec/cm^2 at the partial oxygen pressure of P_{O2} torr and $k \sim 1$. Putting these relationships into Eq. (4.37), a minimum oxygen pressure for the reactive sputtering is given by the following relationships:

Eq. (4.38)
$$P_c \cong N_c /(3.5 \times 10^{20})$$
$$P_s \cong N_s /(3.5 \times 10^{20})$$

where P_c is the minimum oxygen pressure for the cathode-surface reaction, and P_s the minimum oxygen pressure for the substrate-surface reaction. Taking $N_c = 3 \times 10^{16}$ atoms/sec/cm^2, and $N_s = 1 \times 10^{16}$ atoms/sec/cm^2 at a deposition speed of 120 Å/min, we have $P_c = 8.5 \times 10^{-5}$ torr and $P_s = 3 \times 10^{-5}$ torr.[40]

Since a TiO_2 film is produced when the cathode surface layer is fully oxidized to TiO_2 by oxygen absorption, stoichiometric TiO_2 films can be produced when $P_{O2} > 3 \times 10^{-5}$ torr. On the other hand, since metallic Ti films may be produced when neither cathode surface nor the sputtered film are fully oxidized, the metallic Ti films can be produced when $P_{O2} << 3 \times 10^{-5}$ torr.

In reactive sputtering, the reaction at the cathode is called an *oxide mode* and the reaction at the substrate is called a *metallic mode*. The deposition rate is significantly affected by the surface condition of the target and/or the mode of the reactive sputtering. Deposition rates will show the hysteresis curve with the variation of the flow rates of the reactive gas. Typical hysteresis curves were clearly observed by Hata for YSZ thin films in the magnetron sputtering from a crucible Zr/Ti metal target as shown in Fig. 4.42.[41] A typical hysteresis curve is shown in Fig. 4.43. The metallic thin films are deposited below the critical oxygen pressure (*a*). When the partial oxygen pressure increases, the YSZ thin films are deposited through the metallic mode (*b*). In the metallic mode, the deposition rates of the YSZ are higher than those of Y/Zr metal thin films. Above the critical oxygen pressure, the deposition rates abruptly decrease (*c*) and the sputtering mode moves to the oxide mode with low deposition rate (*d*). When the partial oxygen pressure decreases, the sputtering still keeps the oxide mode (*e*). The sputtering deposition will return to the original metallic mode at the oxygen pressure below the critical oxygen pressure, when the oxide layer of the target surface is fully removed (*f*).

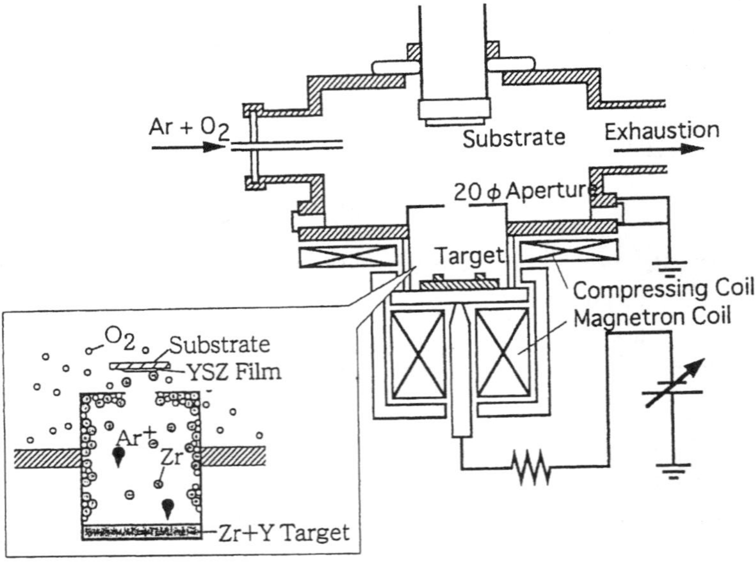

Figure 4.42. The schematic drawing of the reactive sputtering system with an attached glow discharge mass spectrometer.[41]

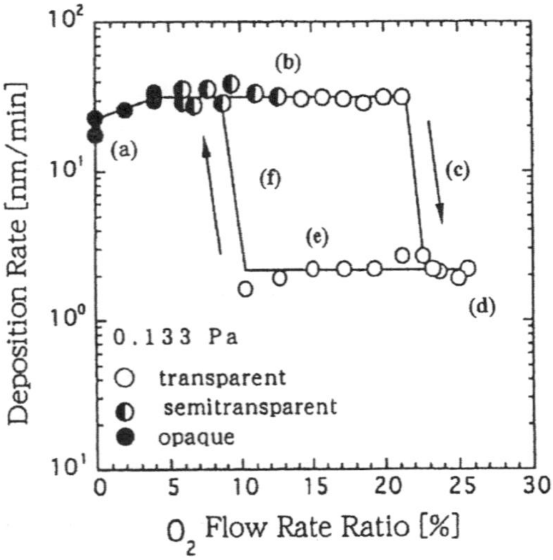

Figure 4.43. Deposition rate vs O_2 flow rate curve showing typical hysteresis.[41]

Generally, we use the mixed gas of Ar and reactive gas for the reactive sputtering. The concentration of the reactive gas is generally 5% to 50%. Table 4.7 shows the typical discharge gas for sputtering deposition. These sputtering gases are uniformly fed to the system through a stainless steel pipe with fitting components.

Table 4.7. Sputtering Gases and Typical Impurity Levels

Gas	Class	Purity (%)	P (kg/cm^2)	Impurities, (ppm)							Dew point (°C)
				N_2	Ar	O_2	CH_4	CO	CO_2	NO_x	
N_2	pure S	99.9998	150			<.5	<.1	<.5	<.1	<.01	<-70
	pure A	>99.9995	150			<.5	<.5				<-70
	pure B	>99.9995	150			<1					<-70
	stand.	99.999	150			<5					<-65
Ar	pure A	<99.9995	150	<3		<.2	<1		<1		<-70
	pure B	<99.999	150			<.2	<1		<1		<-70
	stand.	99.998	150	<10		<2					<-65
O_2	pure A	>99.99	50	<20	<50		<30	<10	<10	<10	<-70
	pure B	>99.9	150	<1k	<1k		<30				<-60
	stand.	99.6	150	<400	<400						<-65
H_2	pure	>99.99999	150				<.05				<-70
	1st cls	>99.99	150								
NH_3	pure	99.995		<10		<5	<5			H_2O <10 ppm	
	1st cls	>99.99	2~10								
CH_4	pure	<99.95	50								
	1st cls	>99	100								
H_2S		>99	7~30								

Nippon Sanso Catalog G 1(80-6) 3000 T, G 2(81-4) 4000M

The presence of nonuniformity in reactive gas density in the system results in nonuniformity in the chemical composition of resultant films. Stable gas flow with controlled partial pressure of the reactive gas is required for reactive sputtering. A precise variable-leak valve controls the gas flow. Typical commercial valves are shown in Fig. 4.44. An electrically controlled mass-flow controller, as shown in Fig. 4.45, also controls the gas flow. In the sputtering system, the mass-flow rate is smaller than the CVD system. The flow rates are typically 1–100 sccm. In these flow controllers the flow rate shows the nitrogen equivalent value. In order to get the absolute flow rate of each sputtering gas, the measured values should be revised. Figure 4.46 shows photographs of the gas flow controller and the vacuum components for the sputtering system.

Outline Drawing inches (mm)

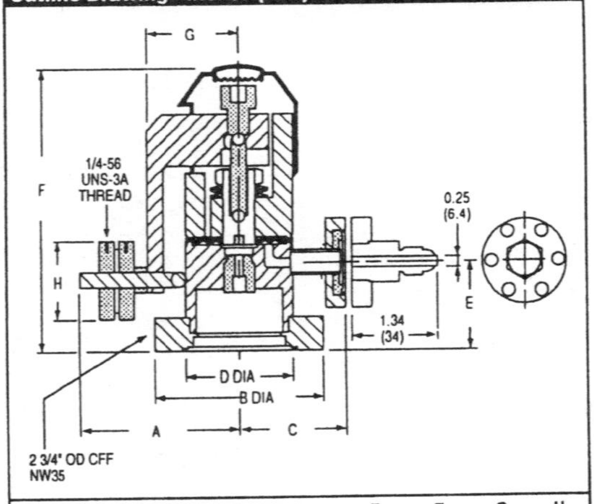

	A	B	C	D	E	F	G	H
Inches	2 21/32	2 3/4	1 3/4	1 3/4	1 5/16	4 1/2	1 17/32	1 1/4
mm	67	70	44	44	33	114	39	32

Technical Specifications

Controlled Leak Rate

10^{-10} torr-l/sec (minimum)

Vacuum Range

Atmosphere to below 10^{-11} torr (mbar)

Leak Rate

No leak detectable on a helium mass spectrometer leak detector with sensitivity of 1×10^{-10} std cc/sec

Bakeable To

450° C

Figure 4.44. Typical commercially available variable leak valve includes a movable piston. *(Varian catalog.)*

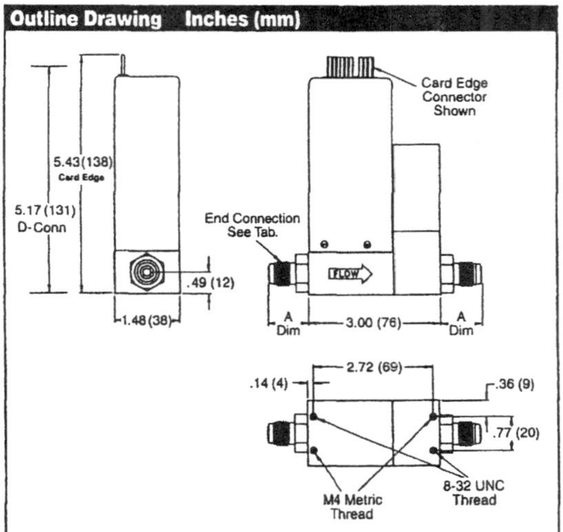

Fitting Options		
Type (all ¹/₄ in.)	"A" DIM	Overall Length
4 VCR**	.94 (24)	4.88 (124)
¼ in. Swagelok**	.72 (18)	4.44 (113)
4 VCO**	.78 (20)	4.56 (116)

Figure 4.45. Electrically controlled mass flow controllers. *(Varian catalog.)*

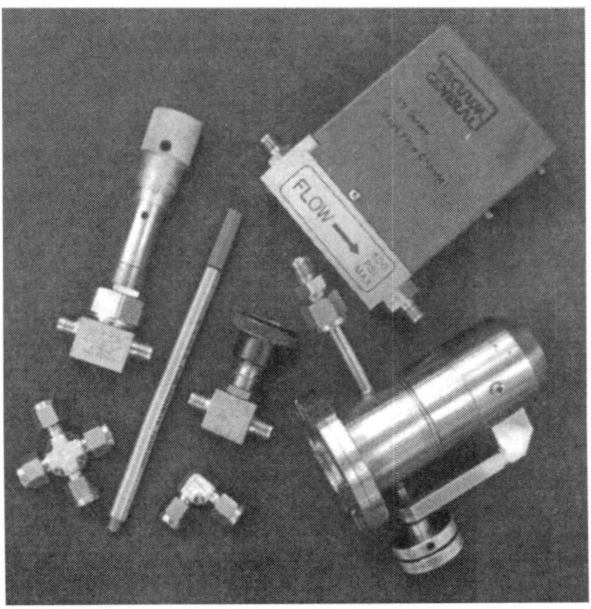

Figure 4.46. Photograph of vacuum components for sputtering system.

4.3.3 Thickness Distribution

The thickness distribution of sputtered films is governed by several factors including the angular distribution of sputtered particles, collisions between sputtered particles and gas molecules, and the construction of the target shield. However, the thickness distribution is estimated by the simple assumption that the angular distribution obeys the cosine law similar to vacuum evaporation, and the collisions are neglected for the sputtered particles.

In the case of the planar sputtering system shown in Fig. 4.47, the thickness distribution is estimated for a disk cathode target (first equation) and a ring cathode target (second equation):

$$\frac{d}{d_0}=\frac{1+(S/h)^2}{2(S/h)^2}\left\{1-\frac{1+(l/h)^2-(S/h)^2}{\sqrt{\left[1-(l/h)^2+(S/h)^2\right]^2+4(l/h)^2}}\right\}$$

Eq. (4.39)

$$\frac{d}{d_0}=\left[1+(S/h)^2\right]^2\frac{1+(l/h)^2+(S/h)^2}{\left\{\left[1-(l/h)^2+(S/h)^2\right]^2+4(l/h)^2\right\}^{3/2}}$$

where d_0 is the center thickness, d is the thickness at a center distance l, s is the radius of the disk and ring cathode, and h is the evaporation distance.[42] The thickness distribution for conventional diode sputtering is estimated by the disk target system. For the planar magnetron, the ring target system is used for estimating the thickness distribution. Figures 4.48 and 4.49 show typical results for a disk target and a ring target, respectively.

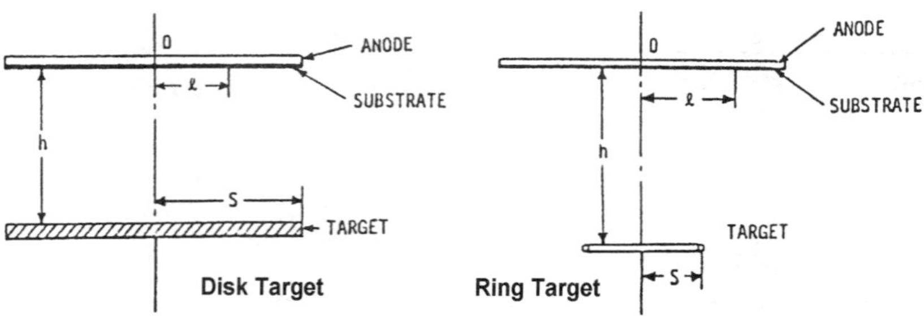

Figure 4.47. Construction of diode sputtering systems using a planar electrode with a disk target and a ring target.

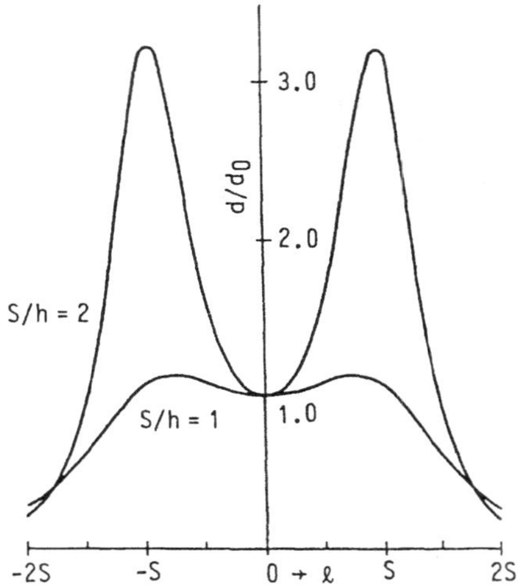

Figure 4.48. Thickness distribution for the disk target.

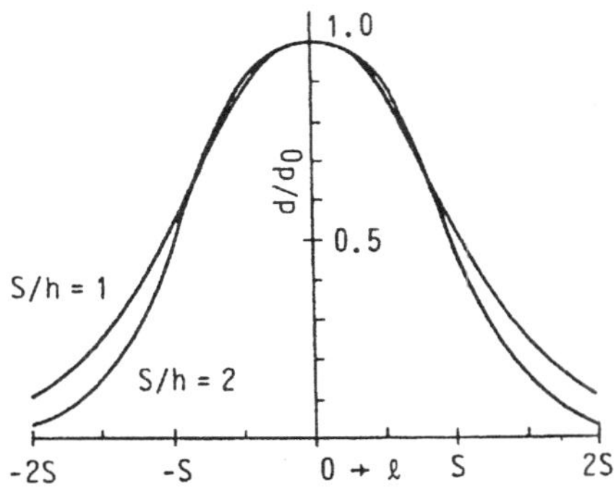

Figure 4.49. Thickness distribution for the ring target.

The thickness distribution is governed by the construction of the magnetic core. Figure 4.50 shows a modified planar magnetron with double erosion rings.[43] The magnetron cathode comprises a rare-earth magnetic core. The magnetic flux density of the magnetic core is 0.4 T on the surface. A permalloy sheet covers the magnetron cathode. The permalloy sheet confines the leakage magnetic field. The permalloy sheet produces an additional magnetic circuit that surrounds the magnetic core accompanied by the additional transverse magnetic field at the outer side of the disk target. Figure 4.51 shows the magnetic flux distribution calculated by the finite element method (FEM) using ANSYS code. The spatial distributions

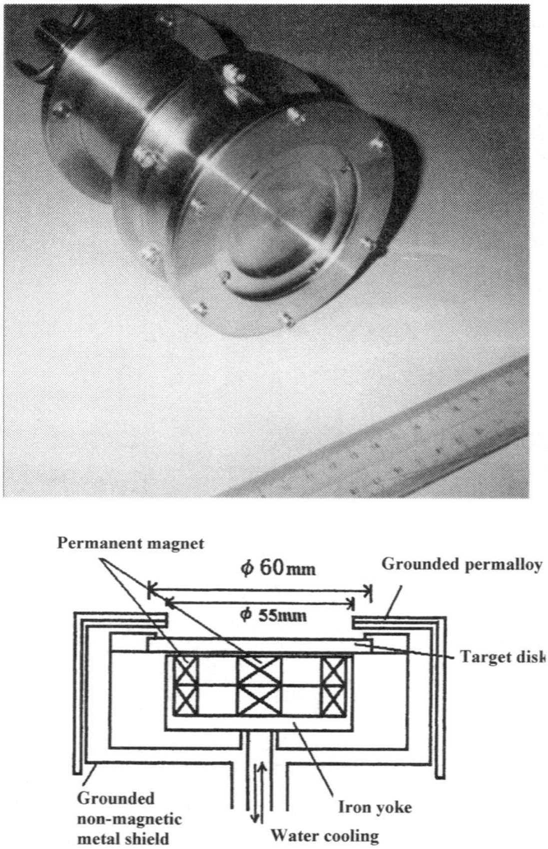

Figure 4.50. Schematic view of the magnetron sputtering cathode with a photograph.

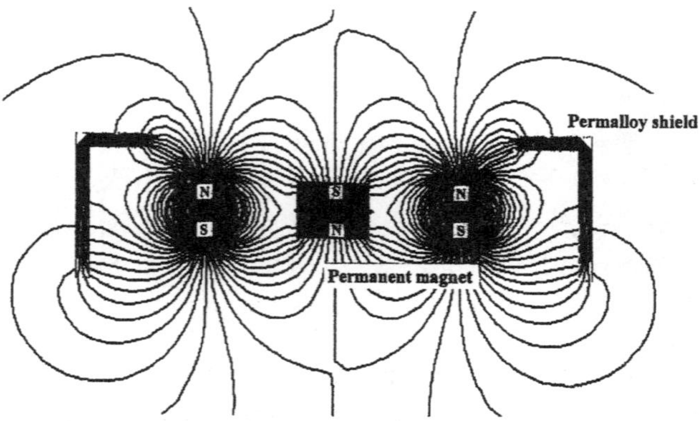

Figure 4.51. Typical distribution of magnetic flux calculated by the FEM.

of the radial magnetic field with and without an additional permalloy magnetic circuit are shown in Fig. 4.52. The additional permalloy magnetic circuit enhances the radial magnetic field at the outer region of the target and produces a second erosion ring. Thus, the modified planar magnetron cathode shows the double erosion rings as shown in Fig. 4.53. The following relation estimates the film thickness distribution, where r_1 and r_2 are the radii of the inner ring and outer ring, respectively:

Eq. (4.40)
$$\frac{d}{d_0} = \frac{(r_1 k_1 + \alpha r_2 k_2)}{(r_1 k_{10} + \alpha r_2 k_{20})}$$

where k_1, k_2, k_{10}, and k_{20} indicate

$$k_{1,2} = \frac{1 + (\ell/h)^2 + (r_{1,2}/h)^2}{\left\{\left[1 - (\ell/h)^2 + (r_{1,2}/h)^2\right]^2 + 4(\ell/h)^2\right\}^{3/2}}$$

$$k_{10,20} = \frac{1}{\left[1 + (r_{1,2}/h)^2\right]^2}$$

and α denotes the ratios of sputtering rate $\Gamma r_2 / \Gamma r_1$ at r_2 (outer ring) and r_1 (inner ring). The double rings will enhance the sputtering at the outer region of the target and improve the uniformity of the film-thickness distribution. The α will be a function of magnetic field B_1 and B_2 at the double rings. The α is empirically expressed by $\alpha = (B_2/B_1)^{1/2}$

Typical thickness distribution with and without the permalloy sheets is shown in Fig. 4.54.[44]

The thickness distribution described is calculated under the assumption that the sputtered atoms will arrive at the substrates without colliding with sputtering gas molecules. Actually, the sputtered atoms will collide with the sputtering gas. The thickness distribution will be governed by the degree of the collisions with the gas molecules. The trajectories of sputtered particles have been studied using a Monte-Carlo simulation. It showed that in the case of a planar magnetron sputtering with a target-substrate distance of 10 to 20 mm, the trajectory was a straight line at a low-gas-pressure region of 0.1 to 1 Pa. The sputtered particles must experience collisions with gas molecules at the high-gas-pressure region of 10 Pa.[45]

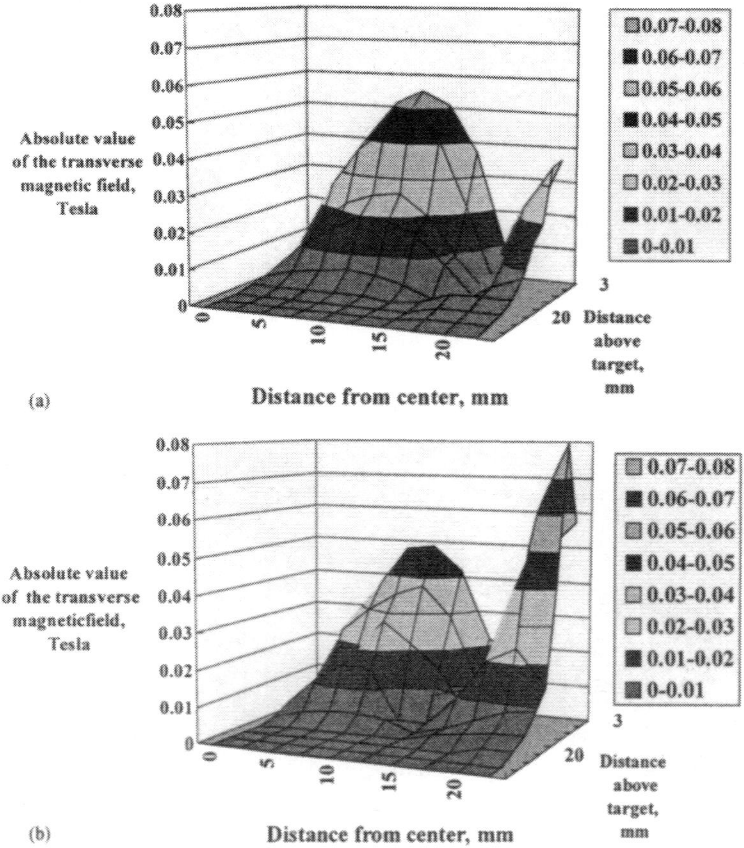

Figure 4.52. Radial distribution of transverse magnetic flux density for the planar magnetron cathode; (*a*) without permalloy sheet, (*b*) with permalloy sheet.

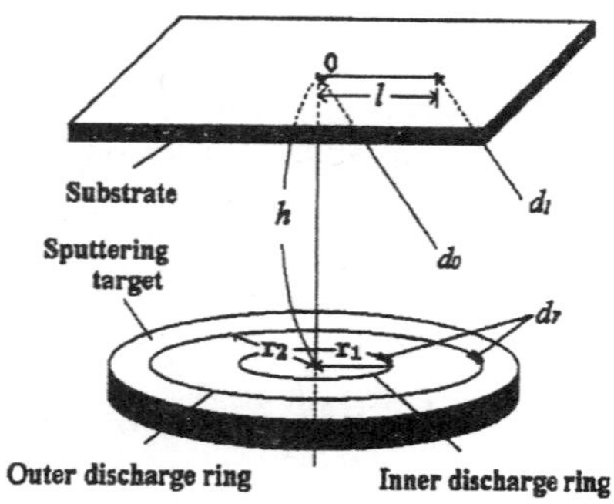

Figure 4.53. Double erosion rings at the modified planar magnetron cathode under the confined magnetic flux.

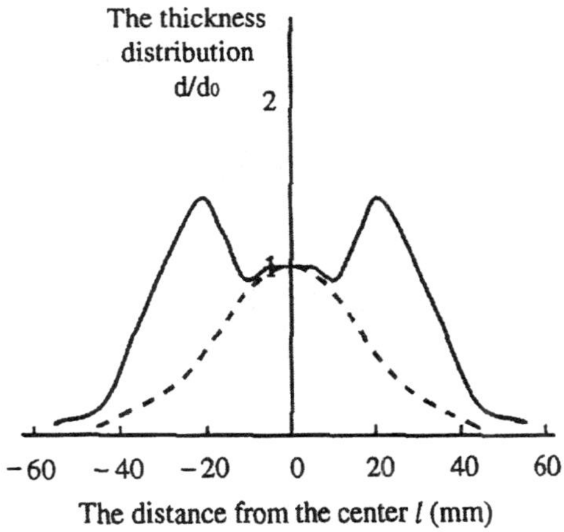

Figure 4.54. Typical radial distributions of the film thickness for Ar- Pt sputtering with the permalloy sheet (dotted line, without the permalloy sheet).

4.3.4 Substrate Temperature

The temperature of the substrate surface is important, yet difficult to control. In conventional sputtering systems, the substrate is mounted on a temperature-controlled substrate holder. However, the heat of the target heats the surface of the substrate. Moreover, bombardment by high-energy secondary electrons also heats the substrate. In the rf diode system, the temperature of the substrate rises up to 700°C without additional substrate heating, as shown in Fig. 4.55. In order to reduce the effect of the heat, the surface of the target must be cooled. Bombardment by the secondary electrons is avoided by negatively biasing the substrate.

It is noted that the temperature rise of the substrate depends on the type of sputtering system. The temperature rise at magnetron is lower than that of the rf diode, since the secondary electrons from the target are trapped by the transverse magnetic field near the surface of the target. The temperature rise at the magnetron system for lab scale is less than 300°C.

4.3.5 Off-Axis Sputtering; Facing-Target Sputtering

In the sputtering system, the highly energetic electrons irradiate the substrates and/or the growing surface of the thin films during deposition.

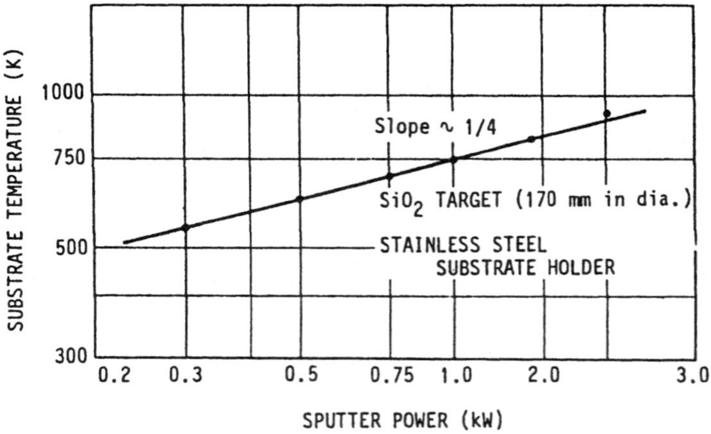

Figure 4.55. Temperature rise of substrates during deposition for rf sputtering system.

The irradiation of the high-energy particles during the deposition of $BaTiO_3$ and ZnO has been studied by Tominaga.[46] A schematic diagram of the measuring system for the high-energy neutral atoms is shown in Fig. 4.56. He suggests that the bombardment of high-energy neutral oxygen atoms on the growing surface of ZnO resputters and suppresses the preferred *c*-axis orientation of the ZnO thin films. The effects of the bombardment are obvious at the position facing the target. In the planar magnetron cathode the effect of the bombardment is obvious at the position facing the ring-erosion area, as shown in Fig. 4.57 [47]

The off-axis sputtering shown in Fig. 4.58 reduces the effects of the irradiation of the high-energy particles. The substrates are settled at the outside of the discharge plasma. The thickness distribution of thin films deposited by off-axis sputtering will be larger than that of on-axis sputtering. A rotating substrate holder with a metal shadow mask is used for the reduction of the thickness distribution of the off-axis sputtering. Under a suitable design, the thickness distribution is less than 2% for substrates of 100 × 100 mm in an rf sputtering system using a 5-inch target.[48]

A sputtering system with two facing targets, proposed by Hoshi and Naoe, and shown in Fig. 4.59, is a modification of off-axis sputtering.[49] The two-facing-targets system provides a high deposition rate for the ferromagnetic thin films of permalloy.[50]

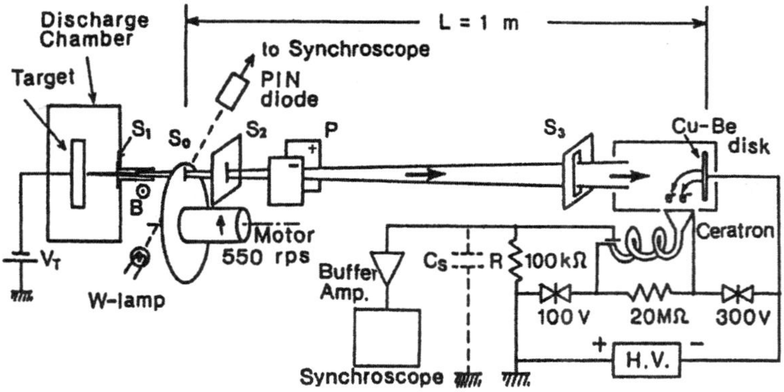

Figure 4.56. Schematic diagram of measuring system for time-of-flight of neutral particles.[46]

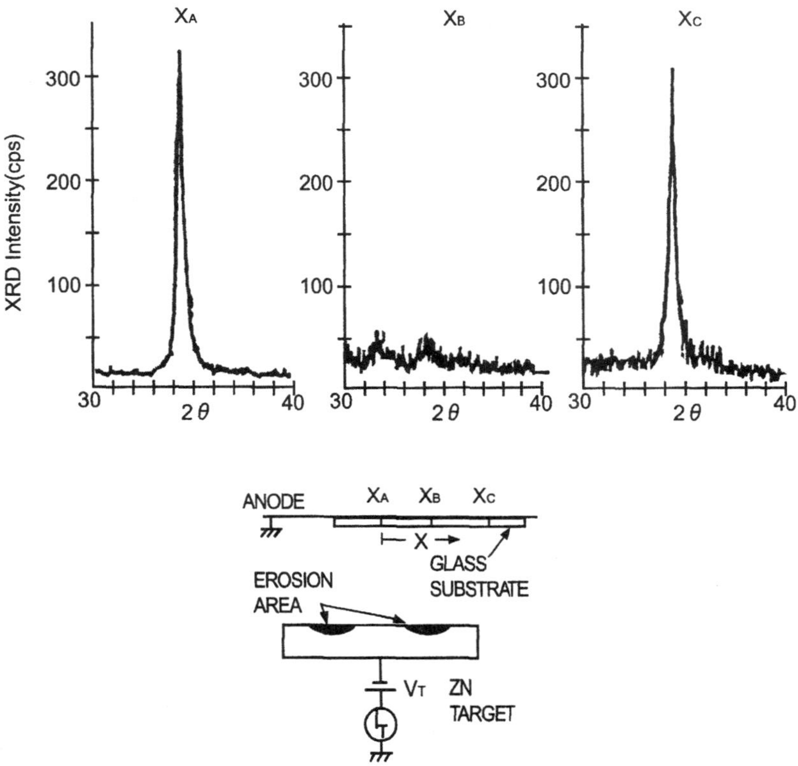

Figure 4.57. X-ray diffraction (XRD) intensity of ZnO thin films at different substrate position; X_A (center), X_B (middle), X_C (edge).[47]

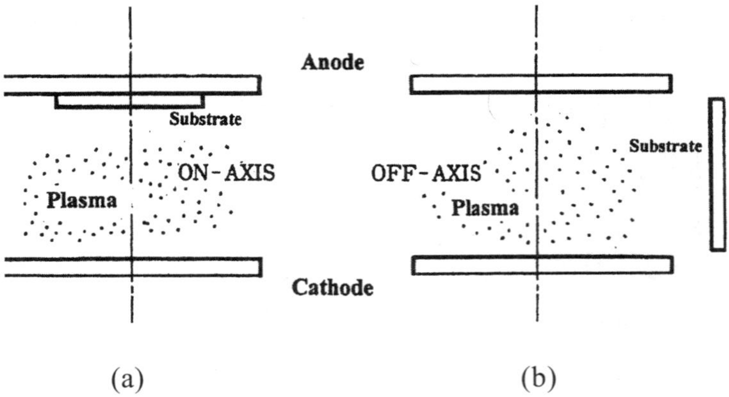

Figure 4.58. On-axis (*a*) and off-axis (*b*) arrangement of the sputtering system.

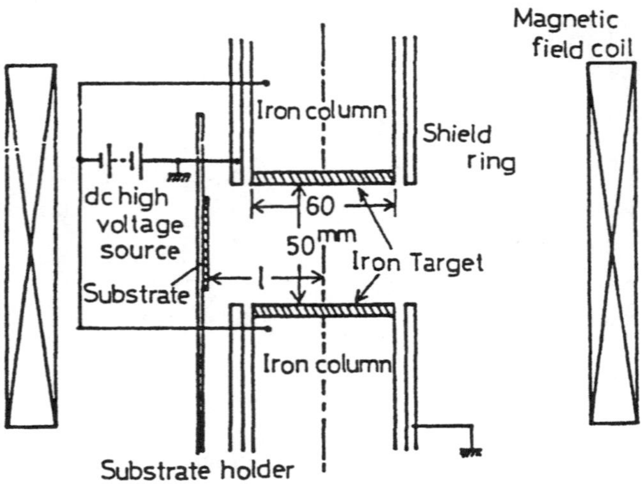

Figure 4.59. Construction of facing target sputtering system.[49]

4.3.6 Monitoring

It is important to monitor sputtering conditions during deposition in order to control the properties of the resultant films. The following parameters should be monitored:

1. The gas composition in the sputtering chamber
   —residual gas
   —total/partial pressure of the sputter gas
2. The sputtering discharge
   —discharge voltage, current, power
3. The substrate temperature
4. The thickness of the sputtered films
5. The film structure

4.3.6.1 Gas Composition

The total gas pressure of the sputtering chamber is measured by a conventional vacuum gauge at the operation range of 10^{-2} to 10 Pa. The quadrupole mass analyzer and the optical spectrometer are two widely used methods of monitoring the composition of the sputter gas. Figure 4.60 shows the construction of the quadrupole mass analyzer. It is composed of an ion source chamber, a focusing electrode, quadrupole electrodes, and an ion collector. The electrodes are 15 to 30 cm long and the pairs of opposing rods are connected to dc and rf voltage supplies. When the sputtering gas is introduced into the ion source chamber, the sputtering gas is ionized and the resulting ions are accelerated along the axis between rods.

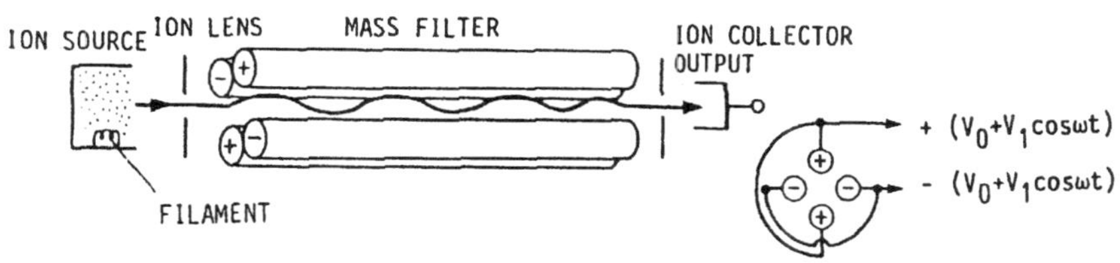

Figure 4.60. Construction of quadrupole mass analyzer.

Since the probability of the number of ions that pass through the quadrupoles and reach the ion collector is governed by the mass number of the sputtering gas and the value of the rf voltage, the mass number is determined by the sweeping of the rf voltage. The photograph of the quadrupole mass analyzer is shown in Fig. 4.61. Since the working pressure of the quadrupole tube is lower than the sputtering pressure, the turbo molecular pump is connected to the mass analyzer.

The other popular technique for monitoring sputtering gas is the optical spectrometric method. The emission intensity of a gas species is characteristic of the ionized species in the sputtering gas. The spatial distribution of the ionized species can be estimated by introduction of the optical fiber. Information about the ionized species informs us of the intensity of sputtered atoms. This enables one to monitor the sputtering rate. The optimum conditions for reactive sputtering can be maintained by the optical spectrometric method. For example, sputtering from an In-Sn target in an oxidizing atmosphere by monitoring the optical spectrum of In ions can easily produce the transparent conductive film, ITO. The optical spectrometric method is also used for monitoring the end point of plasma etching.

4.3.6.2 Sputtering Discharge

In an rf or dc sputtering discharge for lab scale, the power density of the target is on the order of 1 to 5 W/cm^2; at the sputtering voltage

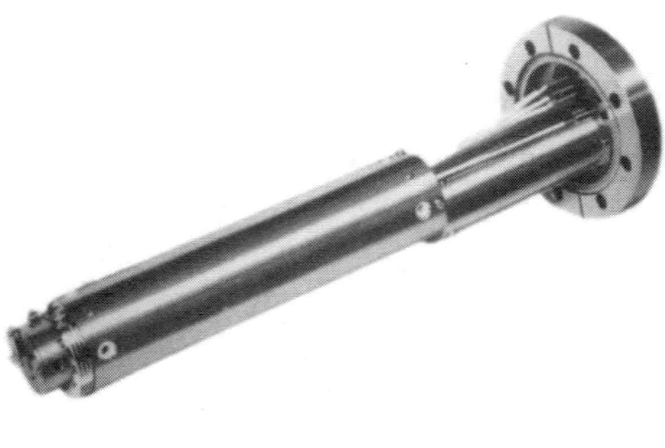

Figure 4.61. Analysis tube of quadrupole mass analyzer.

of 2,000 V, the sputtering current is 0.5 to 2.5 mA/cm^2. At the disk target of 20 cm in radii, the sputtering power is 400 to 2,000 W, the sputtering current 200 to 1,000 mA, and the impedance of the sputtering discharge 2 to 10 kΩ. For pilot and/or production scale, the sputtering discharge shows much lower impedance: taking target size: 1000 mm × 200 mm, power density: 5 W/cm^2, and the sputtering voltage: 500 V, the sputtering power and the impedance become 10 kW and 250Ω, respectively. A conventional high-impedance voltage probe and a low-impedance current meter for dc sputtering can easily measure these discharge parameters. An inductively coupled current probe is used for medium-frequency sputtering. However, the discharge parameters cannot be exactly measured for rf sputtering. In the conventional rf sputtering system, the rf power meter is inserted between the matching circuit and the rf power supply. The power loss at the target will be included in the measured values of the rf power meter.

4.3.6.3 Plasma Parameters

Langmuir probes are used for monitoring discharge or plasma parameters including spatial distribution of potential, electron density, and electron temperature. The construction of the Langmuir probe is shown in Fig. 4.62.[51] The Langmuir probe consists of Mo or W electrodes inserted into the plasma. The plasma parameters are estimated by the current-voltage curve of these electrodes. The typical current-voltage characteristics are shown in Fig. 4.63.

When the electrode is negatively biased against the plasma potential (a negative probe), a positive space charge accumulates around the probe and an ion sheath appears. The electric field in the ion sheath repels the electrons from the plasma. For the negative probe, the probe current I_p consists of the ion current I_+ from the plasma. The I_p will, if the Maxwellian rule holds, be given by

Eq. (4.41) $$I_p = I_+ AeN_+\sqrt{\frac{kT_+}{2\pi M}}$$

where A is the probe area, e is the electron charge, N_+ is the ion density, T_+ is the ion temperature, M is the mass of ion, and k is the Boltzmann constant. Thus the ion density in the plasma will be estimated from the I_p at negative probe voltages (region I).

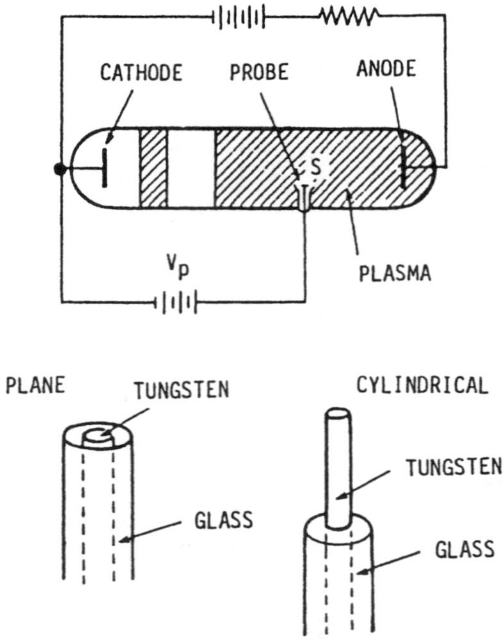

Figure 4.62. Construction of Langmuir probe.

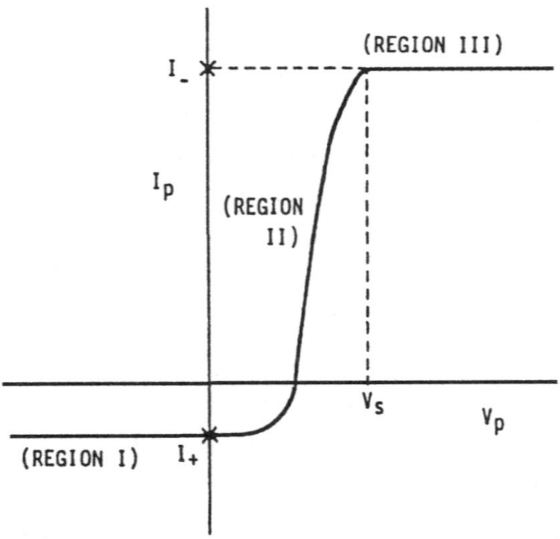

Figure 4.63. Current-voltage characteristics of the Langmuir probe.

When the probe potential is slightly negative to the plasma potential V_s, high-energy electrons will flow into the probe against the retarding bias in the ion sheath. The probe current will be given by

Eq. (4.42) $$I_p = I_+ + I_-$$

where I_- denotes the electron current (region II). The electron current, I_-, is expressed by

Eq. (4.43) $$I_- \cong AeN_-\sqrt{\frac{kT_-}{2\pi m}}\exp\left[-\frac{e(V_p - V_s)}{kT_-}\right]$$

where N_- is the electron density, T_- is the electron temperature, m is the electron mass, and V_p is the probe potential.

Equation (4.43) suggests that when $V_p = V_s$, the I_p becomes the electron current I. We have

Eq. (4.44) $$I_p \cong I_- = AeN_-\sqrt{\frac{kT_-}{2\pi m}}$$

when $V_p > V_s$ and the probe current shows a constant value of I_- (region III).

The electron density will be estimated from the I_p at region III. The space potential V_s will be determined by the transition point from region II to region III.

It is thought that the probe characteristics strongly depend on the construction and surface properties of the metal electrode. The presence of the magnetic field significantly affects the probe characteristics.[52] Also note that the plasma should not be disturbed by the introduction of the probes. The collisions in the probe sheath are negligible. In other words, the following conditions should be kept for probe measurements:

- The size of probe should be smaller than the mean free path of the ions and electrons.
- The thickness of the probe sheath should be smaller than their mean free path.

Under these conditions, the probe current I_p is expressed by the relationship

Eq. (4.45)
$$I_p = AeN\int_{0,u}^{\infty} uf(u)du$$

where N is the density of the charged particles, u is the velocity of the charged particles perpendicular to the probe surface, and $f(u)$ is their velocity distribution. The lower limit of u is to be taken at zero for the accelerating field ($V > 0$) and $u = [2e(V_s - V_p)/m]^{1/2}$ for the retarding field ($V < 0$). In Eqs. (4.40) to (4.43), the velocity distribution is considered to be the Maxwellian rule.

Equation (4.42) suggests that

Eq. (4.46)
$$\frac{d(\log I_-)}{dV_p} = \frac{-e}{kT_-}$$

The log I_- versus V_p plots show the linear properties, and their slope will give the electron temperature T_-.

When we use the Langmuir probe, several discrepancies to the probe theory appear. For instance, the probe currents in regions I and III will not show saturation. This will result from the variation of the effective probe area due to the change of the ion sheath by the probe voltage. The presence of the magnetic field will alter the probe characteristics. In a magnetic field in the order of 100 to 500 gauss, its effects on ion currents will be negligible. The probe characteristics will affect the electron currents since the electrons will be fixed to the magnetic line of force. In order to avoid the effects of the magnetic field, the probe surface should be perpendicular to the magnetic line of force. When the probe surface is parallel to the magnetic line of force, the electron current to the probe is strongly reduced.

Also note that when the magnetic field is superposed onto the discharge, the charged particles show the E×B drift motion. The probe current will increase due to the drift motion. The probe current under the drift motion I_p' is given by

Eq. (4.47)
$$I_p' = I_p\left(1 + \frac{\alpha^2}{2} - \frac{\alpha^4}{16}\right)$$

where α denotes the ratio of the drift velocity to the thermal velocity.

4.3.6.4 Substrate Temperature Monitoring

The substrates are mounted on the substrate holder. The electric controller controls the temperature of the substrate holder. The measurement system of the substrate temperature is shown in Fig. 4.64. The direct measurement of the substrate temperature is done by measuring the infrared radiation energy using a pyrometer. The temperature of the surface or back side of the substrates can be monitored by a pyrometer if the emissivity of the substrates is known. The emissivity of the substrate is measured at the dummy substrates with the aid of temperature measurement using a thermocouple. Since the emissivity of the growing film surface will change during the deposition, the direct measurement at the substrate surface will include the error due of the unknown emissivity. Accurate measurement of the substrate temperature is achieved by the in situ monitoring of emissivity from the back side of the substrates.[53] The emissivity ε is measured by measuring the reflectance ρ, using the relation $\varepsilon = 1 - \rho$. The in-situ measurement of reflectance is done by irradiating infrared laser light onto the backside of the substrates through the optical fiber. For Si substrates, in-situ monitoring of the wafer temperature is achieved with an accuracy of $\pm$ 1° for a temperature range of 300–1250°C.

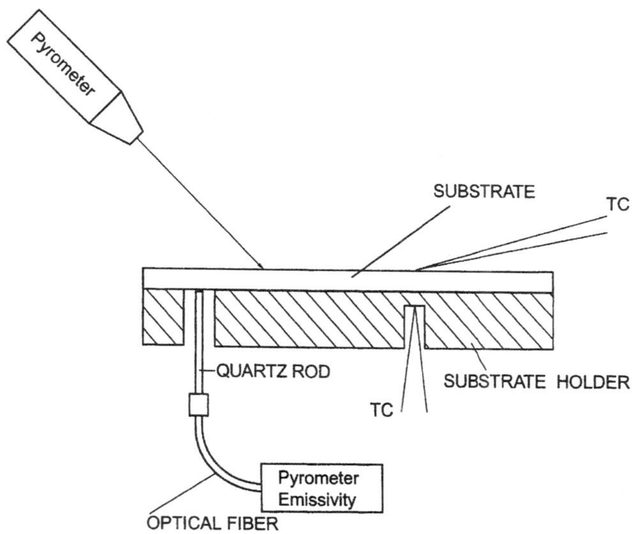

Figure 4.64. Measurements of substrate temperature.

The other method of the measuring substrate temperature is with the use of a thermocouple. A thermocouple mounted on the substrate surface measures the substrate temperature. If a thermocouple cannot be mounted on the substrate surface, dummy substrates can be used for measuring substrate temperature: the temperature differences between the substrate and the substrate holder are measured for the dummy substrates. The temperature of the substrate holder estimates the substrate temperature. If the substrates are closely mounted onto the substrate holder, the temperature of the substrate holder controls the substrate temperature.

4.3.6.5 Thickness Monitoring

Several methods are proposed for thickness monitoring during vacuum deposition: crystal oscillation, resistance monitoring, capacitance monitoring, and optical monitoring. Among these methods, the optical interferometric method is one of the most useful for monitoring the thickness of sputtered films. Typical examples of this method are shown in Fig. 4.65. Helium-neon laser light is introduced into the sputtering chamber and the laser light irradiates the surface of the substrate during sputtering. When optically transparent films are deposited, the reflected light from the surface of the substrate and the sputtered film will show interference with each other. The thickness is monitored by the periodic properties of the reflected light intensity.

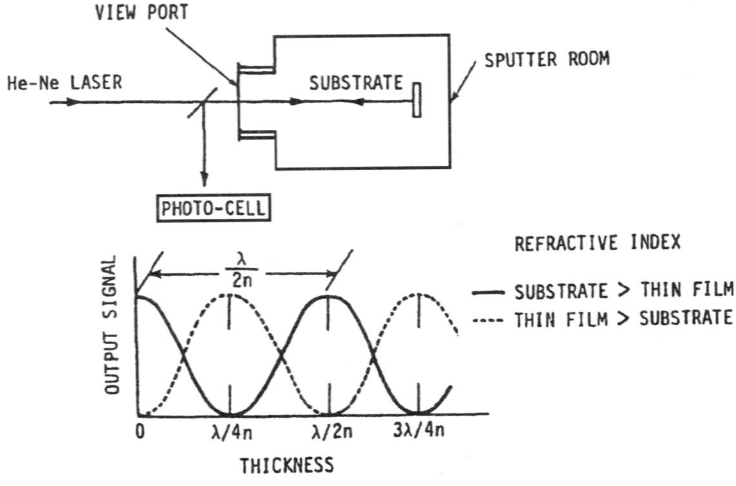

Figure 4.65. Optical interferometric method for monitoring the thickness during the deposition.

An infrared laser will monitor semitransparent films having a narrow band gap. This kind of monitoring method will give the information about the surface roughness during the deposition since the reflected light intensity will decrease for the rough surface of the sputtered films.

4.3.6.6 Film Structure

In-situ monitoring of the crystal structure of sputtered thin films is also important for the deposition of thin films with a controlled structure on an atomic scale. In-situ monitoring of the growth process of epitaxial thin films in the sputtering system is achieved by RHEED analysis. The RHEED system can not be used in a conventional diode sputtering system, since the working pressure is too high for the stable operation of the electron gun in the RHEED. The stable operation of RHEED monitoring is achieved in a low working pressure. The ion beam sputtering system shown in Fig. 4.66 is useful.[54]

The ellipsometric method is also used for in-situ monitoring of the film growth process. Betrand studied the in-situ monitoring of the growth of SiO_2 thin films in the plasma CVD.[55] Non-invasive infrared measurements can be performed using in-situ monitoring during film growth and provide a response that is directly related to the film microstructure, due to vibrations of chemical bonds. The spectroscopic ellipsometry shows the possibilities of the in-situ monitoring of the thin film growth process and/or the chemical reaction of the adatoms on the substrates for complex compound materials like perovskite.[56] Figure 4.67 shows a multi-target sputtering system designed for the in-situ measurement of spectroscopic ellipsometry during film growth.

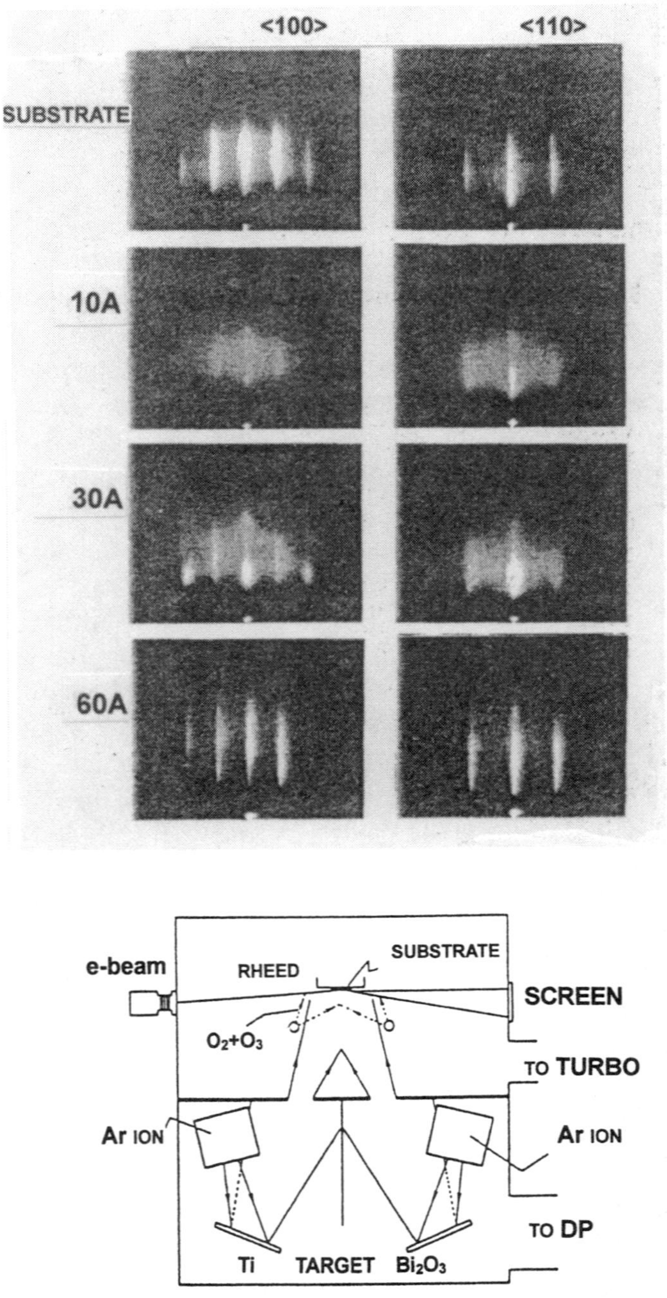

Figure 4.66. Schematic diagram of the dual ion beam sputtering system for preparation of $Bi_4Ti_3O_{12}$ thin films with RHEED patterns.[54]

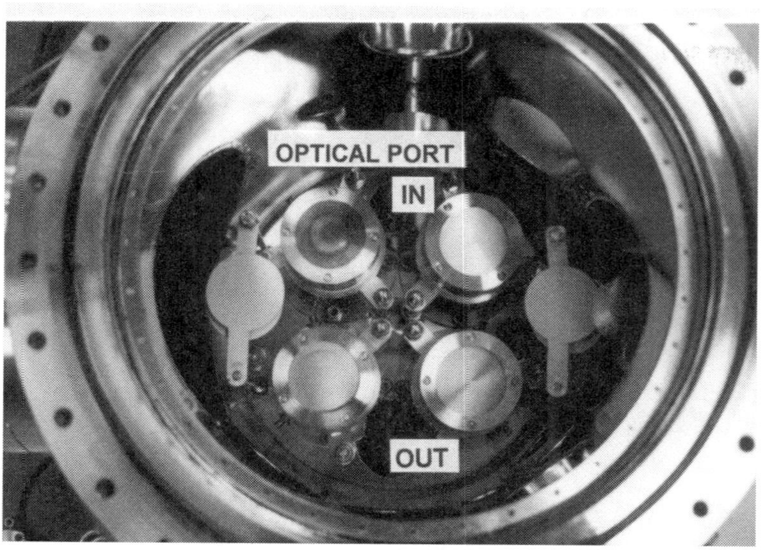

Figure 4.67. Multi-target sputtering system designed for in-situ spectroscopic ellipsometry.

REFERENCES

1. Wasa, K., and Hayakawa, S., *Jpn. J. Electrical Engineering*, 85:130 (1965)
2. Brown, S. C., *Basic Data of Plasma Physics*, MIT Press, p. 258, Cambridge, MA (1959)
3. Von Engel, A., and Steenbeck, M., *Elektrische Gasentladunger* II:68, J. Springer, Berlin (1932)
4. Francis, G., *Ionization Phenomena in Gases*, p. 124, Butterworth Publ. (1960)
5. Wasa, K., and Hayakawa, S., *Proc. IEEE*, 55:2179 (1967)
6. Kay, E., *J. Appl. Phys.*, 34:760 (1962)
7. Wasa, K., and Hayakawa, S., Jpn. Patent 558,099 (1966); U. S. Patent 3,528,902 (Sep. 1970)
8. Goeckner, M. J., Goree, J. A., and Sheridan, T. E., Jr., *IEEE Trans. Plasma Sci.,* 19:301 (1991)
9. McClanahan, E. D., and Laegreid, N., *Sputtering by Particle Bombardment* III. (R. Behrisch and K. Wittmaack, eds.), p. 339, Springer-Verlag, Berlin (1991)

10. Robertson, J. K., and Clapp, C. W., *Nature*, 132:479 (1933)
11. Davidse, P. D., and Maissel, L. I., *J. Appl. Phys.*, 37:754 (1966)
12. Maissel, L. I., and Glang, R., (eds.) *Handbook of Thin Film Technology*, p. 4-35, McGraw-Hill, New York (1970)
13. Penning, F. M., U.S. Patent 2,146,025 (Feb. 1935)
14. Kay, E., *J. Appl. Phys.*, 34:760 (1963)
15. Gill, W. D., and Kay, E., *Rev. Sci. Instrum.*, 36:277 (1965)
16. Wasa, K., and Hayakawa, S., *J. Phys. Soc. Jpn.*, 21:738 (1966)
17. Hayakawa, S., and Wasa, K., *J. Phys. Soc. Jpn.*, 20:1692 (1965)
18. Wasa, K., and Hayakawa, S., *J. Phys. Soc. Jpn.*, 20:1219 (1965)
19. Wasa, K., and Hayakawa, S., *IEEE Trans. Parts Materials Packaging*, PMP-3:71 (1967)
20. Wasa, K., and Hayakawa, S., *Rev. Sci. Instrum.*, 40:693 (1969)
21. Wasa, K., and Hayakawa, S., *Thin Solid Films*, 52:31 (1978)
22. Wasa, K., and Hayakawa, S., Jpn. Patent 642,012 (1967)
23. Bergström, J., Holmberg, S., and Lenert, B., *Arkv Fysik,* 25:49(1963)
24. Clarke, P. J., U. S. Patent 3,616,450 (Oct. 1971)
25. Chapin, J. S., *Res./Dev.,* 25:37 (1974)
26. Thornton, J. A., *J. Vac. Sci. Technol.* 15:171(1978)
27. Hoffman, V., *Solid State Technol.,* p. 57 (Dec. 1976)
28. Rastogi, R. S., Vankar, V. D., and Chopra, K. L., *Rev. Sci. Instrum.* 58:1505 (1987)
29. Sundgren, J. E., Hultman, L., Hakansson, G., Brirch, J., and Petrov, I., *Mater. Res. Soc. Symp. Proc.* 268:71 (1992)
30. Sproul, W. D., *Vacuum*, 51:641(1998)
31. Yamashita, M., *J. Vac. Sci. Technol. A,* 7:151 (1989); Morita, T., Yamamoto, T., Kurauchi, T., and Matsuura, M., *Proc. 4th Int. Symp. on Sputtering and Plasma Processes: ISSP1997*, Kanazawa, p. 405 (1997)
32. Chopra, K. L., and Randlett, M. R., *Rev. Sci. Instr.*, 38:1147 (1967)
33. Matsuoka, M., and Ono, K., *Oyo Buturi*, 57:1301 (1988)
34. Este, G., and Westwood, D., *J. Vac. Sci. Technol.*, A6:1845 (1988)
35. Brauer, G., Bruch, J., Szczyrboeski, G., and Teschner, G., *The 1st Int. Workshop on Pulsed Plasma Surface Technologies*, Abstracts: A2, San Diego (1998)
36. Capps, N., Carter, G., McDonough, G., Roche., G., and Scholl, R., *Proc. 6th Int. Symp. on Sputtering and Plasma Processes: ISSP 2001*, Kanazawa (2001)

37. Schiller, S., and Kirchoff, V., *The 1st Int. Workshop on Pulsed Plasma Surface Technologies*, Abstracts:A1 (San Diego, 1998)
38. Hanak, J. J., *J. Vac. Sci. Technol.*, 8:172 (1971)
39. Hada, T., Wasa, K., and Hayakawa, S., *Thin Solid Films*, 7:135 (1971)
40. Wasa, K., and Hayakawa, S., *Microelectron. Reliab.*, 6:213 (1967)
41. Hata, T., Nakano, S., Masuda, Y., Sasaki, K., Haneda, Y., and Wasa, K., *Vacuum*, 51:583(1998)
42. Maissel, L. I., and Glang, R., (eds.) *Handbook of Thin Film Technology*, p. 1–57, McGraw-Hill, New York (1970)
43. Ai, R., Ichikawa, Y., and Wasa, K., *Vacuum*, 59:466 (2000)
44. Matsui, N., Ai, R., Unno, A., and Wasa, K., *Proc. 6th Int. Symp. on Sputtering and Plasma Processes: ISSP 2001,* Kanazawa (2001)
45. Lugsheider, E., Knotek, O., Löffler, F., Schnaut, U., and Eckert, P., *Mat. Res. Soc. Symp. Proc.*, 389:359(1995)
46. Tominaga, K., Iwamura, S., and Shintani, Y., *Jpn. J. Appl. Phys.*, 21:688 (1982)
47. Tominaga, K., Kume, M., and Yuasa, T., *Jpn. J. Appl. Phys.*, suppl. 24-3:28 (1985)
48. Wasa, K., U.S. Patent 4,288,307 (Sep. 1981)
49. Hoshi, Y., Naoe, M., and Yamanaka, S., *Jpn. J. Appl. Phys.*, 16:1715 (1977)
50. Kadokura, S., and Naoe, M., *Vacuum*, 51:683 (1998)
51. Langmuir, I., and Blodgett, K. B., *Phys. Rev.*, 22:347 (1923) ibid., 24:49 (1924)
52. Hayakawa, S., and Wasa, K., *Nat. Tech. Rept.*, 1.8:419 (1962)
53. ULVAC Tech. J., p. 30 (Sep. 1999)
54. Ichikawa, Y., Adachi, H., Setsune, K., and Wasa, K., *Appl. Surface Sci.*, 60/61:749 (1992)
55. Betrand, N., Drevillon, B., and Bulkin, P., *J. Vac. Sci. Technol.*, A16(1):63 (1998)
56. McKinstry, Susan-Trolier, private communication (1995)

5

Deposition of Compound Thin Films

Many of the technological advances in LSI (large-scale integration) through ULSI (ultra large-scale integration) integrated circuits can be traced to advances in thin-film processing techniques. These advances have allowed the development of many kinds of thin-film electronic devices including thin-film transistors, surface acoustic devices, high-precision resistors, solar cells, magnetic and/or optical memory, LCD (liquid crystal display) and plasma displays, and a variety of sensors and actuators.

Most of these thin films are widely used not only for information devices but also for energy and environmental systems like ecological buildings. Table 5.1 shows typical thin-film materials used for these applications. Compound thin films are also important for these practical applications, as seen in the table.

Compound thin films are deposited by several methods, as shown in Fig. 5.1. The basic method is the direct deposition of a compound source. Thermal evaporation, electron beam evaporation, and sputtering are used for this. Pulsed laser ablation is a simple method of directly depositing compound materials, while thermal evaporation and sputtering are widely used for production. Multisource deposition is used to obtain a higher accuracy of chemical composition. Molecular beam epitaxy (MBE) and chemical vapor deposition (CVD) are reliable deposition processes for finely controlling the chemical composition. The CVD process is widely used in the production of Si and III-V semiconducting devices.

Table 5.1. Thin Film Materials and Applications

Application		Materials
Electronics	electrodes, interconnections	Au, Al, Cu, Cr, Ti, Pt, Mo, W, Al/Si, Pt/Si, Mo/Si.
	resistors	Cr, Ta, Re, TaN, TiN, NiCr, SiCr, TiCr, SnO_2, In_2O_3.
	dielectrics	AlN, BN, Si_3N_4, Al_2O_3, BeO, SiO, SiO_2, TiO_2, Ta_2O_5, HfO_2, PbO, MgO, Nb_2O_5, Y_2O_3, ZrO_2, $BaTiO_3$, $LiNbO_3$, $PbTiO_3$, PLZT, ZnS.
	insulators	Si_3N_4, Al_2O_3, SiO, SiO_2, TiO_2, Ta_2O_5.
	magnetics	Fe, Co, Ni, Ni-Fe, Te-Fe, GdCo.
	superconductors	Nb, NbN, Nb_3Sn, Nb_3Ge, Nb_3Si, La-Sr-Cu-O, Y-Ba-Cu-O, Bi-Sr-Ca-Cu-O.
	semiconductors	Ge, Si, Se, Te, SiC, ZnO, ZnSe, CdSe, CdTe, Cds, PbS, PbO_2, GaAs, GaP, GaN, Mg/Co/Ni/O.
	passivations	Si_3N_4, SiO, SiO_2.
Energy environments	solar cell, battery	SiO_2, TiO_2, SnO_2, In_2O_3, Al_2O_3, AlN, Fe-Si, ZnCuTe, TiS_2.
	sensors, catalysis	$LiTaO_3$, $LiMn_2O_4$, V_2O_5, YSZ, LaSrMn.
Instruments	hardening	Al, Zn, Cd, Cr, Ti, Ta, W, TiN, TiC, SiC.
	coating	Ag, Au, Al, TiC.

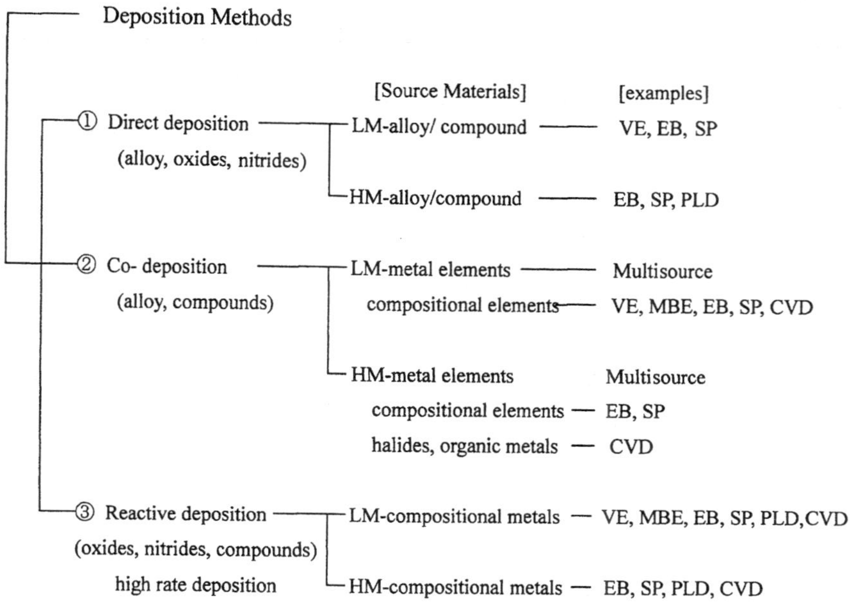

Figure 5.1. Deposition methods of alloys and compound materials. LM: Low melting temperature; HM: high melting temperature; VE: thermal evaporation; MBE: molecular beam epitaxy; EB: electron beam evaporation; PLD: pulse laser deposition; SP: sputtering (rf for insulators, dc for conductors); CVD: chemical vapor deposition.

Recent progress in sputtering enables the production of thin films of new ceramics of complex compounds such as $PbTiO_3$, PLZT [(Pb, La)(Zr, Ti)O_3], and high-temperature oxide superconductors. Semiconducting thin films of II-VI and III-V groups such as ZnSe and GaAs can also be deposited by sputtering, although their semiconductive properties should be further improved to meet the requirements of semiconducting devices. Alloy thin films of silicides, such as Mo-Si deposited by magnetron sputtering, are used for making the Schottky barriers of MOS devices. Reactive sputtering at pulsed dc and/or medium frequency ac is now indispensable in industry for the high-rate deposition of metal oxides. The sputtering process is widely used in practice for the optical coating of large glass plates.

The sputtering process can be used to successfully deposit thin films of high-temperature superconductors composed of layered perovskites. It is noteworthy that the sputtering process provides a layered structure with finely controlled structural properties that are very similar to those deposited by the MBE process. The sputtering process shows a high potential for the deposition of compound thin films.

The physical and chemical properties of compound thin films need to be closely controlled. During typical glow-discharge sputtering, films grow under the influence of the irradiation of energetic particles in the plasma; thus, in order to control the properties of deposited films, we must first understand the influences of sputtering parameters on the structure of the thin films. These structural properties and the conditions that primarily govern them are described below.

Structural Properties:

1. Chemical composition
2. Crystal phase and crystallinity
3. Microstructure
4. The surface and interface structures
5. Stresses in the films

Composition: Control of the chemical composition of thin films is essential. The chemical composition is governed by the following:

1. Flux composition of the sputtering source
2. Flux composition of the sputtered adatoms
3. Flux composition of impurity and/or residual gas atoms
4. Sticking coefficient of the adatoms
5. Surface reaction of the adatoms
6. Interdiffusion between the adatoms and the substrates
7. Mixing of the adatoms in the growing films

Crystal Properties: After we know the chemical composition of the compound thin film, we should consider its crystal properties. The crystal properties are governed by the following:

1. Film composition; stoichiometry, off-stoichiometry
2. Impurity and/or foreign atoms
3. Energy of adatoms and/or irradiation of energetic particles
4. Surface mobility of adatoms
5. Surface chemical properties of substrates
6. Interface properties between thin films and substrates
7. Phase separation

Once we understand the structural properties of the sputtered thin films, it is essential to know the relationship between these properties and the sputtering parameters. The sputtering parameters for the deposition of controlled film properties are listed in Table 5.2. The way we find the relationship between the sputtering parameters and the resultant film properties is also important.

Table 5.3 lists the deposition processes for compound thin films.[1]–[129] It includes not only the sputtering processes but also thermal evaporation, CVD, and laser ablation.

It is common for multiple authors to report different deposition conditions even for the same deposition method. This is true especially for the deposition of thin compound films since the characterization methods of the deposition processes and/or the resultant thin films are not well established, except for the semiconducting processes for Si and/or III-V materials. For example, the values of growth temperature and/or working gas pressure listed in Table 5.3 will, in part, include the values measured near the substrate, not necessarily at the substrate position. If we understand the quality of the data listed in Table. 5.3, these data will be useful for finding the sputtering deposition factors that influence film properties.

For sputtering deposition, we need a sputtering stand and sputtering targets. Table 5.4 shows examples of commercial cathode targets for sputtering deposition.

Table 5.2. Thin Film Structural Properties and Sputtering Parameters

Film Properties	Key Sputtering Parameters
Composition	Target composition
Crystal phase	Deposition rate (sputtering power, spacing)
Microstructure	Deposition temperature
film	Post-anneal temperature
surface	Sputtering atmosphere (total pressure, partial pressure)
interface	Substrate materials (glass, crystal, misfit degree)
Defects	Substrate-position/substrate-bias voltage
	Cathode, anode geometry
	Sputtering mode (metal, oxides)
	Cooling rate, cooling atmosphere

Table 5.3. Deposition Methods for Compound Thin Films

Materials	Structure*	Deposition Method**	Substrate	Deposition Conditions: Substrate Temp. (°C)	Dep. Rate (μm/hr)	Misc.	Film Properties***	Ref.
CdS	PC	VE	fused quartz	≈ 200	≈ 2	CdS source	hex. c-axis orientation; SAW velocity: ≈1700 m/s	1
ZnS	SC	VE	Si (100)	275	unknown	ZnS powder, alumina crucible	$r_{41} = 6.73 \times 10^{-13}$ m/v (6328 Å) cubic	2
	SC	RF-SP	NaCl (100)	20	≈ 18–54	ZnS target, Ar	cubic phase	unpubl.
ZnSe	SC	RF-SP	NaCl (100)	290	0.35	ZnSe target, Ar	cubic phase	3
	SC	MBE	GaAs (100)	≈ 300–400	0.09	Zn, Se sources, Zn/Se ≈ 1–30	PL spectra (4.2 K): 2.7982 eV, Ga-bond exciton	4
	SC	CVD	GaAs (100)	≈ 300–350	2	Dimethylzinc H_2Se	PL spectra (77 K): narrow peak at 447 nm	5
Al_2O_3	a	RF-MSP	Si	≈ 160–300	≈ 0.9–2.1	Al_2O_3 target, $Ar/O_2 = 1$	$\varepsilon^* \approx 9.96$, $n_o \approx 1.61$–1.66, dielectric strength: 4×10^6 V/cm	6
SiO_2	a	CVD	fused quartz	≈ 900–1100	≈ 6–150	reaction gas, $SiCl_4+O_2$	optical wave guide loss: 4.5 dB/cm (1.15 μm), 6.4 dB/cm (6328 Å)	7
	a	RF-MSP	glass	<130	1.2	SiO_2 target, Ar	sputter gas pressure: 1×10^{-3} torr, RF power: 4 kW	8

(cont'd.)

* a: amorphous, PC: polycrystal, SC: single crystal.

** ARE: activated reactive evaporation; CVD: chemical vapor deposition; EB: electron beam; IBS: ion beam sputter; IP: ion plating; LA: laser ablation, MBE: molecular beam epitaxy; MSP: magnetron sputter; P-CVD: plasma-CVD; SP: sputter; VE: vacuum evaporation.

*** SAW: surface acoustic wave; r_{41}: electro-optic constant; PL: photo-luminescence; ε^*: relative dielectric constant; n_o: optical index.

Table 5.3. *(cont'd.)* Deposition Methods for Compound Thin Films

Materials	Structure*	Deposition Method**	Substrate	Deposition Conditions			Film Properties***	Ref.
				Substrate Temp. (°C)	Dep. Rate (μm/hr)	Misc.		
TiO_2	a	DC-MSP	glass	RT ≈ 200	1.4	Ti target, $Ar/O_2 = 85/15$	$n_o \approx 2.5$ (0.5 μm)	8
ZnO	PC	RF-SP	fused quartz	≈ 250–320	≈ 2–3	ZnO target, $Ar/O_2 = 1$	c-axis orientation: $\sigma < 1°$	9
	SC	RF-SP	c, R-sapphire	600	0.2	ZnO target, $Ar/O_2 = 1$	$\rho \approx 2.4 \times 10^3$ Ωcm (c-sapphire), $\rho \approx 70$ Ωcm (R-sapphire) $\mu_H \approx 2.6$–28 cm^2/volt·sec	10
	SC	RF-SP	R-sapphire	400	≈ 0.11–0.25	ZnO(Li_2CO_3) target, $Ar/O_2 = 1$	SAW velocity: ≈5160 m/s, SAW coupling (k^2) ≈ 3.5%	11
	SC	CVD	R-sapphire	200	unknown	Diethylzinc + O_2, RF power: 80 W	[110] orientation: $\sigma < 0.8°$	12
$Bi_{12}GeO_{20}$ (BGO)	PC	RF-SP	glass	≈ 100–350	≈ 0.2–0.6	BGO target	$n_o = 2.6$	13
$Bi_4Ti_3O_{12}$	SC	RF-SP	Pt (001)	700	0.36	$Bi_4Ti_3O_{12}$ target (Bi rich)	$\varepsilon^* = 120$, $P_s = 48$ μC/cm^2, ferroelectrics	14–15

(cont'd.)

* a: amorphous, PC: polycrystal, SC: single crystal.

** ARE: activated reactive evaporation; CVD: chemical vapor deposition; EB: electron beam; IBS: ion beam sputter; IP: ion plating; LA: laser ablation, MBE: molecular beam epitaxy; MSP: magnetron sputter; P-CVD: plasma-CVD; SP: sputter; VE: vacuum evaporation.

*** n_o: optical index; SAW: surface acoustic wave; P_s: polarization; ρ: resistivity; μ_H: hall mobility; ε*: relative dielectric constant.

Table 5.3. *(cont'd.)* Deposition Methods for Compound Thin Films

Materials	Structure*	Deposition Method**	Substrate	Deposition Conditions			Film Properties***	Ref.
				Substrate Temp. (°C)	Dep. Rate (μm/hr)	Misc.		
$Bi_{12}TiO_{20}$	SC	RF-SP	BGO	425	0.5	$Bi_{12}TiO_{20}$ powder target	Optical waveguide loss: 15 dB/cm (6328 Å)	16
Bi_2WO_6	PC	RF-SP	glass	sputter: RT, anneal: 200	0.4	Bi_2WO_6 target	$n = 2.5$, ferroelectrics	17
$(In_2O_3)_{0.8}$ $(SnO_2)_{0.2}$ (ITO)	PC	RF-MSP	glass	130	≈ 1	ITO target, $Ar + O_2$	$\rho \approx 10^3$ Ωcm, $n \approx 10^{21}/cm^3$, $\mu \approx 10$ cm/volt × sec	18
		RF-MSP	glass	40	(sputter power) 200 W, φ100 target	ITO target, Ar: 4×10^{-3} torr	$R/\square \approx 10$–$100\ \Omega/\square$	unpubl.
$K_3Li_2Nb_5O_{15}$ (KLN)	SC	RF-SP	$K_2BiNb_5O_{15}$ (KBN) sapphire	≈ 600–700	0.2	KLN target, (K, Li rich)	Ferroelectrics, $\varepsilon^* = 140$, $T_c = 460°C$ $n_o = 2.277$ (6328 Å)	19
	a	RF-SP	fused quartz	RT	0.38	LNA target, $Ar/O_2 = 1$	$\varepsilon^* \approx 10^4$ (≈ 200–300°C, 1 kHz)	20
$LiNbO_3$ (LN)	SC	RF-SP	c-sapphire	500	0.025	LN target, $Ar + O_2$	Ferroelectrics, $n_o = 2.32$, optical waveguide loss: 9 dB/cm (6328 Å)	21
	SC	LPE	c-$LiTaO_3$	850	thickness: 3 μm	Li_2O-V_2O_5 flux	$n_o = 2.288$, optical waveguide loss 11 dB/cm (6328 Å)	22

(cont'd.)

* a: amorphous, PC: polycrystal, SC: single crystal.

** ARE: activated reactive evaporation; EB: electron beam; IBS: ion beam sputter; IP: ion plating; LA: laser ablation, MBE: molecular beam epitaxy; MSP: magnetron sputter; P-CVD: plasma-CVD; SP: sputter; VE: vacuum evaporation; LPE liquid phase epitaxy.

*** ρ: resistivity; R/□: sheet resistivity (Ω/□); ε*: relative dielectric constant; μ: mobility; n, n_o: optical index.

Table 5.3. *(cont'd.)* Deposition Methods for Compound Thin Films

Materials	Structure*	Deposition Method**	Substrate	Deposition Conditions			Film Properties***	Ref.
				Substrate Temp. (°C)	Dep. Rate (μm/hr)	Misc.		
$PbTiO_3$	SC	CVD	(001) $SrTiO_3$ $KTaO_3$	550	≈ 0.3–0.4	$Pb(C_2H_5)_4$ $Ti(OC_3H_7)_4$	domain patterns spontaneous strains	23
	PC	RF-SP	Pt	610	≈ 0.34–0.3	PbO/TiO_2 target, $Ar + O_2$	$\varepsilon^* \approx 200$ (RT)	24
	SC	RF-SP	c-sapphire	620	≈ 0.3–0.6	$PbTiO_3$, power, $Ar + O_2$	$\varepsilon^* = 60$ (RT)	unpubl.
	SC	MSP	(001) $SrTiO_3$	600	0.18	miscut substrate, sputter gas, $Ar/O_2 = 20/1$, 0.5 Pa	terrace structure, width: ≈ 10–200 nm, height: ≈ 1–3 nm, FWHM = 0.5°	25
PLT	SC	RF-SP	(100) MgO	≈ 600–700	≈ 0.18–0.48	PLT 18/100 target, $Ar + O_2$	$\varepsilon^* \approx 700$ $n_o \approx 2.3$–2.5 (6328 Å)	26

(cont'd.)

* a: amorphous, PC: polycrystal, SC: single crystal.
** ARE: activated reactive evaporation; CVD: chemical vapor deposition; EB: electron beam; IBS: ion beam sputter; IP: ion plating; LA: laser ablation, MBE: molecular beam epitaxy; MSP: magnetron sputter; P-CVD: plasma-CVD; SP: sputter; VE: vacuum evaporation.
*** FWHM: full width, half maximum; ε^*: relative dielectric constant; n_o: optical index.

Table 5.3. *(cont'd.)* Deposition Methods for Compound Thin Films

Materials	Structure*	Deposition Method**	Substrate	Deposition Conditions			Film Properties***	Ref.
				Substrate Temp. (°C)	Dep. Rate (μm/hr)	Misc.		
PZT	SC	CVD	(001) $SrTiO_3$	700	thickness: ≈ 0.1–0.5 μm	$Pb(C_2H_5)_4$ $Zr(OC_4H_9)_4$ $Ti(OC_3H_7)_4$	unfatigued capacitor using $SrRuO_3$ electrodes:10^{12} switching cycles	27
	PC	RF-SP	fused quartz SUS	>500	2	PZT 52/48 target, $Ar + O_2$	Ferroelectrics, $\varepsilon^* \approx 751$ (RT), $P_s \approx 21.6\ \mu C/cm^2$, $T_c = 325°C$, $n_o = 2.36$	28
	SC	MSP	(100) MgO	500	1.2	target: PZT + 20 mol% PbO, Pt coating (70 nm)	ferroelectricity: $P_s \approx 50\ \mu C/cm^2$, piezoelectricity: $d_{31} \approx 100 \times 10^{-12}$ m/V	29
PLZT	PC	RF-SP	fused quartz Pt	sputter: 500, anneal: 650–700	0.2–0.4	PLZT 7/65/35 target, $Ar + O_2$	$\varepsilon^* \approx 1000–1300$ $T_c \approx 170°C$	30
	SC	RF-SP	c-sapphire $SrTiO_3$ (100)	700	0.4	PLZT 9/65/35 target, $Ar + O_2$	$n_o \approx 2.49$ (6328 Å)	31
WO_3	a	VE	glass	100	thickness: 1.4 μm	WO_3 powder source	transmittance > 70%	32

(cont'd.)

* a: amorphous, PC: polycrystal, SC: single crystal.
** ARE: activated reactive evaporation; CVD: chemical vapor deposition; EB: electron beam; IBS: ion beam sputter; IP: ion plating; LA: laser ablation, MBE: molecular beam epitaxy; MSP: magnetron sputter; P-CVD: plasma-CVD; SP: sputter; VE: vacuum evaporation.
*** T_c: Curie temperature; P_s: polarization; d_{31}: piezoelectric coefficient; ε^*: relative dielectric constant; n_o: optical index.

Table 5.3. *(cont'd.)* Deposition Methods for Compound Thin Films

Materials	Structure*	Deposition Method**	Substrate	Deposition Conditions			Film Properties***	Ref.
				Substrate Temp. (°C)	Dep. Rate (μm/hr)	Misc.		
AlN	PC	RF-SP	glass	≈ 200–300	0.6	AlN target, Ar	ρ ≈ 2000 μΩcm	33
	PC	DC-MSP	glass	320	1.3	Al target, Ar/N_2	c-axis orientation, FWHM ≈ 2.9–5.4°	34
	SC/PC	RF-MSP	c-sapphire, glass	≈ 50–500	≈ 0.2–0.8	Al target, Ar/N_2	FWHM = 1° (sapphire), FWHM = 3° (glass)	35
	SC	RF-SP	c, R-sapphire	1200	0.5	Al target, NH_3	SAW velocity ≈ 5500 m/s, SAW coupling (k^2) ≈ 0.05–0.02%	36
	SC	CVD	R-sapphire	1200	3	reaction gas: $(CH_3)Al + NH_3 + H_2$	SAW velocity ≈ 6100 m/s, SAW coupling (k^2) ≈ 0.8%	37
CrN	PC	ARE	glass	≈ 30–450	unknown	reaction gas: NH_3	$T_s > 200$°C: Cr_2N, $T_s > 400$°C: CrN	38
BN	PC	ARE	stainless steel, glass, Si, NaCl	450	9	source: H_3BO_3, reaction gas: NH_3	Hv ≈ 2128 kg/mm^{-2}, cubic boron nitride, optical gap: 3.64 eV	39

(cont'd.)

* a: amorphous, PC: polycrystal, SC: single crystal.

** ARE: activated reactive evaporation; CVD: chemical vapor deposition; EB: electron beam; IBS: ion beam sputter; IP: ion plating; LA: laser ablation, MBE: molecular beam epitaxy; MSP: magnetron sputter; P-CVD: plasma-CVD; SP: sputter; VE: vacuum evaporation.

*** ρ: resistivity; FWHM: full width, half maximum; SAW: surface acoustic wave; T_s: substrate temperature; Hv: Vickers hardness.

Table 5.3. *(cont'd.)* Deposition Methods for Compound Thin Films

Materials	Structure*	Deposition Method**	Substrate	Deposition Conditions			Film Properties***	Ref.
				Substrate Temp. (°C)	Dep. Rate (μm/hr)	Misc.		
Si_3N_4	a	P-CVD	Si	250	3	reaction gas: $N_2 + NH_3 + SiH_4$	$n_0 \approx 2.0–2.1$ (6328 Å)	40
	a	RF-MSP	Si	100	1	Si_3N_4 target: Ar	$n_0 \approx 2.1$ (6328 Å)	unpubl.
TiN	PC	DC-MSP	glass	150	≈ 0.6–1.8	Ti target: $Ar/N_2 = 7/3$	ρ = 250 μΩ cm, TCR = 150 ppm/°C	41
	PC	RF-MSP	fused quartz	500	1.5	TiN target: Ar	ρ = 150 μΩ cm, TCR = 300 ppm/°C	unpubl.
WC	PC	RF-MSD	stainless steel	≈ 200–500	0.36	W target: $Ar + C_2H_2$	$Ts \approx 200°C$: WC, W_2C, W_3C mixed phase, $Ts \approx 400°–500°C$: WC single phase	42
				≈ 300–500	≈ 0.36–4.9	W target: $Ar + C_2H_2$	$Hv^* \approx 2365–3200$ kg/mm^2, friction coefficient: ≈0.09 $Hv^* = 3200$ kg/mm^2 (WC)	
B_4C	a	RF-SP	sapphire	450	≈ 0.5	B_4C target: Ar	Hv = 4800 kg/mm^2	43
CrC	PC	RF-MSP	glass	600	≈ 0.2–0.7	Si target: $Ar + CH_4$	IR absorption, 800 cm^{-1} (SiC), 2000 cm^{-1} (SiH)	44

(cont'd.)

* a: amorphous, PC: polycrystal, SC: single crystal.

** ARE: activated reactive evaporation; CVD: chemical vapor deposition; EB: electron beam; IBS: ion beam sputter; IP: ion plating; LA: laser ablation, MBE: molecular beam epitaxy; MSP: magnetron sputter; P-CVD: plasma-CVD; SP: sputter; VE: vacuum evaporation.

*** n_o: optical index; ρ: resistivity; TCR: temperature coefficient of resistance; T_s: substrate temperature; Hv: Vickers hardness.

Table 5.3. *(cont'd.)* Deposition Methods for Compound Thin Films

Materials	Structure*	Deposition Method**	Substrate	Deposition Conditions			Film Properties***	Ref.
				Substrate Temp. (°C)	Dep. Rate (μm/hr)	Misc.		
SiC	PC	RF-SP	glass, alumina	550	≈ 0.5–1	SiC target, Ar	β-SiC, (220) orientation, ρ ≈ 200 Ωcm, B ≈ 2100 K	43, 45
	SC	IP	Si (111)	1000	≈ 0.9–1.8	reaction gas: C_2H_2	β-SiC	46
	SC	CVD	Si (100)	1330	≈ 4–6	reaction gas: $H_2 + SiH_4 + C_3H_8$	β-SiC carbon buffer layer	47
	SC	CVD	Si (100)	1350	unknown	C_3H_8 treatment, source gas: $SiH_4 + C_3H_8 + H_2$	inclined substrates at ≈ 2°–5° toward (011): single-domain cubic phase	48
	SC	MBE	Si (001)	>900	thickness: 1000 Å	source: graphite, Si, C_2H_2 flux: 3×10^{-8} torr	3C phase, {111} twins, a = 4.357 Å FWHM = 1.64°	49
	SC	CVD	Si (001)	1260	thickness: 1.4 μm	SiH_4: 1 sccm, C_3H_8: 0.4 sccm	3C phase, 3×2 surface, no antiphase domain, FWHM = 0.26°	50

(cont'd.)

* a: amorphous, PC: polycrystal, SC: single crystal.
** ARE: activated reactive evaporation; CVD: chemical vapor deposition; EB: electron beam; IBS: ion beam sputter; IP: ion plating; LA: laser ablation, MBE: molecular beam epitaxy; MSP: magnetron sputter; P-CVD: plasma-CVD; SP: sputter; VE: vacuum evaporation.
*** ρ: resistivity; B: thermistor coefficient; FWHM: full width, half maximum.

Table 5.3. *(cont'd.)* Deposition Methods for Compound Thin Films

Materials	Structure*	Deposition Method**	Substrate	Deposition Conditions			Film Properties***	Ref.
				Substrate Temp. (°C)	Dep. Rate (μm/hr)	Misc.		
Diamond	PC	chemical transport reaction (CTR)	diamond	1000	≈ 1–2	source: graphite, atomic hydrogen	$n = 2.38$, $\rho \approx 10^{12}$–10^{13} Ωcm, microhardness: 9500 kg/mm^2	51
	PC	ion beam SP	Si fused quartz	RT	≈ 0.3–0.4	target: graphite, Ar ion beam: 1200 eV, 60 mA, hydrogen ion	TEM observation: cubic diamond particle, $n = 2.8$, $\sigma = 10^{-2}$ S/cm	52
	PC	microwave P-CVD	Si	900	0.3	with seed layer, source gas: $CO + H_2 + B_2H_6$, 4000 Pa	p-type, $\rho = 3\times10^2$ Ωcm, piezoresistivity, $K \approx 1000$ (RT)	53
	PC	P-CVD	Mo Si	1000	≈ 0.2–0.3	100 MHz plasma: ≈ 400–1000 W, source gas: $CH_4 + H_2$, 50 torr	stable plasma, ⟨111⟩ ⟨110⟩ orientation, hexagonal shape, (100) facet	54
GaAs	SC	MBE	GaAs	600	graphite crucible, Ca (1090°C), As (320°C)		carrier density: 2.0×10^{15}/cm^3, $\mu_{300} = 7500$ cm^2/volt·sec	55
	SC	RF-SP	GaAs (100)	≈ 500–625	≈ 0.7–1.2	GaAs target, Ar	$\rho \approx 10^5$–10^6 Ωcm, $\mu_{300} \approx 5000$ cm^2/volt·sec	56
	SC	CVD	GaAs	≈ 580–620	thickness: 6–9 μm	triethylgallium + tertiarybutylasine	carrier density: ≈ 10^{14} /cm^3, $\mu_{45} \approx 209000$ cm^2/volt·sec	57

(cont'd.)

* a: amorphous, PC: polycrystal, SC: single crystal.

** ARE: activated reactive evaporation; CVD: chemical vapor deposition; EB: electron beam; IBS: ion beam sputter; IP: ion plating; LA: laser ablation, MBE: molecular beam epitaxy; MSP: magnetron sputter; P-CVD: plasma-CVD; SP: sputter; VE: vacuum evaporation.

*** n: optical index; ρ: resistivity; μ_{300}, μ_{45}: mobility at 300 K, 45 K.

Table 5.3. *(cont'd.)* Deposition Methods for Compound Thin Films

Materials	Structure*	Deposition Method**	Substrate	Deposition Conditions			Film Properties***	Ref.
				Substrate Temp. (°C)	Dep. Rate (μm/hr)	Misc.		
GaN	SC	CVD	c-sapphire	1040	thickness: 2–3 μm	AlN buffer, TMG, TMA, NH_3, Cp_2Mg, source: LEEBI	p-type semiconductor, $\rho = 35\ \Omega cm$, carrier density: $2 \times 10^{16}\ /cm^3$, $\mu_{300} \approx 8\ cm^2/volt \cdot sec$	58
	SC	CVD	c-sapphire	1035	thickness: 4 μm	GaN buffer TMG, NH_3, Cp_2Mg, source: N_2, LEEBI	p-type semiconductor, $\rho = 2\ \Omega cm$, carrier density: $3 \times 10^{17}/cm^3$, $\mu_{300} = 10\ cm^2/volt \cdot sec$	59
	SC	MBE	c-sapphire SiC	800	2 μm/hr	NH_3 gas source, $\approx (2–5) \times 10^{-5}$ torr, AlN buffer	wurtzite phase, $\rho > 10^4\ \Omega cm$ (400 K), carrier density: $10^{18}\ cm^{-3}$	60
GaSb	SC	MBE	GaAs	450–600	0.36–1	Ar^+ ion bombarded substrate	p-type semiconductor, carrier density: $(4–6) \times 10^{16}/cm^3$, $\mu_{300} = 670\ cm^2/volt \cdot sec$	61
	a	RF-SP	BaF_2 (111)	400	0.15	GaSb target,Ar	unknown	62
InAs	SC	MBE	GaAs	≈ 450–600	≈ 0.36–1	Ar^+ ion bombarded substrate	n-type semiconductor, carrier density ≈ (4–6) × $10^{16}/cm^3$, $\mu_{300} = 1670\ cm^2/volt \cdot sec$	61

(cont'd.)

* a: amorphous, PC: polycrystal, SC: single crystal.

** ARE: activated reactive evaporation; CVD: chemical vapor deposition; EB: electron beam; IBS: ion beam sputter; IP: ion plating; LA: laser ablation, MBE: molecular beam epitaxy; MSP: magnetron sputter; P-CVD: plasma-CVD; SP: sputter; VE: vacuum evaporation.

*** ρ: resistivity; FWHM: full width, half maximum; μ_{300}: mobility at 300 K.

Table 5.3. *(cont'd.)* Deposition Methods for Compound Thin Films

Materials	Structure*	Deposition Method**	Substrate	Deposition Conditions			Film Properties***	Ref.
				Substrate Temp. (°C)	Dep. Rate (μm/hr)	Misc.		
$In_{1-x}Ga_xSb$	SC	RF-SP	BaF_2 (111)	400	0.15	InSb,GaSb target, Ar	$x = 0.36$	62
Nb_3Sn		DC-MSP	sapphire	≈ 650–800	60	Nb_3Sn target, Ar	Superconducting, transition, $T_c \approx 18.3$ K	63
A15 Nb_3Ge	PC	SP	sapphire	700	0.1	sputter gas pressure, 40Pa	$T_c = 21.6$ K, $\Delta T_c = 3.4$	64
	PC	CVD	Nb tape	900	216	$NbCl_5$ $GeCl_4$ H_2	$T_c = 20.1$	65
B1 NbN	PC	SP	sapphire	<350 anneal 600	0.06–0.5	Nb target, $Ar + N_2$	$T_c = 16$ K, $H_{c2} = 28$ T (13 K) $J_c = 8 \times 10^5$ A/cm^2 (0 T, 4.2 K)	66
	PC	CVD	carbon fiber	≈ 1400–1600	≈ 7–11	$NbCl_5$ N_2 H_2	$T_c = 16.4$ K, $H_{c2} = 11$ T, $J_c = 10^6$ A/cm^2 (0 T, 4.2 K)	67
	PC	P-CVD	carbon fiber	1100	≈ 3.6–7	$NbCl_5$ CH_4 NH_3	$T_c \approx 15$–17 K, $H_{c2} = 21$T (4.2 K), $J_c = 10^6$ A/cm^2 (0 T, 4.2 K)	68

(cont'd.)

* a: amorphous, PC: polycrystal, SC: single crystal.

** ARE: activated reactive evaporation; CVD: chemical vapor deposition; EB: electron beam; IBS: ion beam sputter; IP: ion plating; LA: laser ablation, MBE: molecular beam epitaxy; MSP: magnetron sputter; P-CVD: plasma-CVD; SP: sputter; VE: vacuum evaporation.

*** T_c: superconducting transition temperature; $\Delta T_c = [T_{co}$ (onset temp.) - $T_{R=0}]$; H_{c2}: critical magnetic field; J_c: critical current.

Table 5.3. *(cont'd.)* Deposition Methods for Compound Thin Films

Materials	Structure*	Deposition Method**	Substrate	Deposition Conditions			Film Properties***	Ref.
				Substrate Temp. (°C)	Dep. Rate (μm/hr)	Misc.		
$PbMo_6S_8$	PC	SP	Mo	anneal ≈ 750–1100	thickness: 0.2 μm	MoS_2 Mo PbS	$T_c = 14$ K, $J_c = 8\times10^4$ A/cm^2 (0 T, 4.2 K)	69
	PC	SP	sapphire	anneal 1000	thickness: 1 μm	Mo sputter Pb/MoS_2 1000°C	$T_c = 14$ K, $J_c = 10^6$ A/cm^2 (0 T, 4.2 K)	70
$BaPb_{1-x}Bi_xO_3$	SC	MSP	(100) $SrTiO_3$	700	unknown	$Ar/O_2 = 1$	$T_c = 10.5$ K, $\Delta T_c << 1$ K, $x = 0.3$ $\mu_H = 55$ cm^2/volt·sec, $x = 0$–0.2, $x = 0.35$	71
LSCO K_2NiF_4	PC	SP	sapphire MgO	660 820	0.8	Ar, Ar/O_2	$T_c \approx 40$ K, (001) orientation, $T_{R=0} \approx 24$ K, crystallization, $T_s > 450$°C	72
	SC (001)	M-SP	(100) $SrTiO_2$	800	unknown	$n = 6.28\times10^{21}$ cm^{-3} $\mu_H = 1.3$ cm^2/volt·sec $\rho \approx 25$ μΩcm (T_c)	$T_c \approx 14$ K (as sp.), 22 K (anneal) post anneal: 800°C, 8 hr	73
	PC	SP	YSZ	≈ 400–450	0.125	Target cosputter: LSC, LBC, LSC, LCaC	$T_c \approx 45$ K, $T_{R=0} \approx 17$ K $J_c = 150$ A/cm^2 (0 T, 4.2 K) postanneal in O_2: ≈ 760–850°C, 15 hr	74

(cont'd.)

* a: amorphous, PC: polycrystal, SC: single crystal.

** ARE: activated reactive evaporation; CVD: chemical vapor deposition; EB: electron beam; IBS: ion beam sputter; IP: ion plating; LA: laser ablation, MBE: molecular beam epitaxy; MSP: magnetron sputter; P-CVD: plasma-CVD; SP: sputter; VE: vacuum evaporation.

*** T_c: superconducting transition temperature; J_c: critical current; $T_{R=0}$: transition temperature at $R = 0$; $\Delta T_c = [T_{co}$ (onset temp.) - $T_{R=0}]$; μ_H: mobility.

Table 5.3. *(cont'd.)* Deposition Methods for Compound Thin Films

Materials	Structure*	Deposition Method**	Substrate	Deposition Conditions			Film Properties***	Ref.
				Substrate Temp. (°C)	Dep. Rate (μm/hr)	Misc.		
LSCO K_2NiF_4	SC (001)	MSP	(100) $SrTiO_3$	600	0.6	Ar	$T_c = 34$ K, $\Delta T \approx 3$ K, $T_{R=0} = 25$ K, $J_c \approx 30$ A/cm^2 (0 T, 4.2 K), anneal: 900°C, 72 hr	75
	PC	VE (EB)	sapphire	RT	unknown	La_2O_3, Sr, Ca O_2 reactive vapor deposition (1×10^{-4} torr) anneal	$T_c > 30$ K, $T_{R=0} \approx 15$ K, energy gap: ≈ 20–30 mV	76
YBCO $YBa_2Cu_3O_{7-\delta}$	PC	VE (EB)	MgO sapphire (100) $SrTiO_3$	450	unknown	anneal: ≈ 900–950°C in O_2	$T_c \approx 97$ K, $T_{R=0} \approx 87$ K $J_c = 7 \times 10^5$ A/cm^2, (0 T, 77 K) $J_c = 5 \times 10^7$ A/cm^2 (0 T, 4.2 K)	77
	PC	EB	(100) $SrTiO_3$	≈ 200–870	0.5–1 μm	O_2 reactive vapor deposition, anneal: ≈ 650–850°C	≈ 10%–90% R-transition, 88 K, $\rho \approx 0.5$ mΩ cm (T_c), a-axis orientation (a >> c) $J_{c\parallel} = 2 \times 10^6$ A/cm^2 (0 T, 4.2 K) $J_{c\perp} = 9 \times 10^4$ A/cm^2 (0 T, 78 K)	78

(cont'd.)

* a: amorphous, PC: polycrystal, SC: single crystal.

** ARE: activated reactive evaporation; CVD: chemical vapor deposition; EB: electron beam; IBS: ion beam sputter; IP: ion plating; LA: laser ablation, MBE: molecular beam epitaxy; MSP: magnetron sputter; P-CVD: plasma-CVD; SP: sputter; VE: vacuum evaporation.

*** T_c: superconducting transition temperature; $T_{R=0}$: transition temperature at $R = 0$; $\Delta T_c = [T_{co}$ (onset temp.) - $T_{R=0}]$; J_c: critical current; ρ: resistivity.

Table 5.3. *(cont'd.)* Deposition Methods for Compound Thin Films

Materials	Structure*	Deposition Method**	Substrate	Deposition Conditions			Film Properties***	Ref.
				Substrate Temp. (°C)	Dep. Rate (μm/hr)	Misc.		
YBCO $YBa_2Cu_3O_{7-\delta}$	SC	MSP	(110) $SrTiO_3$	700	thickness: 0.7 μm	sputter Ar/O = 1, 8 × 10^{-2} torr, anneal in O_2, 920°C, 2 hr	$T_{R=0} \approx 84$ K, $\Delta T_c = 6$ K, $J_{c\parallel} = 1.8 \times 10^6$ A/cm^2, (0 T, 77 K), $J_{c\parallel} = 3.2 \times 10^4$ A/cm^2, (0 T,4.2 K), $\rho_{c\parallel}$ =0.5, $\rho_{c\perp} = 16$ mΩ cm (T_c)	79
	PC	MSP	R-sapphire	200	0.9	sputter, Ar, anneal	$T_{R=0} \approx 70$ K, $T_{co} \approx 94$ K	80
	SC	MSP	(100) MgO	650	thickness: 1 μm	in situ deposition	$T_{R=0} \approx 86$ K, $\Delta T = 3$ K, $\rho = 0.5$ mΩ cm (T_C), Er-Ba-Cu-O	81
	PC	LA (KrF excimer laser)	(110) $SrTiO_3$ R-sapphire	450	1 Å/pulse KrF 1 J/pulse	anneal in O_2, 900°C	$T_c \approx 95$ K, $T_{R=0} \approx 85$ K ($SrTiO_3$), $\Delta T = 2$ K, $T_{R=0} \approx 75$ K (sapphire), $\Delta T = 12$ K	82
	PC	EB	sapphire, YSZ	RT	≈ 1–1.8	Y, Ba, Cu multilayer, anneal in O_2, ≈ 800–850°C	$T_c \approx 94$ K, $T_{R=0} \approx 72$ K (YSZ), $T_{R=0} \approx 40$ K (sapphire)	83

(cont'd.)

* a: amorphous, PC: polycrystal, SC: single crystal.

** ARE: activated reactive evaporation; CVD: chemical vapor deposition; EB: electron beam; IBS: ion beam sputter; IP: ion plating; LA: laser ablation, MBE: molecular beam epitaxy; MSP: magnetron sputter; P-CVD: plasma-CVD; SP: sputter; VE: vacuum evaporation.

*** T_c: superconducting transition temperature; $T_{R=0}$: transition temperature at $R = 0$; $\Delta T_c = [T_{co}$ (onset temp.) - $T_{R=0}]$; J_c: critical current; ρ: resistivity.

Table 5.3. *(cont'd.)* Deposition Methods for Compound Thin Films

Materials	Structure*	Deposition Method**	Substrate	Deposition Conditions			Film Properties***	Ref.
				Substrate Temp. (°C)	Dep. Rate (μm/hr)	Misc.		
BSCCO Bi-Sr-Ca-Cu-O system	PC c-axis orientation	MSP	(100) MgO, (100) $SrTiO_3$	700	0.48	anneal in O_2, ≈ 800–900°C	$T_c \approx 115$ K $T_{R=0} \approx 104$K $\rho = 50\ \mu\Omega$cm (T_c) $J_c > 20 \times 10^4$ A/cm^2, (0 T, 77 K)	84
	PC c-axis orientation	MSP co-sputter	(100) MgO	≈ RT	thickness: 0.4 μm	target: Bi, SrCa, Cu, anneal in O_2, 865°C	$T_{R=0} \sim 104$ K, $B_{c2}^{\perp} \approx 8.5$ T/K, $B_{c2}^{\parallel} \approx 0.56$ T/K	85
	PC c-axis orientation	EB	(100) MgO	≈ RT	thickness: 0.2 μm	source: Bi, SrF_2, CaF_2, Cu, anneal in O_2, ≈ 850–890°C	c = 30.605 Å, $T_c \approx 110$ K, $T_{R=0} \approx 60$ K	86
	PC c-axis orientation	EB	(110) (100) $SrTiO_3$	≈ RT	0.3 thickness: ≈ 0.5 μm	source: Bi, SrF_2, CaF_2, Cu, anneal in wet O_2, 860°C, 5 min	$T_{R=0} \approx 80$ K, $J_c = 1 \times 10^6$ A/cm^2 (0 T, 4.2 K), $\rho \approx 150\ \mu\Omega$cm ($T_{co}$)	87
	PC c-axis orientation	LA (ArF excimer)	(100) MgO	RT	0.1 thickness: ≈ 0.2 μm	target:BSCOO, eximer laser, 10 J/cm^2 anneal in air, 890°C, 1 min	$T_c \approx 80$ K (c = 30.8 Å), $T_c \approx 120$ K (c = 36.8 Å)	88

(cont'd.)

* a: amorphous, PC: polycrystal, SC: single crystal.

** ARE: activated reactive evaporation; CVD: chemical vapor deposition; EB: electron beam; IBS: ion beam sputter; IP: ion plating; LA: laser ablation, MBE: molecular beam epitaxy; MSP: magnetron sputter; P-CVD: plasma-CVD; SP: sputter; VE: vacuum evaporation.

*** T_c: superconducting transition temperature; $T_{R=0}$: transition temperature at $R = 0$; $B_{c2}^{\perp}$: critical magnetic field; J_c: critical current; ρ: resistivity.

Table 5.3. *(cont'd.)* Deposition Methods for Compound Thin Films

Materials	Structure*	Deposition Method**	Substrate	Deposition Conditions			Film Properties***	Ref.
				Substrate Temp. (°C)	Dep. Rate (μm/hr)	Misc.		
BSCCO Bi-Sr-Ca-Cu-O system	PC	IBS	(100) MgO ZrO_2	≈ RT	thickness: ≈ 0.3–1 μm	anneal in O_2, 825–880°C	$Bi_2Sr_{2.09}Ca_{0.75}Cu_{2.44}O_X$, $T_c \approx 85$ K, $T_{R=0} \approx 75$ K, $Bi_4Sr_{3.12}Ca_{3.03}Cu_{4.56}O_X$, $T_c \approx 110$ K, $T_{R=0} \approx 63$–67 K	89
	PC c-axis orientation	MSP layer-by-layer deposition	(100) MgO	650	≈ 100 Å/hr, thickness: 200–400 Å	multitarget: Bi, SrCu, CaCu, anneal in O_2, 855°C, 5 hr	$Bi_2Sr_2Ca_1Cu_2O_X$: $T_{co} \approx 80$ K, $Bi_2Sr_2Ca_2Cu_3O_X$: $T_{co} \approx 110$ K, $Bi_2Sr_2Ca_3Cu_4O_X$: $T_{co} \approx 90$ K	90
	PC c-axis orientation	CVD	(100) MgO	910	8 thickness: 8 μm	source: bismuth alkioxide, β-diketonate, chelates of Sr, Ca, Cu	$T_{R=0} \approx 78$ K	91
TBCCO Tl-Ba-Ca-Cu-O system	PC c-axis orientation	EB	YSZ	<50	thickness: 0.7 μm	source: Tl, Ba, Ca, Cu dep. in O_2, anneal in air, 850°C, 5 min	$Tl_2Ba_2Ca_1Cu_2O_X$: $T_c \approx 100$ K (Meissner), $T_{R=0} \approx 97$ K, $J_c \approx 7.1 \times 10^5$ A/cm^2 (0 T, 76 K)	92

(cont'd.)

* a: amorphous, PC: polycrystal, SC: single crystal.

** ARE: activated reactive evaporation; CVD: chemical vapor deposition; EB: electron beam; IBS: ion beam sputter; IP: ion plating; LA: laser ablation, MBE: molecular beam epitaxy; MSP: magnetron sputter; P-CVD: plasma-CVD; SP: sputter; VE: vacuum evaporation.

*** T_c: superconducting transition temperature; $T_{R=0}$: transition temperature at $R = 0$; T_{co}: transition temperature for onset; J_c: critical current.

Table 5.3. *(cont'd.)* Deposition Methods for Compound Thin Films

Materials	Structure*	Deposition Method**	Substrate	Deposition Conditions			Film Properties***	Ref.
				Substrate Temp. (°C)	Dep. Rate (μm/hr)	Misc.		
TBCCO Tl-Ba-Ca-Cu-O system	PC c-axis orientation	MSP	YSZ	≈ RT	thickness: 0.4 μm	target: Tl, BaCa, Cu, anneal in O_2, 850°C, 4 min.	$T_c \approx 110$ K, $T_{R=0} \approx 96$ K, $J_c \approx 2\times10^4$ A/cm^2 (0 T, 4.2 K)	93
	PC c-axis orientation	SP	(100) MgO $SrTiO_3$ YSZ	≈ RT	0.48 thickness: ≈ 1–4 μm	anneal in O_2 / Tl 880–900°C in sealed quartz	$Tl_2Ca_1Ba_2Cu_2O_X$: $T_{R=0} \approx 100$ K, $Tl_2Ca_2Ba_2Cu_3O_X$: $T_{R=0} \approx 110$ K $Tl_2Ca_2Ba_3Cu_4O_X$: $T_{R=0} \approx 120$ K, $J_c \approx 1.5 \times 10^4$ A/cm^2 (77 K)	94
	PC c-axis orientation	SP	(100) MgO $SrTiO_3$	RT	0.2 thickness: ≈ 0.2–1.0 μm	anneal in O_2 or air, 800–880°C	Rocking curve width: 0.32°, $Tl_2Ba_2Ca_1Cu_2O_8$: $T_{co} \approx 102$ K, $J_c \approx 10^4$ A/cm^2 (90 K), $Tl_2Ba_2Ca_2Cu_3O_{10}$: $T_{co} \approx 116$ K, $J_c \approx 10^5$ A/cm^2 (100 K)	95
	PC c-axis orientation	MSP	(100) MgO	200	0.6	anneal in O_2 / Tl, ≈ 850–900°C	$Tl_2Ba_2Ca_1Cu_2O_X$: $T_{R=0} \approx 102$ K $J_c \approx 1.2\times10^5$ A/cm^2 (77 K), $Tl_2Ba_2Ca_2Cu_3O_X$: $T_{R=0} \approx 117$ K, $Tl_2Ba_2Ca_3Cu_4O_X$: $T_{R=0} \approx 113$ K	96–98
TBCCO Tl-Ba-Ca-Cu-O system	PC	off-axis MSP	(100) LaAlO3 NdGaO3	535–565	450 Å/hr	in situ growth, target: $Ba_2CaCu_2O_x$, Tl_2O vapor	$Ta_2Ba_2CaCu_2O_7$ phase, $T_c \approx 80$ K (onset), $T_{R=0} \approx 40$ K, FWHM = 0.2–0.5°	99

(cont'd.)

* a: amorphous, PC: polycrystal, SC: single crystal.

** ARE: activated reactive evaporation; CVD: chemical vapor deposition; EB: electron beam; IBS: ion beam sputter; IP: ion plating; LA: laser ablation, MBE: molecular beam epitaxy; MSP: magnetron sputter; P-CVD: plasma-CVD; SP: sputter; VE: vacuum evaporation.

*** T_c: superconducting transition temperature; $T_{R=0}$: transition temperature at $R = 0$; J_c: critical current; ρ: resistivity; FWHM: full width, half maximum.

Table 5.3. *(cont'd.)* Deposition Methods for Compound Thin Films

Materials	Structure*	Deposition Method**	Substrate	Deposition Conditions			Film Properties***	Ref.
				Substrate Temp. (°C)	Dep. Rate (μm/hr)	Misc.		
PSCCO Pb-Sr-Ca-Cu-O system	PC c-axis orientation	MSP	(100) MgO	550	0.72	in situ growth, target: Pb_5Sr_2Ln $Ca_{0.4}Cu_3O_x$ Ar gas sputter	$Pb_2CuSr_2(Ln,Ca)Cu_2O_x$, Pb-3212 phase, $T_c \approx 75$ K, $T_{R=0} \approx 60$ K	100
	SC	LA (excimer)	(001) $LaAlO_3$	610	thickness: 0.2 μm	target: $Pb_{16}Sr_2Y_{0.5}$ $Ca_{0.5}Cu_3O_x$ O_2: 300 mtorr	PbSrYCaCuO Pb-1212 phase $T_c \approx 90$ K, $T_{R=0} \approx$ 75 K	101
	PC c-axis orientation	LA (ArF excimer)	(100) MgO	550	unknown	target: $Pb_2Sr_2Ca_{0.5}$ $Y_{0.5}Cu_3O_8$ O_2: 0.1 Pa	$Pb_2(Sr,A)_{2+n}Cu_{2+n}O_{6+2n}$ ($n = 1, 2$)	102
	SC	MBE	(001) $SrTiO_3$	650	thickness: 0.05–0.1	Pb, Sr, Eu, Ce, Cu metals, ozone: 2×10^{-4} Pa	$Pb_2CuSr_2(Eu,Ce)_nCu_2O_y$ ($n =$ 3–8), Pb-32n2 ($n = 3$–8) phase	103
	SC	MBE	(001) $LaAlO_3$	≈ 550	unknown	multiple electron-gun, Pb:Sr:Cu = 1:2:1, ozone oxidation	$PbSr_2CuO_{5+x}$, Pb-1201 phase, $T_c \approx 40$ K	104

(cont'd.)

* a: amorphous, PC: polycrystal, SC: single crystal.

** ARE: activated reactive evaporation; CVD: chemical vapor deposition; EB: electron beam; IBS: ion beam sputter; IP: ion plating; LA: laser ablation, MBE: molecular beam epitaxy; MSP: magnetron sputter; P-CVD: plasma-CVD; SP: sputter; VE: vacuum evaporation.

*** T_c: superconducting transition temperature; $T_{R=0}$: transition temperature at $R = 0$.

Table 5.3. *(cont'd.)* Deposition Methods for Compound Thin Films

Materials	Structure*	Deposition Method**	Substrate	Deposition Conditions			Film Properties***	Ref.
				Substrate Temp. (°C)	Dep. Rate (μm/hr)	Misc.		
HBCCO Hg-Ba-Ca-Cu-O system	PC c-axis orientation	MSP	(100) $SrTiO_3$	RT	thickness: 4000 Å	film: Hg:Ba:Cu = 0.7:2:1.3, annealed in $N_2 + O_2$ (1%) 670°C	$HgBa_2CuO_4$, Hg-1201 phase, $T_{R=0} \approx 93$ K $J_c \approx 2\times10^4$ A/cm^2 (0T, 80 K)	105
	PC c-axis orientation	LA	(100) $SrTiO_3$	RT	thickness: 0.5–2 μm	HgO/BaCaCuO multilayer, annealed with pellet, 750–810°	$HgBa_2CaCu_2O_{6+x}$ Hg-1212 phase, $T_{R=0} \approx 124$ K $J_c \approx 5\times10^4$ A/cm^2 (0T, 77 K)	106
	PC c-axis orientation	LA (KrF excimer)	(100) $SrTiO_3$	RT	thickness: 500–8000 Å	HgO/ReBaCa CuO multilayer, annealed with pellet, 725–800°	$HgBa_2Ca_2Cu_3O_y$ Hg-1223 phase, $T_{R=0} \approx 127.5$ K, $J_c \approx 1.5\times10^6$ A/cm^2 (0 T, 77 K)	107
	PC c-axis orientation	LA	(100) $SrTiO_3$	RT	thickness: 0.25 μm	two target: HgO/BaCaCuO, annealed with pellet, 800°, 1 hr	$HgBa_2CaCu_2O_{6+x}$ Hg-1212 phase, $T_c \approx 120$ K, $J_c > 10^7$ A/cm^2 (0T, 5 K), $J_c \approx 10^5$ A/cm^2 (0T, 110 K)	108
	PC c-axis orientation	MSP	(100) $SrTiO_3$	550	thickness: 4000 Å	In-situ growth, target: $Hg_{2.5}Ba_2CuO_{5.5}$ $Po_2 < 0.1$ Pa	in situ $HgBa_2CuO_4$ Hg-1201 phase c-axis orientation $T_c \approx 75$ K (onset)	109

(cont'd.)

* a: amorphous, PC: polycrystal, SC: single crystal.

** ARE: activated reactive evaporation; CVD: chemical vapor deposition; EB: electron beam; IBS: ion beam sputter; IP: ion plating; LA: laser ablation, MBE: molecular beam epitaxy; MSP: magnetron sputter; P-CVD: plasma-CVD; SP: sputter; VE: vacuum evaporation.

*** T_c: superconducting transition temperature; $T_{R=0}$: transition temperature at $R = 0$; J_c: critical current.

Table 5.3. *(cont'd.)* Deposition Methods for Compound Thin Films

Materials	Structure*	Deposition Method**	Substrate	Deposition Conditions			Film Properties***	Ref.
				Substrate Temp. (°C)	Dep. Rate (μm/hr)	Misc.		
$(Sr,Nd)CuO_2$	SC	MSP	(100) $SrTiO_3$	550–600	thickness: 3000 Å	target: $(Sr,Nd)Cu_{1.3}O_y$, Ar: 0.5 Pa	metastable structure, infinite-layer phase, c = 3.4 Å, $T_c \approx 16$ K (onset)	110
$Ba_2Sr_2Cu_4O_{8+x}$	SC	LA	(100) $SrTiO_3$	600	unknown	$SrCuO_2$ - $BaCuO_2$, O_2: 200 mtorr	new structure, c = 23.75 Å $T_c \approx 70$ K (onset)	111
$Ba_2Ca_3Cu_4$-CO_3O_8	SC	LA (KrF excimer)	(001) $LaAlO_3$	625	thickness: 1500 Å	target: BaCaCuO pellet, $O_2 + CO_2$ (6%) 20 Pa	metastable structure, a-axis orientation, $T_{R=0} \approx 60$ K	112
(CO_3, Cu)-Ba_2CuO_x	SC	MSP	(100) $SrTiO_3$	530	thickness: 700 Å	target: $Ba_2Cu_{1.5}O_{3.5}$, Ar + CO_2 gas	new structure, C-1201 phase superconductor, c = 8.3 Å, $T_c \approx 50$ K (onset)	113
$BaCuO_2$	SC	LA (KrF excimer)	(100) $SrTiO_3$	500–630	thickness: 400 Å	target: $BaCuO_2$, laser: 1 J/cm^2, O_2: 1×10^{-5} torr	new structure, infinite-layer phase, c = 4.08 Å	114
$Sr_{n+1}Cu_nO_{2n+1}$	SC	LA	(100) $SrTiO_3$	500–520	thickness: 300–400 Å	two target: $SrCuO_y$, Sr, NO_2: 10^{-5} mbar	metastable structure, $n = 5$, $T_c \approx 45$ K (onset)	115

(cont'd.)

* a: amorphous, PC: polycrystal, SC: single crystal.

** ARE: activated reactive evaporation; CVD: chemical vapor deposition; EB: electron beam; IBS: ion beam sputter; IP: ion plating; LA: laser ablation, MBE: molecular beam epitaxy; MSP: magnetron sputter; P-CVD: plasma-CVD; SP: sputter; VE: vacuum evaporation.

*** T_c: superconducting transition temperature; $T_{R=0}$: transition temperature at $R = 0$.

Table 5.3. *(cont'd.)* Deposition Methods for Compound Thin Films

Materials	Structure*	Deposition Method**	Substrate	Deposition Conditions			Film Properties***	Ref.
				Substrate Temp. (°C)	Dep. Rate (μm/hr)	Misc.		
Ba_2CuO_4	SC	EB	(001) $SrTiO_3$	575–625	Ba: 0.6 Cu: 0.07	flux ratio: Ba:Cu ~ 3:2, $O_2 + O_3$ (5–10%) $\approx 6\times10^{-5}$ torr	new structure, 214 phase, c = 14.6 Å $T_c \approx 70$ K (onset)	116
Fe_3O_4 (magnetite)	PC	SP	Si	500	thickness: 5000 Å	reactive dc sputtering	a = 8.39 Å, $\rho \approx 35$ mΩ cm (RT), MR = 1.7% (0.5T, RT)	117
CrO_2	SC	CVD	(100) TiO_2	390–450	thickness: 500 Å – 1.2 μm	source: CrO_3 powder, 280°C, O_2 flow	$T_c \approx 390$–395°C, FWHM = 0.1°, M = 650 emu/cc (5K), 2 μ_B/Cr ion	118
$Nd_{0.7}Sr_{0.3}MnO_z$	SC	LA	(100) $LaAlO_3$	700	thickness: 2000–3000 Å	target: $Nd_{0.7}$-$Sr_{0.3}MnO_{3-x}$, N_2O: 300 mtorr	113 structure, $\sigma/\sigma_0 = 1700$ (70K, 6T), memory effect	119
$(La,Ca)MnO_{3-x}$	SC	LA	(100) $SrTiO_3$	700	thickness: 1000 Å	laser pulse: 3 J/cm^2 O_2: 300 mtorr	$(La_{0.67}Ca_{0.33})MnO_{3-x}$, 113 structure, MR = 14000% (8 T,140 K), $T_c = 140$–204 K	120
$La_2Ca_1Mn_2O_7$	PC a-axis orientation	MSP	(001) MgO $SrTiO_3$	700–750	thickness: 1000–2000 Å	Ar sputtering, cooling in 10 torr O_2	$La_{2-2x}Ca_{1+2x}Mn_2O_7$, 327 structure, 90° domain MR = 93% (1 T, 133 K), $T^{max}_r = 140$ K	121

(cont'd.)

* a: amorphous, PC: polycrystal, SC: single crystal.

** ARE: activated reactive evaporation; CVD: chemical vapor deposition; EB: electron beam; IBS: ion beam sputter; IP: ion plating; LA: laser ablation, MBE: molecular beam epitaxy; MSP: magnetron sputter; P-CVD: plasma-CVD; SP: sputter; VE: vacuum evaporation.

*** T_c: superconducting transition temperature; ρ: resistivity; MR: magnetoresistance; M: magnetization; FWHM: full width, half maximum.

Table 5.3. *(cont'd.)* Deposition Methods for Compound Thin Films

Materials	Structure*	Deposition Method**	Substrate	Deposition Conditions			Film Properties***	Ref.
				Substrate Temp. (°C)	Dep. Rate (μm/hr)	Misc.		
$(La,Sr)_3Mn_2O_7$	PC c-axis orientation	LA (KrF excimer)	(100) $SrTiO_3$	>900	thickness: 3000 Å	target: $La_{1.2}Sr_{1.8}Mn_2O_7$ $Po_2 < 100$ mtorr	327 structure, $T_c \approx 120$ K, MR = 99.25 % (80K, 12T), M = 2 μ_B/Mn ion (5K)	122
Bi-Sr-Co-O	SC	MBE	(001) MgO Nd:$YAlO_3$	640–800	unknown	crucible: Bi: Ta Sr: PBN Co: MgO/PBN, ozone (98%)	$Bi_2Sr_2CoO_{6+x}$: insulator (E_a = 0.2 eV), $Bi_2Sr_3Co_2O_{9+x}$: metallic behavior	123
$La_{0.5}Sr_{0.5}CoO_{3-x}$	SC	LA (KrF excimer)	(001) $SrTiO_3$	650	thickness: ≈ 1 μm	laser pulse: 2.0 J/cm^2, O_2: 1–10^{-3} mbar	FWHM = 0.5–0.8°, MR = 16 % (0.2 T, 81 K), ρ = 250 Ωcm (81 K)	124
Sr_2FeMoO_6	SC	LA (KrF excimer)	(001) (111) $SrTiO_3$	900	thickness: 1000 Å	target: SrFeMoO, $Po_2 = 10^{-6}$ torr	ferromagnetic ($T_c > 400$ K), M ≈ 2–2.5 μ_B/f.u. (5 K), MR = 20% (7 T, 5 K)	125
La_2FeCrO_6	SC	LA (ArF excimer)	(111) $SrTiO_3$	unknown	thickness: 600–1100 Å	target: $LaFeO_3$ and $LaCrO_3$, unit-by-unit	ordered strucutre (1/1 unit), ferromagnetic ($T_c \approx 375$ K), M = 2 μ_B / TM ion	126
PtMnSb	PC a-axis orientation	MSP	(100) MgO	500	50 Å/min	multitarget Pt (9W), Mn (24W), Sb (21W)	M = 450 emu/cc (RT), dot patterning perpendicular magnetization	127
NiMnSb	PC	SP	(001) MgO (001) Si	150–300	1.4 Å/min	target: Ni:Mn:Sb = 1:1:1	in-situ x-ray scattering	128

* a: amorphous, PC: polycrystal, SC: single crystal.

** ARE: activated reactive evaporation; CVD: chemical vapor deposition; EB: electron beam; IBS: ion beam sputter; IP: ion plating; LA: laser ablation, MBE: molecular beam epitaxy; MSP: magnetron sputter; P-CVD: plasma-CVD; SP: sputter; VE: vacuum evaporation.

*** T_c: superconducting transition temperature; ρ: resistivity; MR: magnetoresistance; M: magnetization; FWHM: full width, half maximum.

Table 5.4. Sputtering Targets for Electronics (Mitsubishi Materials)

Function	Composition	Purity
Targets for Conductors	Cu	6N
	Ni	4N
	Al, Al-Si	5N
	Al-Cu-Ti	5N
	Al-Si-Cu	5N
	W-Si	5N
	Mo-Si	5N
	Ti-Si	5N
	Ni-Si	4N
	Ti	4N5 ~ 6N
	Ti-W	5N
	W, Mo	5N
	Ru	5N
	Co	5N
Targets for Reflective Films	Au, Au Alloy	4N
	Ag, Ag Alloy	4N
Targets for Transparent Conductive Films	ITO	-
Targets for Wear-Resistance	SiAlON	-
EL (Electroluminescence) Targets	ZnS system	-
	CaS system	
	SrS system	
PDP (Plasma Display) Targets	MgO	3N
Semiconductor Targets	B doped Si	
	P doped Si	
Dielectric Targets	BaSrTiO	3N
	BaTiO	
	SrTiO	5N
	PbZrTiO	4N
	PbLaZrTiO	
Targets for Magnetic Records	Co-Cr-Ta	3N
	Co-Cr-Pt	3N
	Co-Pt + α	3N
	Fe-Co, Ni-Fe	3N
Targets for Optical Records (Chalcogen-type targets)	Te-Ge-Sb	4N
	Te, Se, Sb	4N
	Sb-Se, Bi-Te	4N
	In-Ag-Sb-Te	4N
Targets for Magneto-optical Records (Rare-earth-metal/transition-metal type targets)	Tb-Fe-Co	3N
	TbFeCoCr	3N
	Gd-Fe-Co	3N
	Dy-Fe-Co	3N
Superconductive Targets	Ba-Y-Cu-O	3N

Several attempts have been made to provide a more controlled deposition environment for the thin-film growth process. Irradiation with charged particles during film growth is one of the most promising methods to achieve close control of film growth. Diamond crystallites can be formed by ion-beam sputtering from a graphite target under irradiation of a proton beam onto the film-growth surface.[130] The irradiation may cause (1) heating of the surface of the thin films and/or substrate, (2) rearrangement of adatoms, (3) recrystallization, (4) a change in the level of defects or vacancies, and (5) a chemical reaction between irradiated particles and adatoms. The effects of irradiation are estimated by calculating collisions between the irradiated particles and adatoms using the three-dimensional Monte Carlo cascade code for sputtering (TRIMSP).[131]–[132]

For specific applications, the electric, physical, chemical, mechanical, and/or optical properties of thin films are evaluated. Further, evaluation of the short-term and long-term stability of the deposition process and the resultant sputtered film properties helps our understanding of their use in practice. Examples of the sputtering deposition of thin compound films are described in this chapter. The examples are mostly laboratory scale for making new materials.

The sputtering process uses different growth conditions for the metallization of Si ICs and the high-rate coating of large window glass than for new materials. The growth rate for new materials is not as high, since a lower rate is necessary for the growth of high-quality crystals.

5.1 OXIDES

5.1.1 ZnO Thin Films

Zinc oxide (ZnO) single crystals show a wurtzite hexagonal structure (Fig. 5.2). These ZnO crystals are known as piezoelectric materials with a large electromechanical coupling factor and a low dielectric constant.[133] Typical physical properties of ZnO are listed in Table 5.5.

Owing to these excellent piezoelectric properties, thin films of ZnO are used for making ultrasonic transducers in high-frequency regions. In the past fifteen years, many workers have investigated fabrication processes for ZnO, including sputter deposition, chemical vapor deposition,[134] and ion plating.[135] Among these processes, sputter deposition is the most popular for ZnO thin-film deposition.

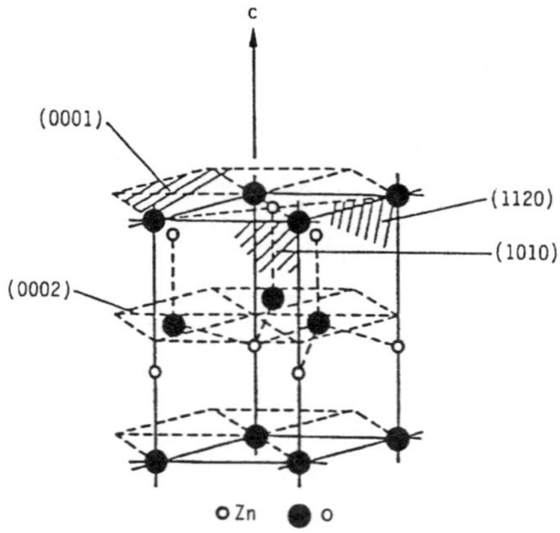

Figure 5.2. Crystal structure of ZnO single crystal.

Table 5.5. Physical Properties of ZnO Single Crystal

Crystal System	Wurtzite (6 mm)
Space group	$P6_3$ mc
Lattice constant	a = 3.24265 Å, c = 5.1948 Å
Sublimation point	1975 ± 25°C
Hardness	4 moh
Dielectric constants	$\varepsilon^S_{11} = 8.55$, $\varepsilon^S_{33} = 10.20 \times 10^{-11}$ F/m
Density	5.665×10^3 kg/m^3
Thermal expansion coefficient	$\alpha_{11} = 4.0$, $\alpha_{33} = 2.1$ ($\times 10^{-6}$/°C)
Optical transparency	0.4–2.5 μm
Refractive index	$n_0 = 1.9985$, $n_e = 2.0147$ (λ = 6328 Å)
Electrooptic constant	$r_{33} = 2.6$, $r_{13} = 1.4$ ($\times 10^{-12}$ m/V, λ = 6328Å)

5.1.1.1 Deposition of ZnO

Polycrystalline thin films. Polycrystalline ZnO thin films with a c-axis orientation are among the most popular piezoelectric thin films. A typical sputtering system for the deposition of c-axis-oriented ZnO films is shown in Fig. 5.3. The basic system consists of a planar diode sputtering device. Zinc metal or ZnO ceramic is used as the cathode (target). Sputtering is done with a mixed gas of Ar and O_2. A dc high voltage is supplied for the Zn metal or ZnO ceramic target. The electrical conductivity of the ZnO ceramic target should be higher than 10^3 mho in order to keep the dc glow discharge between the electrodes. When rf high voltage is used for sputtering (i.e., rf-diode sputtering), a high-resistance ZnO target can be used.

The ZnO ceramic target for rf sputtering is prepared as follows: First, ZnO powder (purity > 99.3%) is sintered in air at 800 to 850°C for one hour. The sintered powder is then pressed at about 100 kg/cm^2 into the form of the target (typically a disk) and is finally sintered at 930°C for two hours. The resultant ZnO ceramic is not completely sintered until the ceramic is mechanically shaped into the final form of the target. The high-conductivity ZnO ceramic for dc sputtering is made by sintering the ZnO at a higher temperature of around 1300°C.

Polycrystalline ZnO films with a c-axis orientation are commonly deposited on a glass substrate by dc or rf sputtering when the temperature of the substrate is kept at 100 to 200°C. Figure 5.4 shows their reflection electron diffraction (RED) patterns and a typical cross section observed by SEM. The c-axis-oriented films consist of a so-called fiber structure and the c-axis is oriented normally to the substrate plane.

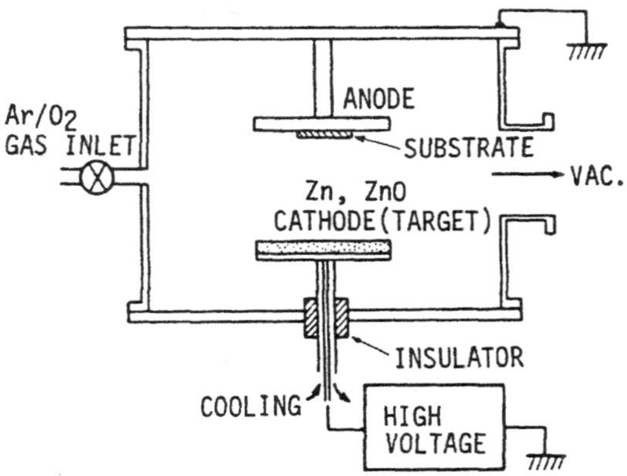

Figure 5.3. Typical sputtering systems for the deposition of ZnO thin films.

In general, the c-axis orientation is frequently observed in these deposited films. This is reasonably well understood since the c plane of the ZnO crystallites corresponds to the densest packed plane, and the growth mechanism of the sputtered ZnO thin Bravais' empirical law for crystal growth will govern films.

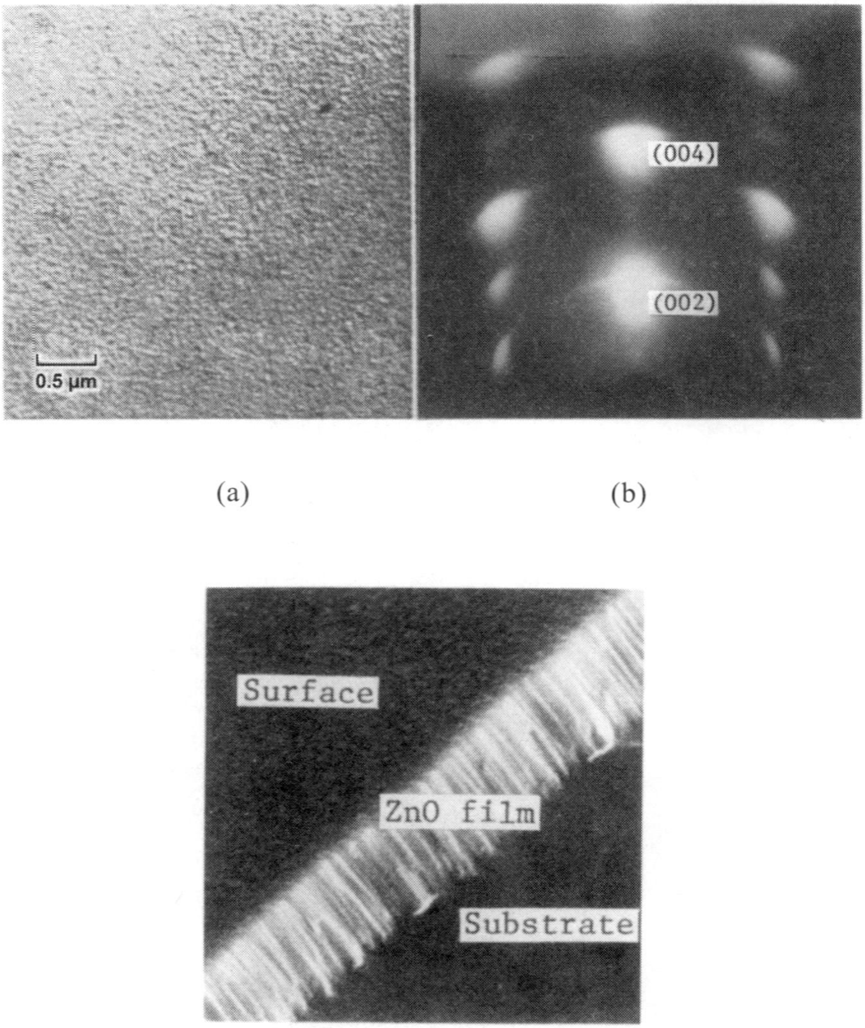

(a) (b)

(c)

Figure 5.4. (*a*) Surface structure, (*b*) RED pattern, and (*c*) SEM image of ZnO film of c-axis orientation, 2 µm thick, prepared on a glass substrate.

The reflection electron diffraction patterns of ZnO films of different film thickness on glass substrates are shown in Fig. 5.5.[136] The (002) orientation is clearly observed for film thicknesses of more than 500 Å. The angular spread of (002) arcs decreases with thickness. The typical half-angular spread was within 7.5–9 degrees for films 0.5–10.0 µm thick. The mean inclination of the c-axis from the substrate normal was within 3 degrees. The lattice constant c_o of the film is 5.23 to 5.24 Å. The c_o is slightly longer than the bulk single-crystal value (c_o = 5.2066 Å, a_o = 3.2497 Å).

(a)
(b)
(c)

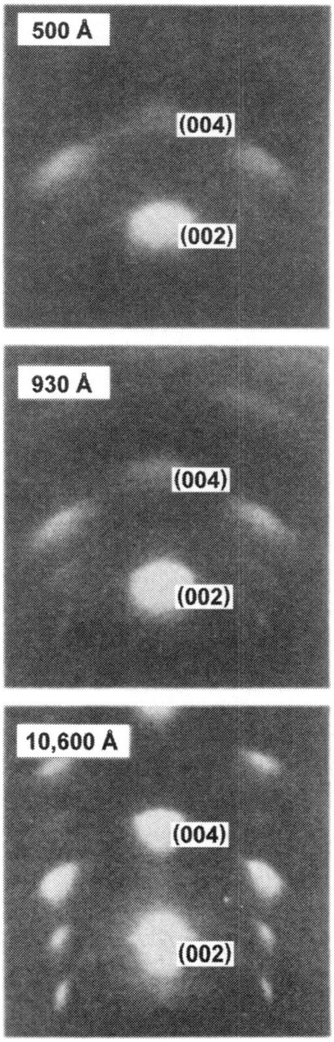

Figure 5.5. Typical electron diffraction patterns of sputtered ZnO thin films on glass substrates for various film thickness: (*a*) 500 Å, (*b*) 930 Å, (*c*) 10,600 Å.

Sputtering parameters. Detailed studies of ZnO thin-film growth suggest that the structure of sputtered ZnO films depends on various sputtering parameters including deposition rate, substrate temperature, gas pressure and composition, residual gas, and target composition. In some cases, the growth of ZnO films does not obey Bravais' empirical law and does not show a c-axis orientation.

Deposition rate and substrate temperature. The deposition rate and the substrate temperature will drastically influence the crystal structure of deposited films. Figure 5.6 shows the optimum conditions for the deposition of c-axis-oriented films by various sputtering systems.[137] It shows that the optimum condition of the substrate temperature is 100 to 200°C for a deposition rate below 1 μm/hr with the rf-diode sputtering system. In the magnetron sputtering system, the optimum condition shifts to a higher substrate temperature with a higher deposition rate; typically the substrate temperature is 300 to 400°C and the deposition rate is 1 to 5 μm/h.[137]

A high deposition rate with a low substrate temperature and/or a low deposition rate with a high substrate temperature frequently cause mixed orientation of the c-axis and the a-axis in sputtered films. A typical RED pattern of a ZnO film with mixed orientation is shown in Fig. 5.7.

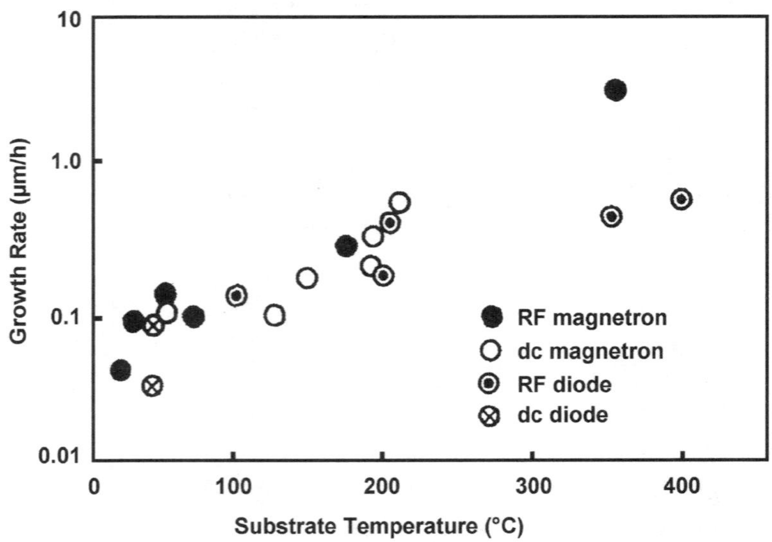

Figure 5.6. Optimum sputtering conditions for the deposition of c-axis-oriented film on a glass substrate for various sputtering systems.

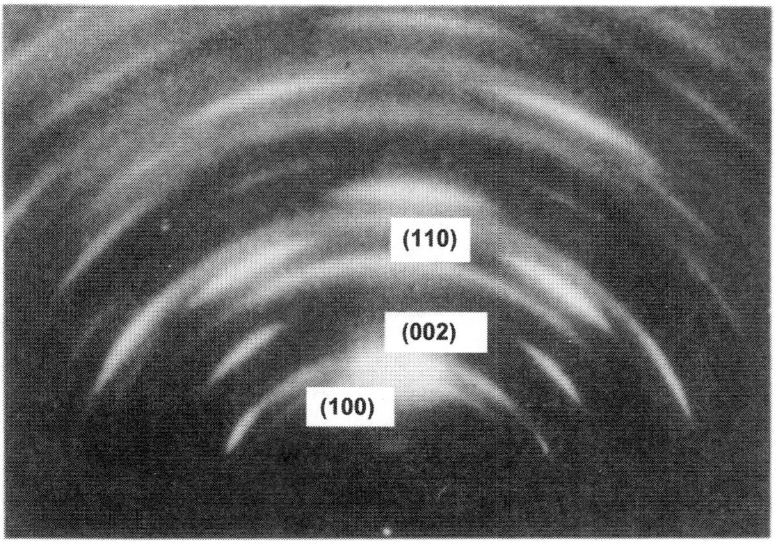

Figure 5.7. Electron diffraction pattern of ZnO thin film showing mixed orientation.

Sputtering gas composition and gas pressure. Westwood reported that an optimum partial oxygen pressure existed in the mixed gas of (Ar + O_2) for making c-axis-oriented ZnO films.[138] The optimum content of O_2 was reported to be 67% at a total sputtering gas pressure of 3.5×10^{-2} torr. The optimum oxygen pressure suggests that favorable oxidization at the surface of the cathode target and/or deposited film is necessary for obtaining c-axis-oriented ZnO films. When the sputtering gas pressure is reduced below 1×10^{-3} torr, a different feature is observed.[139] The c-axis orientation is suppressed.

Table 5.6 shows the crystallographic orientation for different gas pressures. Two types of orientation are observed. One is the c-axis orientation (normal orientation, $c_\perp$). The other is the c-axis parallel to the substrate (parallel orientation, $c_\parallel$); in this case, either the $\langle 110 \rangle$ or $\langle 100 \rangle$ axis is normal to the film surface. Figure 5.8 gives a composite plot of the crystallographic orientation, substrate temperature, and deposition rate for ZnO sputtered films at low gas pressure (1×10^{-3} torr), showing that the parallel orientation is predominant. In contrast, the normal orientation is predominant at high gas pressures (greater than 3×10^{-2} torr).

The crystalline properties of the resultant films are influenced by the degree of oxidation of both the surface of the target and the film during deposition. The degree of oxidation at the film surface increases with increasing substrate temperature. Under low oxygen partial pressure, a high substrate temperature is necessary for obtaining a favorable oxidation. Figure 5.9 shows the optimum sputtering conditions for deposition of c-axis-oriented films at various gas pressures and substrate temperatures. It stands to reason that lowering the gas pressure shifts the optimum substrate temperature to higher values.

Table 5.6. Crystallographic Orientation of ZnO Thin Films for Different Sputtering Gas Pressure

Sample No.	Total Sputtering Pressure* (10^{-3} torr)	Substrate Temperature (°C)	Condensation Rate (μm/h)	Thickness (μm)	Crystal-lographic Orientation
1-0	1	40	0.03	0.1	$c_\perp$
1-1	1	40	0.1	0.25	$c_\perp$
1-2	1	40	0.12	0.36	$c_\perp$, $c_\parallel$
1-3	1	150	0.1	0.3	$c_\parallel$
1-4	1	150	0.7	0.3	$c_\parallel$
1-5	1	200	0.7	0.3	$c_\parallel$
1-6	1	270	0.07	0.2	$c_\parallel$
1-7	1	270	0.6	0.3	$c_\parallel$
1-8	5	200	0.2	0.1	$c_\parallel$
1-9	30	100	0.15	0.15	$c_\perp$
1-10	30	200	0.15	0.15	$c_\perp$
1-11	60	200	0.2	0.2	$c_\perp$
1-12	100	200	0.15	0.15	$c_\perp$

*Al + O_2, 50% O_2

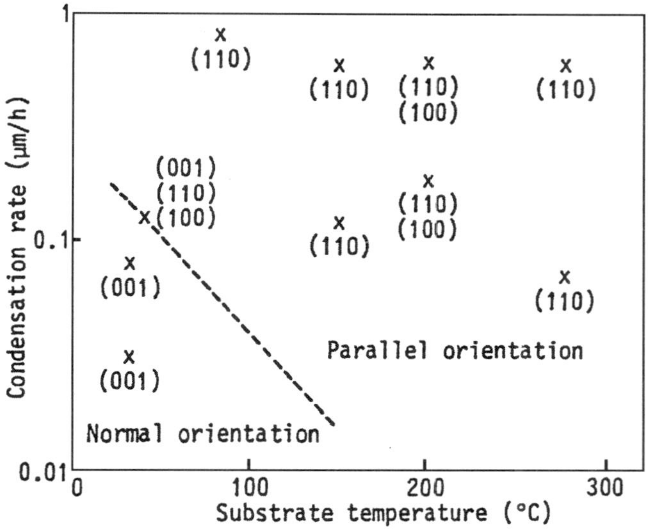

Figure 5.8. Variation of crystallographic orientation for ZnO thin films with deposition rate and substrate temperature sputtered in low gas pressure (1×10^{-3} torr).

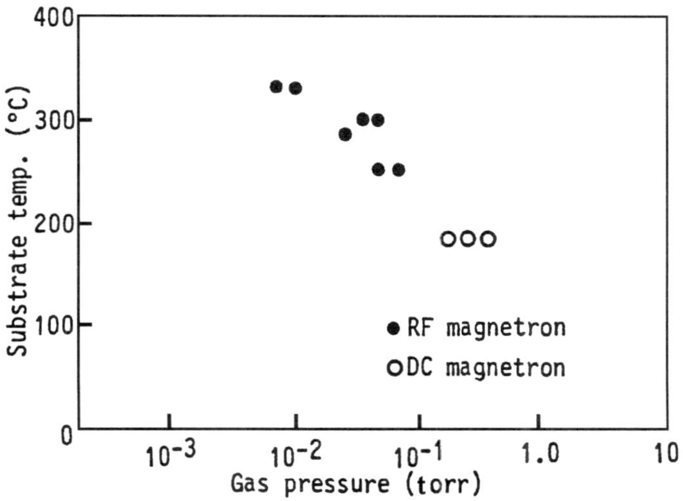

Figure 5.9. Optimum sputtering conditions for the deposition of c-axis-oriented ZnO films for various sputtering gas pressure.

Target composition and Purity. A sintered ZnO target is more suitable for c-axis-oriented films than a Zn metal target. Foster pointed out that the presence of organic vapor, such as methane, in the sputtering atmosphere reduced the growth of c-axis orientation and induced the a-axis orientation.[140]

It is also interesting to note that the addition of foreign atoms, such as aluminum copper, during film growth changes the crystallographic orientation of the sputtered films. Table 5.7 indicates the typical change of orientation with aluminum and copper. The proportions of aluminum and copper were determined by chemical analyses. Typical photographs (electron micrographs) and reflection electron diffraction patterns are shown in Fig. 5.10. The results are summarized as follows: the addition of aluminum enhances the growth of parallel orientation (c-axis parallel to the film surface, $c_{\parallel}$), and the increase of aluminum results in an increase of parallel orientation intensity. For a favorable amount of aluminum, the normal orientation (c-axis normal to the film surface, $c_{\perp}$) disappears and the parallel orientation remains.

Table 5.7. Crystallographic Orientation of ZnO Thin Films with Admixed Foreign Atoms

Sample No.	Total Sputtering Pressure* (10^{-3} torr)	Contents of Foreign Metals (at. %)	Substrate Temperature (°C)	Condensation Rate (μm/h)	Thickness (μm)	Crystal-lographic Orientation
2-0		-				$c_{\perp}$
2-1	1	6 (Al)	40	0.1	0.3	$c_{\parallel}$
2-2		-				$c_{\parallel}$
2-3		0.1 (Al)				$c_{\parallel}$
2-4	1	1.3 (Al)	200	1.2	0.6	$c_{\parallel}$
2-5		6 (Al)				$c_{\parallel}$
2-6		-				$c_{\perp}$
2-7		0.04 (Al)				$c_{\perp} + c_{\parallel}$
2-8	30	5 (Al)	200	0.075	0.15	$c_{\parallel}$
2-9		-				$c_{\parallel}$
2-10	1	0.013 (Cu)	200	0.9	0.45	$c_{\parallel}$
2-11		0.5 (Cu)				$c_{\perp} + c_{\parallel}$

*Al + O_2, 50% O_2

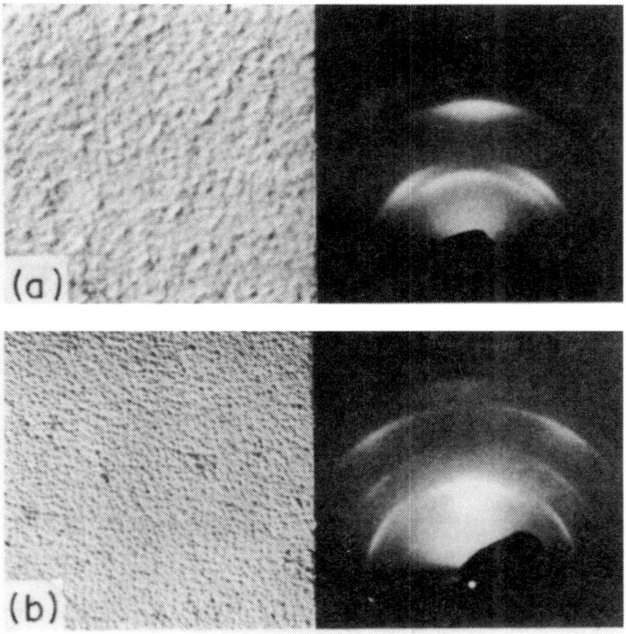

Figure 5.10. Electron micrographs of ZnO thin films of 0.3 μm thick with admixed foreign atoms: (*a*) with Al of 6 atm.%, (*b*) with Cu of 0.5 atm%.

Copper has the opposite effect on crystallographic orientation. It enhances the normal orientation. The enhancement of the normal c-axis orientation is observed for Mn (not shown).

Electrode configuration and substrate position. The electrode configuration of the sputtering system also affects the crystallographic orientation. Two types of rf sputtering systems are shown in Fig. 5.11. One is a conventional planar electrode system, Fig. 5.11a, and the other is a hemispherical electrode system, Fig. 5.11b. The glass substrates are placed on a substrate holder behind the anode. Thin films of ZnO with a c-axis orientation are deposited under the conditions shown in Fig. 5.6. However, the configuration affects the degree of the crystalline orientation.

Figure 5.12 shows typical x-ray diffraction patterns from ZnO thin films on glass substrates prepared by the two sputtering systems. It shows that ZnO thin films prepared by the planar system at substrate position "1" exhibit very poor c-axis orientation.

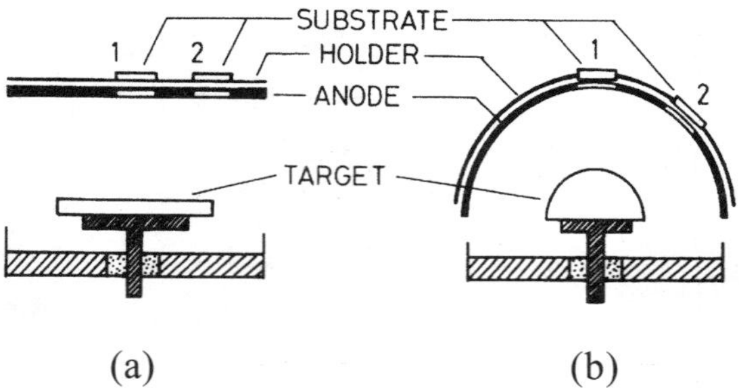

Figure 5.11. Electrode configuration of two sputtering experiments: (*a*) conventional planar system and (*b*) hemispherical system.

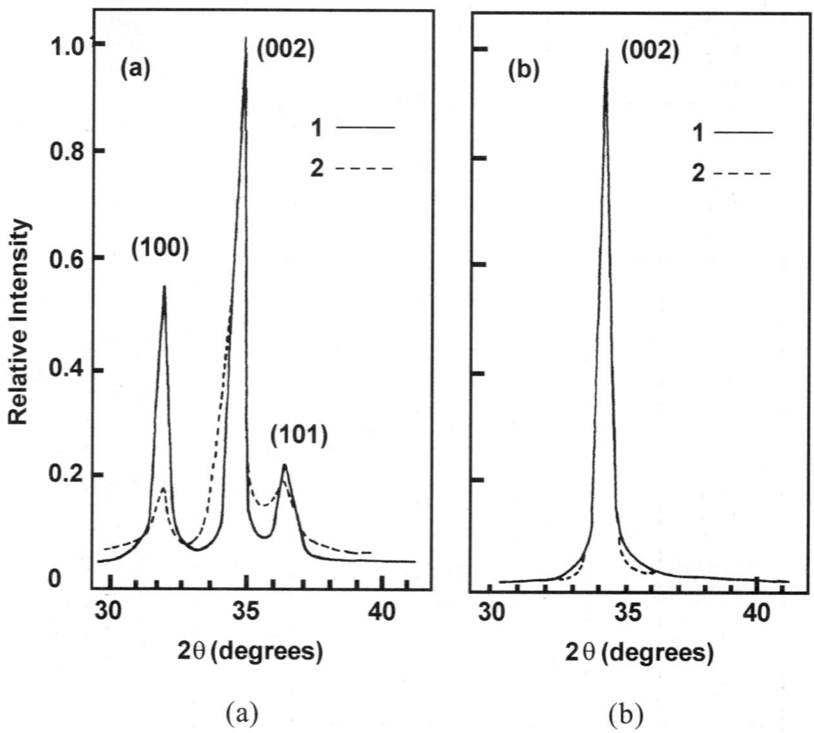

Figure 5.12. X-ray diffraction patterns from ZnO thin films on glass substrates prepared by (*a*) planar system and (*b*) by hemispherical system.

The degree of c-axis orientation of the film prepared at position "2" increases somewhat but (100) and (101) peaks still remain in the diffraction pattern. Such variation of c-axis orientation with the substrate position is also observed in dc-sputtered ZnO thin films.[141] In addition, it has been found that the degree of the c-axis orientation is strongly affected by a slight change of sputtering conditions. On the other hand, thin films prepared by the hemispherical system normally exhibit excellent c-axis orientation regardless of the substrate position as shown in Fig. 5.11b. From the x-ray rocking-curve analysis, the standard deviation of c-axis orientation is found to be less than 3 degrees for ZnO thin films prepared by the hemispherical system. In this system, the distribution of incident angles of the sputtered particles at the substrate surface is considered to be much narrower than in the planar system. This probably causes formation of a beam-like flow of sputtered particles, including Zn atoms, onto the substrate, resulting in the growth of highly oriented ZnO thin films. The reduction of highly energetic sputtered particles (i.e., O^- ions and/or O atoms from the ZnO cathode) is considered a geometrical effect.[142] The irradiation effects of the highly energetic O^- ions and/or O atoms are reduced during the reactive sputtering using the metal Zn cathode.

Since the hemispherical system shows uniformity in film thickness and crystal orientation, it is useful for the production of ZnO thin films. Figure 5.13 shows the construction of an rf sputtering system designed for production. In this system, diameters of the ZnO target and substrate holder (anode) are 70 mm and 220 mm, respectively. Sixty substrate wafers of 25 mm square can be loaded on the substrate holder. The holder is rotated for further improvement in uniformity of film thickness. The sputtering is controlled to flow uniformly at the whole target surface, so the variation in a single wafer $(h_0 - h_r)/h_0$ is simply given by the relationship

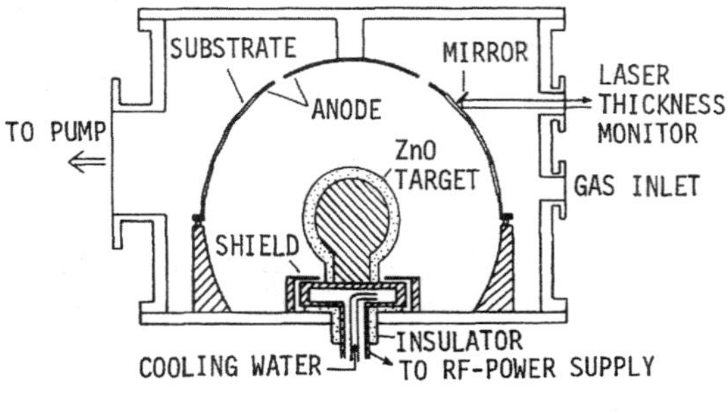

Figure 5.13. Schematic diagram of the hemispherical sputtering system for ZnO thin film deposition.

Eq. (5.1)
$$\frac{(h_0 - h_r)}{h_0} = \frac{3}{2}\left(\frac{r}{L}\right)^2$$

where h_0 is the thickness at the center of the wafer, h_r is the thickness at distance r from the center of the wafer, and L is the space between the wafer and the target center. Taking, $L = 11$ mm, and $r = 12.5$ mm, the maximum thickness variation in the wafer is ±1%. The thickness variation between different substrate wafers is less than 1%, which is governed by the geometrical accuracy of the electrode system. A laser interference device monitors the thickness of the ZnO thin films during deposition. A photograph of the sputtering chamber with the ZnO target is shown in Fig. 5.14.

Ion plating, cluster ion-beam deposition, and chemical vapor deposition are also used for the deposition of ZnO thin films. The difference in the growth mechanism of ZnO thin films in these deposition processes is not yet well understood. However, it is very interesting that the ion plating process shows nearly the same optimum condition as the sputtering process in growth rate and substrate temperature indicated in Fig. 5.9.[135] This suggests that the growth mechanism of ZnO thin films in ion plating resembles that of the sputtering process. Recent experiments

Figure 5.14. Photograph of the sputtering chamber of the hemispherical system with ZnO target.

suggest that the ECR (electron cyclotron resonance) plasma CVD process can also be applied to the deposition of c-axis-oriented ZnO thin films.[143]

Single crystal films. Two types of piezoelectric ZnO single-crystal films are epitaxially grown on sapphire single-crystal substrates by rf sputtering.[144] Epitaxial relations between these ZnO thin films and the sapphire (Al_2O_3) substrates are determined as follows:

Eq. (5.2)
$$(0001)ZnO//(0001)Al_2O_3, (11\bar{2}0)ZnO//(10\bar{1}0)Al_2O_3$$
$$(11\bar{2}0)ZnO//(01\bar{1}2)Al_2O_3, (0001)ZnO//(0\bar{1}11)Al_2O_3$$

A ZnO target is used for the sputter deposition of epitaxial films as well as the deposition of polycrystalline films. The key sputtering parameters for making epitaxial films are the deposition rate and the substrate temperature. Figure 5.15 shows typical variations of their crystal properties with the sputtering conditions. Epitaxial single-crystal films are grown at substrate temperatures of 400 to 600°C. Typical electron micrographs with electron diffraction patterns for these ZnO films are shown in Fig. 5.16. The epitaxial films are very smooth and no texture is observed. The lattice constant c_o of the film on (0001) sapphire is 5.210 Å and a_o of the film on

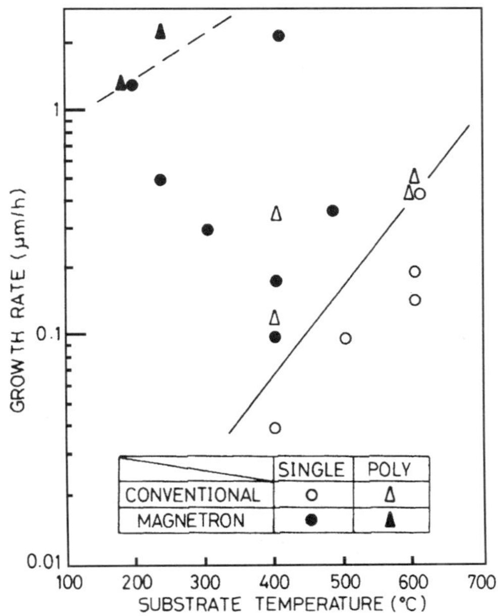

Figure 5.15. Crystalline structures of ZnO films sputtered at various conditions.

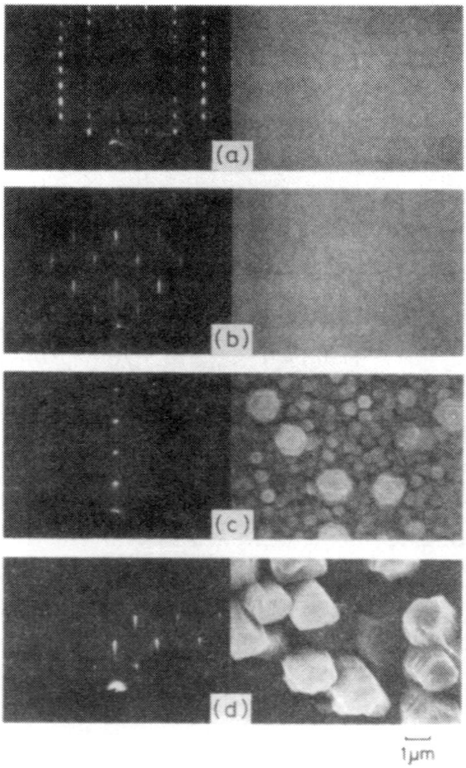

Figure 5.16. Electron micrographs and electron diffraction patterns of ZnO films on sapphire sputtered at a substrate temperature of 600°C with various deposition rates. The orientation of the sapphire substrates and deposition rates are (*a*) (0001) and 0.2 μm/h, (*b*) (01$\bar{1}$2) and 0.2 μm/h, (*c*) (0001) and 0.4 μm/h, and (*d*) (01$\bar{1}$2) and 0.44 μm/h.

(0112) sapphire is 3.264 Å. The c_0 of the film is almost equal to the standard (+0.065%), but a_0 is greater than the standard (+0.44%). This phenomenon can be explained by stress resulting from the difference in thermal expansion characteristics of films and substrates. The thermal expansion coefficients across the c-axis for ZnO and sapphire show almost the same values: 5.5 ppm/°C for ZnO and 5.42 ppm/°C for sapphire; therefore little stress is induced through cooling down after deposition for the (0001)/ZnO/(0001)Al_2O_3 structure, where the c-axis of both ZnO and sapphire are perpendicular to the surface. On the other hand, the thermal expansion coefficient along the c-axis for ZnO is considerably smaller than that of sapphire; ZnO shows 3.8 ppm/°C and sapphire shows 6.58 ppm/°C. In the (1120)ZnO/(0112)Al_3O_3 structure, the c-axis of ZnO is parallel to the surface, and that of sapphire is nearly parallel to the surface. This suggests that the compressive stress parallel to the surface is likely to be present in the film after deposition.

The electrical resistivity of these epitaxial ZnO thin films is as low as 10^2 to 10^3 Ωcm. Strong piezoelectric properties with high electrical resistivity are observed when the sputter deposition is conducted from Li-doped ZnO targets. The Li-doped ZnO targets are prepared by the addition of Li_2CO_3 before sintering ZnO to a level of 0.5 to 2%. As-sputtered ZnO thin films doped with Li show resistivities of 10^3 to 10^4 Ωcm. Resistivity increases by two orders in magnitude after postannealing in air at 600°C for thirty minutes. Figure 5.17 shows the variation of resistivity with the concentration of Li_2O_3 in the target.

The surface flatness of these epitaxial films strongly depends on the growth temperature. Lowering the growth temperature results in a smooth surface. Detailed studies of epitaxial growth of ZnO single-crystal films by sputtering suggest that rf-magnetron sputtering allows a decrease in the growth temperature and improves the surface smoothness of the epitaxial ZnO thin films. Typical sputtering conditions for ZnO single films are listed in Table 5.8. The growth temperature of ZnO single crystal films in magnetron sputtering is as low as 200°C.[145] The most popular technique for preparing single-crystal ZnO thin films is sputtering, although chemical vapor is also used.[146] In chemical-vapor deposition, oxidation

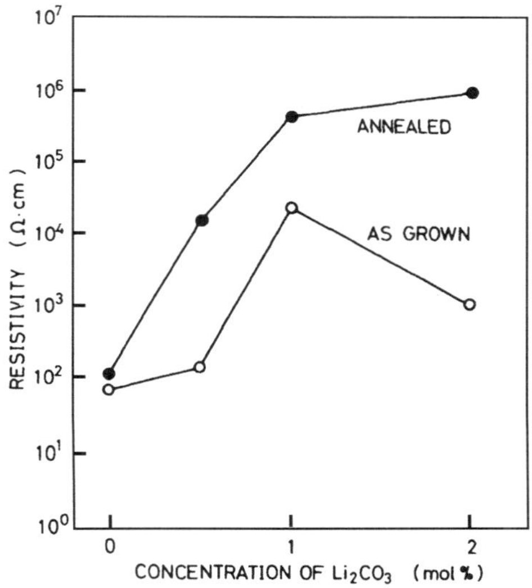

Figure 5.17. Variation of the resistivity of the sputtered ZnO thin films with concentration of Li_2CO_3 in the sputtering target.

Table 5.8. Typical Sputtering Conditions for the Preparation of ZnO Single Crystal Films

	RF-diode system	RF-magnetron system
Target dimension	30 mm ϕ	85 mm ϕ
Target-substrate spacing	25 mm	50 mm
Sputtering gas	Ar + O_2 (1:1)	Ar + O_2 (1:1)
Gas pressure	6.7 Pa	0.67–1.3 Pa
RF power	1–10 W	25–200 W
Substrate temperature	400–600°C	180–480°C
Growth rate	0.04–0.5 μm/h	0.1–2.2 μm/h

of Zn vapor or of zinc oxides or zinc halides is used for the reaction. The epitaxial temperature is around 650° to 850°C. The temperature is lowered in plasma-enhanced chemical depositions. The resistivity of the epitaxial films is then as low as 1 Ωcm. The crystal properties and their surface smoothness are improved by the deposition of a thin ZnO sputtered layer on the substrate as a buffer layer.[147]

5.1.1.2 Electrical Properties and Applications

Longitudinal/shear mode couplings. The piezoelectric properties of ZnO thin films are evaluated by measuring the frequency dependence of electrical admittance in a ZnO film transducer. The construction of the ZnO thin-film transducer and typical characteristics are shown in Fig. 5.18. The transducer is composed of the c-axis-oriented ZnO film of about 5-μm thickness sputtered onto the end of a fused-quartz rod (15 mm long and 5 mm in diameter), which is precoated with a thin Cr/Al base electrode. A counterelectrode of gold thin film (1 mm in diameter, 0.2 μm thick) is deposited in a vacuum onto the ZnO thin film.

The phase velocity of the bulk acoustic wave of the longitudinal-mode is calculated from the relationship $v_p = 2f_r d$, where f_r is a frequency at the resonance, and d is the ZnO film thickness. The longitudinal-mode electromechanical coupling factor k_t is calculated from admittance characteristics using the equation

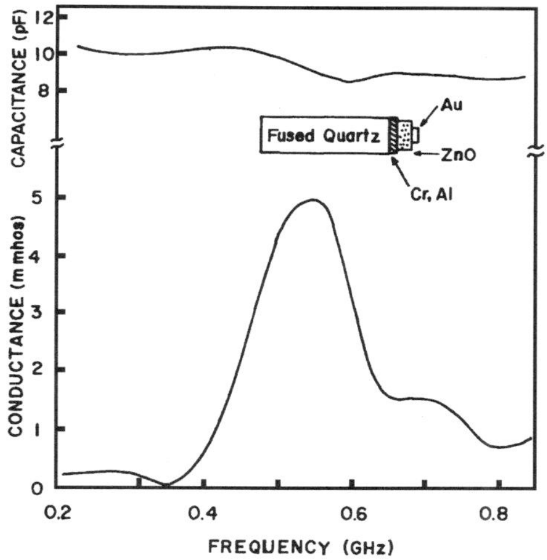

Figure 5.18. Frequency dependence of the admittance characteristics of a sputtered ZnO thin film transducer. ZnO film thickness, about 5 mm.

Eq. (5.3)
$$k_t = \frac{\pi G_A X_C Z_M}{4Z_T}$$

where G_A is the conductance above the background at antiresonance, and X_C is the transducer capacitive reactance, also at the antiresonant frequency.[148] The acoustic impedance of the fused quartz propagation medium and ZnO transducer material are, respectively, $Z_M = 1.58 \times 10^7$ kg/m^{2}sec, and $Z_T = 3.64 \times 10^7$ kg/m^{2}sec. The admittance is measured by using a network analyzer. The results show that the coupling factors, k_t, for thin films with c-axis orientation is $k_t = 0.23$ to 0.24. These coupling factors are 85% to 88% of the bulk single-crystal value. The relation in Eq. (5.3) neglects the effect of the metallic electrodes.

The effect of the electrodes should be included for precise evaluation of k_t.[149] The shear-mode coupling of ZnO thin films is evaluated by measuring the frequency dependence of electrical admittance in a ZnO thin-film transducer, which is composed of a ZnO thin film with parallel orientation.

Thin films of ZnO with parallel orientation are prepared by the addition of aluminum during sputtering.[150] Figure 5.19 shows a typical frequency response of this kind of transducer measured at unmatched and

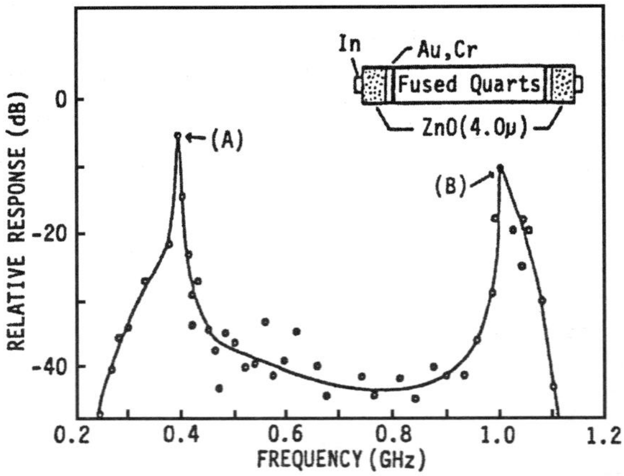

Figure 5.19. Frequency response of a two port, ZnO thin-film, acoustic element with parallel orientation.

untuned conditions. Two peaks, (A) and (B) in Fig. 5.19, are observed. Peak (A) is a resonance point, which corresponds to the excitation of shear-mode elastic waves. Peak (B) may be the third higher overtone of peak (A).

Piezoelectric ZnO thin films are used as bulk acoustic wave (BAW) resonators similar to quartz crystals. A typical structure is shown in Fig. 5.20.[151] The basic structure of the ZnO BAW resonator comprises a sandwich structure: top-electrode/ZnO/base-electrode. An air gap is inserted between the piezoelectric thin films and the substrates to reduce the acoustic load of the substrates as shown in Fig. 5.20a. The fundamental resonant frequency, f_0, is expressed by $f_0 = v/2d$ (half-wavelength mode), where v denotes the longitudinal sound velocity in ZnO thin films and d is the ZnO film thickness. Taking $d = 1$ μm, and $v = 6000$ m/s, f_0 becomes 3 GHz. In the quarter-wavelength mode, the resonant frequency becomes 6 GHz for $d = 1$ μm. These thin-film resonators are promising for use in a very high frequency range of 0.5 to about 10 GHz. Several designs of the resonator are proposed to make solidly mounted resonators without the air gap, as shown in Fig. 5.20b. Figure 5.21 shows the solidly mounted ZnO thin-film resonators.[152] The resonator has an acoustic reflector of multilayers of ZnO and SiO_2 between piezoelectric ZnO films and the substrates. The multilayers reduce the acoustic load of substrates to the piezoelectric ZnO films. Electrical-mechanical coupling of ZnO depends on the crystal axis. The a-axis-oriented, ZnO thin films, epitaxially grown on R-sapphire are used for the fabrication of shear-mode BAW resonators for higher coupling devices.

Another interesting application of ZnO thin-film transducers is an ultrasonic microscope in which the seminal work has been done by Quate.[153] Figure 5.22 shows the construction of the ultrasonic microscope developed by Chubachi.[154] The thin-film transducer easily generates an ultrasonic beam onto the small test sample in a liquid medium.

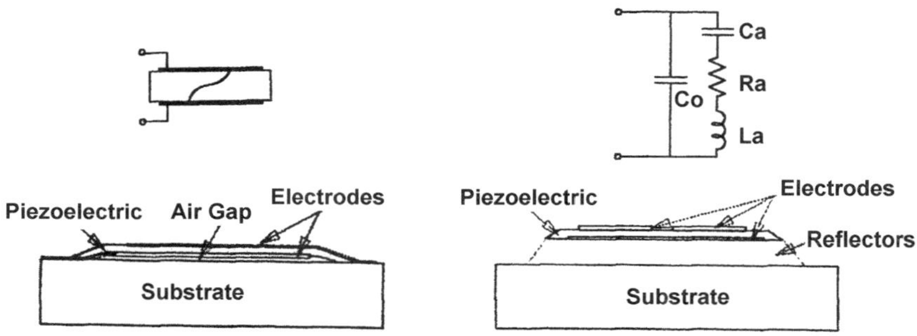

(a) (b)

Figure 5.20. Typical constructions of thin film BAW resonators: (*a*) thin film resonators with air gap, (*b*) solidly mounted resonators with reflectors.[152]

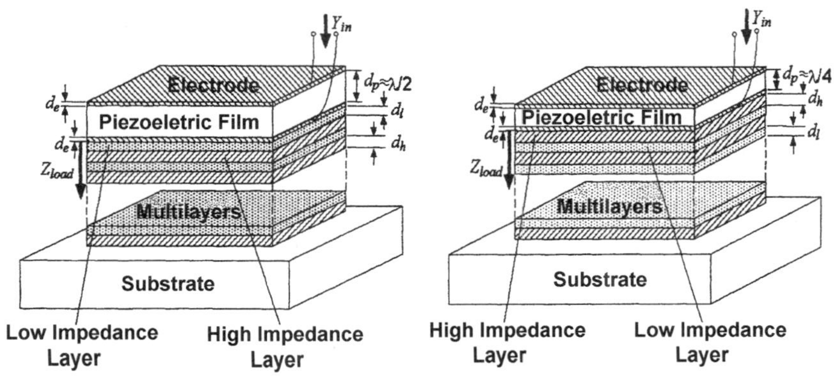

(a) (b)

Figure 5.21. Two configurations of solidly mounted thin film BAW resonators: (*a*) λ/2 mode, (*b*) λ/4 mode.[153]

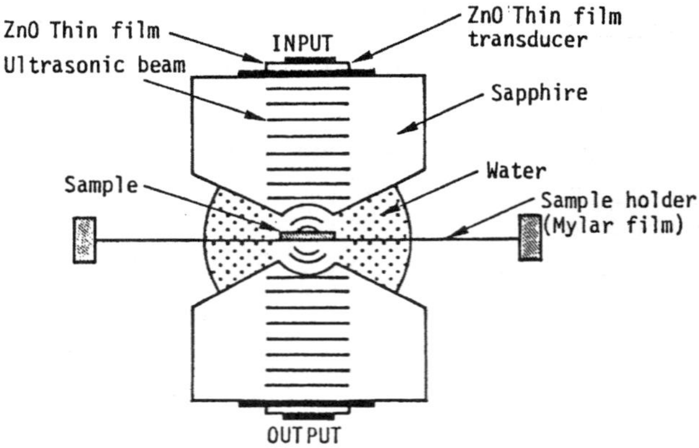

Figure 5.22. Construction of thin film ultrasonic microscope.[154]

Surface acoustic wave (SAW) properties. Many workers have investigated the surface acoustic wave (SAW) properties of ZnO thin films, since they are promising materials for making SAW devices for consumer electronics, communication systems, data processing systems, and acousto-optic devices.[155] Thin-film SAW devices are essentially composed of a layered structure: substrates over-coated by thin piezoelectric ZnO films. Surface acoustic wave properties, which include phase velocity, electromechanical coupling, and propagation loss, are governed by the thickness of the ZnO films, the wavelength of the SAW, and the materials' constants of the ZnO and the substrates.

As shown in Fig. 5.23, there are four types of electrode configurations for the excitation of SAWs in a layered structure in which c-axis-oriented ZnO films are deposited on a glass substrate, and epitaxial single-crystal ZnO thin films are grown on a single-crystal substrate.

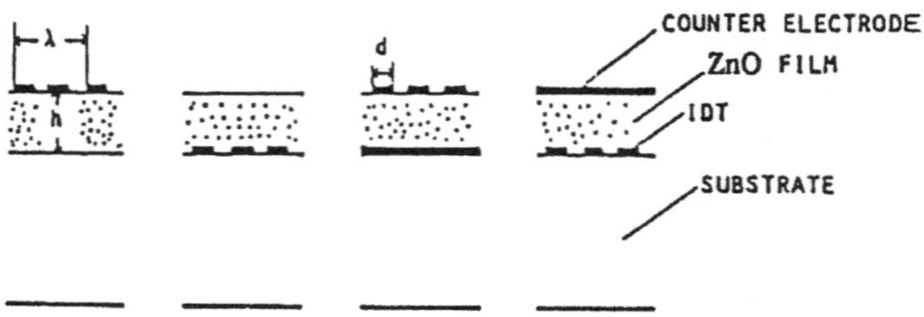

Figure 5.23. Four different configurations of interdigital transducers for the excitation of surface acoustical waves.

Figure 5.24 shows the calculated values of the phase velocity and electromechanical coupling, $k^2 = 2\Delta v/v$ for a layered structure, where v is the vave velocity with an electroded surface and Δv is the change in velocity produced by removing the electrode. The effective coupling, k^2, varies with the ZnO film thickness to wavelength ratio (h/λ). The variations show a double-peaked character for the ZnO/glass structure where the first peak is at $h/\lambda \cong 0.02$–0.03 and the second peak at $h/\lambda \cong 0.5$.[156]

The phase velocity of the SAWs propagating on the layered structure mainly depends on the acoustic properties of the substrate. For a ZnO/glass structure, the phase velocity is approximately 3000 m/s at the first peak, $h/\lambda \cong 0.02$–0.03. Under a high h/λ value ($h/\lambda \cong 0.4$), the phase velocity is around 2700 m/s, which is close to that of bulk ZnO. Note that high values of k^2 with high phase velocity are achieved for ZnO single-crystal films on R-plane sapphire as shown in Fig. 5.25. The maximum value of k^2 is around 5%, which is much higher than bulk ZnO and $LiNbO_3$.

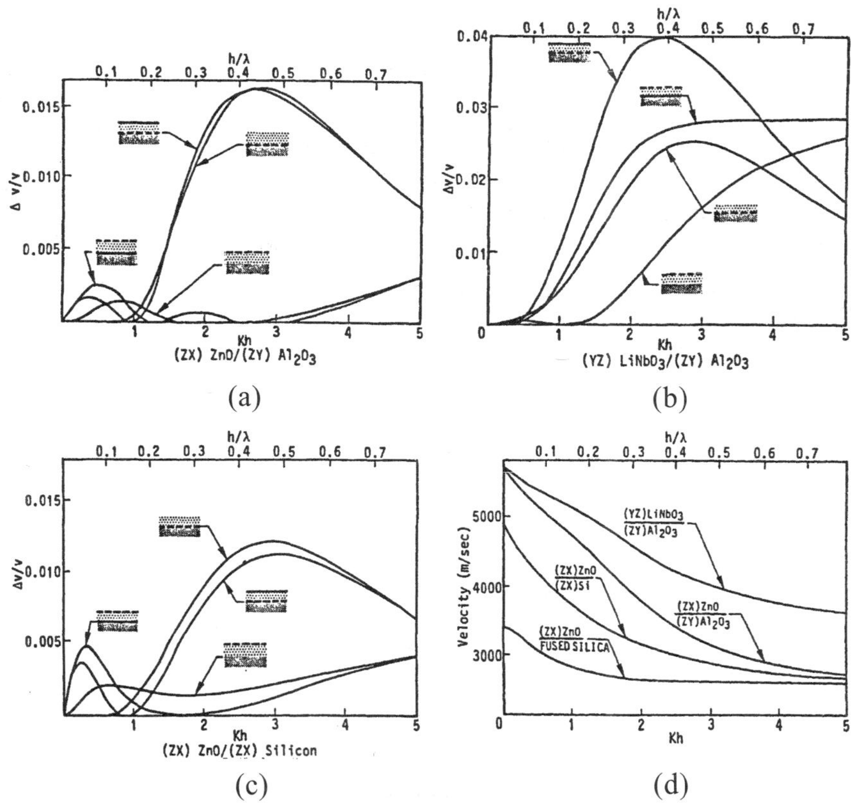

Figure 5.24. SAW properties for (*a*) a (ZX) ZnO film on a (ZY) Al_2O_3 substrate; (*b*) a (YX) $LiNbO_3$ film on a (ZY) Al_2O_3 substrate; and (*c*) a (ZX) ZnO film on a (ZX) silicon substrate. (*d*) The phase velocities are shown for four combinations of film and substrate.

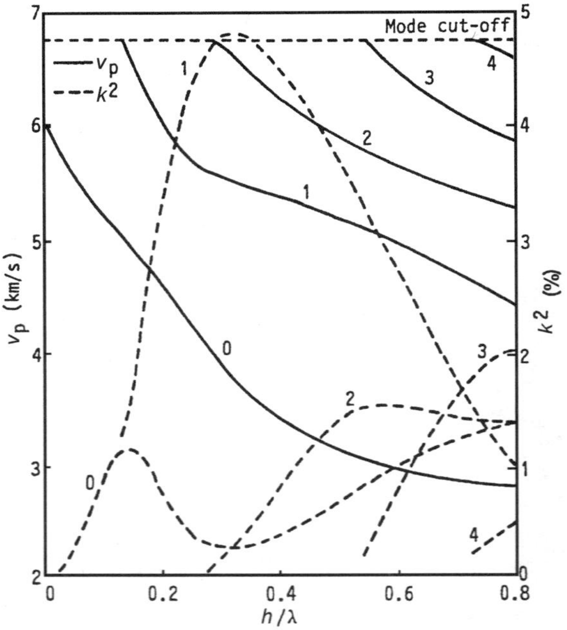

Figure 5.25. Calculated values of V_p and k^2 vs h/λ for $(11\bar{2}0)$-ZnO/$(01\bar{1}2)$ Al_2O_3 structure.

The thin-film SAW devices are integrated in Si devices. Figure 5.26 shows typical theoretical data for ZnO/SiO_2/Si structure. The coupling factor depends on the conductivity of the Si, Fig. 5.26a. The temperature variations of the phase velocity are controlled by the thickness of the SiO_2 layer as expected, Fig. 5.26b.[157]

For ZnO SAW devices on glass substrates, the temperature coefficient of phase velocity, and/or the delay time, depends on the property of the glass substrate. Figure 5.27 shows the temperature coefficient of delay time for various substrates. It shows that a fused quartz substrate gives a small temperature coefficient at high h/λ, while borosilicate glass gives a small temperature coefficient at small h/λ.

The propagation loss for a SAW on a ZnO/glass structure was measured to be 4 dB/cm at 98 MHz for $h/\lambda = 0.03$. This value is rather large in comparison with single-crystal materials, but considerably smaller than that of ceramic materials. Considering the loss value, polycrystalline ZnO films are usable for SAW devices operating at frequencies of up to several hundred MHz. Low loss value has been achieved for epitaxial ZnO single-crystal films.

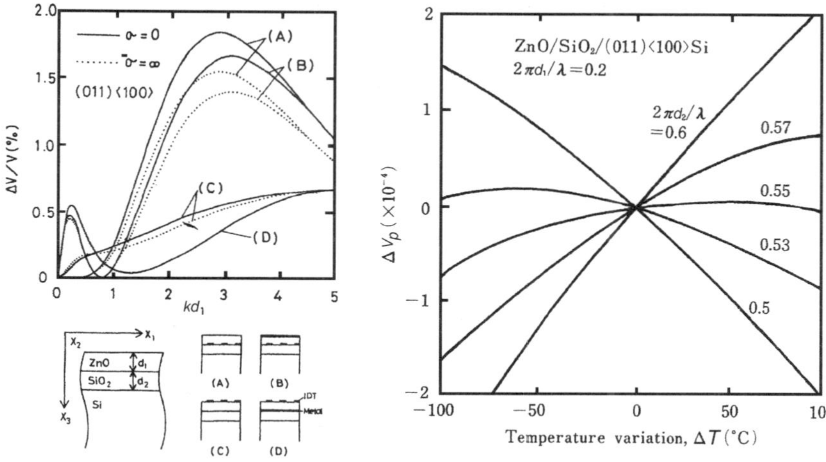

(a) (b)

Figure 5.26. SAW properties of $ZnO/SiO_2/Si$ (011)⟨100⟩ layered system at $kd_1 = 0.2$. (*a*) The coupling factor depends on the conductivity of the Si. (*b*) Temperature variations of the phase velocity are controlled by the thickness of the SiO_2 layer

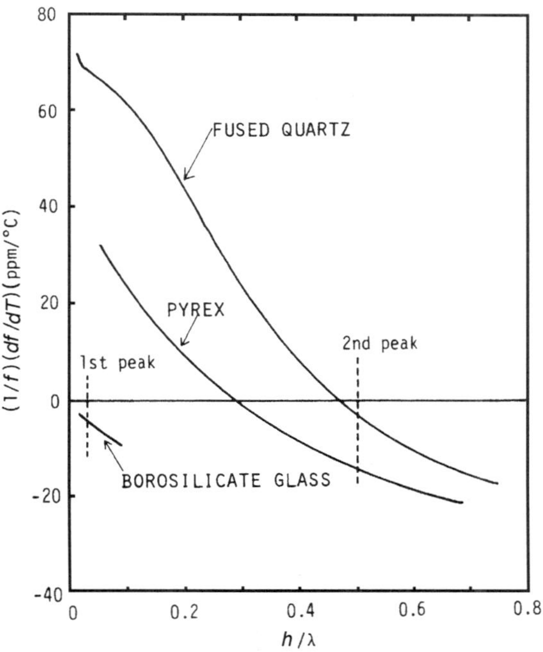

Figure 5.27. Temperature coefficient of delay time in the c-axis-oriented ZnO films for various glass substrates.

Table 5.9 summarizes the physical properties of sputtered ZnO thin films. In ZnO thin-film SAW devices, the SAW coupling factors, phase velocity, and temperature variations of the phase velocity are controlled by the selection of substrate materials. Figure 5.28 summarizes ZnO SAW properties for different substrates. The layered structure shows a thin-film effect. Enhanced coupling is expected in $ZnO/LiNbO_3$/sapphire structures. Zero temperature coefficients with high coupling are obtained in $ZnO/SiO_2/LiNbO_3$ structures.[158]

Table 5.9. Properties of ZnO Thin Films

Deposition Method* (Source)	Substrate	Deposition Temperature (°C)	Structure**	Film Properties***	See Ref. 299
DC-SP (ZnO)	sapphire	75	PC	$k_t \sim 0.25$ (0.4–1.8 GHz)	a
RF-MSP (ZnO)	sapphire	150–300	SC	$V_{SAW} \sim 5300$ m/s, Sezawa	b
RF-SP (ZnO)	FQ		PC	$k^2_{SAW} \sim 0.0025$ ($hk = 1.1$)	c
RF-SP (ZnO)	sapphire	200	PC	$k_t = \sim 0.2$–0.25	d
DC-TSP (ZnO)	FQ	175	PC	$k^2_{SAW} \sim 0.024$	e
DC-TSP (ZnO)	FQ/SiC	350	PC	$V_t \sim 12500$, $V_{SAW} \sim 7250$ m/s	f
RF-MSP (ZnO)	glass	~ 100–200	PC	$k_t = \sim 0.2$–0.24 $V_t = \sim 5900$–6900 m/s $\varepsilon^* = \sim 8$–9	g
DC-SP (Zn, Al)	glass	~ 100–200	PC	$k_s = \sim 0.05$–0.08 $V_s = \sim 2600$–3000 m/s	h
DC-SP (ZnO)	FQ	~ 180–260	PC	$k_t = \sim 0.18$–0.25	i
CVD	sapphire	~ 700–750	PC	$k^2_{SAW} \sim 0.0139$	j
RF-MSP (ZnO)	SiO_2	~ 250–320	PC		k
RF-SP (ZnO)	sapphire	600	SC	~ 2.4×10^3 Ωcm (0001) sapphire $\mu_H \sim 2.6$–28 cm^2/volt·sec	l
RF-MSP (ZnO, Li)	sapphire	400	SC	$V_{SAW} \sim 5160$ m/s $k^2_{SAW} \sim 0.035$	m
ECR (Zn)	FQ	~ 200–500	PC	$k^2_{SAW} \sim 0.05$	b

* SP, diode sputtering; TSP, triode sputtering; HSP, hemispherical sputtering; MSP, magnetron sputtering; IP ion plating, ECR, electron cyclotron resonance sputtering.

** PC, polycrystalline; SC, single crystal.

*** k_t, longitudinal mode coupling; k_s, shear mode coupling; k^2_{SAW}, SAW effectivecoupling ($=2\Delta V/V$); V_t, longitudinal mode phase velocity; V_s, shear mode phase velocity; V_{SAW}, SAW phase velocity.

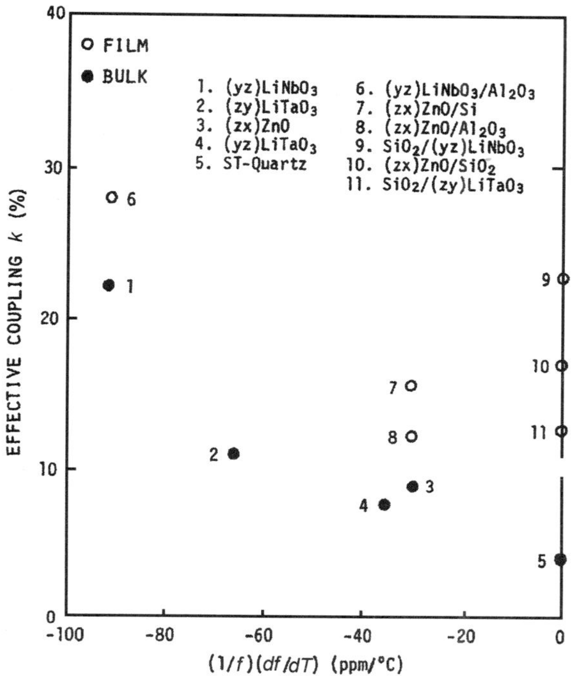

Figure 5.28. SAW properties for various layered structures.

Surface acoustic wave (SAW) devices. Thin-film SAW devices are one of the most interesting thin-film electronic components.[159] The devices include band-pass filters, resonators, voltage-controlled oscillators, and convolvers in frequencies of 10 MHz to the gigahertz range. The ZnO thin-film SAW video intermediate frequency (VIF) filters for color TV sets and video tape recorders (VTR) are now widely in production.[158][160]

Figure 5.29 shows a typical construction and band-pass characteristics of ZnO thin-film SAW VIF filters designed for the NTSC (National Television System Committee) American band. The ZnO thin films are composed of a polycrystalline structure with a c-axis orientation. They are deposited on a borosilicate glass substrate using hemispherical rf sputtering. The interdigital electrodes are made of evaporated Al thin films. The thickness of the ZnO is 1.5 μm corresponding to 3% of the SAW wavelength, which induces strong electromechanical coupling in the layered structure. These ZnO thin-film SAW filters reduce the number of electronic components in color TV sets. The layered structure, ZnO on borosilicate glass, strongly improves the temperature stability of band-pass properties due to the suitable selection of borosilicate glass composition.

Single-crystal ZnO thin films, epitaxially grown on sapphire, are used for making the ZnO thin-film SAW devices for the UHF region since they exhibit a small propagation loss. Such UHF SAW devices are used for various communication systems. The layered structure of ZnO/sapphire yields high electromechanical coupling with high SAW phase-velocity. The high phase-velocity allows the high-frequency operation of the SAW devices. Suitable design of a SAW filter makes it possible to operate at a frequency of 4.4 GHz with an interdigital transducer (IDT) finger width of 0.5 μm. The frequency response is shown in Fig. 5.30.

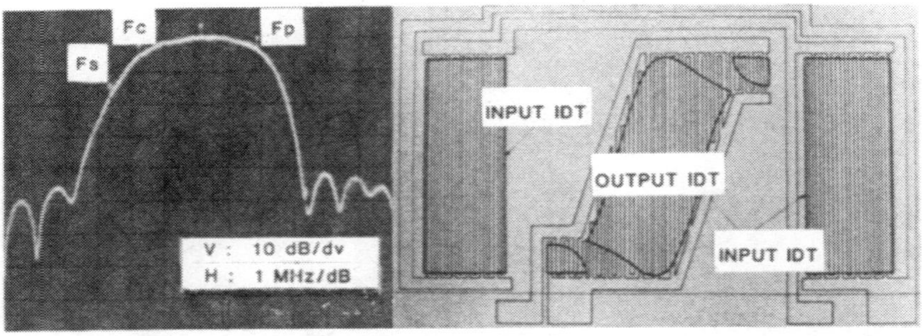

Figure 5.29. Typical configuration and frequency response of ZnO thin-film SAW VIF filters for TV sets. $F_p = 45.75$ MHz.

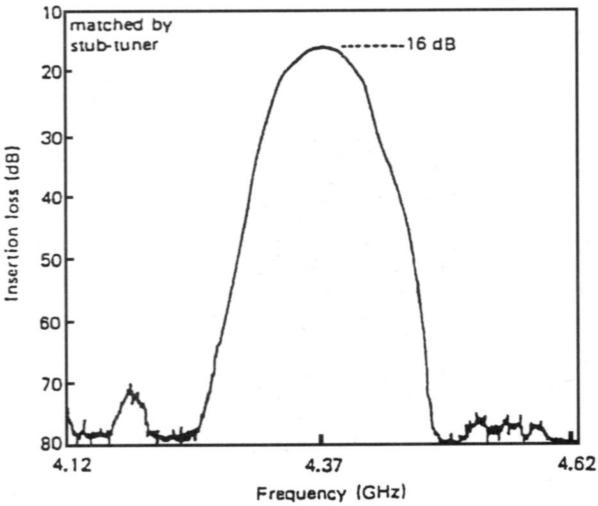

Figure 5.30. Frequency response of a SAW filter using single-crystal ZnO thin film on sapphire.

Single-crystal ZnO thin films are also excellent waveguide materials for planar guided light, and are useful for making an acousto-optic Bragg diffractor as shown in Fig. 5.31a.[161] High-frequency operation of SAWs causes a high diffractive angle. Since the Bragg angle corresponds to the acoustic frequency, this diffractor works as a spectrum analyzer. Figure 5.31b shows the typical operation of the diffractor. Each spot is diffracted light by the acoustic wave of the respective frequency.

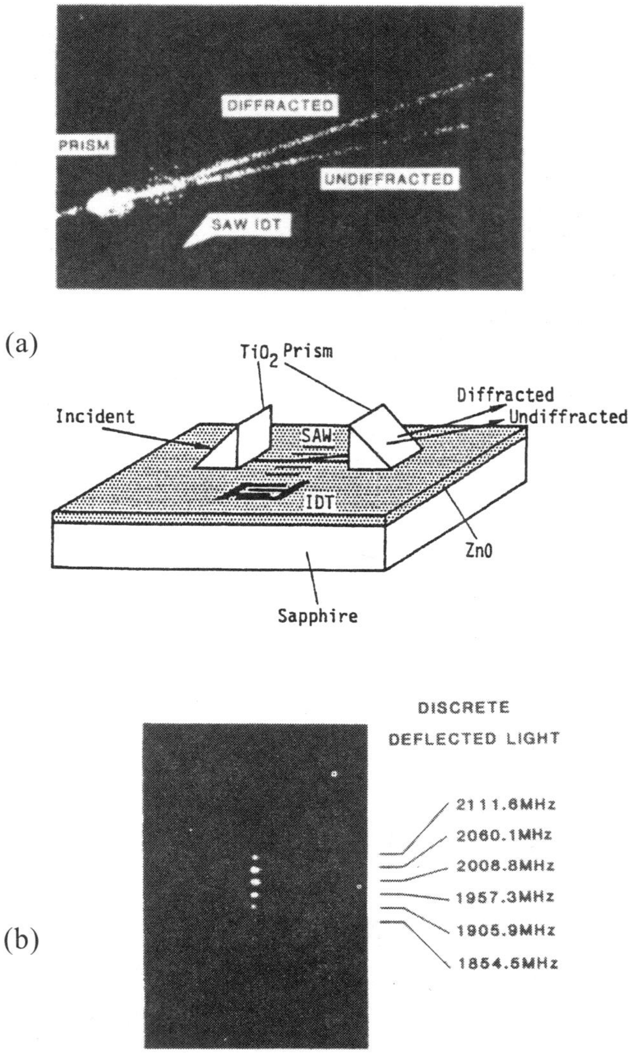

Figure 5.31. (*a*) Schematic illustration and a typical experiment of an A-O deflector using ZnO thin film on sapphire structure. (*b*) Deflected light beam positions as a function of input frequency.

The semiconductive properties of ZnO thin films are used for making ZnO/Si heterojunction photo detectors,[162] ZnO/BiO thin-film varistors,[163] and ZnO/Pt gas sensors with Schottky structures. Recent work suggests that magnetron sputtering can produce transparent ZnO thin films of high-conducting properties. The p-type ZnO, produced by nitrogen doping, shows the possibility of producing a ZnO blue laser.[164] Heavy doping of Li into ZnO makes ferroelectric ZnO.[165] The ZnO thin films are promising materials for future applications.

5.1.2 Sillenite Thin Films

Crystals of the γ-Bi_2O_3 family, called sillenites, are attractive materials with strong electro-optical effects, acousto-optical effects, and piezoelectricity.[166] The sillenites are generally composed of Bi_2O_3 and foreign oxides such as GeO_2, SiO_2, and PbO; for example, $Bi_{12}GeO_{20}$ (BGO), and $Bi_{12}TiO_{20}$ (BTO) are known as sillenites. The crystal properties for BGO are shown in Table 5.10.

Thin films of sillenites are deposited by rf sputtering from a compound target of sillenites. Bismuth shows the highest vapor pressure in the composition and will exceed 10^{-5} torr when the substrate temperature is higher than 450°C. This suggests that stoichiometry will be achieved by sputtering from a target of stoichiometric composition at a substrate temperature below 450°C.

Table 5.10. Physical Properties of Single-Crystal $Bi_{12}GeO_{20}$ are Typical of Sillenites

Crystal system	23(bcc)
Space group	I23
Lattice constant	a = 10.1455 Å
Melting point	935°C
Dielectric constant	$\varepsilon^{s}_{11} = 34.2 \times 10^{-11}$ F/m
Density	9.2 g/cm^3
Refractive index	2.5476
Optical transparency	0.45–7.5 μm
Piezoelectric constant	$\varepsilon_{14} = 0.99$ C/m^2
Acoustic velocity	3.33×10^3 m/s (⟨111⟩ longitudinal wave)

5.1.2.1 Amorphous/Polycrystalline Films

RF sputtering from a sillenite target makes thin films of sillenites with an amorphous and/or polycrystalline structure. Figure 5.32 shows the construction of the sputtering system. Typical sputtering conditions are shown in Table 5.11.[13] The target is made of a ceramic disk with stoichiometric composition sintered at 800°C for four hours in air.

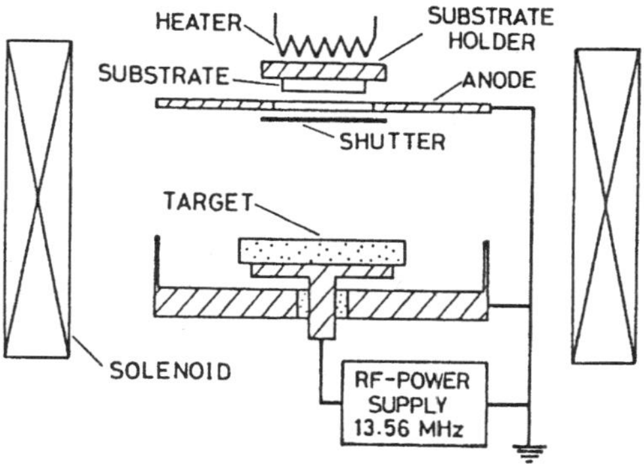

Figure 5.32. Electrode configuration of the sputtering system for the deposition of $Bi_{12}GeO_{20}$ thin films.

Table 5.11. Sputtering Conditions for Amorphous Polycrystalline Sillenite Thin Films

Target dimension	30 mm diameter
Target-substrate spacing	25 mm
Sputtering gas	Ar + O_2 (1:1)
Gas pressure	5×10^{-2} torr
RF power density	≈ 1 W/cm^2
Magnetic field	100 G
Substrate temperature	350–550°C
Growth rate	≈ 0.5 μm/h
Film thickness	1.5–4 μm

The crystallographic structure of sputtered films strongly depends on the substrate temperature during deposition. In the case of BGO films, sputtered films show an amorphous phase at substrate temperatures below 150°C. At substrate temperatures of 150° to 350°C, the sputtered films show a polycrystalline form of a metastable δ phase. The δ phase shows a face-centered cubic structure, which does not exhibit piezoelectricity. At substrate temperatures above 400°C, the γ phase appears. Between 350° to 400°C, mixed phases of δ and γ appear. Typical reflection electron patterns and electron micrographs of sputtered BGO films are shown in Fig. 5.33.

The crystallographic properties of other sputtered sillenite films are generally similar to those of BGO films. Sputtered BLO films, however, show preferential orientation as seen in Fig. 5.34. The direction of the preferred orientation is the (310) axis, perpendicular to the substrate.

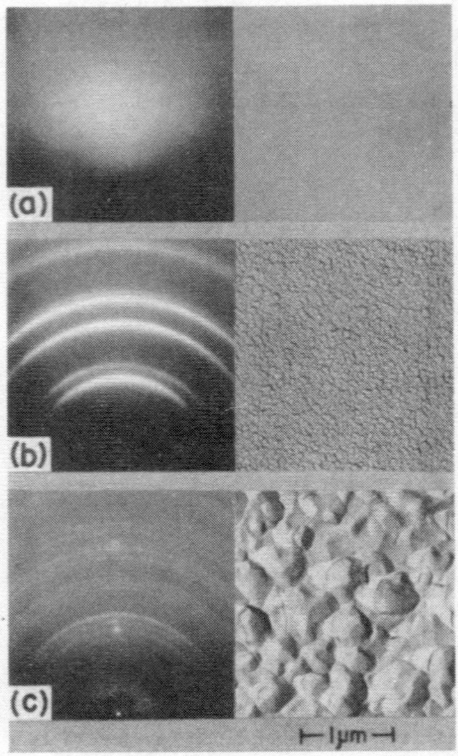

Figure 5.33. Typical reflection electron diffraction patterns and electron micrographs of $Bi_{12}GeO_{20}$ films sputtered onto glass substrates. (*a*) Film sputtered at about 100°C with growth rate of 0.25 μm/h, amorphous state; (*b*) film sputtered at 200°C, 0.20 μm/h, fcc form; (*c*) film sputtered at 400°C, 0.15 μm/h, bcc form.

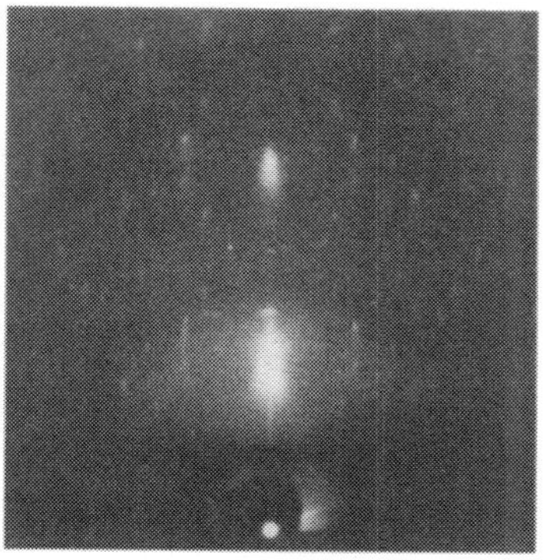

Figure 5.34. Typical electron diffraction pattern of $Bi_{12}PbO_{19}$ sputtered onto a glass substrate.

These sputtered films are semitransparent with a yellow to light brown color. Figure 5.35a shows typical optical absorption spectra for sputtered BGO films.[13] The optical absorption edge for the amorphous films is about 510 μm, the δ phase is 620 μm, and the γ phase is 620 μm. The absorption edge for the sputtered phase films is nearly equal to the bulk single-crystal value of 450 μm.[16] In the infrared region, a broad absorption is observed at 20 μm as shown in Fig. 5.35b. For sputtered phase films, several weak absorptions are superposed on the broad absorption spectrum, which relates to the characterized lattice vibration of γ-phase sillenites.

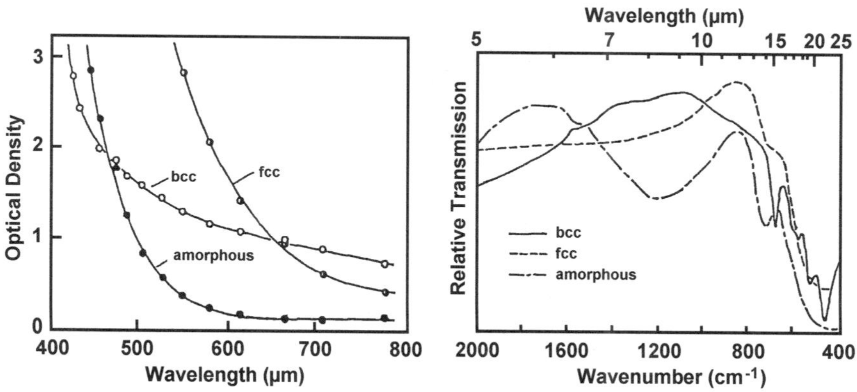

(a) (b)

Figure 5.35. (*a*) Optical absorption spectra in the visible region for three types of sputtered $Bi_{12}GeO_{20}$ films. (*b*) Infrared transmission spectra for three types of sputtered $Bi_{12}GeO_{20}$ films.

5.1.2.2 Single-Crystal Films

Single-crystal films of γ-Bi_2O_3 compounds are epitaxially grown on single-crystal substrates by rf-diode sputtering.[15] One example is the BTO thin films on BGO single-crystal substrates. Table 5.12 shows some crystal properties of BTO and BGO. The structure BTO/BGO provides a waveguide for optical integrated circuits since the refractive index of BTO is greater than that of BGO. The small lattice mismatch (≈0.3%) between BTO and BGO leads to the growth of excellent quality, single-crystal BTO on BGO substrates. These substrates are sliced from a single crystal, parallel to the (110) plane, and then polished by 0.3-μm alumina powder. The epitaxial growth is carried out by the rf-diode sputtering system. Since the epitaxial temperature of γ-Bi_2O_3 compound thin films is in the range of 400° to 500°C, the reevaporation of Bi from the sputtered films must be considered stoichiometric. Table 5.13 shows the Bi/Ti atomic ratios, determined by electron probe microanalysis (EPMA), for films sputtered at substrate temperatures of 400° and 500°C from three different targets. It shows that the compositions of the films sputtered at 400°C are almost equal to that of the respective targets. Films sputtered at 500°C show a remarkable decrease of the Bi/Ti atomic ratio due to the reevaporation of Bi. Their crystalline structures for various substrate temperatures are summarized in Fig. 5.36. Single-crystalline γ-phase films are epitaxially grown from the stoichiometric target at 400° to 450°C. From the Bi-rich target ($9Bi_2O_3 \cdot TiO_2$), γ-phase films are obtained at temperatures above 450°C. From the Ti-rich target ($4Bi_2O_3 \cdot TiO_2$), mixed phases of the γ phase and $Bi_4Ti_3O_4$ appear. Similar mixed phases are also observed in the sputtered films from stoichiometric targets at substrate temperatures above 500°C.

Single-crystal films obtained from stoichiometric targets are transparent with a smooth surface, as shown in Fig. 5.37. The films act as waveguides. Figure 5.38 shows a guided beam of He-Ne laser light (6328 Å), which is fed into the film by a rutile prism. The film thickness is 2.4 μm and the waveguide mode is TE_o.

Table 5.12. Properties of $Bi_{12}TiO_{20}$ and $Bi_{12}GeO_{20}$

	Crystal Structure	Lattice Constant (Å)	Melting Point (°C)	Refractive Index
$Bi_{12}TiO_{20}$	bcc, space group I23	10.176	~930	2.5619
$Bi_{12}GeO_{20}$	bcc, space group I23	10.1455	~935	2.5476

Table 5.13. Atomic Ratios of Bi/Ti Sputtered Films Determined by EPMA

Substrate Temperature (°C)	Target Composition		
	$9Bi_2O_3 \cdot TiO_2$ (Bi/Ti = 18)	$6Bi_2O_3 \cdot TiO_2$ (Bi/Ti = 12)	$4Bi_2O_3 \cdot TiO_2$ (Bi/Ti = 8)
400	17.3	12.4	7.1
500	13.3	6.2	5.3

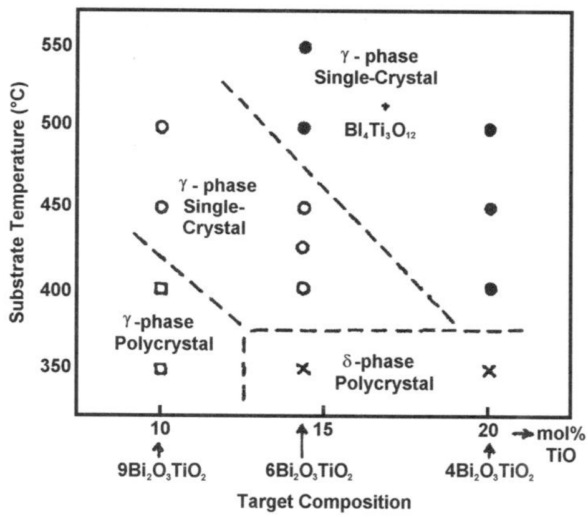

Figure 5.36. Crystalline structures of sputtered films.

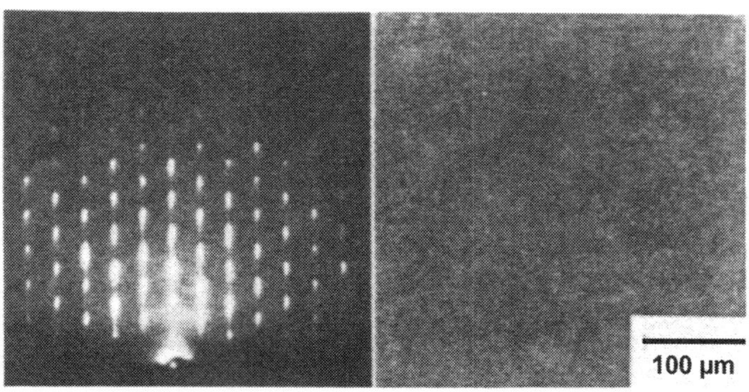

Figure 5.37. Typical electron diffraction pattern and photomicrograph of epitaxially grown $Bi_{12}TiO_{20}$ onto $Bi_{12}GeO_{20}$ substrate.

Figure 5.38. Guided beam of He-Ne laser light on the $Bi_{12}TiO_{20}/Bi_{12}GeO_{20}$ structure.

5.1.3 Perovskite Dielectric Thin Films

The perovskite structure observed in ABO_3-type compounds such as $BaTiO_3$ has ferroelectricity similar to the γ-Bi_2O_3 family.[167] Figure 5.39 shows a typical perovskite crystal structure. Thin films of the perovskites, including $BaTiO_3$, $PbTiO_3$, PZT ($PbTiO_3$-$PbZrO_3$), and PLZT [(Pb, La)(Zr, Ti)O_3], have been studied in relation to making thin films of dielectrics, pyroelectrics, piezoelectrics, and electro-optic materials.

Thin films of perovskites with a polycrystalline or single-crystal structure are deposited by rf sputtering from a sintered ceramic target. Substrates for deposition of polycrystalline films are Pt or fused quartz. Single crystals of MgO, $SrTiO_3$, sapphire, and spinel are used for epitaxial growth of single-crystal perovskite films. The growth temperature of the perovskite structure is 600° to 700°C; thus, the substrate temperature must be higher than 600°C during the deposition. However, sputtering is sometimes done at a low substrate temperature of 300°C and the resultant film is postannealed in air at 600° to 750°C to form the polycrystalline structure of perovskite.

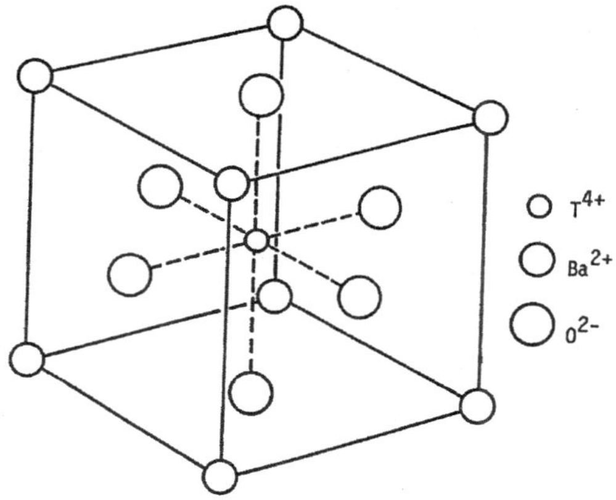

Figure 5.39. Crystal structure of perovskite ($BaTiO_3$).

5.1.3.1 $PbTiO_3$ Thin Films

Deposition. Thin films of perovskite materials including $PbTiO_3$, PZT, and PLZT have been prepared by sputtering and/or by chemical vapor deposition.

Bickley and Campbell have previously deposited mixed films of PbO and TiO_2 by reactive sputtering from a composite lead titanium cathode in an oxidizing atmosphere. They used a conventional dc-diode sputtering system, and the mean permittivity of the resultant films was 33 when chemical composition was $PbTiO_3$. This value, however, is much smaller than that of a true $PbTiO_3$ compound. The as-grown films gave little information about the formation of $PbTiO_3$. In order to synthesize $PbTiO_3$, the substrate temperature must normally be more than 600°C.

The structure and dielectric properties of these sputtered films may vary with the sputtering system. The growth of $PbTiO_3$ will be achieved even at low substrate temperatures of 200°C or less by magnetron sputtering deposition under low working pressure.[168] Typical sputtering conditions are listed in Table 5.14.

As shown in Table 5.14, a composite lead-titanium cathode was used for deposition. During sputtering, PbO and TiO_2 are codeposited onto the substrate, and mixed films of PbO and TiO_2 are fabricated. The chemical composition of the sputtered films is controlled by the ratio of

Table 5.14. Typical Sputtering Conditions for the Deposition of $PbTiO_3$ Thin Films by a dc Magnetron System

Sputtering system	dc magnetron
Target	Pb-Ti composite
Sputter gas	$Ar/O_2 = 1$ (6×10^{-4}torr)
Substrate	glass
Substrate temperature	150–300°C
Deposition rate	30–600 Å/min

Pb/Ti in the composite lead-titanium cathode. Figure 5.40 shows the dielectric properties of the mixed films for various chemical compositions. It shows that the permittivity maximum observed in the chemical composition of $PbTiO_3$ is higher than the permittivity of PbO or TiO_2. The temperature variation of the permittivity also shows the maximum at about 490°C, as indicated in Fig. 5.41, which is the value expected for $PbTiO_3$. These electrical properties suggest that $PbTiO_3$ is synthesized in mixed films prepared by magnetron sputtering.[169]

The PbO and TiO_2 phases were detected by electron diffraction analysis of the mixed films. The remainder is amorphous $PbTiO_3$. The population of $PbTiO_3$ is estimated from

Eq. (5.4)
$$X_{PT} = \frac{\log\left(\dfrac{\varepsilon_M}{X_{TiO_2}\,X_{PbO}}\right)}{\log\left(\dfrac{\varepsilon_{PT}}{X_{TiO_2}\,X_{PbO}}\right)}$$

if we assume that Lichteneker's empirical logarithmic mixing rule is established between ε_{PbO}, ε_{TiO_2}, and ε_{PT}, where these values are the permittivities of PbO, TiO_2, $PbTiO_3$ and the sputtered mixed films, respectively, and X_{PbO}, X_{TiO_2}, and X_{PT} are the proportions by volume of PbO, TiO_2, and $PbTiO_3$, respectively, so that $X_{PbO} + X_{TiO_2} + X_{PT} = 1$. Putting $\varepsilon_M \approx 120$, $\varepsilon_{PbO} \approx 25$, and $\varepsilon_{TiO_2} \approx 60$, we have $X_{PT} \approx 0.7$. This estimate suggests that 70% of the sputtered film is $PbTiO_3$.

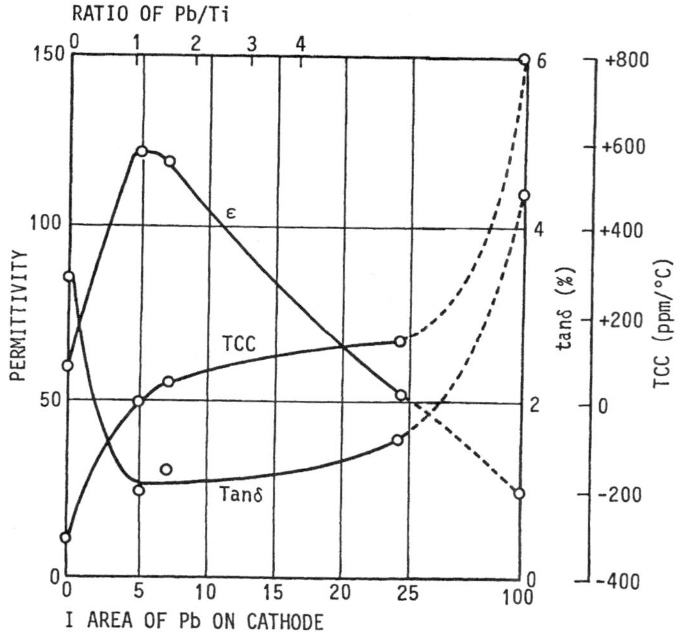

Figure 5.40. The dielectric properties of PbTiO films 3000 Å thick on 7059 glass deposited by a magnetron sputtering system (measured at 1 MHz and room temperature).

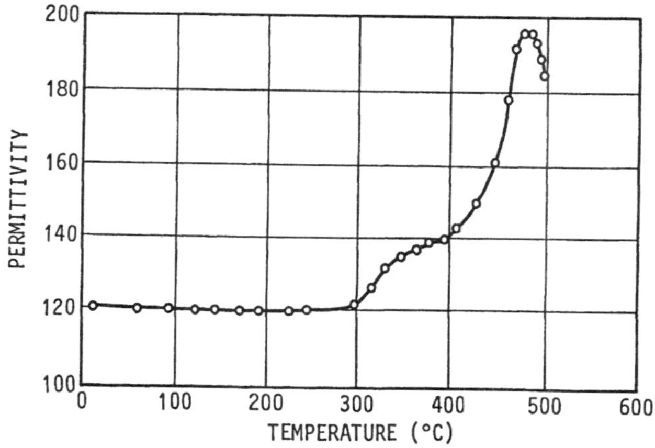

Figure 5.41. The temperature variation of the permittivity of PbTiO films 3000 Å thick on 7059 glass deposited by a magnetron sputtering system (measured at 1 MHz).

In magnetron sputtering, the oxides of the cathode metals (e.g., PbO and TiO_2) are initially formed at the cathode surface. These oxides will be cosputtered, and some fraction may be sputtered as molecules of PbO and TiO_2, which are deposited on the substrates. These sputtered molecules have an energy of 1–10 eV when they strike the substrates since they suffer few collisions with gas molecules in transit and have approximately the same energy as when they were removed from the cathode surface. The activation energy in the chemical reaction $PbO + TiO_2 \rightarrow PbTiO_3$ is on the order of 1 eV; thus when these sputtered PbO and TiO_2 molecules collide with each other on the substrates, $PbTiO_3$ can conceivably be synthesized even at low substrate temperatures. In conventional sputtering (i.e., higher pressure rf diode), the reaction between PbO and TiO_2 might not be possible at low substrate temperatures because the energy of the sputtered PbO and TiO_2 molecules is greatly reduced on the substrate due to collisions between the sputtered molecules and gas molecules in transit.

The rf-magnetron sputtering system is also employed for the preparation of the $PbTiO_3$ thin films.[170] Typical sputtering conditions are shown in Table 5.15. A powder of $PbTiO_3$, which is put in a stainless steel dish, is used as the target. X-ray diffraction analyses suggest that the $PbTiO_3$ thin films deposited on a cooled glass substrate (liquid N_2 at approximately room temperature) are amorphous structures (a-$PbTiO_3$) with Pb crystallites. Lead crystallites make the film electrically conductive [$\sigma > 10\ (\Omega cm)^{-1}$]. They disappear upon annealing the film above 220°C. The formation of Pb crystallites may be attributed to extremely rapid quenching of adsorbed atoms on the substrate in the sputter deposition process. The surface migration of the adsorbed atoms is not activated on the cooled substrate. The adsorbed atoms are condensed so rapidly that Pb crystallites (whose free energy is closer to that of the vapor) are formed.

Table 5.15. Typical Sputtering Conditions for the Deposition of $PbTiO_3$ Thin Films by RF-Magnetron Sputtering

Sputtering system	RF-Magnetron
Target	$PbTiO_3$ compounds
Sputter gas	$Ar/O_2 = 1$ (5×10^{-3} torr)
Substrate	Glass, sapphire, MgO, $SrTiO_3$
Substrate temperature	Liquid N_2 – 700°C
Deposition rate	50–70 Å/min

Heating the substrate activates the surface migration of the adsorbed atoms on the substrate. The $PbTiO_3$ deposited on the heated substrates (> 200°C) did not contain Pb crystallites. The x-ray diffraction pattern of amorphous $PbTiO_3$ deposited at 200°C is shown in Fig. 5.42. The diffraction pattern is a halo and suggests that the film exhibits a uniform amorphous configuration (which contained no crystallites). Activating the surface migration can cause the disappearance of Pb crystallites and the formation of uniform amorphous configurations.

Further activation of the surface migration may cause the formation of crystalline configurations. When the substrate is heated above 200°C during the deposition, the film is partially crystallized. Figure 5.43 shows the x-ray diffraction pattern of the film deposited at 500°C. The observed peaks can be indexed by pyrochlore $Pb_2Ti_2O_6$ (ASTM card 26-142). The films deposited above 500°C are a polycrystalline mixture of pyrochlore and perovskite. Figure 5.44 shows the schematic phase diagram of the sputter-deposited $PbTiO_3$ film.

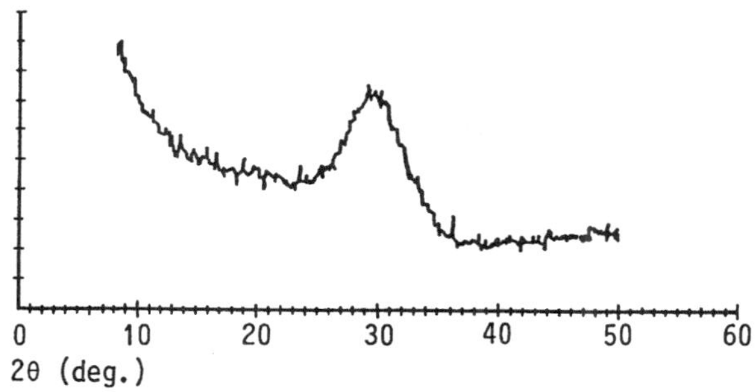

Figure 5.42. X-ray diffraction pattern of a-$PbTiO_3$ deposited at 200°C.

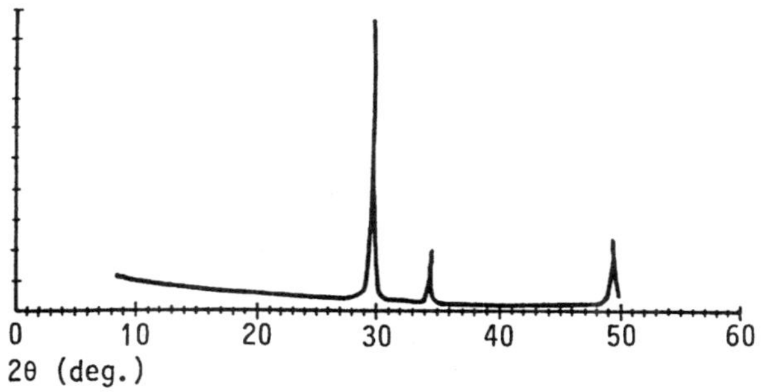

Figure 5.43. X-ray diffraction pattern of an a-$PbTiO_3$ film deposited at 500°C.

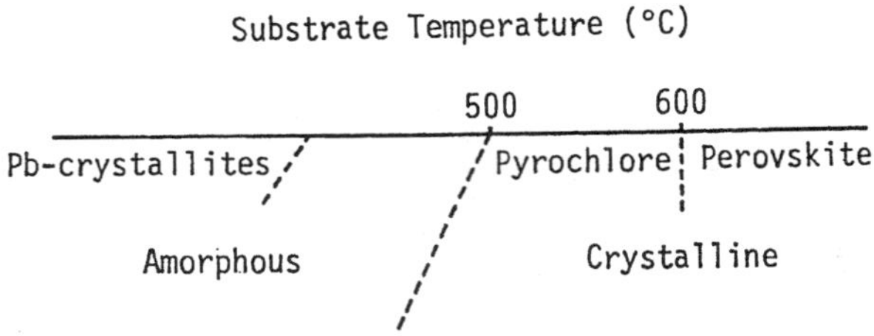

Figure 5.44. Schematic phase diagram of sputter-deposited $PbTiO_3$.

Figure 5.45 shows the x-ray diffraction pattern of an amorphous $PbTiO_3$ film annealed at 600°C. All of the x-ray diffraction peaks can be indexed by perovskite $PbTiO_3$ (ASTM card 6-0452). No diffraction peak attributed to pyrochlore is observed. The amorphous $PbTiO_3$ film changes to polycrystalline perovskite when it is annealed above 520°C, whereas the film deposited at 600°C is a polycrystalline mixture of pyrochlore and perovskite. The pure perovskite $PbTiO_3$ film is more easily prepared by annealing the amorphous $PbTiO_3$ film.

Figure 5.46 shows the x-ray diffraction pattern of an amorphous $PbTiO_3$ film annealed at 480°C (below the crystallization temperature). This film is partially crystallized. A mixture of perovskite and pyrochlore can index the diffraction peaks. Two kinds of crystallites grow below the crystallization temperature, whereas only the perovskite crystallites grow above the crystallization temperature. We may consider that the sputter-deposited amorphous $PbTiO_3$ contains crystallites of perovskite-like and/or pyrochlore-like microstructures.

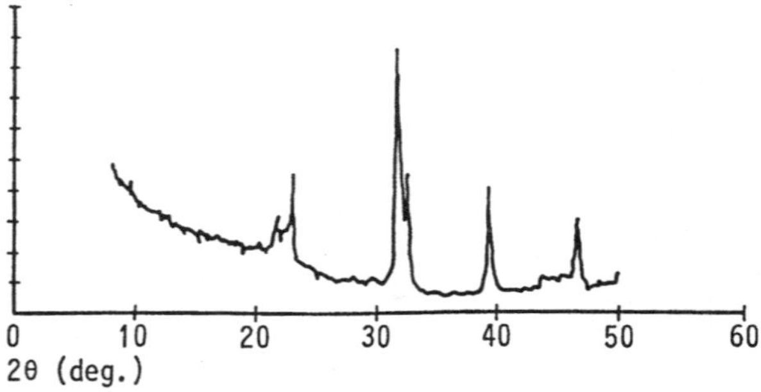

Figure 5.45. X-ray diffraction pattern of an a-$PbTiO_3$ film annealed at 600°C (above crystallization temperature).

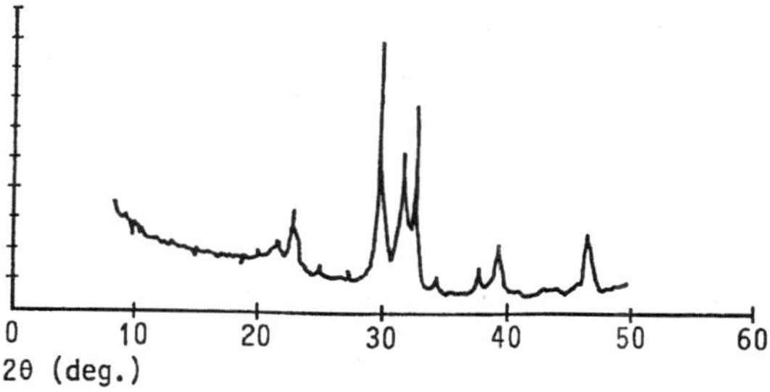

Figure 5.46. X-ray diffraction pattern of an a-$PbTiO_3$ film annealed at 480°C (below the crystallization temperature).

Polycrystalline or single-crystal thin films of $PbTiO_3$ are deposited at substrate temperatures higher than the crystalline temperature. Figure 5.47 shows a typical x-ray reflection spectra obtained from the target powder and the epitaxial $PbTiO_3$ thin films on a c-axis sapphire sputtered at 620°C. The epitaxial relationship is

Eq. (5.5) $$(111)PbTiO_3 \,//\, (0001)\text{sapphire}$$

A typical RED pattern and optical transmission spectrum of $PbTiO_3$ thin films epitaxially grown on sapphire are shown in Figs. 5.48 and 5.49, respectively. The (100)$PbTiO_3$ thin films are epitaxially grown on the (100) surface of MgO and/or $SrTiO_3$ single-crystal substrates.

Polycrystalline films with a perovskite structure are grown at a substrate temperature of 450° to 600°C. The polycrystalline films show a preferred orientation of (110) which corresponds to the most densely packed plane.

Since the vapor pressure of Pb becomes high at the epitaxial temperature of about 600°C, the resultant $PbTiO_3$ thin films often show a deficiency of Pb. In order to keep the stoichiometric composition in the sputtered $PbTiO_3$ thin films, a multitarget sputtering system is used for the deposition.[171] The schematic illustration of the multitarget sputtering system is shown in Fig. 5.50.

Separate magnetron cathodes are equipped and metal targets of each component are placed on them. The targets are then focused on the substrate. The normal line of the substrate makes an angle of 30 degrees to that of each target. The sputtering rate of each target is individually controlled by a dc power supply. Introducing a mixed gas of oxygen and argon creates reactive sputtering. The details of the multitarget sputtering apparatus and sputtering conditions are summarized in Table 5.16.

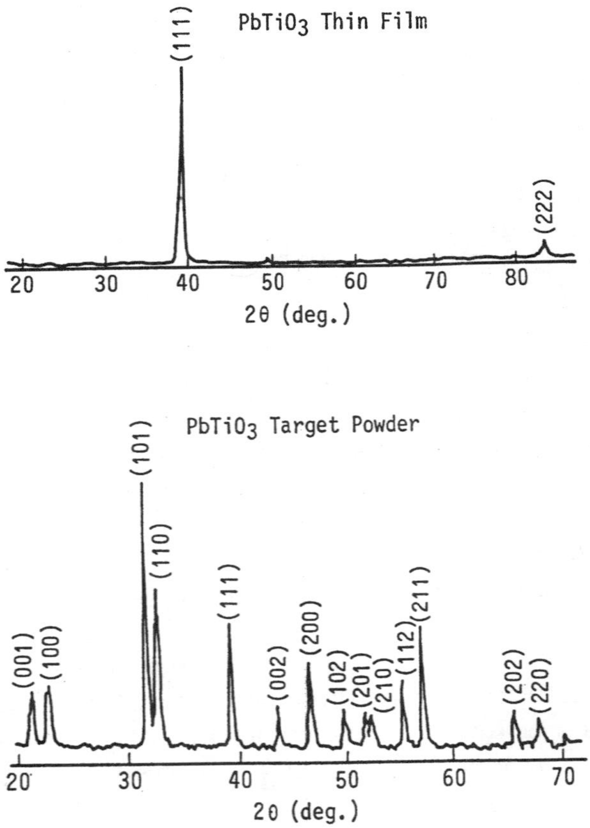

Figure 5.47. Typical x-ray reflection spectra from $PbTiO_3$ powder and $PbTiO_3$ single-crystal films on (0001) sapphire.

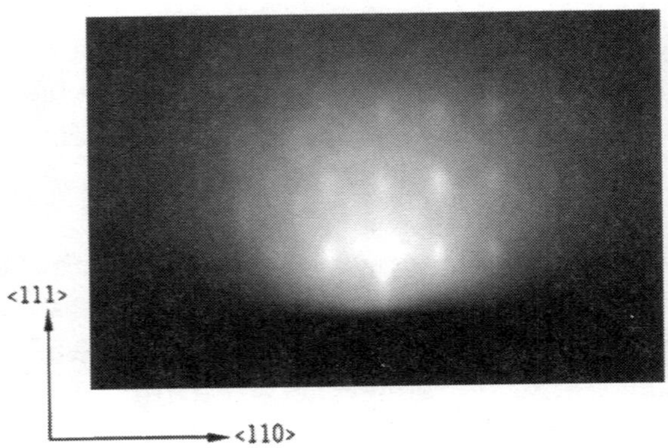

Figure 5.48. A typical electron pattern of $PbTiO_3$ thin film, epitaxially grown on (0001) sapphire.

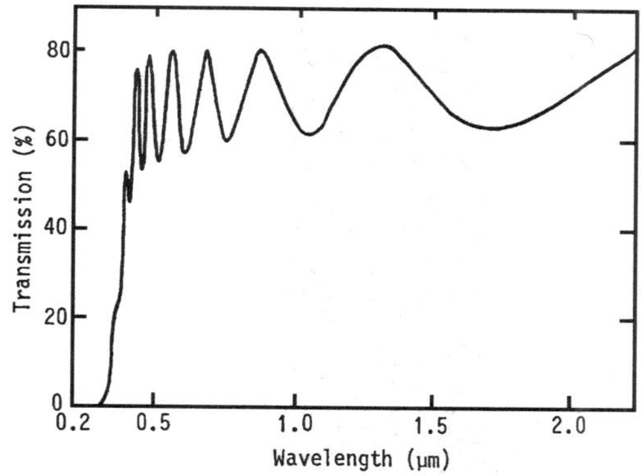

Figure 5.49. Optical transmission spectrum of epitaxial $PbTiO_3$ thin films about 0.4 μm thick.

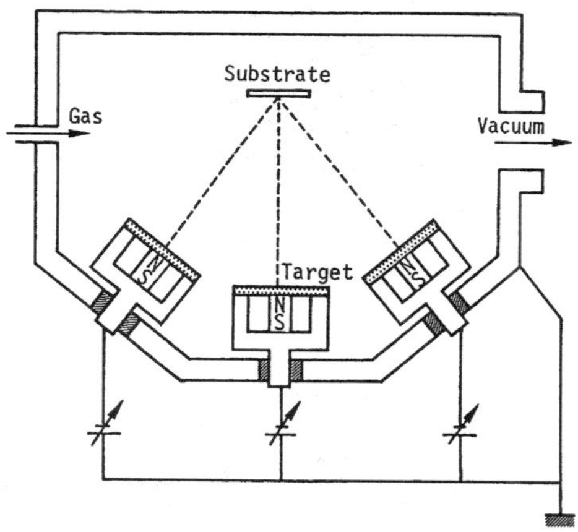

Figure 5.50. Schematic illustration of multitarget system.

Table 5.16. Sputtering Conditions For a Multitarget System

Target	Pb, Ti metal
Target diameter	60 mm
Target-substrate spacing	100 mm
Sputtering gas	Ar/O_2 = 1/0–1/1
Gas pressure	0.5–10 Pa
Input power	Pb: 0–15 W, Ti: 0–200 W

In general, the sputtering rate of the metal target decreases when the oxygen partial pressure increases. When the oxygen partial pressure is high, the surface of the target is oxidized and the sputtering rate decreases markedly. The effect of the oxygen partial pressure on the sputtering rate for Pb and Ti targets is shown in Fig. 5.51. The dependence of the sputtering rate upon oxygen gas pressure differs greatly between the Pb and Ti targets. The sputtering rate decreases more rapidly for the Ti target than the Pb target. The deposition rate of lead oxide is much larger than that of titanium oxide. In order to provide the same deposition rate for Ti and Pb in a fully oxidizing atmosphere, the sputtering conditions are fixed at an oxygen partial gas pressure of 1 Pa, with input power of ≈ 5–15 W for the Pb target and ≈ 100–200 W for the Ti target, respectively.

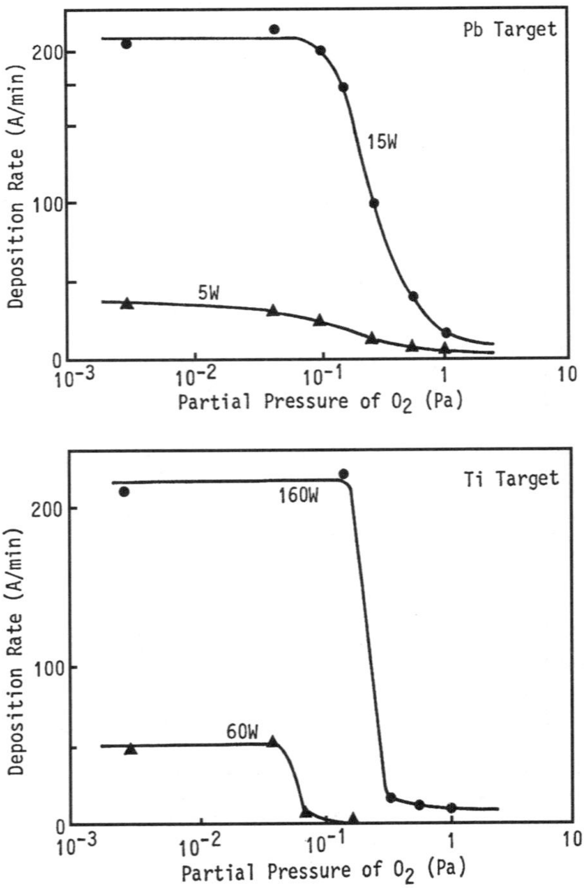

Figure 5.51. The relation between the oxygen partial pressure and the deposition rate for a Pb-Ti target.

Figure 5.52 shows the composition and crystallinity of films obtained at various substrate temperatures and incident Pb/Ti ratios. The composition and crystallinity of the films were examined by EPMA and x-ray diffraction methods, respectively. The compositional Pb/Ti ratios of thin films grown at 50°C can be regarded as Pb/Ti ratios incident to the substrate, since reevaporation is negligible. The epitaxial growth of the perovskite structure is found at high substrate temperatures. When the substrate temperature is 700°C, excess Pb (greater than Pb/Ti of 1.1) seemed to reevaporate from the substrate. In the event that only the Pb target is sputtered, film growth was not observed at 700°C. This indicates that the affinity between Pb and Ti prevents the evaporation of Pb as seen in the sintering of $PbTiO_3$ ceramics. The crystallinity of the epitaxial thin films is evaluated by x-ray diffraction peak intensity of (111). Figure 5.53 shows the x-ray peak intensity of the films as a function of the incident Pb/Ti ratio. The crystallinity of the films seems to improve as the incident Pb/Ti ratio approaches about 1.1. Though the compositional Pb/Ti ratio of the films at the incident Pb/Ti ratios of 2.3, 1.6, and 1.2 are nearly equal at a substrate temperature of 700°C, the x-ray peak intensities of these three films are obviously different. An optimum incident ratio of Pb to other components does exist for preparing epitaxial thin films of good quality.

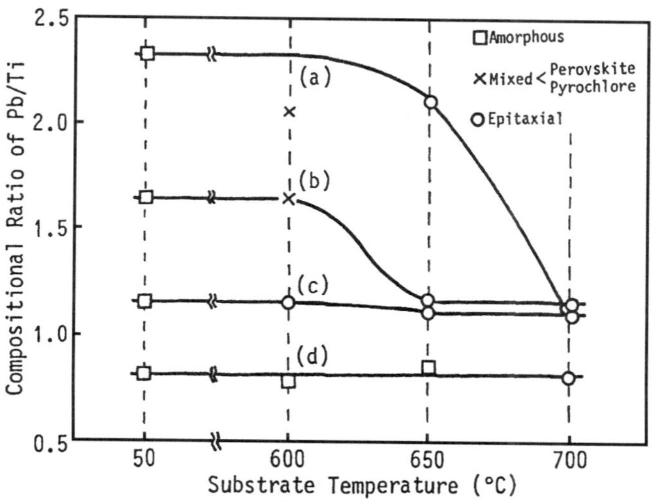

Figure 5.52. Composition and crystallinity of the films obtained at various substrate temperatures and incident Pb/Ti ratios. The incident Pb/Ti ratios are (*a*) 2.3, (*b*) 1.6, (*c*) 1.2 and (*d*) 0.8.

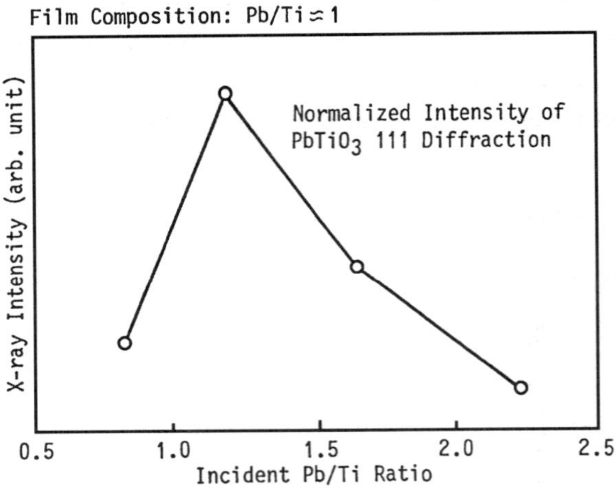

Figure 5.53. The relation between the incident Pb/Ti ratio and x-ray diffraction peak intensity of epitaxial films prepared at substrate temperatures of 650°C.

Electrical properties. Figure 5.54 shows the temperature dependence of the dielectric constant of as-deposited amorphous $PbTiO_3$ film deposited at 300°C. Two anomalies are observed at 520° and 480°C. The crystallization temperature is 520°C and one of the anomalies is attributed to it. Figure 5.55 shows the temperature dependence of the dielectric constant of the annealed (above 520°C) and crystallized film. An anomaly, which is caused by the phase transition of ferroelectric perovskite $PbTiO_3$, is observed at 480°C. The dielectric anomaly (at 480°C) of the as-deposited amorphous $PbTiO_3$ is attributed to the phase transition of ferroelectric perovskite $PbTiO_3$. This dielectric anomaly of the as-deposited $PbTiO_3$ is larger and sharper than that of the crystallized one. The as-deposited amorphous film partially crystallizes and contains perovskite $PbTiO_3$ crystallites.

It is reported that the roller-quenched amorphous $PbTiO_3$ and its annealed platelets are in a "heavily pressed state," and the dielectric anomaly, which is caused by the phase transition of ferroelectric perovskite $PbTiO_3$, shifted to a lower temperature. The sputter-deposited amorphous $PbTiO_3$ and its annealed film exhibit this anomaly at the same temperature, and no temperature shift is observed. Therefore, the perovskite $PbTiO_3$ crystallites, which grow in the sputter-deposited amorphous $PbTiO_3$, are considered to be in a "stress-free state." This property is the most significant difference between sputter-deposited and roller-quenched amorphous $PbTiO_3$. Scanning electron microscopy suggests that sputter-deposited amorphous $PbTiO_3$ contains many voids. These voids may compensate for the stress. This stress-free growth of the perovskite $PbTiO_3$ crystallites enables us to study the pure "grain-size effect" of ferroelectrics.

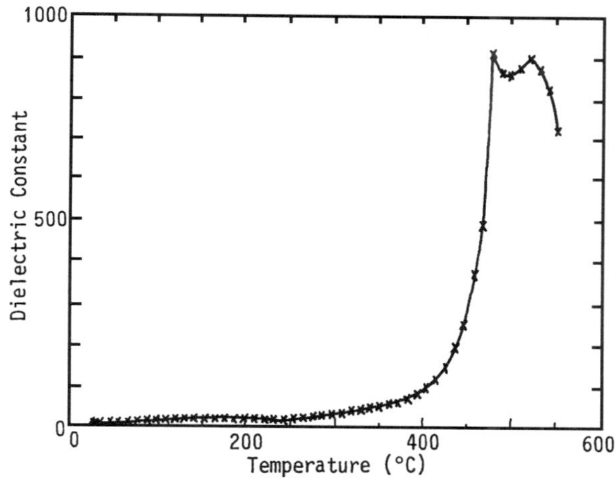

Figure 5.54. Temperature dependence of the dielectric constant of an a-$PbTiO_3$ film deposited at 300°C. The film thickness was about 1.5 μm. The measuring ac (10 kHz) voltage is 0.1 V rms.

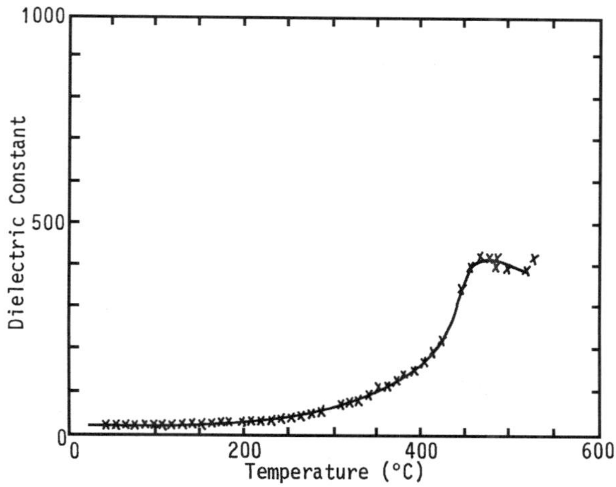

Figure 5.55. Temperature dependence of the dielectric of a $PbTiO_3$ crystallized film (measured at 10 kHz).

The dielectric anomaly in $PbTiO_3$ thin films is clearly observed in epitaxially grown films. A typical result is shown in Fig. 5.56.[172] This anomaly is observed at 490°C, which corresponds to the anomaly temperature for bulk $PbTiO_3$.

The piezoelectric properties of $PbTiO_3$ thin films have been studied in detail by Kushida and Takeuchi for c-axis-oriented films.[173] Measurements of the electromechanical coupling factor, k_t, show c-axis-oriented $PbTiO_3$ films were formed on patterned Pt electrode films embedded in the $SrTiO_3$ single-crystal seeded lateral overgrowth. The structure of the sample is shown in Fig. 5.57. The substrate is a (100) $SrTiO_3$ single-crystal plate. The impedance characteristics of the Au/$PbTiO_3$/Pt/$SrTiO_3$ structure are evaluated by the Mason equivalent circuit shown in Fig. 5.58.

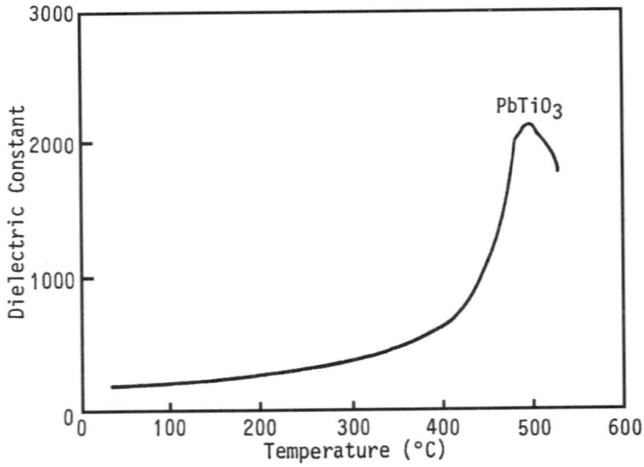

Figure 5.56. Temperature dependence of the dielectric constant of $PbTiO_3$ thin film, about 0.4 μm thick, epitaxially grown on (0001) sapphire (measured at 100 kHz).

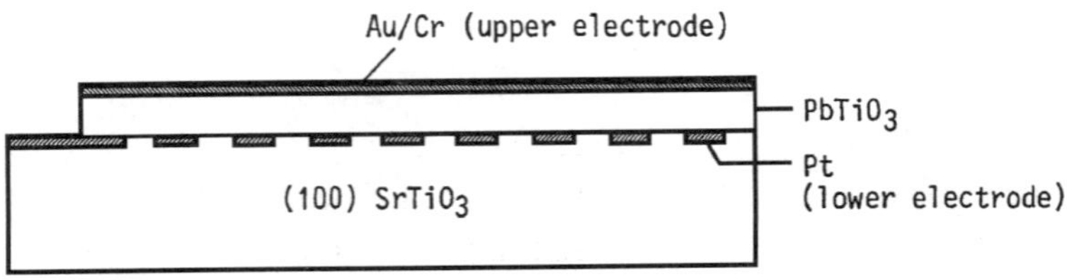

Figure 5.57. Structure of the sample for the measurements of the piezoelectric coupling factor for thickness vibration, k_t.[173]

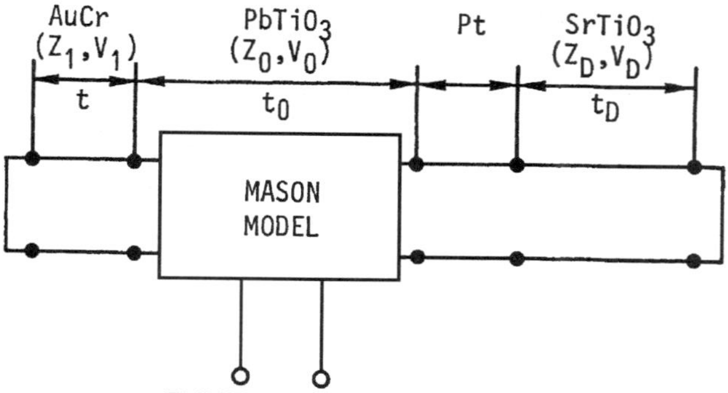

Figure 5.58. Equivalent circuit model used in analysis of the composite resonator.[173]

In the Mason circuit, the electromechanical coupling factor, k_t, is deduced from the following impedance formula:

Eq. (5.6) $$S = \frac{1}{j\omega C_0}\left\{1+\frac{k_1^2}{\Theta_0}\left[\frac{2(1-\cos\Theta_0)+(z_1+z_D)\sin\Theta_0}{(1+z_1 z_D)\sin\Theta_0-(z_1+z_D)\cos\Theta_0}\right]\right\}$$

where C_0 is the constant strain capacitance of the piezo layer, $k_1^2 = h_{33}{}^2\varepsilon_{33}/c_{33}$ is the electromechanical coupling constant of the $PbTiO_3$ film (h is the piezoelectric constant, ε is the dielectric constant, and c is the elastic stiffness), $z_1 = Z_1 \tan\Theta_1 / z_0$, $z_D = Z_D \tan\Theta_D / z_0$, and Z_0, Z_1, and Z_D are the acoustic impedances, $\Theta_0 = 2\pi f t_0/u_0$ is the acoustic phase in the piezo film, $\Theta_D = 2\pi f t_D/u_D$ is the acoustic phase in the substrate, and $\Theta_1 = 2\pi f t_1/u_1$ is the acoustic phase in the electrode.

The value k_t is evaluated from impedance measurements, shown in Fig. 5.59. The impedance characteristics show the resonant properties of a composite bulk wave resonator. The impedance measurements suggest that the c-axis-oriented $PbTiO_3$ thin films, 1–2 μm thick, exhibit a $k_t \approx 0.8$ at a frequency below 350 MHz. This value is extremely large for piezoelectric thin films and is comparable to the value obtained for a $PbTiO_3$ single crystal.[173]

Thin films of $PbTiO_3$ are also known as the pyroelectrics materials. Figure 5.60 shows the construction of a thin-film pyroelectrics sensor.[174] A small heat capacity with high pyroelectricity increases the pyroelectrics current and decreases Johnson noise, and makes for an excellent pyroelectrics sensor. Table 5.17 summarizes the electrical properties of $PbTiO_3$ thin films.

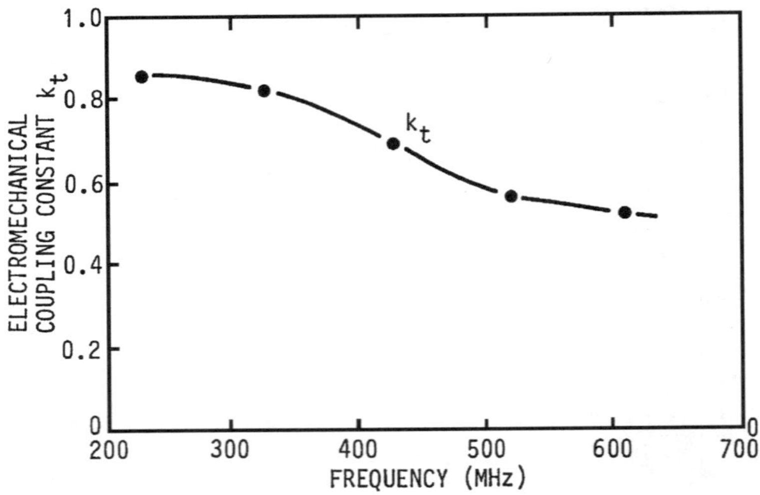

Figure 5.59. The frequency variation of the calculated electromechanical coupling constant k.[173]

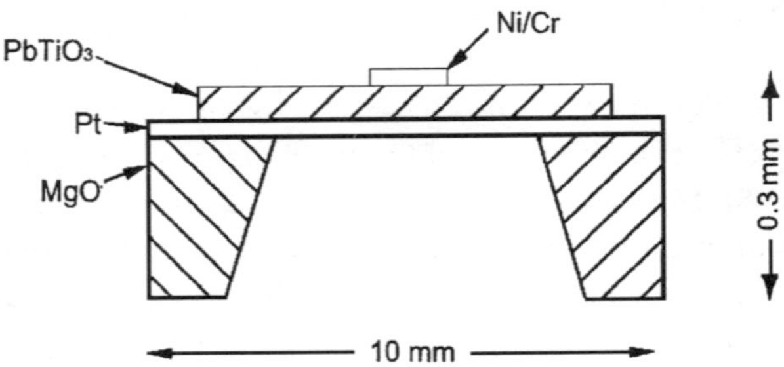

Figure 5.60. Construction of a typical $PbTiO_3$ thin film pyroelectric sensor for IR detection.

Table 5.17. Properties of $PbTiO_3$ Thin Films

Deposition* Method	Substrate	Deposition Temp. (°C)	Structure**	Film Properties	Ref.
DC-MSP	glass	200	a	$\varepsilon^* = 120$ (RT) $T_c = 490$°C	79
RF-MSP	glass	liquid N_2	a	$\varepsilon^* = 800$ (200°C)	80
RF-MSP	sapphire	580	pc	$\varepsilon^* = 370$ (RT) $T_c = 490$/°C	34
RF-MSP	Pt	630	pc	$\varepsilon^* = 110$ (RT) $k_t = 0.8$	83
RF-MSP	Pt	575	pc	$\varepsilon^* = 97$ (RT) Pyroelectric Coefficient $\gamma = 3\times10^{-8}$ C/cm²K	84

* MSP: magnetron sputtering
** a: amorphous; pc: polycrystalline

5.1.3.2 PLZT Thin Films

Deposition. Figure 5.61 shows the phase diagram of PLZT ($x/y/x$), $[(Pb_{1-x}, La_x)(Zr_yTi_{1-y})_{1-x/4}O_3]$ ceramics.[175] Thin films of PLZT are prepared by sputter deposition similar to the deposition of $PbTiO_3$ thin films.

Since the vapor pressure of Pb becomes high at the epitaxial temperature, the resultant films often show a deficiency of lead components. Achieving stoichiometry in epitaxial films is much more important for the deposition of complex compounds like PLZT. Figure 5.62 shows the typical XMA spectra obtained from a PLZT (9/65/35) thin film and from the ceramic target. The composition of the sputtered PLZT films is roughly estimated from the spectra. The content of Pb decreases with the increase of substrate temperature as described before. In order to compensate for the lead deficiency, excess PbO, 5 to 10 mole %, should be added to the target. Also note that the degree of Pb deficiency will strongly depend on the target conditions and type of sputtering system. Typical results are shown in Fig. 5.63. Magnetron sputtering with a powder target shows the smallest Pb deficiency. The ratio of Zr/Ti is close to the target in magnetron sputtering as indicated in Table 5.18. The magnetron discharge permits a lowering of the sputtering gas pressure, which helps to lower the epitaxial temperature. This enables one to keep the stoichiometric composition in epitaxial films. Typical sputtering conditions for PLZT thin films are shown in Table 5.19. Sapphire is used for the substrate. With respect to crystal orientation, the (0001) plane, the c-plane of sapphire, is suitable for epitaxial growth when one considers the atomic configuration. The plane has a normal threefold axis to the plane and is the same symmetry as the (111) plane of PLZT. Their atomic configurations are shown in Fig. 5.64. The average distances between oxygen atoms are 2.75 Å for the sapphire and 2.8 Å for the PLZT. The lattice mismatch is about 2%. Their epitaxial relationship is as follows:

$$(111)\text{PLZT} \,//\, (0001)\text{sapphire}$$

Eq. (5.7) $$[1\bar{1}0]\text{PLZT} \,//\, [10\bar{1}0]\text{ sapphire}$$

The crystal orientation of the epitaxial PLZT thin film on the sapphire substrate is schematically shown in Fig. 5.65.

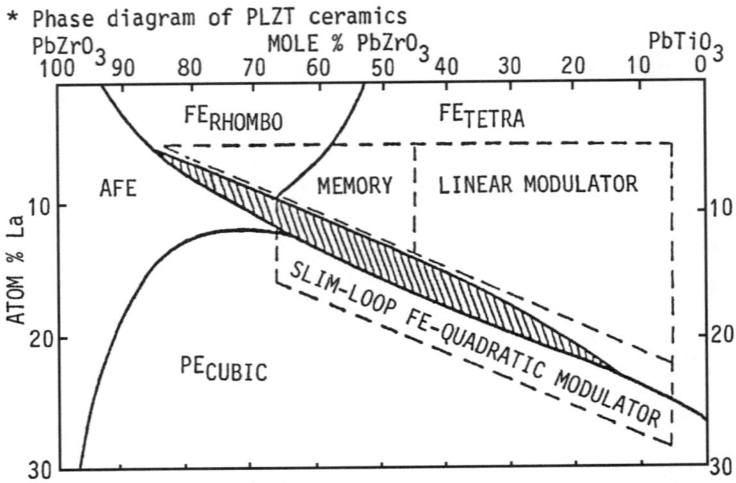

Figure 5.61. Phase diagram of PLZT ceramics.[175]

(a)

(b)

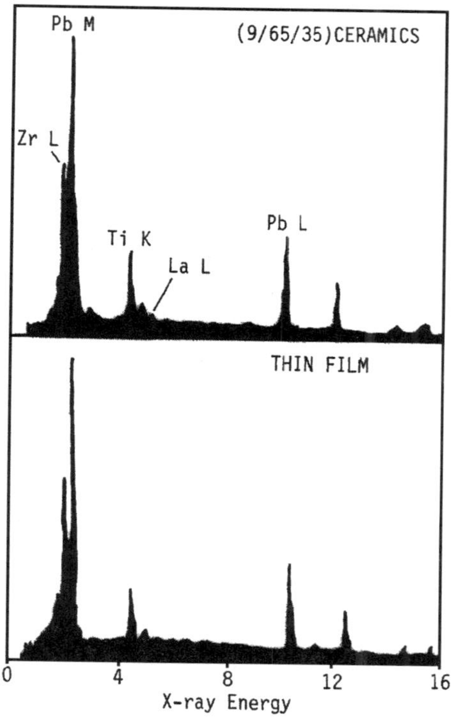

Figure 5.62. (*a*) XMA patterns of a PLZT (9/65/35) ceramic target, (*b*) and sputtered thin film.

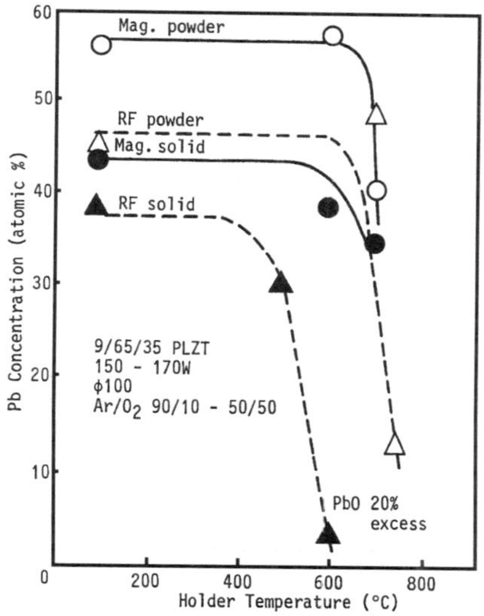

Figure 5.63. Pb concentration vs substrate holder temperature in the compound thin films sputtered from PLZT (9/65/35) target.

Table 5.18. The Atomic Ratio Zr/Ti in Sputtered PLZT (9/65/35) Thin Films for Various Sputtering Conditions*

Sputtering Method	Target	Zr/Ti Ratio
Rf-diode	Powder Plate	73/27 73/27
Rf-magnetron	Powder Plate	64/36 64/36

*Target Zr/Ti = 65/35, substrate temperature = 700°C.

Table 5.19. Typical Sputtering Conditions for the Deposition of PLZT Thin Films by RF-Magnetron Sputtering

Target	PLZT Powder
Target diameter	100 mm
Substrate	Sapphire (0001)
Target-substrate spacing	35 mm
Sputtering gas	Ar (60%) + O_2 (40%)
Gas pressure	0.5 Pa
Substrate temperature	500–700°C
RF power	150–250 W
Deposition rate	60–100 Å/min

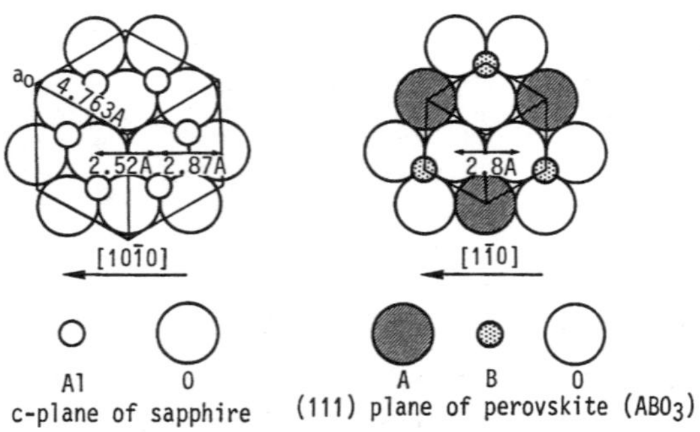

Figure 5.64. Planar atomic arrangements of c-plane of sapphire and the (111) plane of cubic perovskite (ABO_3).

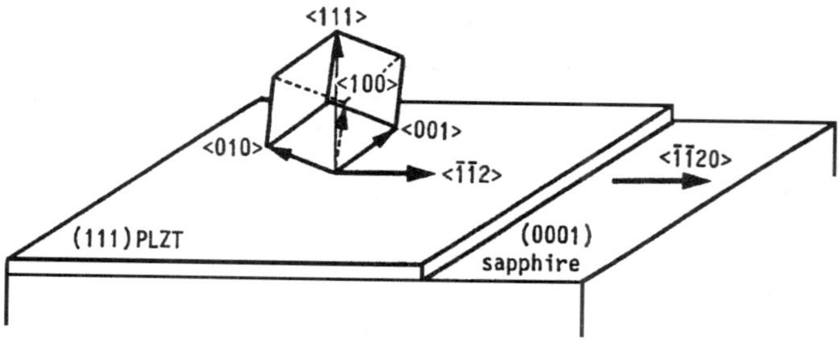

Figure 5.65. Crystal orientation of the epitaxial (111) PLZT film on (0001) sapphire.

Figure 5.66 indicates the crystalline structures of films deposited at various conditions. It shows that the film structure depends primarily on the substrate temperature and is only slightly affected by the growth rate. At substrate temperatures lower than 550°C, a metastable pyrochlore structure appears. The general formula of the pyrochlore is $A_2B_2O_7$, and its films have an intense yellow color. The figure also shows that epitaxial films with a perovskite structure are obtained at substrate temperatures higher than about 550°C. Epitaxial perovskite films are colorless. Figure 5.67 shows a typical RHEED pattern of the epitaxial PLZT film.

The composition of deposited films also depends on the substrate temperature and is independent of growth rate. The solid line (a) in Fig. 5.68 shows the compositional Pb/Ti ratio as a function of the substrate temperature at the growth rate of 80 Å/min. The epitaxial perovskite films obtained at above 550°C are almost stoichiometric, while the pyrochlore films obtained at below 550°C are markedly Pb rich. The excess Pb content in the film is considered to prevent the epitaxial growth of the perovskite structure. These considerations are experimentally confirmed as shown by the dashed line (b) in Fig. 5.68. The minimum substrate temperature for epitaxial growth is decreased to 450°C. It is observed that the composition of the PLZT films is close to stoichiometric when a target with less Pb is used. Thus, epitaxial film growth can be observed even at lower substrate temperatures. This result seems to disagree with Ishida's work using conventional diode sputtering where a Pb-rich target was required.[31] In the magnetron sputtering system, however, the discharge plasma is located primarily near the target, so that Pb in the target is considered to have evaporated as compared to conventional diode sputtering.

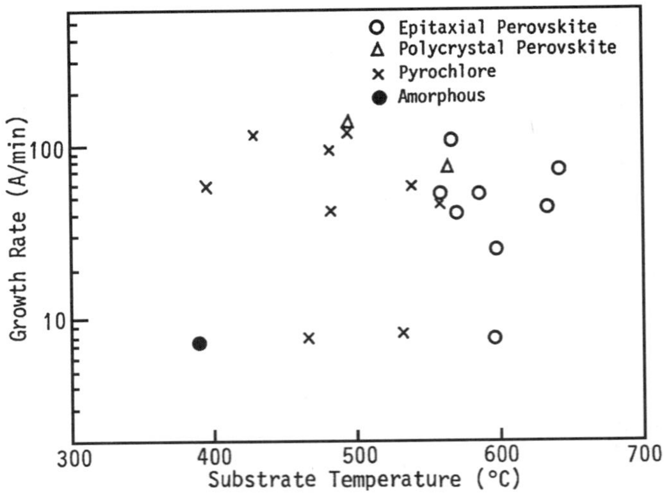

Figure 5.66. Crystalline structures of films deposited on sapphire at various substrate temperatures and growth rates.

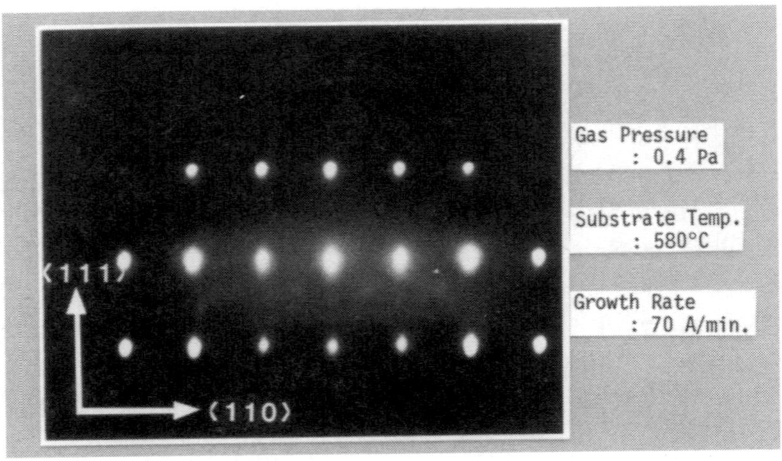

Figure 5.67. RHEED pattern of an epitaxial PLZT (28/0/100) thin film. The thickness is 0.4 μm.

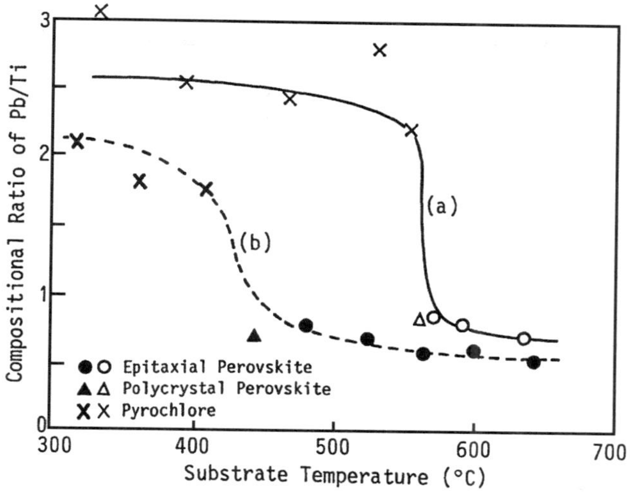

Figure 5.68. Compositional Pb/Ti ratio and structure of films at various substrate temperatures using PLZT (28/0/100) target (*a*) and, Pb-reduced PLZT (28/0/100)-target (*b*). The growth rate is about 80 Å/min.

In order to keep the correct stoichiometric composition in sputtered PLZT thin films, the multitarget sputtering system described in Fig. 5.50 is much more useful than the conventional rf-magnetron sputtering system.[176] It is also interesting that an artificial superlattice structure composed of multilayers of different ferroelectric materials can be made by the multitarget sputtering system. Figure 5.69 shows the construction of the multitarget sputtering system for the deposition of the ferroelectric superlattice PLT-PT structure. Targets of each element are separately positioned facing the substrate. Reactive cosputtering is carried out by introducing a mixed gas of argon and oxygen. The dc power supply for each target is controlled by a desktop computer (HP-9835A) and the input power of each target is varied periodically. Typical sputtering conditions are shown in Table 5.20.

Note that the input power for the Pb target for depositing the PLT layer is larger than for the PT layer despite the fact that the relative composition of Pb in PLT is smaller than in PT. This phenomenon is due to the reevaporation of Pb occurring during the growth of PLT because of the weak affinity between Pb and La atoms. Thus, excessive incident Pb content is required. The deposition times for PLT and PT layers are kept equal and the ratio of thickness (PLT:PT) is about 3:2 at these conditions. The superlattice films with a total thickness of about 3000 Å are prepared by varying the period of PLT-PT deposition from 120 to 300 Å/sec.

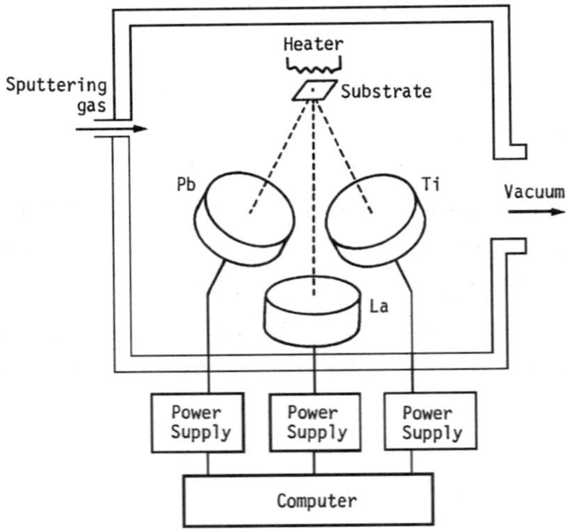

Figure 5.69. Schematic illustration of a multitarget sputtering system for the preparation of a PLT-PT superlattice structure.

Table 5.20. Typical Sputtering Conditions for the Deposition of PLZT Thin Films by a Multitarget Deposition System

Target	Pb, Ti and La metal
Target diameter	60 mm
Substrate	Sapphire c-plane
Substrate temperature	700°C
Target-substrate spacing	100 mm
Sputtering gas	$Ar/O_2 = 2/1$
Gas pressure	3 Pa
Input power:	
PLT layer	Pb: 15 W La: 22 W Ti: 180 W
PT layer	Pb: 9 W Ti: 180 W

RHEED analysis suggests that the sputtered films are epitaxially grown with the relationship (111)perovskite || (0001)sapphire and [$10\bar{1}$]perovskite || [$10\bar{1}0$]sapphire. Figure 5.70 shows the x-ray diffraction pattern of a film grown with a deposition period of 300 seconds. Diffraction from the (111) plane of a perovskite structure with a pair of satellite peaks is observed. This suggests the presence of a superlattice structure in the sputtered film.

Figure 5.71 shows the x-ray diffraction patterns of the (111) peak as a function of deposition period. When the deposition period is as long as 3000 seconds, the (111) peak is divided into (111) peaks of PLT layer and PT layer as shown in Fig. 5.71a. However, for a film grown for the deposition period of 600 seconds, a new strong peak appears between PLT (111) and PT (111) peak, and some satellite peaks around the center peak areobserved as seen in Fig. 5.71b. These satellite peaks are the first (Θ_1^+, Θ_1^-) and second (Θ_2^+, Θ_2^-) diffracted peaks caused by the superlattice

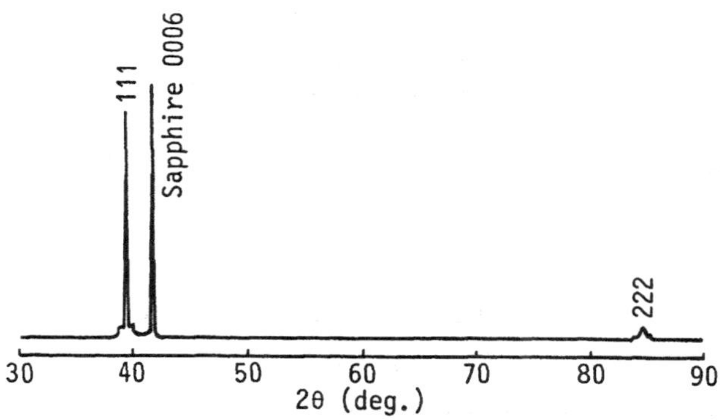

Figure 5.70. X-ray diffraction pattern for the PLT-PT superlattice film on sapphire grown during a deposition period of 300 sec.

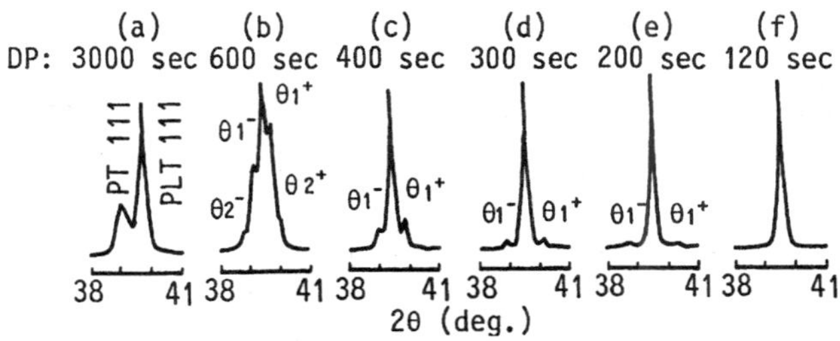

Figure 5.71. X-ray diffraction patterns of (111) peak for the films grown during various deposition periods.

structure with a modulation wavelength of 330 Å. The modulation wavelength Λ is calculated by the equation

Eq. (5.8)
$$\Lambda = \frac{\lambda}{\sin\Theta_1^+ - \sin\Theta_1^-}$$

where λ is the wavelength of the x-ray and Θ_i^+, Θ_i^- are the Bragg angles of the ith satellite peaks. As the deposition period becomes shorter, the wavelength becomes shorter and the intensity of the satellite peaks decreases. For films grown for a deposition period of 120 seconds, no satellite peaks are observed. This result indicates that interdiffusion of each layer occurs and the modulation structure disappears. The relation between the deposition period and the modulation wavelength is plotted in Fig. 5.72. Clear linearity is observed in the figure. The modulation wavelength can be strictly determined by the deposition period. These results suggest that the multitarget sputtering system is useful for the deposition of ferroelectric compound thin films with controlled chemical composition and crystal structures. Figure 5.73 shows a typical SEM image and RHEED pattern of an epitaxial $(Pb_{0.77}La_{0.23})Ti_{0.94}O_3$ thin film prepared by the multitarget sputtering system.

The transmission spectrum of a PLZT (28/0/100) film is shown in Fig. 5.74. These films are transparent from the visible to the near-infrared region. The refractive indexes of PLZT films varies with the Pb content in the film and are in the range of 2.4–2.7 at 0.633 μm.

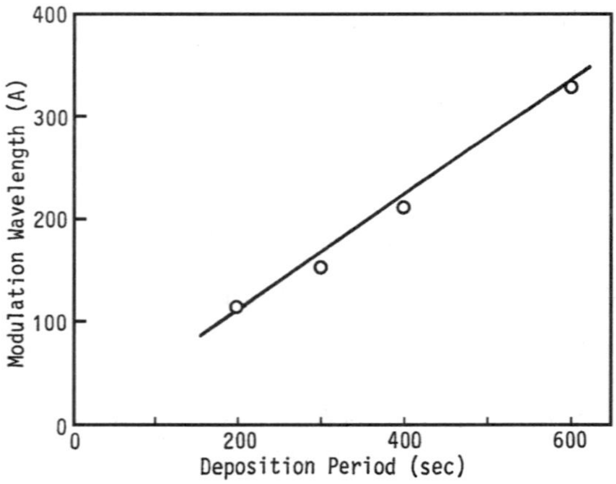

Figure 5.72. The relation between the deposition period and the modulation wavelength.

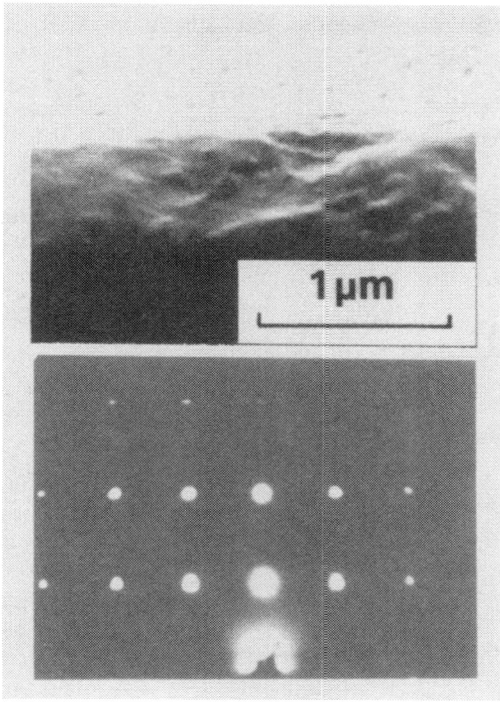

Figure 5.73. Typical SEM and RHEED images of (Pb, La)TiO_3 thin film.

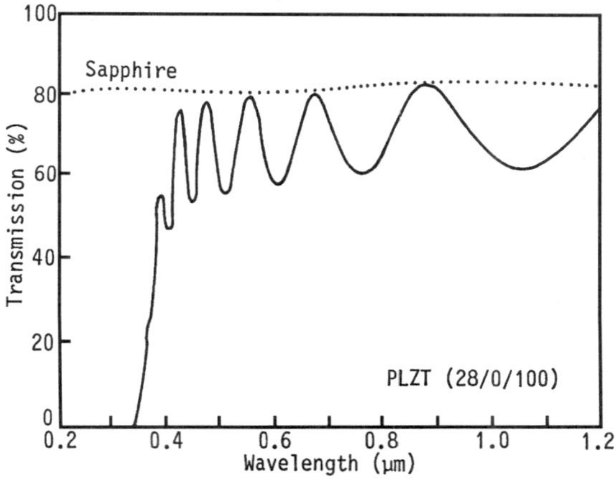

Figure 5.74. Optical transmission spectrum of an epitaxial PLZT (28/0/100) film about 0.4 µm thick.

Electrical properties. Electrodes made of the sandwich structure shown in Fig. 5.75 are used for measuring the dielectric properties of sputtered PLZT films. The PLZT thin films are deposited on a sapphire substrate over-coated by a TiN thin-film electrode. The sputtered films exhibit the polycrystalline form and show the dielectric anomaly similar to ceramics. However, a broad transition is observed in the temperature-permittivity characteristics as shown in Fig. 5.76.

Dielectric properties of epitaxial PLZT thin films are evaluated by the deposition of comb Al electrodes on the surface of the PLZT thin films as shown in Fig. 5.77. In the structure, the thin-film dielectric constant of PLZT is calculated from the measured capacitance, *C,* of the Al comb electrodes on top of the film using the following approximation:[177]

Eq. (5.9)

$$C = Knl\left\{(\varepsilon_s + 1) + (\varepsilon_f - \varepsilon_s)\left[1 - \exp\left(-\frac{4.6h}{L}\right)\right]\right\} \text{ (farad)}$$

$$K = \left[6.5\left(\frac{D}{L}\right)^2 + 1.08\left(\frac{D}{l}\right) + 2.37\right] \times 10^{-12}$$

where ε_s is the relative dielectric constant of the substrate ($\varepsilon_s = 10$), h is the thin-film thickness ($h = 0.33$ μm), L is the center-to-center spacing between adjacent electrodes ($L = 6$ μm), n is the number of electrode strips ($n = 160$), l is the length of the fingers of electrodes ($l = 720 \times 10^{-6}$ m), D is the width of each finger ($D = 3$ μm), and K is the constant given by the structure of electrodes ($K = 4.53 \times 10^{-12}$).

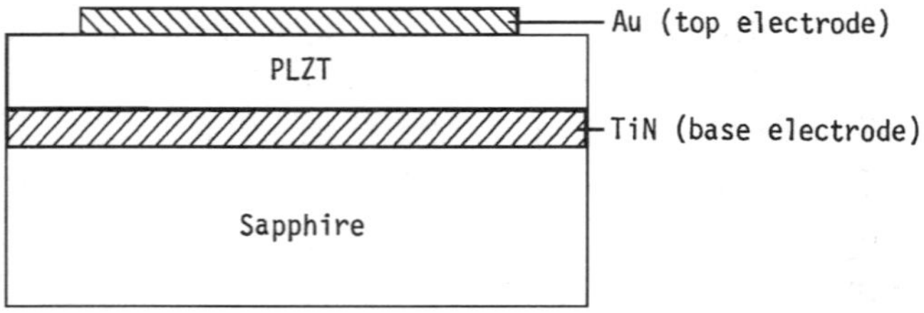

Figure 5.75. Structure of electrodes for the measurements of dielectric properties of PLZT thin films.

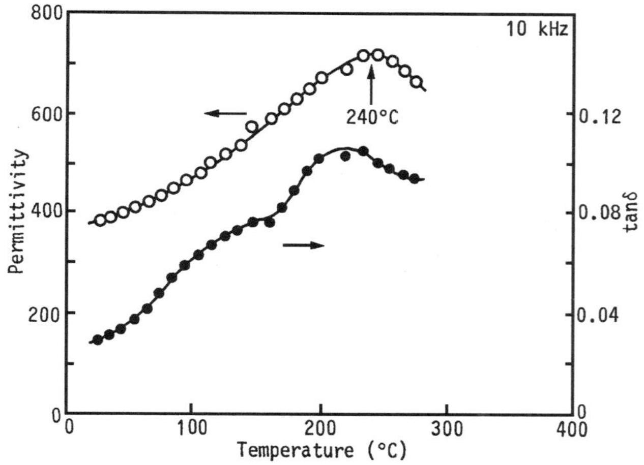

Figure 5.76. Temperature dependence of dielectric properties for PLZT (9/65/35) film.

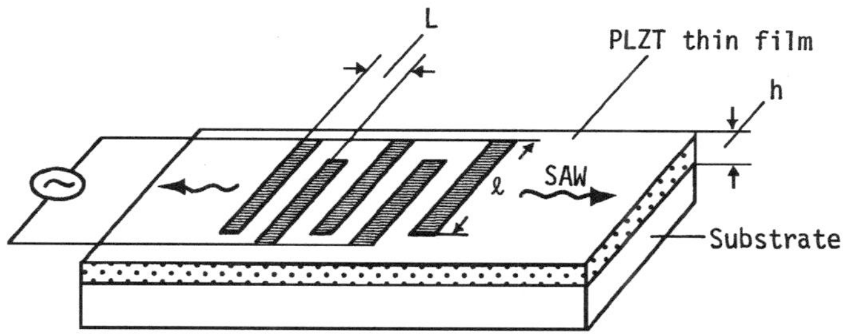

Figure 5.77. Comb electrodes for the measurement of dielectric properties of epitaxial PLZT thin films.

Figure 5.78 shows the temperature dependence of the relative dielectric constant measured for films with various compositions. The peaks of the dielectric constants correspond to the Curie temperatures (T_c) since the P-E (polarization-electric field) hysteresis measured using Sawyer-Tower circuit disappeared at temperatures above the peaks.[176] Typical P-E hysteresis curves of PLZT thin films measured at various temperatures are shown in Fig. 5.79. Note that the dielectric constant maximum for PLZT (9/65/35) film shows a broad temperature dependence similar to PLZT (9/65/35) ceramic, but the T_c of the film is approximately 100°C higher than that of the ceramic. Table 5.21 indicates the dielectric properties of PLZT thin films compared with those of ceramics of various chemical composition.

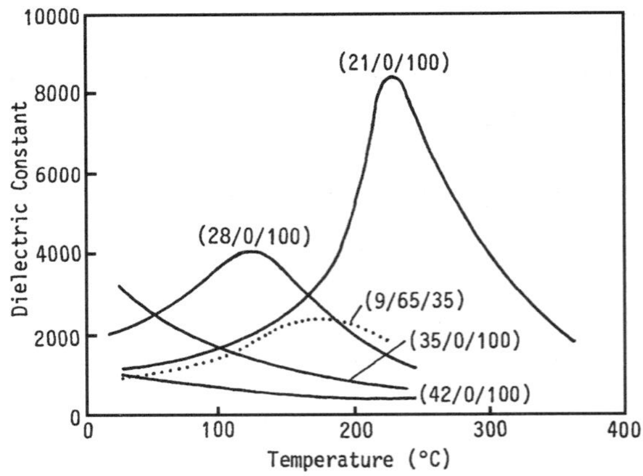

Figure 5.78. Temperature dependence of the relative dielectric constant of PLZT about 0.4 μm thick. The measuring frequency is 100 kHz.

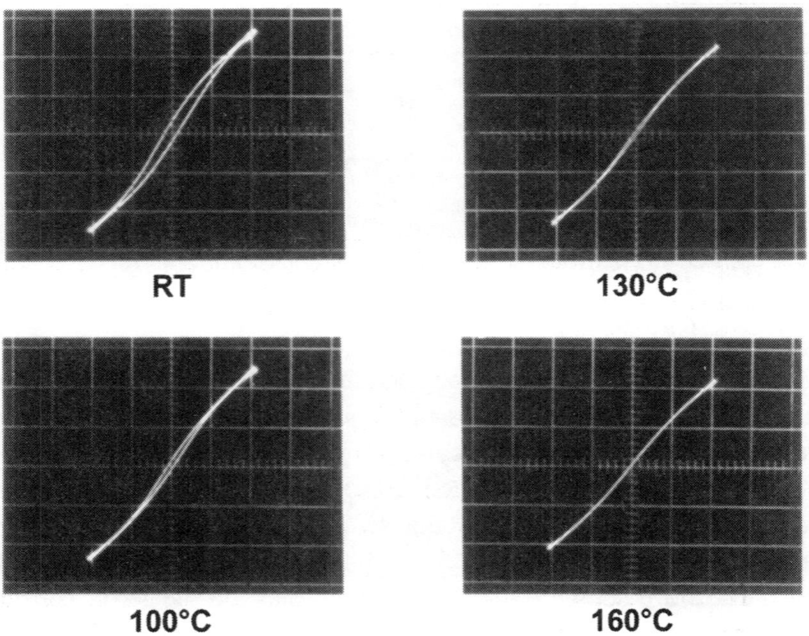

Figure 5.79. Temperature dependence of P-E hysteresis loop of PLZT (28/0/100) thin film at 60 Hz. Horizontal scale: 5 V/division. Vertical scale: 3×10^{-9} C/division.

Table 5.21. Dielectric Properties of (111) PLZT Thin Films on (0001) Sapphire Prepared by Sputter Deposition

Target Composition	Thin Films		Ceramics	
	ε*	T_c (°C)	ε *	T_c (°C)
$PbTiO_3$	370	490	230	490
PLZT (0/65/35)	450	275		365
PLZT (7/65/35)	480	260	1570	150
PLZT (9/65/35)	710	240	4650	85
PLZT (11/65/35)	630	220	4100	70
PLZT (14/65/35)	380	220	1450	50
PLZT (14/0/100)	600	290	1200	220
PLZT (21/0/100)	1300	225	2000	100
PLZT (28/0/100)	1800	120	2000	-100
PLZT (42/0/100)	1100			

* Measured at 10 kHz, RT.

It is understood that the T_c of PLZT (*x*/0/100) ceramics will increase with the decrease of La content.[178] Similar phenomena have been observed in sputtered PLZT thin films. The T_c of PLZT (*x*/0/100) films and ceramics are compared in Fig. 5.80. The T_c's of thin films is higher than those of ceramics. Under the assumption that the relationship between T_c and the La content of the thin films agrees well with that of ceramics, the disagreement of T_c between films and ceramics may be due to the compositional difference of the thin films. These assumptions are confirmed by chemical analyses of the sputtered PLZT thin films. Typical composition differences between the PLZT target and the sputtered thin films are shown in Fig. 5.81.

The piezoelectric properties of PLZT thin films are evaluated by excitation of a surface acoustic wave (SAW). An as-grown state of epitaxial PLZT thin film has three equivalent anisotropic axes along the edges of a pseudocubic lattice, which makes an angle of about 35 degrees to the film plane. Poling treatment is done as follows: First, Al electrodes are fabricated with 1-mm gaps on the film surface. Then the temperature is elevated to 200°C, which is higher than the Curie temperature. Next, the sample is gradually cooled from 200°C by applying a voltage of 2 kV as shown in Fig. 5.82a. The direction of the applied electric fields is parallel to the $(\bar{1}\bar{1}2)$ plane of PLZT. The polarization of this region in the thin film will be uniformly arranged to (001). In order to excite and detect the SAW, interdigital transducers (IDTs) of Al are made by a lift-off method on the polarized regions as shown in Fig. 5.82b. The period of the IDT finger is 12 μm and the pair number of the fingers is 80. The SAW will propagate along the $\langle\bar{1}\bar{1}2\rangle$ direction of PLZT.

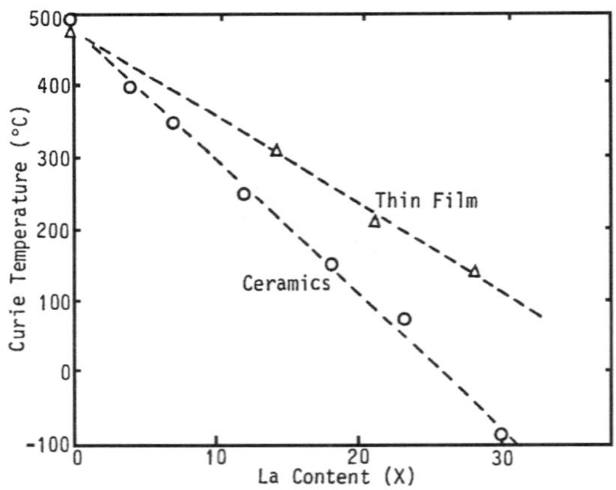

Figure 5.80. Curie temperature vs lanthanum content of ceramics and thin films deposited by PLZT (*x*/0/100) targets.

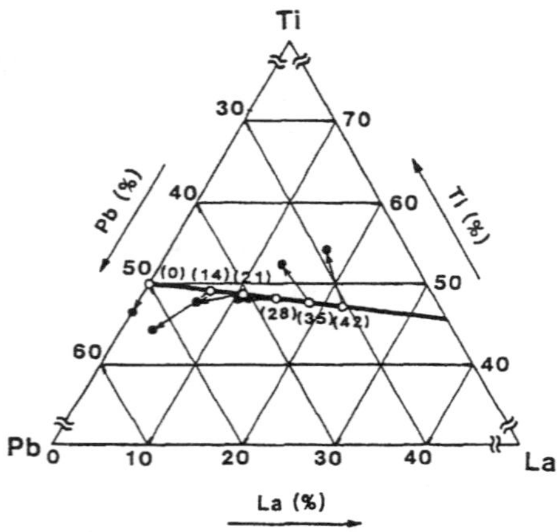

Figure 5.81. Composition of target and sputtered PLZT thin films: single target sapphire substrate at a substrate temperature of 650°C; o: target; •: films; Zr = 0.

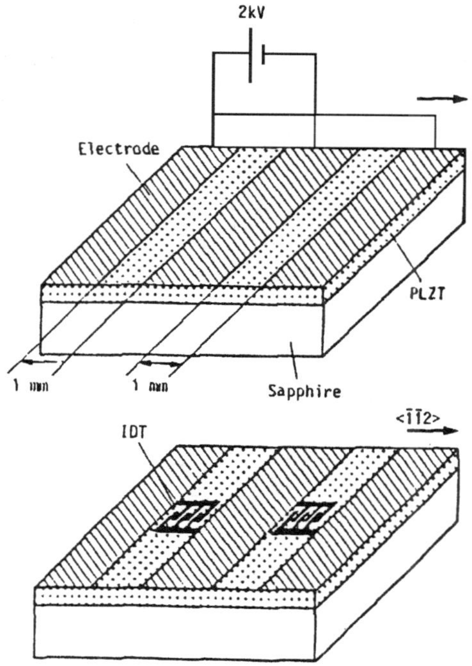

Figure 5.82. Fabrication procedure of PLZT SAW devices.

Figure 5.83 shows the Smith chart plot of the impedance characteristic of the IDT. The normalized thickness of the film is $Kd = 0.42$, where K is the wave number of the SAW and d is the film thickness. Two modes exist whose center frequencies are 405 MHz and 455 MHz. These modes are the fundamental (0th) and the higher-order modes of the SAW, respectively. The electromechanical coupling constant, k^2, is evaluated by Smith's relationship

Eq. (5.10) $$k^2 = \frac{\pi^2 f_o C_T R_q}{2N}$$

where N is the pair number of the fingers of the IDT, C_T is the capacitance of the IDT, f_o is the center frequency, and R_q is the measured radiation resistance.[179] The coupling constant, k^2, is calculated to be about 0.85% for the 0th mode of the SAW, which is a relatively large value.

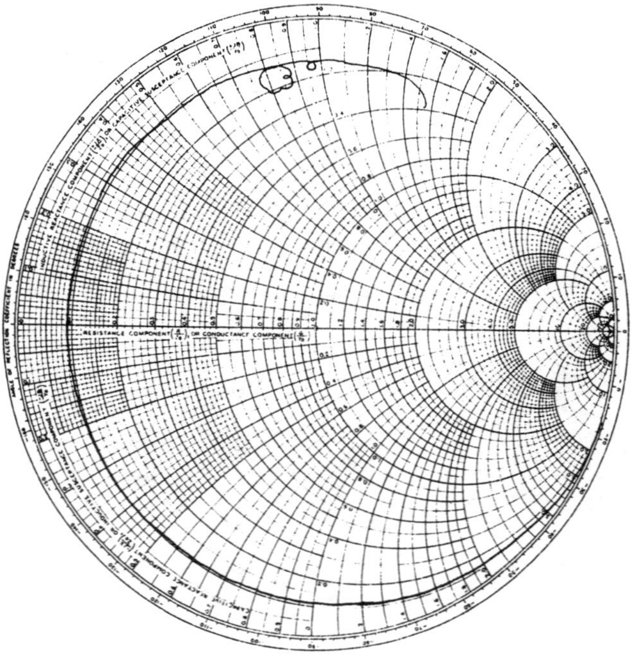

Figure 5.83. Smith chart impedance pattern for the IDT fabricated on the PLZT thin film. Frequency range is 350–450 MHz.

The phase velocity, V_p, and the coupling constant, k^2, of the SAW for various PLZT thicknesses are shown in Fig. 5.84. The open circle indicates the 0th mode SAW and the solid circle indicates the 1st mode. Since the physical parameters of the present PLZT are unknown, we have calculated the SAW properties using the parameters of $BaTiO_3$ and $PbTiO_3$ for comparison.[180][181] These are similar perovskite-type ferroelectric materials. The epitaxial relationships and the propagating direction of the SAW were similar to those in the experiment. The coupling constant, k^2, is expressed as follows:

Eq. (5.11) $$k^2 = 2F\frac{\Delta V}{V_p}$$

where V_p is the SAW velocity, ΔV is the perturbation of velocity, and F is the filling factor. The calculation is carried out by substituting 1 for F as

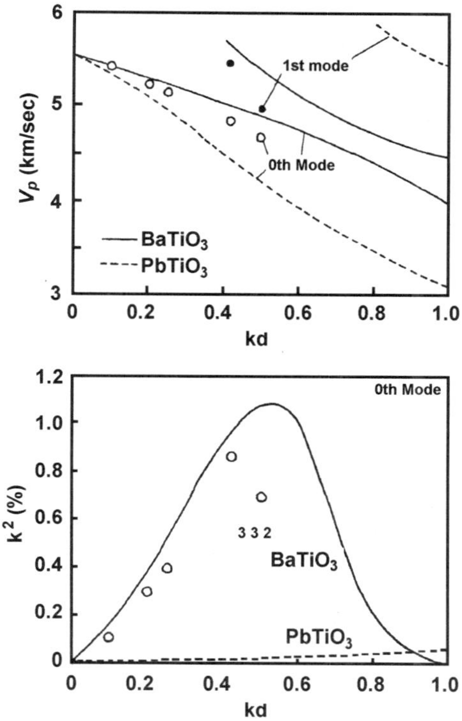

Figure 5.84. The phase velocity and coupling constant of SAWs for various K_d. Dots represent the experimental results for PLZT/sapphire. Lines represent the calculation for $BaTiO_3$/sapphire and $PbTiO_3$/sapphire.

usual. The results are shown in the figure by solid lines for $BaTiO_3$ and broken lines for $PbTiO_3$. Although the composition of the present PLZT is similar to $PbTiO_3$, the SAW properties show characteristics close to $BaTiO_3$. The piezoelectric effect of the PLZT thin film seems as strong as that of $BaTiO_3$.

The PLZT films also show high electro-optic properties. The electro-optic properties are evaluated by measuring the variations of the birefringence shift with the electric fields. The measurement system is shown in Fig. 5.85. A polarized He-Ne laser (6328 Å) will pass through the PLZT thin films perpendicularly between a pair of 1-mm-gap electrodes deposited on the film surface. The polarized laser light is oriented 45 degrees with respect to the direction of the applied electric field. The phase retardation shift, $\Delta\Gamma$, of the laser light in passing through the thin films is small since the film thickness of the PLZT is less than 1 μm. A $\lambda/4$ plate and

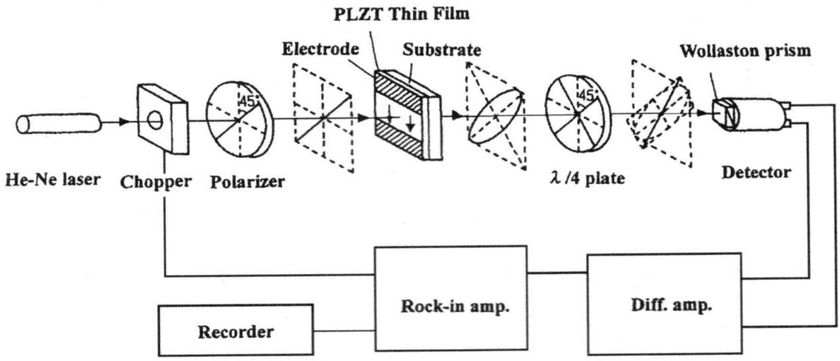

Figure 5.85. Experimental setup for electrooptic effect measurement.

a Wollaston prism are used for the measurement of the small shift in phase retardation. The birefringence shift, $\delta\Delta n$, is estimated by the relation

Eq. (5.12)
$$\delta\Delta n = \frac{\lambda_0 \Delta\Gamma}{2\pi d}$$

where n is the refractive index of the thin films, d is the film thickness, and λ_o is the wavelength of the laser light.

Figure 5.86 shows a typical birefringence shift of the PLZT thin films as a function of the electric field, E. In Fig. 5.86 a, b, and c, the characteristic shows a nearly quadratic effect with small hysteresis. Large electro-optic effects are observed for PLZT (28/0/100) and PLZT (9/65/35) thin films, similar to the bulk PLZT ceramics. The quadratic electro-optic coefficient, R, is shown by

Eq. (5.13)
$$\delta\Delta n = -\frac{1}{2} n^3 R E^2$$

The value of R is 0.6×10^{-16} (m/V)^2 for PLZT (28/0/100) thin films and 1.0×10^{-16} (m/V)^2 for PLZT (9/65/35) thin films. The electro-optic coefficient of as-sputtered PLZT (14/0/100) and PLZT (21/0/100) thin films is very small. After in-plane poling treatment, the PLZT thin films show a linear electro-optic effect as seen in Fig. 5.86(d). The linear electro-optic coefficient, r, is defined by the relation

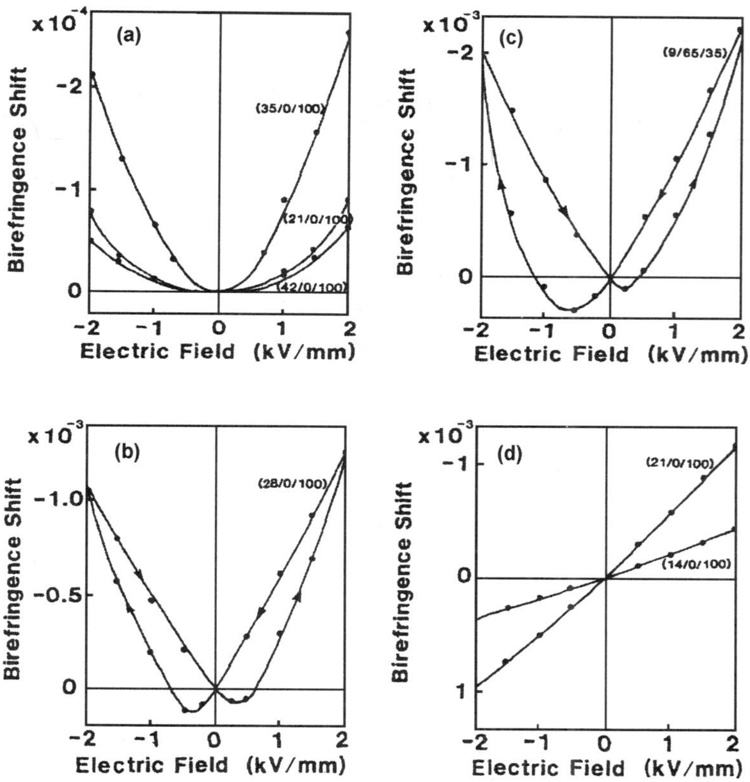

Figure 5.86. The effective birefringence shift as a function of transverse electric field for various PLZT films, each about 0.4 μm thick. (*a*) PLZT (35/0/100), (21/0/100), (42/0/100); (*b*) PLZT (28/0/100); (*c*) PLZT (9/65/35); (*d*) PLZT (21/0/100), (14/0/100).

Eq. (5.14)
$$\delta\Delta n = -\frac{1}{2} n^3 r E$$

The linear electro-optic coefficient, *r*, is 0.28×10^{-10} m/V for PLZT (14/0/100) and 0.81×10^{-10} m/V for PLZT (21/0/100) thin films. The chemical composition of the thin films is slightly different from the bulk ceramics as indicated in Fig. 5.81. The composition shift will lower the electro-optic coefficient of the thin films.[171] However, the value *r* of the PLZT (21/0/100) thin films is still several times higher than that of bulk $LiNbO_3$ single crystals. Table 5.22 summarizes the electrical properties of the PLZT thin films.

Table 5.22. Properties of PZT, PLZT Thin Films

Materials	Deposition method* (source)	Substrate	Deposition temp. (°C)	Structure**	Film properties	See Ref. 300
PZT	EB	SiO_2	350 (postanneal at 700°C)	PC	$\varepsilon^* \approx$ 100 (RT), $P_s \approx$ 4.2 μC/cm^2, $T_c \approx$ 340°C	a
	RF-SP (PZT 52/48)	SiO_2	> 500	PC	$\varepsilon^* \approx$ 751 (RT), $P_s \approx$ 21.6 μC/cm^2, $T_c \approx$ 325°C , n_0 = 2.36	b
PLT	RF-SP (PLT 18/100)	MgO	≈ 600–700	SC	$\varepsilon^* \approx$ 700 (RT), $n_0 \approx$ 2.3–2.5 (6328Å)	c
	RF-MSP (PLT 21/100)	sapphire	580	SC	$\varepsilon^* \approx$ 1300 (RT), $T_c \approx$ 225°C	d
	RF-MSP (PLT 28/100)	sapphire	580	SC	$\varepsilon^* \approx$ 2000 (RT) , $T_c \approx$ 150°C, $n_0 \approx$ 2.4–2.7 (6328Å), electrooptic coefficient: $R \approx 0.6\times10^{-16}$ m^2/V^2 (6328 Å), SAW coupling: k^2_{SAW} = 0.85% (Kd = 0.4)	d
PLZT	RF-SP (7/65/35)	SiO_2	500, post-annealing ≈ 650–700°C	PC	$\varepsilon^* \approx$ 1000–1300 (RT), $T_c \approx$ 170°C	e
	RF-SP (9/65/35)	sapphire $SrTiO_3$	700	SC	$n_0 \approx$ 2.49 (6328 Å)	f
	RF-MSP (9/65/35)	sapphire	580	SC	$\varepsilon^* \approx$ 710, $T_c \approx$ 240°C, $R \approx 1\times10^{-16}$ m^2/V^2 (6328 Å)	d

* EB: electron beam deposition; SP: diode sputtering; MSP: magnetron sputtering.
** PC: polycrystalline; SC: single crystal.

Optical switches. Light-beam switching in a four-port channel waveguide has been achieved using an electro-optic modulation of epitaxially grown PLZT thin film on sapphire.[182]

Figure 5.87 shows the typical configuration of channel waveguide switches. The switches are composed of a total internal reflection (TIR) structure.[183] The thickness of the PLZT thin film is 3500 Å. The PLZT thin film is prepared by rf-planar magnetron sputtering from a target of sintered PLZT (28/0/100) powder. The four-port channel waveguides are formed by ion-beam etching. The intersecting angle is 2.0 degrees. The width of each channel waveguide is 20 μm. A pair of parallel metal electrodes, separated from each other by 4 μm, is deposited at the center of the intersection region. In the absence of a switching voltage, the incident guided-light beam from port 1 encounters no refractive index fluctuation between the parallel electrodes and propagates straight to port 3. When the switching voltage performs due to a refractive index fluctuation, the guided-light beam is reflected to port 4.

Photographs of the transmitted and reflected light beam are shown in Fig. 5.88. The switching voltage is less than 5 V.[184] The time response of the TIR switch is shown in Fig. 5.89. The time response goes up to 26.5 GHz, and is mainly governed by the configuration of the electrodes. The traveling-wave electrodes will realize a high frequency operation of more than 100 GHz.[185] These types of switches can be used for high-speed multiplexers and demultiplexers for an optical LAN system.[186]

Since a variety of thin-film optical devices including lasers, light detectors, acoustic deflectors, and other micro-optical elements can be deposited on the same sapphire substrates, the PLZT/sapphire layered structure has the potential for making novel integrated optic circuits.[187]

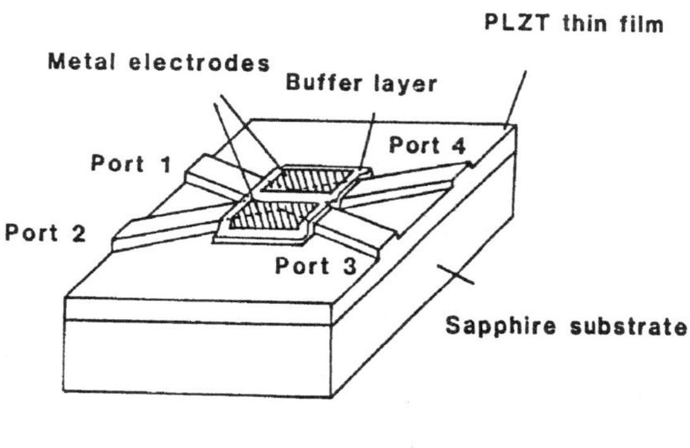

Figure 5.87. Configuration of optical TIR switches comprising a PLZT thin-film/sapphire layered structure.

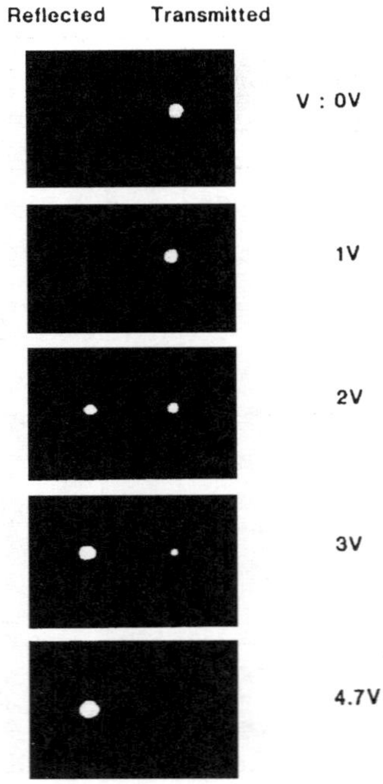

Figure 5.88. Transmitted (port 3) reflected light beam intensity (port 4) at the PLZT thin film TIR switch.

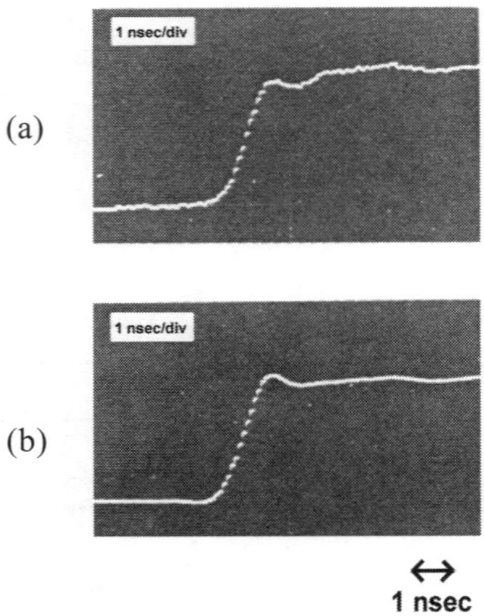

Figure 5.89. The time response of the PLZT, TIR switch: (*a*) applied voltage; (*b*) reflected light.

5.1.4 Perovskite Superconducting Thin Films

Extensive work has been done on high-temperature superconducting ceramics since Bednorz and Müller discovered the La-Ba-Cu-O compound system with the transition temperature of $T_c = 30$ K.[188] The K_2NiF_4 structure is found to be the major phase. Numerous compositions have been studied in an effort to raise the T_c. The replacement of Ba ions by smaller Sr ions in the La-Ba-Cu-O compound system has elevated the T_c.[189] The oxygen-deficient perovskite Y-Ba-Cu-O system with $T_c = 90$ K has been developed by Chu.[190] These compounds are composed of rare-earth elements. Extensive research on superconductors has led to several new high-T_c oxide materials, including rare-earth-free, high-T_c oxides of Bi-Sr-Ca-Cu-O and Tl-Ba-Ca-Cu-O systems with T_c exceeding 100 K, developed by Maeda and Hermann, respectively.[191]–[192] These high-temperature oxide superconductors, which are composed of a copper oxide layer, are shown in Table 5.23.

Historically, oxide superconductors have been known since 1964. Table 5.24 shows these oxide superconductors. These oxides are composed of perovskite $SrTiO_{3-\delta}$ with a low transition temperature of 0.55 K. The history of high-T_c superconductors of perovskite goes back to $BaPb_{1-x}Bi_xO_3$ with a T_c of 13 K, proposed by A. W. Sleight in 1974.[193]

Table 5.23. High T_c Superconducting Oxides

A_2BO_4 $(La_{1-x}M_x)_2CuO_4$	M: Ca, Sr, Ba	T_c: 20–40 K
$A_3B_3O_{7-\delta}$ $Ba_2LnCu_3O_{7-\delta}$	Ln: Y, Nd, Sm, Eu, La, Lu, Gd, Dy, Ho, Er, Tm, Yb	T_c: 90 K
$T_2A_2BO_x$, $T_2A_3B_2O_y$, $T_2A_4B_3O_z$	T: Bi, A: Sr, Ca, B: Cu T: Tl, A: Ba, Ca, B: Cu	T_c: 80–120 K
Miscellaneous	$(Nd_{0.8}Sr_{0.2}Ce_{0.2})_2CuO_4$	T_c: 27 K
	$(Nd_{1-x}Ce_x)_2CuO_4$ ($x \approx 0.07$)	T_c: 25 K
	$(Ba_{1-x}M_x)BiO_3$ ($x \approx 0.4$) M = K, Rb	T_c: 30 K

Table 5.24. Traditional Oxide Superconductors

Materials	T_c (K)	Structure	Date
$SrTiO_{3-\delta}$	0.55	perovskite	1964
TiO	2.3	NaCl	1964
NbO	1.25	NaCl	1964
M_xWO_3	6.7	tungsten bronze	1964
$Ag_7O_3^+X^-$	1.04	clathrate	1966
($X^- = NO_3^-, F^-, BF_4^-$)			
$Li_{1+x}Ti_{2-x}O_4$	13.7	spinel	1973
$BaPb_{1-x}Bi_xO_3$	13	perovskite	1975

The crystal structures of high-T_c perovskite-related copper oxide superconductors are shown in Fig. 5.90. In the perovskite compound ABO_3, the A-site and the B-site elements could be substituted by other kinds of cations. High oxidizations are achieved by substituting the B-side with 3d-transition metals accompanied by the generation of a mixed-valence state. These features are observed in perovskite-related, high-T_c superconductors.

The basic properties of copper oxide superconductors have been reviewed by several workers[194] and are shown in Table 5.25, with accompanying phase diagrams. Note that this type of oxide often shows a large oxygen nonstoichiometry. Figure 5.91 shows the δ-log $P(O_2)$ curves for various ambient temperatures in the $A_3B_3O_{7-\delta}$ system.[195] It is seen that the superconducting orthorhombic phase is stable at temperatures below 600°–650°C. Lattice parameters of the high-T_c superconductors are shown in Table 5.26.

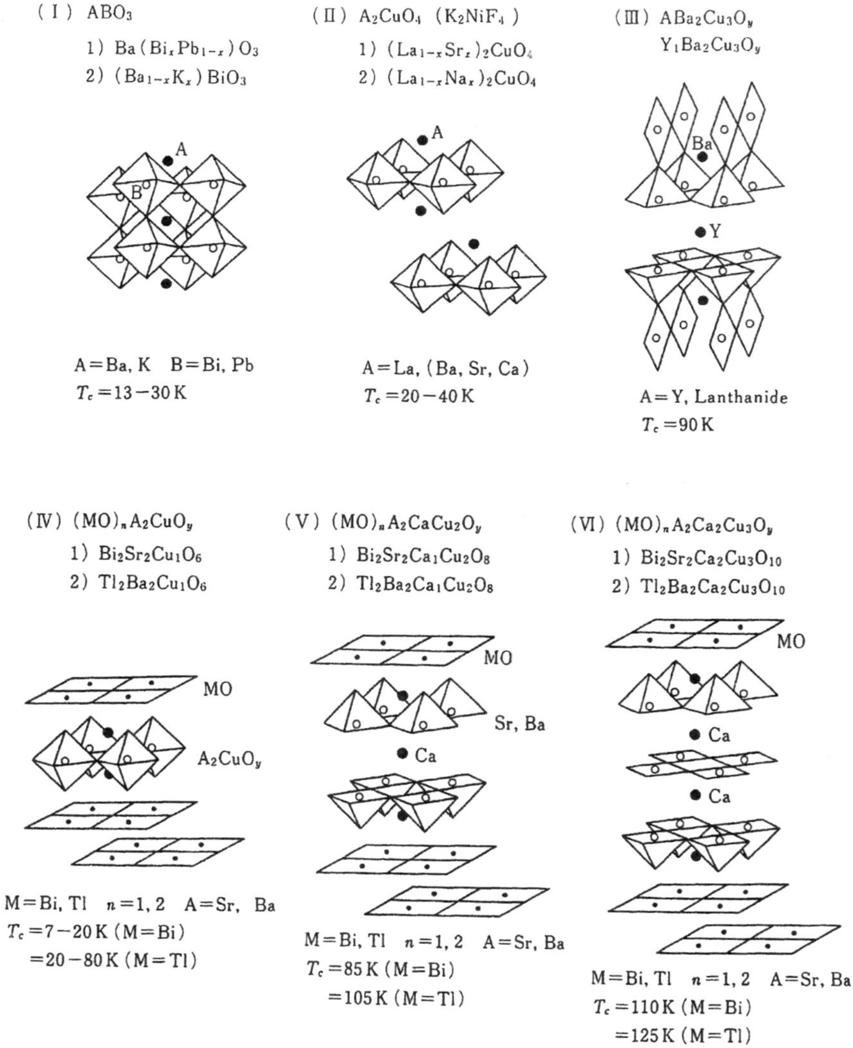

Figure 5.90. Crystal structures of oxide superconductors.

Table 5.25. Lattice Parameters and Phase Diagrams of High-T_c Superconductors

$YBa_2Cu_3O_{6.9}$, Ortho.
a=3.8218(7) Å,
b=3.8913(7) Å,
c=11.677(2) Å

$h\ k\ l$	2θ (deg)
2 1 3	58.856
1 1 6, 1 2 3	58.245
1 1 5	51.483
2 0 0	47.582
0 0 6, 0 2 0	46.675
1 1 3	40.411
1 0 4	38.798
1 1 0, 1 0 3	32.854
0 1 3	32.559
1 0 2	27.898
0 1 2	27.57
1 0 0	23.273
0 0 3	22.843

$Bi_2Sr_2CuO_6$, Tetra
a=3.8097 Å,
c=24.607 Å

$h\ k\ l$	2θ (deg)
2 1 5	57.209
2 0 8	56.853
2 1 3	55.039
2 0 0	47.736
1 1 8	44.715
0 0 12	44.172
1 0 7	34.714
1 1 0	33.254
1 0 5	29.647
0 0 8	29.021
1 0 3	25.770
0 0 6	21.656
0 0 2	7.155

$Bi_2Sr_2CaCu_2O_6$, Tetra
a=3.812 Å,
c=30.66 Å

$h\ k\ l$	2θ (deg)
2 1 7	57.970
2 1 5	55.898
1 0 15	50.678
2 0 0	47.627
1 0 13	45.205
1 1 10	44.757
1 1 4	35.235
1 1 0	33.196
1 0 7	31.091
0 0 10	29.154
1 0 5	27.535
1 0 3	24.910
0 0 8	23.231
0 0 2	5.784

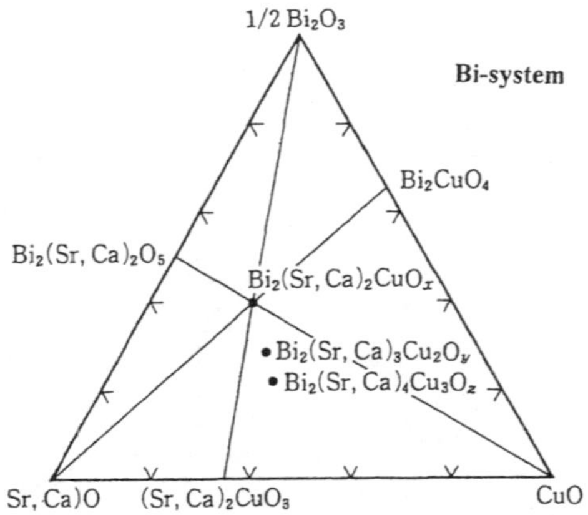

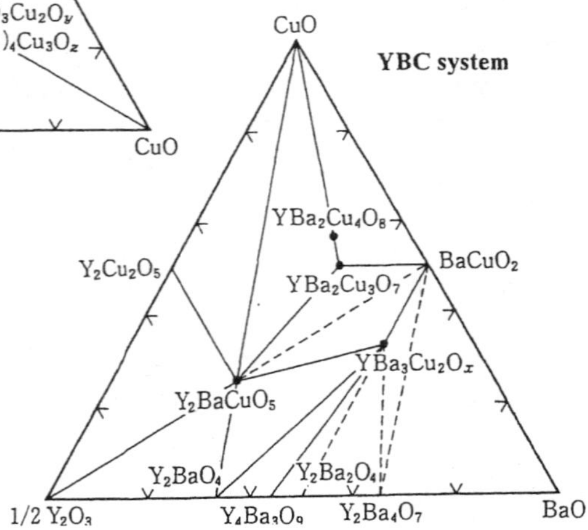

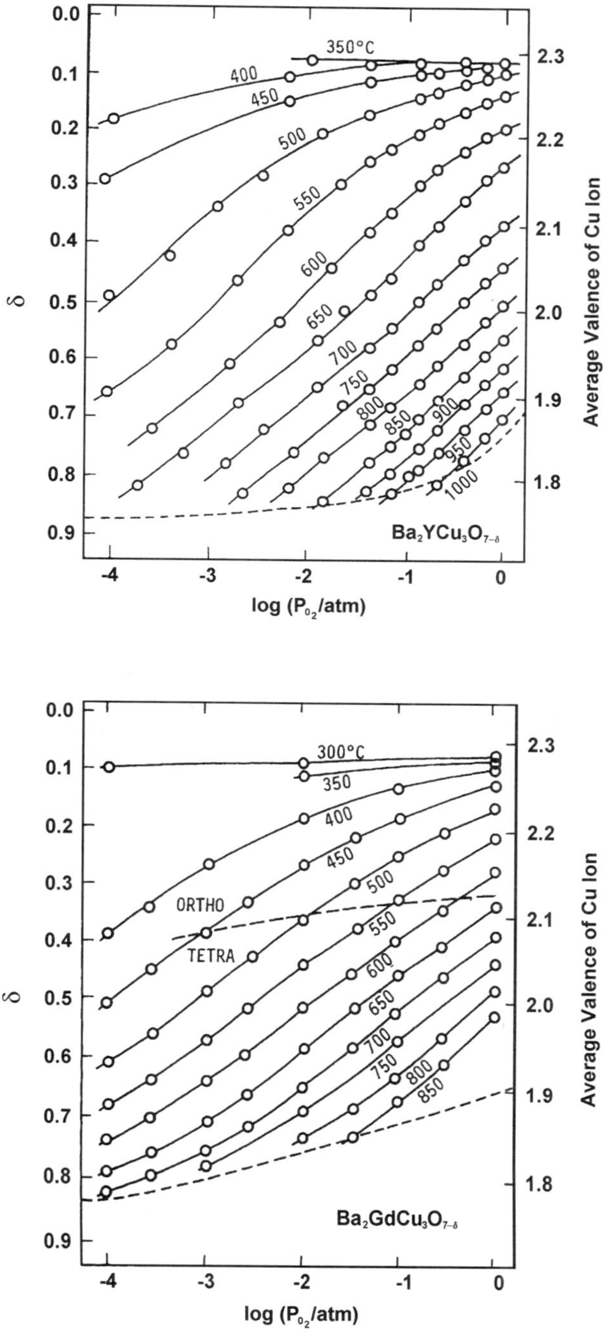

Figure 5.91. Oxygen deficiency and average valence of Cu ion as a function of log $P(O_2)$.[195]

Table 5.26. Material Parameters for High T_c Superconductors[194]

Material	T_c (K)	ε (nm)	Λ (nm)	κ (Λ/ξ)	
LaSr	37		250		Aeppli (1987)
	36	3	330	110	Finnemore (1987)
		1.8			Kobayashi (1987)
	34		200		Kossler (1987)
	38	1.3	210	160	Orlando (1987)
				~75	Renker (1987)
				~40	Takagi (1987)
	37		230		Wappling (1987)
LaSr (0.096)	38	2.0			Nakao (1987)
LaSr (0.1)		2.6			Kobayashi (1987)
	35	1.3	210	160	Orlando (1987)
	39	2.0	100	50	Uchida (1987)
LaSr (0.3)	35	3.2			Murata (1987)
$YBa_2Cu_3O_{7-\delta}$				180	Bezinge (1987)
			22.5		Felici (1987)
		1.5	120	80	Gottwick (1987)
				70	Grant (1987)
	95	2.7(‖)			Hikita (1987)
	95	0.6(⊥)			Hikita (1987)
	84		130		Kossler (1987)
	89	1.7			Orlando (1987)
	89	3.4(‖)	26(⊥)	7.6(‖)	Worthington (1987)
	89	0.7(⊥)	125(‖)	37(⊥)	Worthington (1987)
			400		Zuo (1987)
$YBa_2Cu_3O_{6.9}$	92.5	2.2	140	65	Cava (1987)
$Y_{0.4}Ba_{0.6}CuO_{3-\delta}$	89	1.4			Murata (1987)
BiSrCaCuO		4.2(‖)			Hikita (1988)
		0.1(⊥)			Hidaka (1988)

Coherence lengths (ξ), penetration depths (Λ), and their ratios (Ginzburg-Landau Parameter) $\kappa = (\Lambda/\xi)$. The notation used is $(La_{1-x}M_x)_2CuO_{4-\delta} = LaM(x)$. Several values are given of the coherence length $\xi_{\parallel} = \xi_{ab}$ in the Cu-O planes and $\xi_{\perp} = \xi_c$ perpendicular to these planes are given.

5.1.4.1 Studies of Thin Film Processes

Thin films of perovskite-related, high-T_c superconductors have been successfully synthesized. This is due to previous studies on dielectric thin films of perovskite-type oxides[33] and/or superconducting thin films of the perovskite-type, $BaPb_{1-x}Bi_xO_3$,[71] and the alloy superconductors of the $A_{15}Nb_3Ge$ and B_1NbN type.[64][66]

In the early periods of research on these superconducting films, sputtering and/or electron beam deposition was used for the deposition. The polycrystalline and/or single-crystal thin films of La-Sr-Cu-O were prepared by several workers.[72]–[76] Adachi found that sputtered single-crystal thin films of La-Sr-Cu-O exhibited a transition temperature of T_c = 34 K, which corresponded to the best results for the ceramics.[75] Suzuki found that this kind of superconductor shows a small carrier density of around $n_H = 6.8 \times 10^{21}$ cm^{-3}, similar to the $BaPb_{1-x}Bi_xO_3$ oxides.[73] Naito evaluated the energy gap of La-Sr-Cu-O thin films by infrared reflectance measurements; the values were found to be 20–30 mV.[76]

After the discovery of Y-Ba-Cu-O high-T_c superconductors by Chu, most thin-film studies were shifted to the deposition of Y-Ba-Cu-O thin films. The sputtered films exhibited high transition temperatures of around 90 K, which was close to the value of ceramics. Their critical currents were found to exceed 10^6 A/cm^2 at a liquid nitrogen temperature of 77 K under a zero magnetic field, although critical currents measured in bulk ceramics were less than 10^3 A/cm^2.[79]

Several basic properties in the new high-T_c Y-Ba-Cu-O thin films were evaluated, including the tunneling gap and anisotropy in critical currents and the critical field.[79] In these experiments, the critical field was found to exceed 4 T/K.[196] These extensive studies have suggested that the new high-T_c superconductors have possible applications for electronic devices and/or power systems.

The possibility of lowering the synthesis temperature has been discussed. Sputter deposition has allowed the reduction of the synthesis temperature from 900°C to 600°C by irradiation with an oxygen plasma during deposition.[81] Lowering the synthesis temperature has been found to reduce the mutual diffusion between thin films and the substrates, and stabilizes the interface.[197] Several workers have studied mutual diffusion between thin films and substrates.[198][199] Aluminum atoms in the sapphire substrates are found to easily diffuse into superconducting thin films during postannealing. Buffer layers are used for the reduction of mutual diffusions. Layers of ZrO_2,[200] CaF_2,[197] and Pt[201] are considered buffer layers on Si and sapphire substrates.

A multilayer deposition to keep the stoichiometry was also examined.[202] A new deposition process, pulsed laser deposition (PLD), has also been evaluated for controlled deposition of high-T_c superconductors.[203] Chemical vapor deposition is also considered an available method for deposition of new high-T_c superconductors.[204]

Thin-film deposition of the high-T_c superconductors of Bi-Sr-Ca-Cu-O and/or Tl-Ba-Ca-Cu-O systems has been tried using methods similar to those of La-Sr-Cu-O and/or Y-Ba-Cu-O thin films. These deposition methods are listed in Table. 5.3.

In these deposition methods, the most important problem is keeping the correct composition. The layer-by-layer deposition in an atomic scale proposed by Adachi is one of the most promising methods for the controlled deposition of high-T_c superconductors.[205]

Aside from deposition, extensive studies on passivation and/or microfabrication have been done by several workers.[206][207] ECR plasma CVD can be used for making a stable passivation layer onto high-T_c superconducting thin films due to its low working pressure and low deposition temperatures.[208]

5.1.4.2 Basic Thin Film Processes

The basic processes for the deposition of perovskite thin films are shown in Table 5.27. Thin films of amorphous phase are deposited at the substrate temperature T_s below the crystallizing temperature T_{cr}, which is 500°–700°C for the perovskite-type oxides. In some cases, a different crystal structure appears at substrate temperatures below the T_{cr} for the perovskite structure. For thin films of $PbTiO_3$, the pyrochlore phase appears at substrate temperatures of around 400°C.[209]

Thin films of polycrystalline phase are deposited at $T_s > T_{cr}$. This phase is also achieved by deposition of the amorphous phase followed by postannealing at temperatures above T_{cr}.

Thin single-crystal films are epitaxially deposited on a single-crystal substrate at substrate temperatures above the epitaxial temperature T_e ($T_e > T_{cr}$). The amorphous thin films deposited on the single-crystal substrate will be converted into single-crystalline thin films after postannealing at temperatures above T_e owing to solid-phase epitaxy.

Table 5.27. Basic Process of the Deposition of Perovskite Thin Films

Amorphous phase	$T_s < T_{cr}$
Polycrystalline	$T_s > T_{cr}$ $T_s < T_{cr}$, postannealing
Single crystals (single-crystal substrate)	$T_s > T_e$ $T_s < T_{cr}$, postannealing (solid-phase epitaxy)

Note: T_s, substrate temperature during deposition; T_{cr}, crystallizing temperature during deposition; T_e, epitaxial temperature.

Several kinds of deposition processes are proposed for perovskite-type oxides, including electron beam deposition,[210] laser beam deposition,[211] cathodic sputtering,[212] and chemical vapor deposition.[213] Oxidization is considered necessary for the deposition of high-T_c superconductors. For this purpose, oxygen, ozone, and/or oxygen ions are supplied onto the growing surface of thin films during deposition by the electron beam and/or molecular beam deposition system as shown in Fig. 5.92.

In cathodic sputtering, thin films of perovskites are deposited directly from the compound ceramic target of perovskites in an rf-system. RF-magnetron sputtering is commonly used for deposition from a compound ceramic target. A sintered ceramic plate or sintered ceramic powder is used for the sputtering target. In dc-magnetron sputtering, the metal targets of A-site and B-site elements are sputtered in an oxidizing atmosphere.

Wehner and coworkers have determined that, in a planar diode system, the film composition from a multicomponent target is a function of substrate location, and is usually different from that of the target. The main reason for this is that different atomic species are sputtered with different angular distributions. An additional problem arises with negative oxygen ions that cause presputtering from substrates located opposite the target. These complications disappear when sputtering is performed using spherical targets,[214] although this is generally an impractical means of film formation.

In chemical vapor deposition, metal-organic compounds such as $Y(C_{11}H_{19}O_2)_3$, $Ba(C_{11}H_{19}O_2)_2$, and $Cu(C_{11}H_{19}O_2)_2$ are tentatively used as the source for deposition of YBC thin films. Halides such as $BiCl_3$, CuI, CaI_2, and SrI_2 are used as the source for deposition of BSCC thin films.[215]

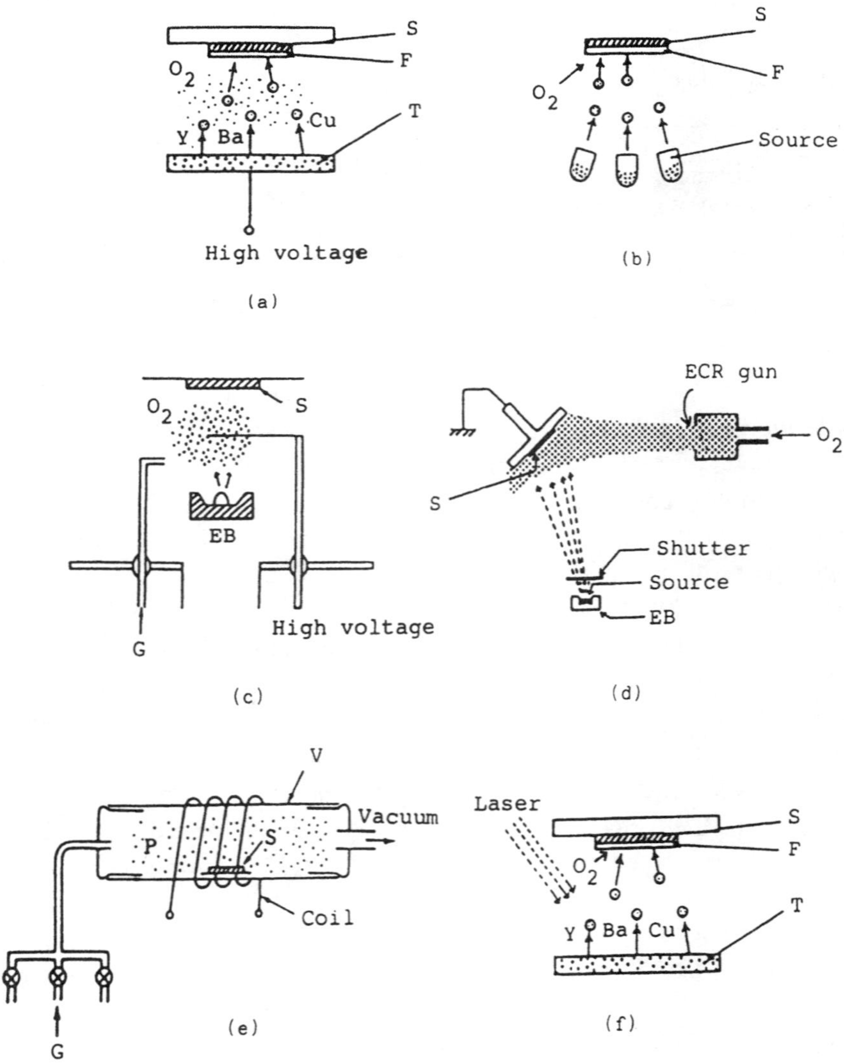

Figure 5.92. Typical deposition systems for the high-T_c superconductors; (*a*) sputtering, (*b*) reactive evaporation, (*c*) activated reactive evaporation, (*d*) ion assisted evaporation, (*e*) plasma CVD, (*f*) reactive laser abrasion. (S: substrate, F: thin film, T: target, V: vac. chamber, EB: electron beam, G: reactive gas source.)

For the deposition of single-crystal films, the selection of the substrate crystal will affect the crystal properties of the resultant films. The crystallographic properties of the crystal substrates used for epitaxial growth of perovskite-type oxides are shown in Table 5.28.[216]

Figure 5.93 shows a typical epitaxial relationship between high T_c superconductors and cubic substrates. The c-axis of the epitaxial films will be perpendicular to the (100) plane of the substrate crystals. The isotropic superconducting currents will flow in the (001) plane of the deposited films. On the (110) plane of the cubic crystal substrates the c-axis of the epitaxial films will lie in the films. Large anisotropy will be expected for the current flow in the (110) plane of the epitaxial films.

It should be noted that most of the crystal substrates exhibit cubic structure. The crystal twin will often be formed in epitaxial films of orthorhombic superconductors, and orthorhombic substrates are important for the reduction of them. Besides crystallographic properties, the possibility of mutual diffusion at the film and substrate interface should be considered in the selection of the substrates.[217]–[218]

A number of experiments have been done for deposition of high-T_c thin films. For the Y-Ba-Cu-O system, three deposition processes are classified as indicated in Table 5.29.

The process for making YBC ceramics is composed of three stages: (1) mixing, (2) annealing for crystallization and sintering, and (3) annealing for the control of oxygen vacancies and/or crystal structure. In the annealing process, the oxidation of copper will be promoted and the density of Cu^{+1} and/or Cu^{+2} decrease and Cu^{+3} density increases.

For YBC thin films, three processes are considered:

- Process (1) is deposition at low substrate temperature followed by postannealing. This process is commonly used for deposition of high-T_c superconducting thin films, since stoichiometric composition of the thin films is relatively easily achieved. Single-crystal thin films are expected to be obtained under the condition of solid-phase epitaxy during the postannealing process. At present, however, the resultant films show the polycrystalline phase.
- Process (2) is deposition at high substrate temperature followed also by a postannealing process.
- Process (3) is deposition at high substrate temperature without a subsequent postannealing process.

Table 5.28. Substrates for the Deposition of High-T_c Superconducting Thin Films

	Crystal System	Structure	Lattice Constants	Thermal Expansion Coefficient (K^{-1})	Dielectric constant
Substrates					
sapphire (α-Al_2O_3)	trigonal	corundum	hex.axes a' = 4.763 c' = 13.003	$\parallel$c 8×10^{-6} $\perp$c 13×10^{-6}	9.9
MgO	cubic	NaCl	a = 4.203	8×10^{-6}	17
$MgAl_2O_4$	cubic	spinel	a = 8.059	7.6×10^{-6}	
YSZ	cubic	fluorite	a = 5.16	10×10^{-6}	27
$SrTiO_3$	cubic	perovskite	a = 3.905	9×10^{-6}	
$LaGaO_3$	orthorhombic	perovskite	a = 5.482 b = 5.526 c = 7.780	10×10^{-6}	25
$LaAlO_3$	pseudo cubic	perovskite	a = 3.792	10×10^{-6}	16
Nd: $YAlO_3$	orthorhombic	perovskite	a = 5.18 b = 5.33 c = 7.37	2.2×10^{-6}	
$NdGaO_3$	orthorhombic	perovskite	a = 5.426 b = 5.502 c = 7.706	3×10^{-6}	22
Si	cubic		a = 5.43	2.6×10^{-6}	12
GaAs	cubic		a = 5.65	6.86×10^{-6}	13
Superconductors					
$La_{1.8}Sr_{0.2}CuO_4$ (LSC)	tetragonal#	K_2NiF_4	a = 3.78 c = 13.23	$10–15\times10^{-6}$	
$YBa_2Cu_3O_x$ (YBC)	orthorhombic	oxygen deficient perovskite	a = 3.82 b = 3.89 c = 11.68	14×10^{-6} 12×10^{-6} 25×10^{-6}	
Bi-Sr-Ca-Cu-O (BSCC)	Pseudo tetra.	Bi-layered structure	a = 5.4 c = 30, 36	12×10^{-6}	
Tl-Ba-Ca-Cu-O (TBCC)	Pseudo tetra.	Bi-layered structure	a = 5.4 c = 30, 36		

Superconductivity at orthorhombic phase.

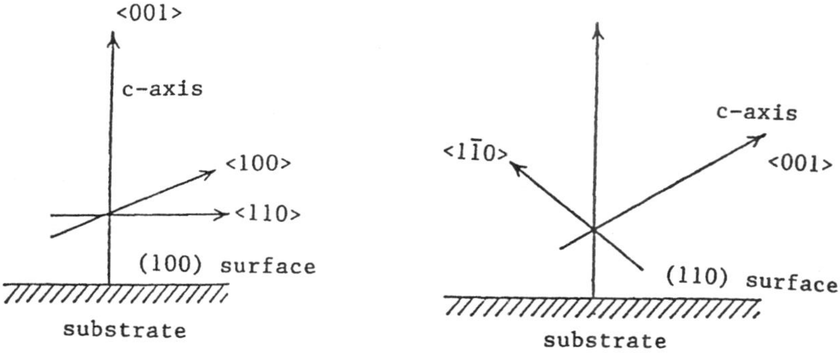

Figure 5.93. Epitaxial relations of the high-T_c superconducting thin films on crystal substrate.

Table 5.29. Fabrication Processes for High-T_c Superconducting Ceramics and Thin Tilms

		Chemical Composition	Crystallization	Oxygen Vacancy and/or Structural Control
Ceramics		mixing	sintering (850°–950°C)	annealing* (850°–950°C)
Thin films	1		annealing (850°–950°C)	annealing* (850°–950°C)
	2	deposition ($T_s > T_{cr}$)		annealing* (400°–950°C)
	3	deposition** ($T_s > T_{cr}$)		

Note: T_s, substrate temperature during deposition; T_{cr}, crystallizing temperature (500°–700°C); *slow cooling; **quenching.

Single-crystal films can be obtained by vapor-phase epitaxy achieved in processes (2) and (3). These considerations may also be adopted for deposition of the LSC (La-Sr-Cu-O) system and the BSCC (Bi-Sr-Ca-Cu-O) and/or the TBCC (Tl-Bi-Ca-Cu-O) system. However, in the LSC system, the oxidation of copper will be promoted by the substitution of La^{+2} site by Sr^{+3} during the postannealing process. In the BSCC and/or TBCC system, the rearrangements of Sr, Ba, and/or Ca will act as the oxidation promoter, as described in Sec. 5.1.4.7.

5.1.4.3 Synthesis Temperature

As indicated in Table 5.29, the synthesis temperature of high-T_c superconducting ceramics is around 850°–950°C. Lowering it is very important not only for scientific interests but also for fabrication of thin-film superconducting devices. Table 5.29 shows that the maximum temperature in the thin-film process may be governed by the postannealing process. In thin-film process (2) for rare-earth $YBa_2Cu_3O_x$ (YBC) superconductors, the synthesis temperature is governed by the postannealing process for control of the oxygen vacancies if the as-deposited thin films are crystallized.

The structural analysis for $YBa_2Cu_3O_x$ ceramics suggests that the structural transition from the nonsuperconducting tetragonal phase ($x < 6.3$) to the superconducting orthorhombic phase ($7 > x > 6.3$) occurs around 700°C, and the latter phase is predominant at annealing temperatures below 600°C as shown in Fig. 5.91.[219]

It is known that when the annealing temperature is near the tetra/ortho transition temperature, superconductors show the ortho-II phase with $T_c = 50$–60 K ($6.7 \geq x > 6.3$). The ortho-I phase with $T_c = 90$ K ($7 > x > 6.7$) is obtained at a lower annealing temperature below 600°C. The stability of the ortho phase is sensitive to the partial pressure of the oxygen.[220]

Similar results are obtained in Gd-Ba-Cu-O (GBC) films.[221] Figure 5.94 shows typical experimental results on the variations of the c-axis lattice parameter with the postannealing temperature for the c-axis-oriented GBC films of the tetragonal phase, c = 11.83–11.84 Å. It shows that the c-axis lattice constant is reduced by postannealing in O_2. The films annealed at 550°–650°C with c = 11.73–11.75 Å may correspond to the ortho-II phase. The high-T_c ortho-I phase with c = 11.71–11.73 Å and $T_c = 90$ K is obtained at an annealing temperature of 350°–550°C. These structural analyses suggest that the synthesis temperature is not governed by the postannealing temperature for YBC films but by the crystallizing temperature.

For rare-earth $La_{2-x}Sr_xCuO_4$ thin films, the as-deposited films show the insufficient oxidation of copper. Oxidation takes place during the postannealing process. The maximum synthesis temperature for LSC thin films may also be governed by the crystallization temperature, since oxidation will take place below it. The crystallization temperature for these rare-earth, high-T_c oxides is around 500°–600°C. For rare-earth-free, high-

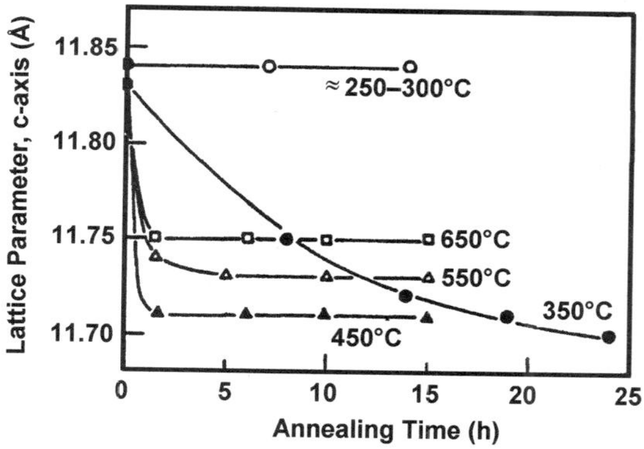

Figure 5.94. Variation of the c-axis lattice parameter of the Gd-Ba-Cu-O thin films with postannealing in O_2 atmosphere at various temperatures.

T_c superconductors, the minimum synthesis temperature may also correspond to the crystallization temperature of perovskites, 500°–700°C.

Note that in YBC thin films, the partial substitution of O-sites by S shows a tendency to decrease the crystallizing temperature accompanied by a sharpened transition.[222]–[223]

5.1.4.4 Low-Temperature Processes, In-Situ Deposition

It is possible that lowering the synthesis temperature for the YBC system can be achieved by deposition at the crystallizing temperature of 500°–700°C followed by postannealing in O_2 at a temperature of 350°–550°C. The maximum temperature for rare-earth, high-T_c film processes is then governed by the crystallizing temperature of 500°–700°C.

Several studies have been done on the low temperature synthesis of rare-earth, high-T_c superconductors. These processes are classified into two types:

(i) *In-situ annealing:* Deposition at a substrate temperature above the crystallizing temperature T_{cr} (500°–700°C), followed by postannealing at a lower temperature (400°–600°C) in O_2.

(ii) *In-situ deposition:* Deposition at a substrate temperature above the crystallizing temperature without any additional postannealing process.

Type (i), in-situ annealing, corresponds to process (2),* that is, postannealing is conducted successively in the deposition equipment without breaking the vacuum. In type (ii), in-situ deposition, without any postannealing process. Type (ii) corresponds to process (3).*

The low-temperature process with either in-situ annealing or in-situ deposition was studied in several deposition processes including sputtering,[224] pulsed laser deposition,[225] and reactive deposition.[226] In these studies, the temperature of the substrates during deposition was 500°–700°C. The annealing was done at around 400°–500°C in O_2. The results showed that the low-temperature process without annealing (in-situ deposition) is readily available for making thin film electronic devices since it achieves the formation of the multilayered structure of high-T_c superconductors.

In the YBC system, the as-deposited films will show the single-crystal phase and exhibit superconductivity without postannealing if enough oxygen is supplied onto the film surface during deposition to oxidize the deposited films, and if the substrate temperature during deposition T_s satisfies the relationship,

Eq. (5.15) $$T_e \leq T_s \leq T_t$$

where T_e denotes the epitaxial temperature, and T_t the transition temperature from the tetragonal to the orthorhombic phase.

In-situ deposition of the YBC system has been attempted by magnetron sputtering. It has also been confirmed that irradiation of oxygen ions and/or plasma onto the deposited film during deposition is important in order to achieve in-situ deposition. Under suitable irradiation of oxygen plasma onto the film surface, excellent superconducting transition temperatures were observed in as-deposited Er-Ba-Cu-O (EBC) thin films. The onset temperatures were 95 K with a zero-resistance temperature of 86 K for films deposited at 650°C.[81]

Several advantages of in-situ deposition have been found such as the smooth surface of the deposited films and the small interdiffusion between deposited superconducting films and substrates.[198] In-situ deposition, however, has exhibited low critical-current density due to the presence of crystal boundaries in the deposited films, which will form weak links. The magnitude of critical currents is strongly affected by the application of an external magnetic field when its direction is parallel to the

---

*See the description of processes (1), (2) and (3) in Sec. 5.1.4.2.

c-axis of the oriented superconducting films.[227] These weak points observed in in-situ deposition may result from imperfect crystallinity of the deposited films, which is essentially improved by refinement of the deposition system.

In rare-earth-free, high-T_c superconductors of the BSCC and/or TBCC systems, several different superconducting phases are simultaneously formed during the postannealing process. Although the basic thin-film processes for rare-earth-free, high-T_c superconductors are essentially the same as those of rare-earth, high-T_c superconductors, the simultaneous growth of the different superconducting phase causes difficulty in the controlled deposition of the single-phase, high-T_c superconducting thin films.

5.1.4.5 Deposition of Rare-Earth, High-T_c Superconductors

The simplest method for making rare-earth, high-T_c films is the deposition of amorphous films by sputtering at a low substrate temperature followed by a postannealing process (process 1). Typical sputtering conditions for YBC films are shown in Table 5.30. These targets were obtained by reacting a mixed powder of Y_2O_3 (99.99%), $BaCO_3$ (99.99%), and CuO (99.9%) in air at 900°C for 8 hours and then sintering at 900°C for 8 hours in air.

The measurement of resistivity was carried out using the standard four-probe technique with gold electrodes fabricated on the surface of the films. The measured current density was about 5 A/cm^2. Samples were fixed to the copper block and the temperature was measured by a Chromel-Au (Fe) thermocouple attached to the copper block. Figure 5.95 shows

Table 5.30. Sputtering Conditions for the Deposition of High-T_c Thin Films (Process 1)

Target	$(Y_{0.4}Ba_{0.6})_3Cu_3O_x$ (100 mm in dia.)
Substrate	$(1\bar{1}02)$ plane of sapphire
Substrate temperature	200°C
Sputtering gas	Ar
Gas pressure	0.4 Pa
RF input power	150 W
Growth rate	150 Å/min

photographs of the samples. The as-sputtered films were insulating and brown in color. After postannealing at 900°C for 1 hour in O_2, the films showed superconductivity as shown in Fig. 5.96.*

---

*High frequency surface impedance Z_s, Hall mobility, and a critical current/critical magnetic field are also important electrical properties of high-T_c superconductors. The Z_s is expressed by $Z_s = \mu_0\sigma\lambda^3\omega^2/2 + j(\mu_0\lambda\omega)$, where μ_0, σ, λ, and ω are absolute permeability of free space, electrical conductivity, London penetration depth, and measurement frequency, respectively. The impedance measurements are conducted by a strip line or a cavity resonator at the frequency range of 1 GHz to 1000 GHz. The surface impedance of YBCO thin films is 10^{-5} to 10^{-6} at 10 GHz and 77 K, which is three orders in magnitude lower than that of copper. (Chin, C. C., Rainville, P. J., Dreman, A. J., Devor, J. S., Steinbeck, J., Dresselhaus, G., and Dresselhaus, M. S., *J. Mater. Res.*, 5:1599 [1990])

The van der Pauw method is used for the Hall measurement. A test sample plate of thickness d has four measurement points, A, B, C, D. For the isotropic conductor, the resistivity, ρ, is obtained by the relation

$$\exp(\pi R_{AB,CD}\, d/\rho) + \exp(-\pi R_{BC,DA}\, d/\rho) = 1$$

where $R_{AB,CD}$ denotes the potential difference between D and C (= $V_D - V_C$) when the unit current passes through from A to B, $R_{BC,DA}$ denotes the potential difference between A and D (= $V_A - V_D$) when unit current passes through from A to D. The Hall mobility $\mu_H (= R/\rho)$ is obtained by (d/B) × ($\Delta R_{BDAC}/\rho$), where the shift of R_{BDCD} becomes ΔR_{BDAC}, under the application of the magnetic field B perpendicular to the sample plate. Typical measurement results are shown in the table below.

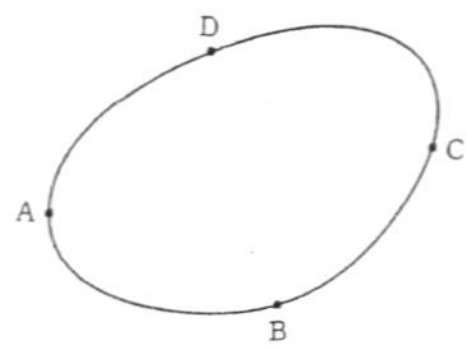

Sample	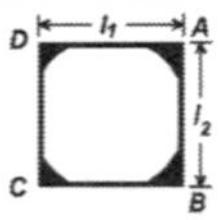	$l_1 \approx l_2 = 1$ cm $d = 2000$ Å	Sputtered Bi-Sr-Ca-Cu-O thin film
Magnetic Field		4.6 kOe	
Current		1 mA, room temperature	
$V_D - V_c = R_{AB,CD}$		37.34 mV	
$V_A - V_D = R_{BC,DA}$		66.08 mV	
$\Delta(V_A - V_C) = \Delta R_{BD,AC}$		0.030 mV	
$\rho \approx 1.04 \times 10^{-3}$		$\exp(-\pi R_{AB,DC}\, d/\rho)$	
$n \approx 3.09 \times 10^{21}\,\text{cm}^{-3}$		$+ \exp(-\pi R_{BC,DA}\, d/\rho) = 1$	
$\mu_H \approx 3.025\ \text{cm}^2/\text{V·s}$		$\mu_{H} = (d/B)\,(\Delta R_{BD,AC}/\rho)$	

The carrier concentration, n, is also evaluated by the measurement of a plasma frequency, ω_p, which is equal to $(4\pi\, ne^2/m^*)^{1/2}$, where m^* is the electron effective mass. The ω_p is measured by optical spectroscopy.

The critical current density, J_c, is directly evaluated by a current transportation method or by a measurement of antimagnetization, M. The J_c is estimated by the Bean's equation $J_c = 30(M/R)$, where R denotes the equivalent radius of the sample. The SQUID system is used for the measurement M.

Figure 5.95. A photograph of a thin-film superconductor for measuring resistive properties.

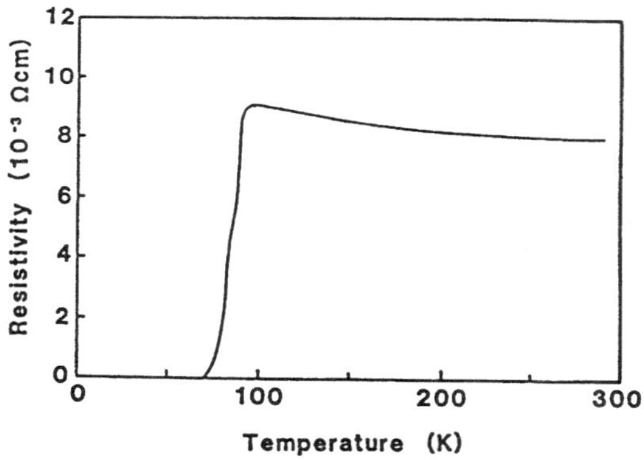

Figure 5.96. Temperature dependence of resistivity for sputtered Y-Ba-Cu-O thin films on (1102) sapphire (process 1).

The onset temperature was 94 K with zero resistivity at 70 K.[80] These annealed YBC films showed a polycrystalline phase with the preferred orientation of (103) crystal axis perpendicular to the substrate, as shown in Fig. 5.97. The (103) surface of YBC corresponds to the closest packed plane. The temperature dependence of resistivity shown in Fig. 5.96 suggests that YBC films are composed of the superconducting orthorhombic phase and the semiconducting tetragonal phase since the resistivity indicates that above the transition temperature, the material is semiconductive. The relatively high resistivity, 2 mΩcm at the transition temperature, may also result from the presence of grain boundaries in the YBC films. The sharp superconducting transition with a single superconducting ortho-I phase is obtained for stoichiometric composition in deposited films.

Improvement of the crystallinity of high-T_c films is achieved by deposition at a higher substrate temperature (process 2). Table 5.31 shows typical sputtering conditions for the improvement of crystalline properties of GBC films. Since the concentrations of Ba and Cu in GBC films are reduced at higher substrate temperatures, the composition of the target is modified so as to achieve stoichiometric composition for sputtered GBC films.[228] Typical superconducting properties of these GBC films are shown in Fig. 5.98. The low resistivity, less than 0.5 mWcm, at the transition temperature corresponds to bulk resistivity was observed for YBC films. The temperature dependence of resistivity was metallic at temperatures above the transition temperature. Similar properties were also obtained by electron beam deposition.[77]

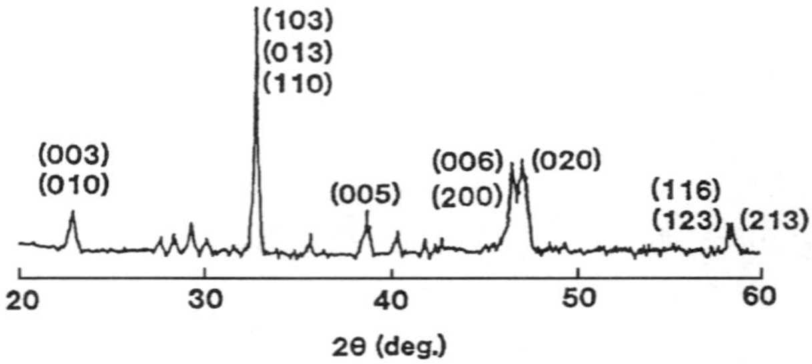

Figure 5.97. Typical x-ray diffraction pattern of sputtered Y-Ba-Cu-O thin film on (100) $SrTiO_3$ (process 1).

Table 5.31. Sputtering Conditions of Low Temperature Deposition for Gd-Ba-Cu-O Thin Films (Process 2)

Target	$GdBa_2Cu_3O_x$ (100 mm in dia.)
Substrate	(100) plane of MgO
Substrate temperature	600° and 750°C
Sputtering gas	$Ar + O_2$ (3:2)
Gas pressure	0.4 Pa
RF input power	130 W
Growth rate	80 Å/min
Target-substrate spacing	25–35nm

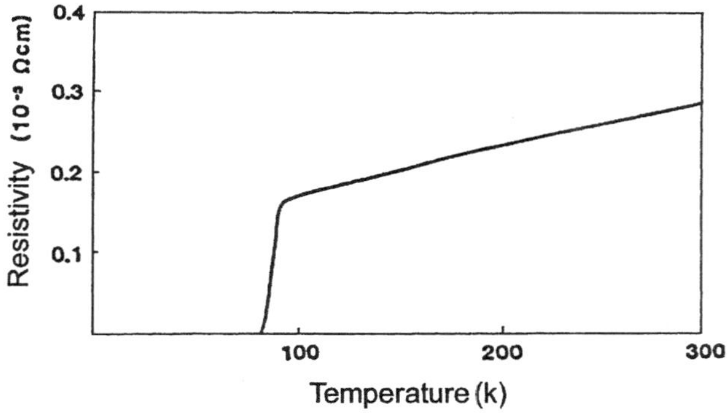

Figure 5.98. Temperature dependence of resistivity for sputtered Gd-Ba-Cu-O thin films on (100) MgO (process 2).

However, epitaxial YBC thin films with single-phase YBC are generally difficult to deposit, since the composition of the films often differs from the stoichiometric value of YBC at substrate temperatures above T_{cr} (500°–700°C). In contrast, epitaxial LSC thin films are easily deposited since LSC is composed of a solid solution.

Typical sputtering conditions for LSC thin films are shown in Table 5.32.[229] The target was stoichiometric $(La_{0.9}Sr_{0.1})_2CuO_4$ and was made by sintering a mixture of La_2O_3 (99.99%), $SrCO_3$ (99.9%), and CuO (99.9%) at 900°C in air for about 8 hours. The as-sputtered films were conductive with a black color similar to the target. Electron-probe x-ray microanalyses showed that the concentrations of La, Sr, and Cu were close to the target composition. The electron diffraction pattern suggested that an excellent single crystal was epitaxially grown on the substrate. However, when LSC films were deposited on (100) MgO, the resultant films showed the polycrystalline phase. This is possibly due to the large lattice mismatch between LSC and MgO.

These as-sputtered films showed semiconductive behavior. The superconductivity was observed after postannealing in air at 900°C for 3 days. Typical electron diffraction patterns and electrical properties of these sputtered films are shown in Fig. 5.99. Single-crystal LSC films grown on (100) $SrTiO_3$ exhibited excellent superconducting properties. The onset temperature was ≈34 K with T_c = 25 K. The narrow transition width, less than 3 K, suggests that these sputtered films are composed of the single phase of layered perovskites, K_2NiF_4.

Table 5.32. Sputtering Conditions (Process 2)

Target	$(La_{0.9}Sr_{0.1})_2CuO_4$ (100 mm in dia.)
Substrate	(100) plane of $SrTiO_3$
Substrate temperature	600°C
Sputtering gas	Ar
Gas pressure	0.4 Pa
RF input power	150 W
Growth rate	100 Å/min

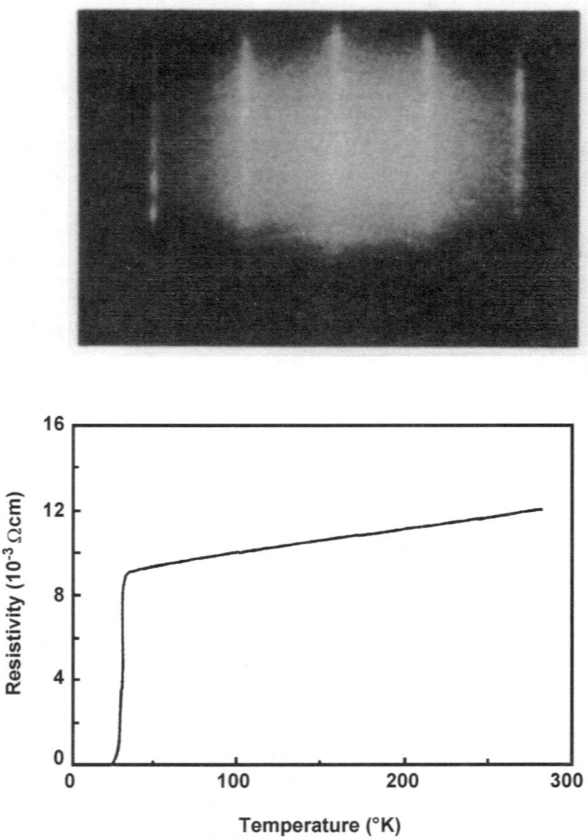

Figure 5.99. Electron diffraction pattern and temperature dependence of resistivity for sputtered La-Sr-Cu-O thin films on (100) $SrTiO_3$ (process 2).

However, these processes still need the troublesome postannealing. Diffusion at the film/substrate interface took place at the postannealing temperatures above 800°–900°C. This causes a broad transition due to mutual diffusion between substrate and the deposited films.[199] A discharge plasma of oxygen in sputtering is considered suitable for oxidation of thin films during deposition. Thus, if the thin films are immersed in the oxygen plasma during deposition, in-situ deposition (process 3) will potentially be achieved.[230]

Sputter deposition with two target–substrate spacings, 35 mm and 40 mm, was used in the preparation of Er-Ba-Cu-O films on MgO. Typical sputtering conditions are shown in Table 5.33. The target was made by sintering the mixture of Er_2O_3, $BaCO_3$, and CuO at 900°C for 20 hours in air. The surface of the substrate was exposed to the discharge plasma with a target spacing of 35 mm. For spacing of 40 mm, the substrate is situated outside of the plasma.

The temperature dependence of resistivity for as-sputtered films is shown in Fig. 5.100. The film made with a spacing of 35 mm showed a sharp superconducting transition with onset at 92 K and $T_{R=o}$ = 86 K. On the other hand, the film with spacing of 40 mm exhibited a much broader superconducting transition, and zero resistance was realized at 57 K. It is considered that the effect of target spacing on superconducting properties results from the difference of oxidation in the films. We can only roughly presume oxidation of the films from the crystalline information obtained. Sufficient oxidation leads surely to the superconducting orthorhombic structure, while oxygen defects cause the semiconducting tetragonal structure.

Table 5.33. Sputtering Conditions (Process 3)

Target	$Er_1Ba_2Cu_{4.5}O$ (100 mm in dia.)
Substrate	(100) MgO and (110) $SrTiO_3$
Sputtering gas	Ar + O_2 (4:1)
Gas pressure	0.4 Pa
RF input power	175 W
Substrate temperature	650°C
Growth rate	70 Å/min

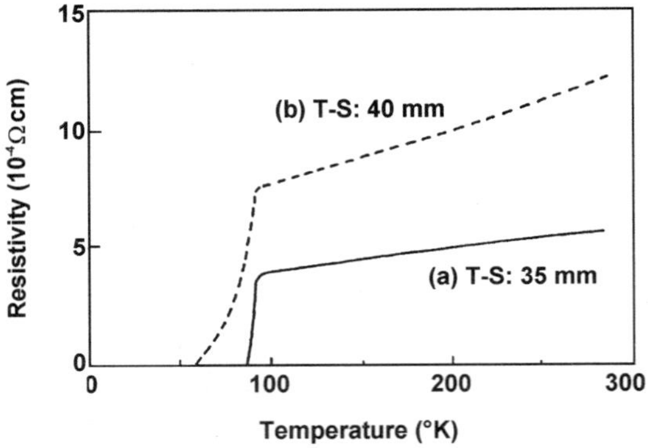

Figure 5.100. Temperature dependence of the resistivity for Er-Ba-Cu-O thin films on (100) MgO (process 3).

Figures 5.101a and 5.101b show the x-ray diffraction patterns of films made with spacings of 35 mm and 40 mm, respectively. The c-axis is primarily oriented perpendicular to the film plane. The crystal system can be discriminated by the lattice constant c, i.e., c = 11.68 Å for the orthorhombic structure (*o*), and c = 11.8–11.9 Å for the tetragonal structure (*t*). The film made with a spacing of 35 mm shows a mixed structure with dominant orthorhombic and minor tetragonal phases. On the other hand, the film with a spacing of 40 mm shows the tetragonal structure. From these results, it is evident that oxidation progresses more for a spacing of 35 mm than for 40 mm. For comparison, the x-ray diffraction pattern of the film $T_{R=0}$ = 55 K, made with a target-to-sample distance of 35 mm and at a higher substrate temperature of 700°C, is shown in Fig. 5.101c. The film shows the tetragonal structure. Since deposition was carried out at a temperature higher than the *t-o* transition, the oxidation in passing through the *t-o* point was not sufficient for quick cooling.

On $SrTiO_3$ (110) substrates, epitaxial films are prepared by the same process. Figure 5.102 shows the RHEED pattern of the epitaxial Er-Ba-Cu-O film. The temperature dependence of resistivity for as-deposited films shows similar characteristics, and zero reactivity is realized below 80 K. The Y-Ba-Cu-O films are also prepared by this process.

These facts suggest the possibility of in-situ deposition (process 3) in Table 5.29, although the in-situ deposited films are not composed of the single phase of the orthorhombic structure.

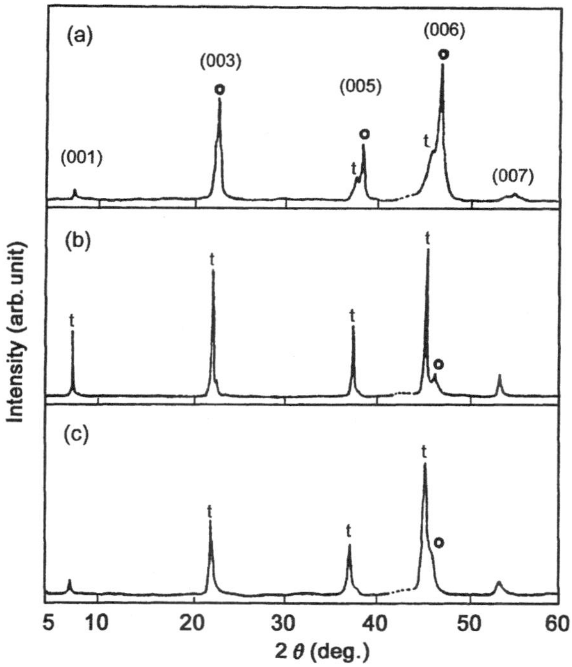

Figure 5.101. X-ray diffraction patterns for Er-Ba-Cu-O thin films on (100) MgO (process 3).

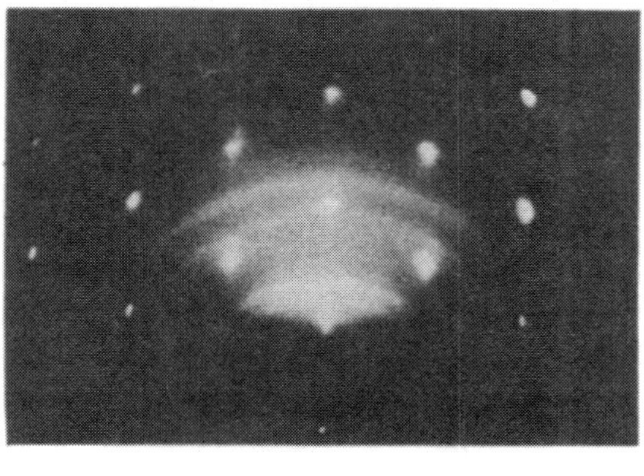

Figure 5.102. Electron diffraction pattern of the epitaxial Er-Ba-Cu-O thin film on (100) $SrTiO_3$ (process 3).

The experiments on Gd-Ba-Cu-O thin films suggest that in-situ postannealing in O_2 at the relatively low temperature of 400°–609°C increases the orthorhombic phase and improves the superconducting properties.[231] The effects of low-temperature postannealing are also verified in pulsed laser deposition.

5.1.4.6 Deposition of Rare-Earth-Free, High-T_c Superconductors

In rare-earth-free, high-T_c superconductors, several superconducting phases are present for different chemical compositions. Typical chemical compositions for the Bi system and Tl system are listed in Table 5.34. Their superconducting properties have not yet been fully explained.

Table 5.34. Rare-Earth-Free High-T_c Superconductors

		T_c (K)	Institute	Date
Bi -system: $Bi_2O_2 \cdot 2SrO \cdot (n\text{-}1)Ca \cdot nCuO_2$				
$Bi_2Sr_2CuO_6$	(2 2 0 1)*	≈ 7–22	Caen Univ. (France) Aoyamagakuin Univ. (Japan)	1987.5
$Bi_2Sr_2CaCu_2O_8$	(2 2 1 2)*	80	National Res. Institute for Metals (Japan)	1988.1
$Bi_2Sr_2Ca_2Cu_3O_{10}$	(2 2 2 3)*	110	National Res. Institute for Metals (Japan)	1988.3
$Bi_2Sr_2Ca_3Cu_4O_{12}$	(2 2 3 4)*	≈ 90	Matsushita Elec. (Japan)	1988.9
Tl -system: $Tl_2O_2 \cdot 2BaO \cdot (n\text{-}1)Ca \cdot nCuO_2$				
$Tl_2Ba_2CuO_6$	(2 2 0 1)*	≈ 20–90	Institute for Molecular Sci. (Japan) Arkansas Univ. (USA)	1987.12
$Tl_2Ba_2CaCu_2O_8$	(2 2 1 2)*	105	Arkansas Univ. (USA)	1988.2
$Tl_2Ba_2Ca_2Cu_3O_{10}$	(2 2 2 3)*	125	Arkansas Univ. IBM (USA)	1988.3
Tl -system: $TlO \cdot 2BaO \cdot (n\text{-}1)Ca \cdot nCuO_2$				
$TlBa_2CaCu_2O_7$	(1 2 1 2)*	≈ 70–80	IBM (USA)	1988.5
$TlBa_2Ca_2Cu_3O_9$	(1 2 2 3)*	≈ 110–116	IBM (USA)	1988.3
$TlBa_2Ca_3Cu_4O_{11}$	(1 2 3 4)*	120	ETL (Japan)	1988.5
$TlBa_2Ca_4Cu_5O_{13}$	(1 2 4 5)*	<120	ETL (Japan)	1988.5

* Numbers indicate a composition; e.g., $Bi_{\underline{2}}Sr_{\underline{2}}Ca_{\underline{1}}Cu_{\underline{2}}O_8 \rightarrow$ (2 2 1 2).

Thin films of the Bi-Sr-Ca-Cu-O system are prepared by rf planar magnetron sputtering similar to YBC films. The target is complex oxides of Bi-Sr-Ca-Cu-O, which is made by sintering a mixture of Bi_2O_3 (99.999%), $SrCO_3$ (99.9%), $CaCO_3$(99%), and CuO (99.9%) at 880°C for 8 hours in air.

It is known that superconducting properties are strongly affected by the substrate temperature during deposition. Figure 5.103 shows typical x-ray diffraction patterns with resistivity-temperature characteristics for Bi-Sr-Ca-Cu-O thin films, about 0.4 μm thick, deposited at various substrate temperatures. It shows that films deposited at 200°C exhibit a $Bi_2Sr_2CaCu_2O_x$ structure with the lattice constant c = 30.64 Å, which corresponds to the low-T_c phase.[232] The films show a zero-resistance temperature of about 70 K; see Fig. 5.103 (top).

When the substrate temperature is raised during deposition of the high-T_c phase with T_c = 110 K, the $Bi_2Sr_2Ca_2Cu_3O_x$ structure with the lattice constant c =36 Å, is superposed on the x-ray diffraction pattern; see Fig. 5.103 (center).[233] At a substrate temperature of around 800°C, a single high-T_c phase is observed. The films show a zero-resistance temperature of 104 K, shown in Fig. 5.103 (bottom).

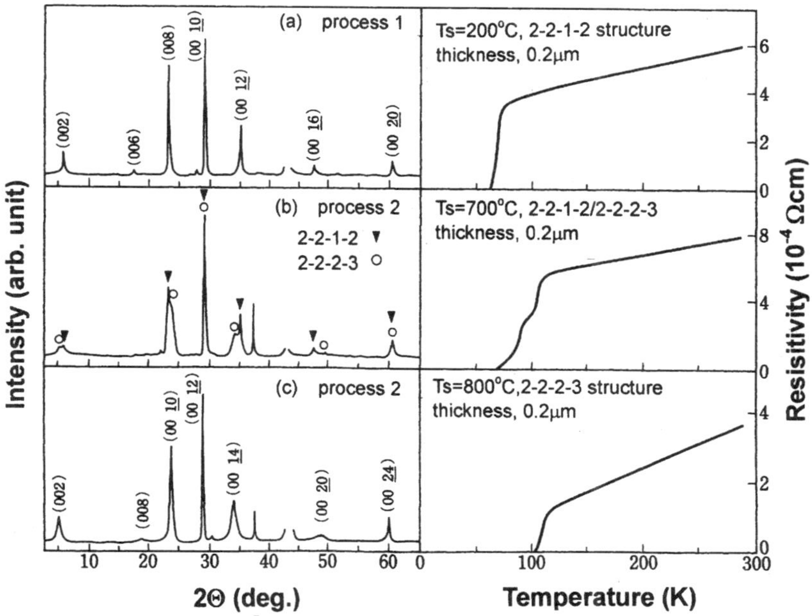

Figure 5.103. X-ray diffraction patterns with resistivity vs temperature for the annealed Bi-Sr-Ca-Cu-O films deposited on (100)MgO substrates.

Typical sputtering conditions are shown in Table 5.35. The target is complex oxides of Bi-Sr-Ca-Cu-O. The compositions are near the 1-1-1-2 ratio of Bi-Sr-Ca-Cu. Processes (1) and/or (2) are used for deposition. Single crystals of (100) MgO are used as substrates. The superconducting properties are improved by postannealing at 850°–900°C for 5 hours in O_2.[234]

It is noted that the formation of these superconducting phases strongly depends on the annealing temperature and chemical composition as shown in Fig. 5.104.[235]

Similar to the Bi-Sr-Ca-Cu-O system, thin films of the Tl-Ba-Ca-Cu-O system are prepared by rf-magnetron sputtering on a MgO substrate. Typical sputtering conditions are shown in Table 5.36. However, their chemical composition is quite unstable during deposition, and the postannealing process is due to the high vapor pressure of Tl. Thin

Table 5.35. Sputtering Conditions for Bi-Sr-Ca-Cu-O Thin Films

Target	Bi:Sr:Ca:Cu = 1–1.7:1:1–1.7:2 (100 mm in dia.)
Sputtering gas	Ar/O_2 = 1–1.5
Gas pressure	0.5 Pa
RF input power	150 W
Substrate temperature	200°–800°C
Growth rate	80 Å/min

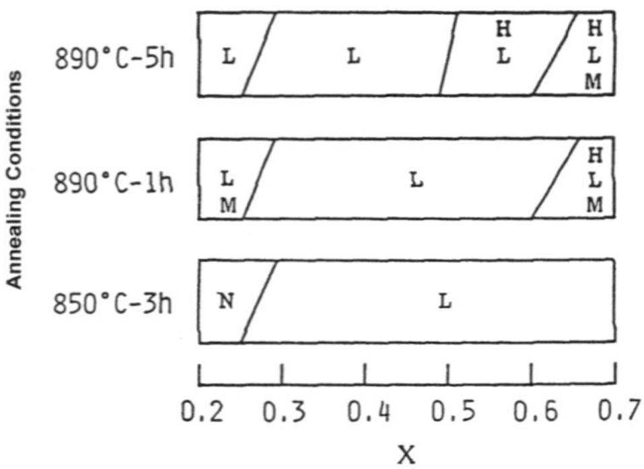

Figure 5.104. Variations of superconducting phase with annealing conditions for $Bi_2(Sr_{1-x}Ca_x)_{n+1}Cu_nO_y$ thin films: *H*, high-T_c phase ($n = 3$); *L*, low-T_c phase ($n = 2$); *M* and *N*, other phases.[235]

films of the Tl system are deposited without the intentional heating of substrates (< 200°C) and are annealed at 890°–900°C in Tl vapor.[92] It is seen that the superconducting phase of the resultant films strongly depends on postannealing conditions.

Figure 5.105 shows typical x-ray diffraction patterns with resistivity-temperature characteristics for Tl-Ba-Ca-Cu-O thin films annealed at different conditions.

Table 5.36. Sputtering Conditions for Tl-Ba-Ca-Cu-O Thin Films

Target	Tl:Ba:Ca:Cu = 2:1–2:2:3 (100 mm in dia.)
Sputtering gas	$Ar/O_2 = 1$
Gas pressure	0.5 Pa
RF input power	100 W
Substrate temperature	200°C
Growth rate	70 Å/min

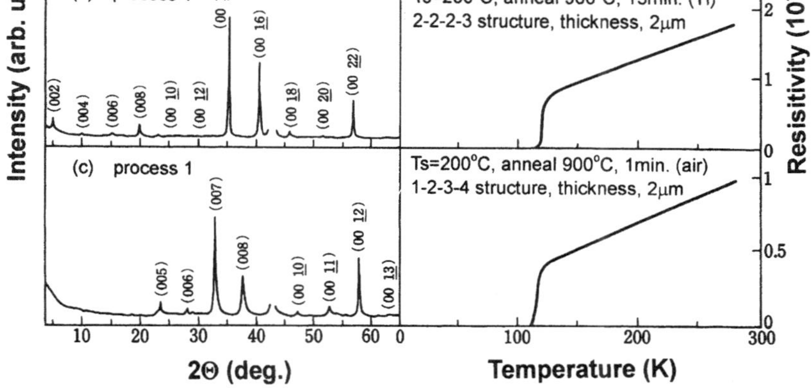

Figure 5.105. X-ray diffraction patterns with resistivity vs temperature for annealed Tl-Ba-Ca-Cu-O films deposited on (100)MgO substrates.

The 0.4-μm-thick film exhibits the low-temperature phase $Tl_2Ba_2CaCu_2O_x$ structure, with the lattice constant c = 29 Å after slight annealing for 1 minute at 900°C (Fig. 5.105a). The 2-μm-thick films annealed at 900°C for 13 minutes show the high-temperature phase $Tl_2Ba_2Ca_2Cu_3O_x$ structure with the lattice constant c = 36 Å (Fig. 5.105b). In specific annealing conditions, the other superconducting phase $TlBa_2Ca_3Cu_4O_x$ structure with the lattice constant c = 19 Å is also obtained (Fig. 5.105c).*

5.1.4.7 Structure and Structural Control

As described in a previous section, thin-film processing of high-T_c superconductors is classified into three processes: deposition at low substrate temperature with postannealing (process 1), deposition at high temperature with postannealing (process 2), and deposition at high temperature without postannealing (process 3).

---

*The high-T_c superconductors include a variety of compounds. The deposition with controlled crystal phase is essential. The no uniformity and/or off stoichiometry will lead a different crystal phase as shown in the table below. The stoichiometric composition is necessary for the deposition with controlled superconducting phase.

Deposition temperature (T_s)	Postannealing temperature (T_{anneal})	Phenomena observed during annealing	Superconducting phase
$T_s < T_{cr}$ process (1)	$T_{anneal} \sim T_{mp}$	melt crystallization segregation oxidization	low T_c (high T_c mixed phase)
$T_s > T_{cr}$ process (2)	$T_{anneal} \sim T_{mp}$	melt crystallization segregation oxidization	low T_c high T_c intergrowth
	$\sim T_{cr}$	crystallization oxidization	single phase**
	$\sim T_{cr}$	oxidization	single phase**
$T_s > T_{cr}$ process (3)		oxidization and crystallization during deposition	single phase
**at stoichiometric composition,		T_{mp}: melting point T_{cr}: crystallizng temp.	

One of the most important problems to be solved for thin-film processing is lowering the synthesis temperature. At present, lowering of the synthesis temperature can be achieved by both process (2) and process (3) for rare-earth, high-T_c superconductors.

In process (2), the lowering of the synthesis temperature can be achieved with low-temperature postannealing at around 400°–600°C. The minimum synthesis temperature is determined by the crystallizing temperature of the high-T_c superconductors in these processes. The crystallizing temperature of YBC, for instance, is around 500°–700°C. It is noted that in process (3), as-deposited films show superconducting properties without any postannealing (i.e., in-situ deposition).

As seen in the x-ray diffraction pattern, the in-situ-deposited films are composed of the orthorhombic phase and the tetragonal phase. The TEM image suggests that these films are composed of small crystallites as shown in Fig. 5.106. The TEM image of the sputtered films also denotes the presence of crystal boundaries as shown in Fig. 5.107. This may reduce the critical current J_c. The J_c for in-situ-deposited Er-Ba-Cu-O thin films is proportional to $(1 - T/T_c)^{1.8}$, which is close to $(1 - T/T_c)^{1.5}$. This indicates that the current transport will be partially governed by the weak link of superconductive regions. At present, high critical current is obtained in process (2) using postannealing.[235]

Although low-temperature-synthesized films are not perfect single crystals, the low-temperature process gives several favorable properties such as the suppression of interdiffusion at the film and substrate interface as shown in Fig. 5.108.

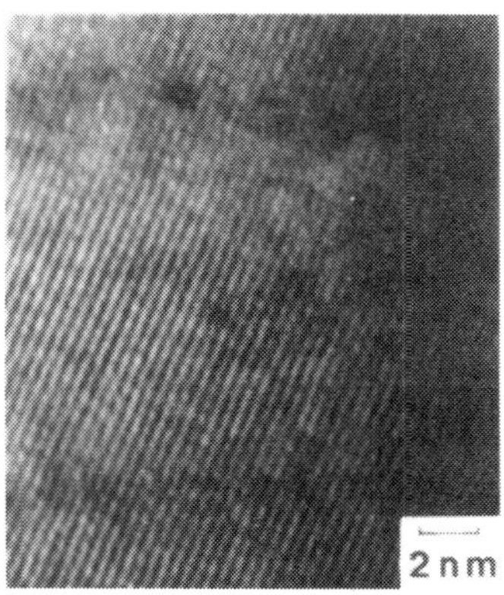

Figure 5.106. TEM image of in-situ deposited Er-Ba-Cu-O thin films on (100) MgO.

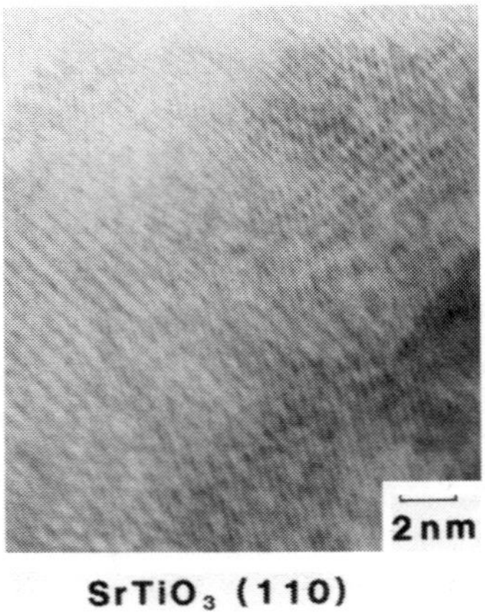

Figure 5.107. TEM image of in-situ deposited Er-Ba-Cu-O thin films on (110) $SrTiO_3$.

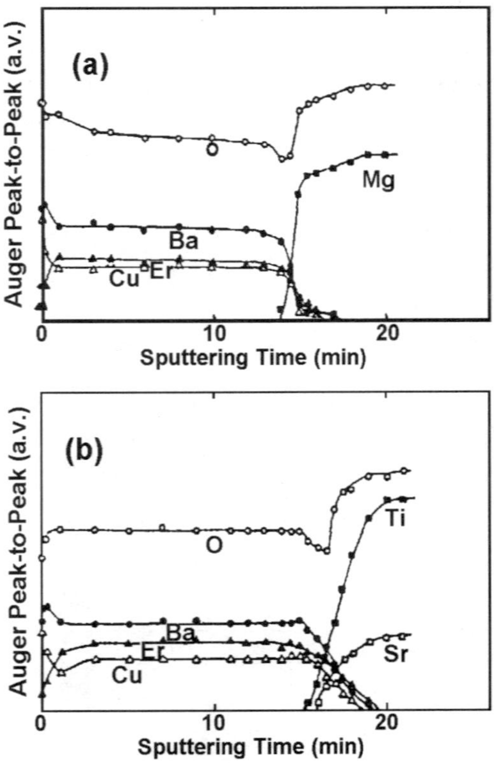

Figure 5.108. Auger depth profile of Er-Ba-Cu-O thin films, 2000 Å thick, deposited on (*a*) (100)MgO substrate, and (*b*) (100)$SrTiO_3$ substrate, by the low-temperature process without postannealing.

As described in Sec. 5.1.4.3 for YBC high-T_c superconductors, the superconducting orthorhombic phase is stabilized during the postannealing process. For rare-earth-free, high-T_c superconductors of the Bi system, the superconducting phases of the sputtered films are controlled by the substrate temperature during deposition; the low-T_c phase of the $Bi_2Sr_2CaCu_2O_x$ system is obtained at the substrate temperature below 600°C, and the high-T_c phase of the $Bi_2Sr_2Ca_2Cu_3O_x$ system is obtained at the substrate temperature above 750°C.

The SEM image suggests that the thin films of the Bi system are composed of mica-like crystallites as shown in Fig. 5.109. The c-axis of the crystallites is perpendicular to the crystal plane. The large crystallites allow for the large critical current.

The critical current density measured for the Bi-Sr-Ca-Cu-O films is as high as 2×10^5 A/cm^2 at 77 K and 6×10^6 A/cm^2 at 4.2 K. The critical current density at 77 K will be governed by the high T_c phase. The current will flow through the current channel presented in the sputtered Bi-Sr-Ca-Cu-O films since the films are composed of a mixture of the high-T_c phase and low-T_c phase. A higher critical current density will be possible in the case of films with a single high-T_c phase. The diamagnetic measurement suggests that the film is composed of 5%–10% of the high-T_c phase.[237] This suggests that the net critical current of the high-T_c phase will be 2×10^6 to 4×10^6 A/cm^2 at 77 K and 6×10^7 to 1.2×10^8 A/cm^2 at 4.2 K.

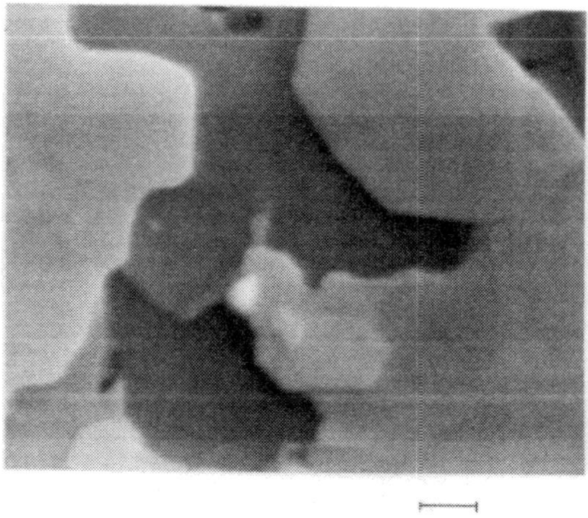

Figure 5.109. SEM image Bi-Sr-Ca-Cu-O of thin films with c-axis orientation.

It is noted that the temperature variations of the J_c are governed by $(1 - T/T_c)^2$.[238] The square power-dependence is different from the 3/2 power dependence predicted by the well-studied proximity junction tunneling model, which is based on the Bardeen-Cooper-Shrieffer (BCS) theory. The presence of the layered structure will cause the square power-dependence. Similar properties are observed in the T1-Ba-Ca-Cu-O films.[239]

In the crystallites of the low-T_c Bi-Sr-Ca-Cu 2-2-1-2 phase, the atomic arrangements are found to be uniform as indicated in the TEM image shown in Fig. 5.110. However, in the crystallites of the high-T_c Bi-Sr-Ca-Cu 2-2-2-3 phase, the crystallites are composed of the different superconducting phases including the Ba-Sr-Ca-Cu 2-2-1-2, 2-2-3-4, and 2-2-4-5 phases, although the resistivity-temperature characteristics correspond to the single superconducting phase of the 2-2-2-3 structure. The presence of the mixed phase is also confirmed by the spreading skirt observed in the x-ray diffraction pattern at the low-angle peak of around $2\Theta = 4°C$.

It is reasonable to consider that the presence of the mixed phases results from the specific growth process of the present rare-earth-free superconducting thin films: The rare-earth-free superconducting thin films may be molten during the annealing process and the superconducting phase will be formed during the cooling cycle.[240]

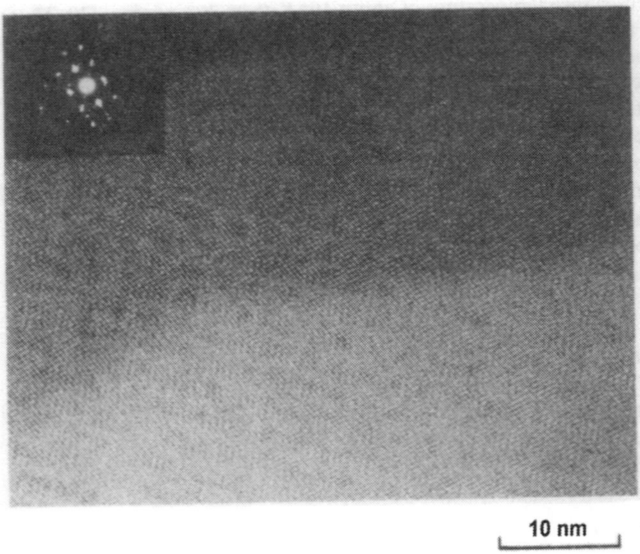

Figure 5.110. TEM image of the sputtered Bi-Sr-Ca-Cu-O thin films of 2-2-1-2 structure.

For the Tl-Ba-Ca-Cu-O system, the sputtered films exhibit rough surface morphology as shown in Fig. 5.111. The critical current is lower than that of the Bi-system.

Figure 5.112 shows XPS measurements for the crystallized Bi-Sr-Ca-Cu-O films. It shows that the annealing process modifies the crystal structure near the Cu-O_2 layer and increases the density of Cu^{3+}. It is also noted that the Sr 3d and/or Ca 2p electron spectra move during the annealing. This implies that some structural changes will appear around Sr and/or Ca sites during the annealing. In contrast, the Bi-O layered structure is stable during annealing.[241] This implies that the single superconducting phase will be synthesized when the Bi-O basic structure is crystallized and the stoichiometric composition is kept for the unit cell of Bi-Sr-Ca-Cu-O.

5.1.4.8 Phase Control by Layer-by-Layer Deposition

The present high-T_c superconductors are composed of layered oxides. If the layered oxides are atomically synthesized by layer-by-layer deposition, the superconducting phase will be closely controlled. These considerations have been successfully confirmed by sputter deposition of the Bi systems in the multitarget sputtering process shown in Fig. 5.113.

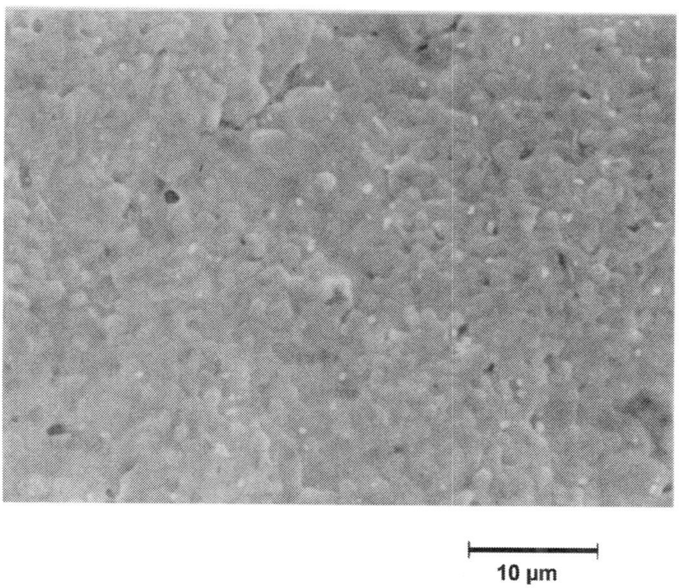

Figure 5.111. SEM image of Tl-Ba-Ca-Cu-O thin films.

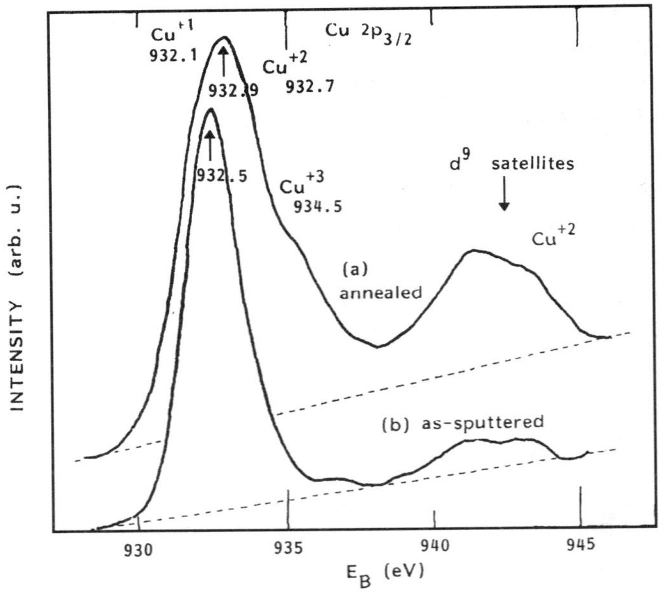

Figure 5.112. XPS spectrum of sputtered Bi-Sr-Ca-Cu-O thin films: (*a*) postannealed in O_2 at 845°C, 300 min; (*b*) as sputtered.

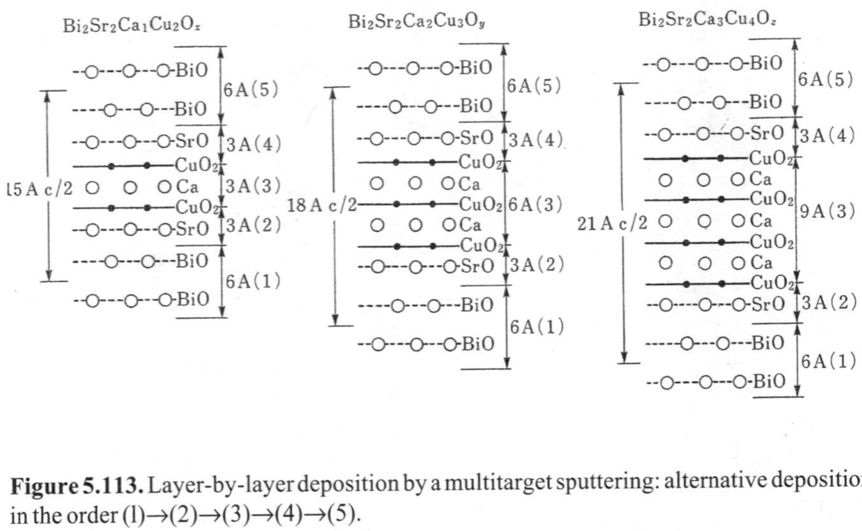

Figure 5.113. Layer-by-layer deposition by a multitarget sputtering: alternative deposition in the order (1)→(2)→(3)→(4)→(5).

The deposition rate was selected so as to deposit the Bi-O, Sr-O, and Cu-O layers in an atomic-scale range. The substrate temperature was kept around the crystallizing temperature of 650°C. X-ray diffraction analyses suggest that as-sputtered films show the Bi-layered oxide structure with broad superconducting transitions. These superconducting properties were improved by postannealing at 850°–900°C in O_2. Figure 5.114 shows typical results for layer-by-layer deposition with postannealing. It is noted that phase control is achieved simply by controlling the amounts of Cu-Ca-O during the layer-by-layer deposition. Experiments show that the T_c does not increase monotonously with the number of the Cu-O layers. In the Bilayer system, the T_c shows the maximum 110 K at three layers of Cu-O, $Bi_2Sr_2Ca_2Cu_3O_x$. At four layers of Cu-O, $Bi_2Sr_2Ca_3Cu_4O_x$, the T_c becomes 90 K.[242]

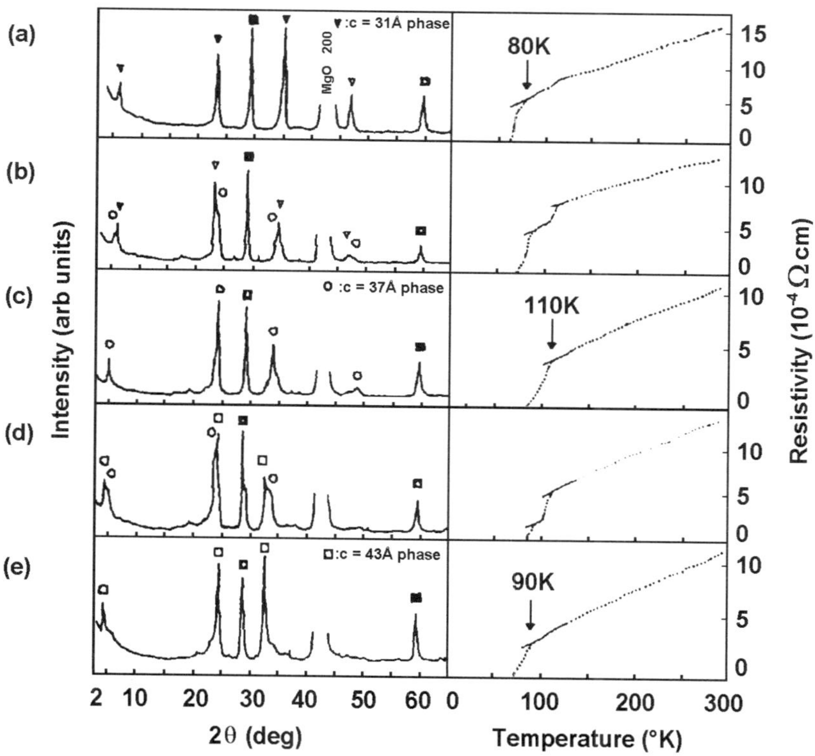

Figure 5.114. X-ray diffraction patterns with resistive-temperature characteristics for phase-controlled Bi-Sr-Ca-Cu-O thin films: (*a*) $Bi_2Sr_2CaCu_2O_x$; (*b*) $Bi_2Sr_2CaCu_2O_x$/ $Bi_2Sr_2CaCu_3O_y$; (*c*) $Bi_2Sr_2Ca_2Cu_3O_y$; (*d*) $Bi_2Sr_2Ca_2Cu_3O_y$/$Bi_2Sr_2Ca_3Cu_4O_z$; and (*e*) $Bi_2Sr_2Ca_3Cu_4O_z$.

Layer-by-layer deposition is one of the most promising processes for fine control of the superconducting phase. Improvements of crystallinity during layer-by-layer deposition will reduce the annealing temperature and allow the low temperature process and/or in-situ deposition. Artificially made layered-oxide superconductors (ALOS) can also be synthesized by layer-by-layer deposition.[242]

5.1.4.9 Diamagnetization Properties

Magnetization properties are measured by an RF-SQUID (superconducting quantum interference device) susceptometer. The operation conditions of an RF-SQUID are listed in Table 5.37. The diamagnetization of the high-T_c oxide superconducting films with c-axis orientation essentially indicates the anisotropy, when the external field is applied both perpendicularly and parallel to the c axis. The high diamagnetization is observed when the film plane is perpendicular to the magnetic field.

Typical diamagnetic hysteresis loops of superconducting oxide films measured at 4.2 K are shown in Fig. 5.115.[243] These loops show the "Lenz Law."[244] When the external field is decreased to a small extent while holding the same field direction, diamagnetization is reversed to the opposite direction within a very short time, keeping its absolute value. Similar hysteresis loops are also observed in single crystals.[245] It is noted that small additions of sulfur into the YBC system increases diamagnetization.[222]

Table 5.37. Operating Conditions of RF-SQUID Susceptor for the Measurement of Diamagnetization Properties

Range of measurements	± 2 emu
Accuracy magnetic flux susceptance	 1×10^{-8} emu√Hz 1×10^{-10} emu/cm^3/√Hz
Applied magnetic field	± 10 kGauss
Sample dimension	5 mm × 5 mm in dia.

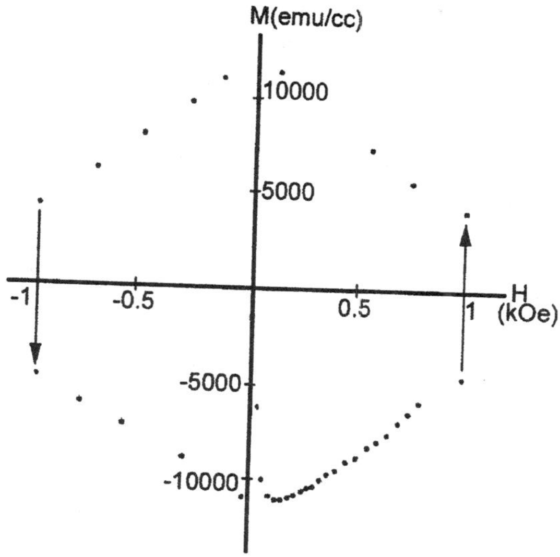

Figure 5.115. Magnetic hysteresis curve for Bi-Sr-Ca-Cu-O thin films measured at 4.2 K.

Temperature variations of the diamagnetization of c-axis-oriented superconducting films are measured both for cooling the specimen in the external field (Meissner effect), and for the warming specimen by applying the field after zero-field cooling (shield effect). Typical results for Bi-Sr-Ca-Cu-O thin films are shown in Fig. 5.116.

By using Bean's formula, the critical current J_c (A/cm^2) becomes

Eq. (5.16) $$J_c = 30\frac{M}{\gamma}$$

where M denotes the diamagnetization (emu/cm^3), and γ (cm) is the effective radius of the sample specimen. For Bi-Sr-Ca-Cu-O thin films with $\gamma = 0.1$ cm, J_c becomes 3.3×10^6 A/cm^2 at 4.2 K, a value very close to the J_c measured by transport methods.[244]

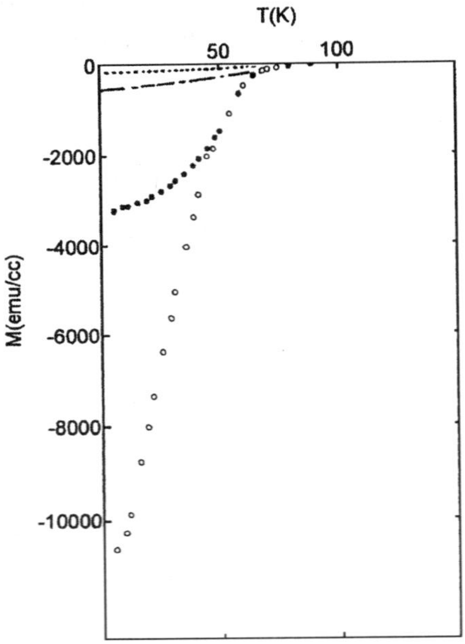

Figure 5.116. The temperature dependence of diamagnetization for Bi-Sr-Ca-Cu-O film; *open circles:* shielding effect in 140 Oe; *closed circles:* shielding effect in 10 Oe; *dash-dot:* Meissner effect in 140 Oe; *dotted line:* Meissner effect in Oe.

5.1.4.10 Passivation of Sputtered High-T_c Thin Films

It is of practical importance to form passivation films on superconducting films in order to reduce environmental influences such as humidity. For such purposes, inorganic insulating films are considered preferable to organic materials. When aluminum oxide films were deposited on Y-Ba-Cu-O thin films with sapphire substrates by rf-magnetron sputtering, the Ba atoms were incorporated in the Al_2O_3 films. In this case, the substrates were exposed to Ar-O_2 plasma. It is considered more desirable to form passivation films without exposing superconducting films directly to the plasma discharge. Film preparation should be performed at the lowest temperatures possible in order to minimize the influence on crystal structures. One promising method for low-temperature deposition is the reactive evaporation method (REM).

Figure 5.117 shows a schematic configuration of the system for electron cyclotron resonance (ECR) REM. It is composed of an ECR plasma source, electron evaporation source (e-gun) and a vacuum chamber where the substrates are placed. The microwave frequency is 2.45 GHz and the magnetic flux density is 875 gauss. Silicon is evaporated using an electron beam gun. For Si-N and Si-O film depositions, N_2 and O_2 gases are introduced, respectively. Silicon evaporation and ECR plasma irradiations are simultaneously performed for Si-N and Si-O formations. Deposition conditions are shown in Table 5.38.

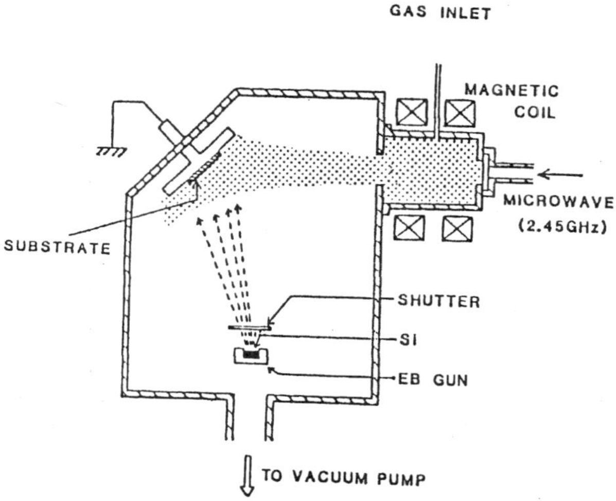

Figure 5.117. A schematic configuration of the plasma assisted electron beam deposition. The oxygen plasma is supplied by the ECR plasma source.

Table 5.38. Typical Deposition Conditions of Si-N and Si-O Passivation Films

Passivation film	Si-N	Si-O
Gas flow rate	N_2 50 sccm	O_2 20 sccm
Gas pressure	1×10^{-3} torr	8×10^{-4} torr
Microwave power	600 W	600 W
Deposition rate (Si)	1–5 Å/s	1 Å/s

Figure 5.118 represents the structure and superconductive properties of the Si-O/Gd-Ba-Cu-O system. The depth distributions of the compositional elements as determined by AES measurement are shown in Fig. 5.118a. It is seen that the compositional elements of Gd-Ba-Cu-O films are not detected in the Si-O film, although the depth distributions of both compositional elements in the Si-O/Gd-Ba-Cu-O system around the interface are slightly complicated. The depth distribution of Si around the interface is not symmetrical in both sides of the Si-O and Gd-Ba-Cu-O films. At the depth where the Si signal is not detected, the signals caused by Gd, Ba, and Cu still increase with depth. It is deduced that the resultant depth distributions are caused by the film coverage over the superconducting Gd-Ba-Cu-O film when the superconducting film surface is not smooth, since the Si-O film formation is essentially carried out obliquely to the Gd-Ba-Cu-O film. Figure 5.118b shows the temperature dependences of resistivities for the Gd-Ba-Cu-O film before and after Si-O film deposition. The onset temperature of the as-deposited Gd-Ba-Cu-O is 88 K and the zero point ($T_{R=0}$) is 57 K. After Si-O film formation, the $T_{R=0}$ increased slightly to 62 K. This may be caused by oxygen ECR plasma exposure to the sample surface during the initial stage of Si-O film deposition.

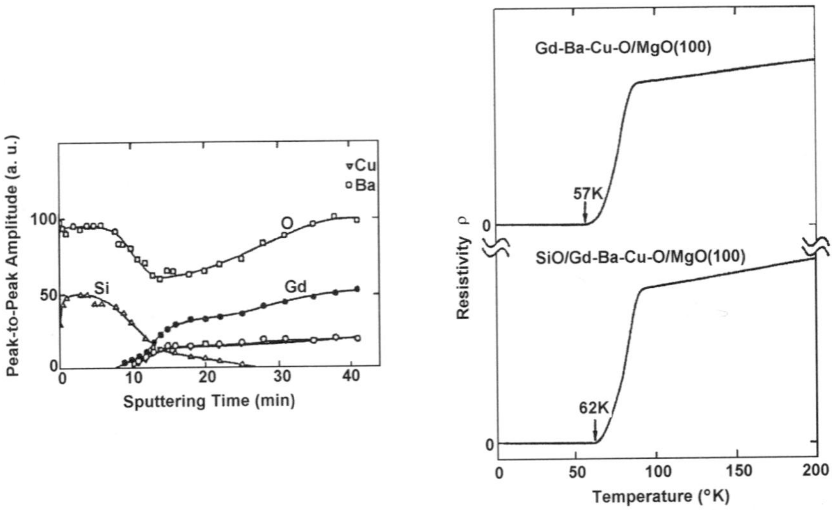

(a) (b)

Figure 5.118. (*a*) Depth distribution of the compositional elements in the Si-O/Gd-Ba-Cu-O system determined by AES measurement, and (*b*) the superconducting properties of sputtered Gd-Ba-Cu-O thin films deposited on (100) MgO substrates with and without overcoating of a Si-O layer.

In the case of Si-N film formation on the superconducting thin film of Gd-Ba-Cu-O, both the values of T_c and $T_{R=0}$ after Si-N film deposition are the same as those of the as-deposited Gd-Ba-Cu-O film as shown in Table 5.39.

Hall measurements suggest that the high-T_c superconductors, LSC, YBC, BSCC, and TBCC show p-type conduction. The postannealing used for achieving superconducting properties will act as a hole injection process. In contrast, the Nd-Ce-Cu-O systems are known as n-type superconductors. The postannealing is done in a reducing atmosphere. It is noted the density of Cu^{+1} will increase during the postannealing process. Table 5.40 shows the physical properties of high-T_c superconducting thin films.

Table 5.39. Influence of Overcoating of Dielectric Layers on Superconducting Properties for Gd-Ba-Cu-O Thin Films

	T_C (K)	$T_{R=0}$ (K)
Virgin Gd-Ba-Cu-O thin film	88	57
Si-O/Gd-Ba-Cu-O thin film	89	62
Si-N/Gd-Ba-Cu-O thin film	89	58

Table 5.40. Typical Superconducting Properties of High-T_c Superconducting Thin Films

	$La_{1-x}Sr_xCuO_4$ $x = 0.05$	$YBa_2Cu_3O_{7-x}$	Bi-Sr-Ca-Cu-O		Tl-Ba-Ca-Cu-O	
			(2 2 1 2)	(2 2 2 3)	(2 2 1 2)	(2 2 2 3)
$Tc_{R=0}$ (K)	30	84	80	104	102	117
$-dH_{C2\parallel}/dT$ (T/K)	6.3	4.6	7.3	7	20	-
$-dH_{C2\perp}/dT$ (T/K)	1.2	0.54	0.36	0.34	0.6	-
$H_{C2\parallel}(0)$ (T)	130	373	440	500	1408	-
$H_{C2\perp}(0)$ (T)	25	44	20	24	42	-
$\xi_\parallel$ (A)	37	27	12.8	11.7	28	-
$\xi_\perp$ (A)	7.1	3.2	2.7	2.6	0.8	-
Anisotropy	5.3	8.4	20	20	31	-

5.1.4.11 Multilayers and Superconducting Devices

The multilayers of superconductors will modify the superconductive properties of each layer. Figure 5.119 shows an example of multilayered superconductor–Bi-2212/insulator–Bi-2201. The multilayers are deposited by multitarget sputtering of Bi-2212 and Bi-2202.[246] The superconducting transition temperature decreases with the decrease of thickness of multilayers as shown in Fig. 5.120.

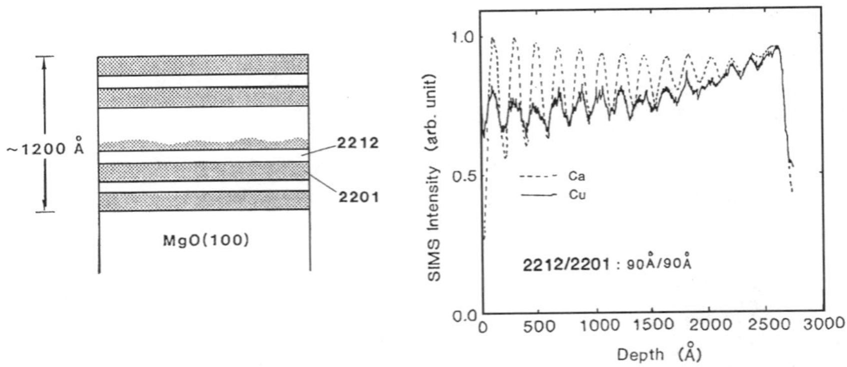

Figure 5.119. Schematic diagram of Bi-2212/Bi-2201 multilayer with depth profile evaluated by SIMS.

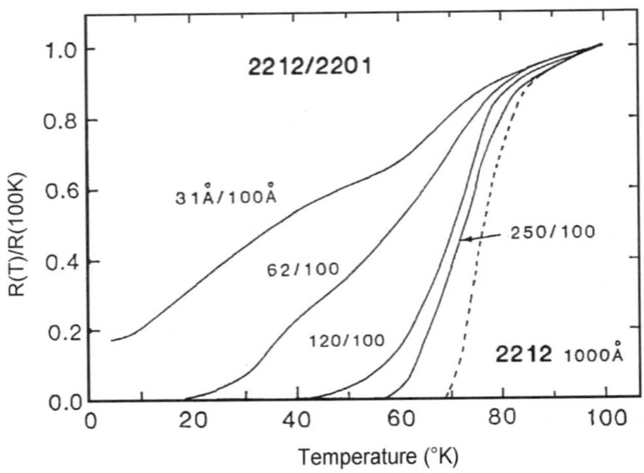

Figure 5.120. Typical temperature dependence of resistivity for B-2212/Bi-2201 multilayers.

Figure. 5.121 shows current-voltage characteristics of sandwich structure, Bi-2212/Bi-2201/Bi-2212. The thickness of the Bi-2201 layer is 60 nm. Under the irradiation of 12 GHz microwaves, clear Shapiro steps are observed at 4.2 K. Since the thickness of a Bi-2201 insulating layer is much larger than the coherent length, a tunnel junction will not be formed in the sandwich structure. The sandwich structure shows the S/N/S-type junction properties governed by the proximity effect.

The simple superconducting electronic devices are low-loss resonators and/or filters of GHz range described in Ch. 1. Figure 5.122 shows the construction and the typical operation of a low-loss filter using hairpin resonators.[247] The sputtered $Tl_2Bi_2CaCu_2O_x$ thin films were used for the hairpin resonators. The substrates are 5-mm thick $LaAlO_3$.

The superconductor reduces the insertion loss as low as 0.8 dB at 77 K. The loss values are 20 dB smaller than gold electrodes.

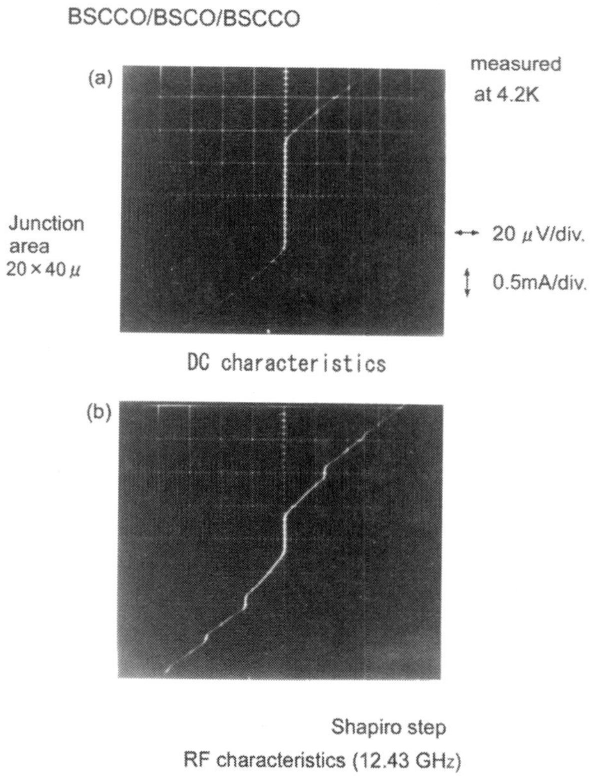

Figure 5.121. Current-voltage characteristics of the Bi-2212/2201/2212 sandwich junction at 4.2 K (vertical: 0.5 mA/div, horizontal: 20 μV/div) (*a*) without rf radiation, (*b*) under rf radiation.

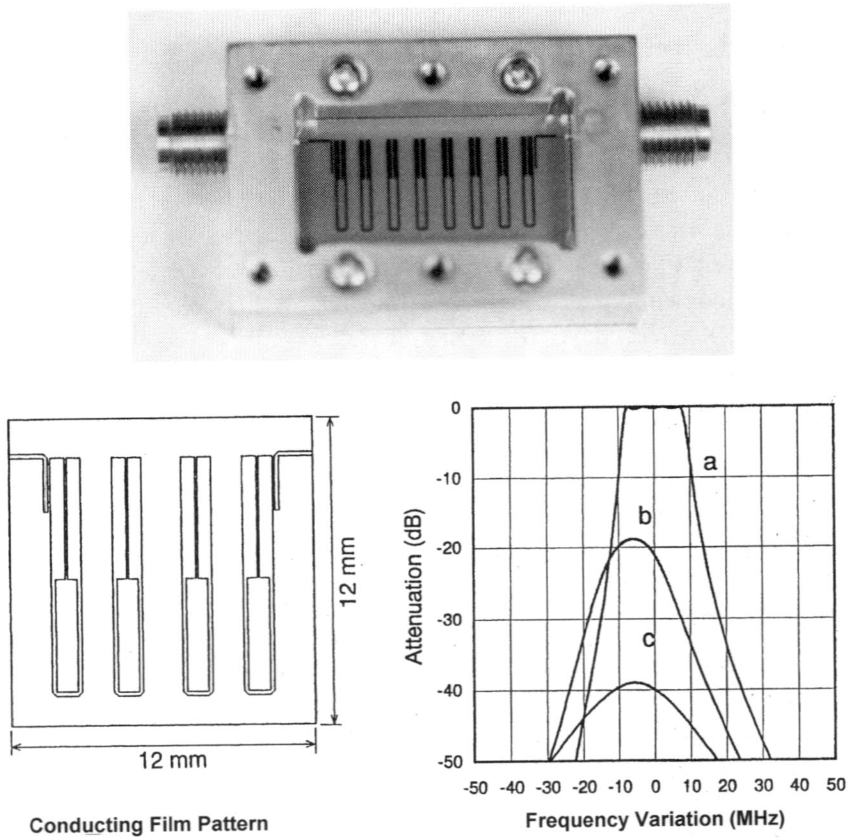

Figure 5.122. A 1.5 GHz four-stage band-path filter (*top*) and details of its conducting film pattern (*bottom left*). Frequency responses are shown (*bottom right*) for various conductors: (line *a*) $Tl_2Ba_2CaCu_2O_x$ thin films measured at 77 K; (line *b*) gold film measured at 77 K; (line *c*) gold film measured at room temperature.

5.1.5 Transparent Conducting Films

Thin films of SnO_2 and In_2O_3 are transparent with high electrical conductivity. These conduction films are prepared by a chemical deposition process, which includes spray coating, or a physical deposition process, a sputtering process, and reactive vacuum evaporation.[248]

Of these processes, the sputtering process gives the most controlled deposition for this type of conducting film. Generally the conducting films are prepared by dc sputtering from an alloy target of In-Sn in an oxygen atmosphere, or by rf sputtering from the compound target of In-Sn oxides. In dc sputtering, the target surface variation changes the electrical conductivity. In order to achieve high reproducibility of the film properties, the partial pressure of oxygen during sputtering deposition should be closely controlled. The as-sputtered films are often annealed in air at 400°C to 500°C to increase their transparency.[249] In contrast to dc sputtering, sputtering of the oxide target produces the conductive transparent films without any postannealing process.

Table 5.41 shows typical sputtering conditions for the deposition of transparent conductive films. The target is a ceramic of indium tin oxides, In_2O_3 with 5% to 10% SnO_2. The rf-magnetron sputtering, where the working pressure is as low as 1 mtorr, allows the deposition of highly conductive films at a low substrate temperature. The addition of oxygen at 10^{-4} to 10^{-5} torr during sputtering increases the crystallinity of the sputtered films and increases the conductivity.[250] Since these sputtered films exhibit high transparency in the visual region and high reflectance in the infrared region, they are used for both liquid crystal display and selective coating in resistively heated solar energy conversion systems.

It is known that the resistivity of these films is around 10^{-3} Ωcm. Lowering the deposition temperature is required for many applications. Recent experiments suggest that the addition of H_2O vapor at 10^{-4} to 10^{-5} torr during deposition reduces resistivity below 5×10^{-4} Ωcm even if the substrate temperature is lower than 200°C.

Table 5.41. Sputtering Conditions for the Deposition of Transparent Conducting Films

Sputtering system	RF planar magnetron
Target	Sintered ITO (100 mm in diameter)
Substrate	Fused quartz
Target-substrate distance	30 mm
Sputter gas	4×10^{-3} torr (Ar)
Substrate temperature	40°C
Sputter power	200 W
Sputter time	5–10 minutes
Sheet resistivity	10–100 Ω/□

5.2 NITRIDES

Most nitrides can be characterized as high-temperature materials that show high mechanical strength. A wide variety of electronic properties, from superconductors to dielectrics, can be found in various nitrides. McLean and his co-workers performed pioneering works on sputter deposition of TaN films for making highly precise thin-film resistors.[251] They were used in touch-tone telephones at that time.

Nitride thin films are easily prepared by sputtering; the vapor pressure of nitrides is generally so low that composition in sputtered films will scarcely shift due to evaporation of one species. A sintered nitride target is used for sputtering in Ar gas, and a metal target is used for sputtering in a nitride-forming atmosphere.

5.2.1 TiN Thin Films

Titanium nitride, TiN, shows a cubic structure of the NaCl type. Thin films of TiN are prepared by sputtering from a TiN powder target in Ar. Table 5.42 shows typical sputtering conditions for the deposition of TiN thin films. These sputtered films show a crystalline structure even at a low substrate temperature.

Table 5.42. Sputtering Conditions for the Deposition of TiN Thin Films

Sputtering conditions		Film properties
Sputter system	RF-magnetron	
Target	TiN sintered powder (stainless target dish, 100 mm dia, is used for powder target materials)	Polycrystal (cubic)
Sputter gas	4×10^{-2} torr (Ar 6N)	
Substrate	Fused quartz	(111) orientation
Sputter power	400 W	d = 4.24–4.25 A
Substrate temp.	500°C	$\rho = 2\times10^{-4}\Omega$cm
Growth rate	1.5 μm/hr	

5.2.2 Compound Nitride Thin Films

Thin films of the Ti-Al-N compound families have high mechanical strength and show a wide range of electrical resistivity. Thin films of Ti-Al-N families are prepared by reactively sputtering from the composite metal target[252] or by direct sputtering of a mixed nitride powder. Table 5.43 shows the sputtering conditions for the deposition of Ti-Al-N films.[253] Their electrical properties are shown in Fig. 5.123. Figure 5.123 also shows the resistive properties of sputtered Ti-Al-N films for various compositions. The TiN films have a resistivity of 150 μΩcm, and a temperature coefficient of resistivity T_{CR} = 300 ppm/°C; for the AlN films, ρ = 2000 μΩcm, and T_{CR} = −400 ppm/°C. It is noted that when AlN/TiN = 1.0, the Ti-Al-N films give zero T_{CR}. For T_{CR} < +100 ppm/°C, ρ becomes 1800 μΩcm, which is one order in magnitude higher than thin films of β-Ta (180 μΩcm) or Ta_2N (290 μΩcm). The Ti-Al-N films are composed of crystalline TiN with amorphous AlN when AlN/TiN = 1. Thin films of Ta-Al-N, Ti-Si-N, and Ta-Si-N also show electrical properties similar to Ti-Al-N films. The addition of Zr to Ti-Al-N makes a ternary compound Ti-Zr-Al-N.[254] The ternary composition expands the range of resistivity with small T_{CR}. These nitride films are useful for making precise thin-film resistors and thin-film heaters for thermal printer heads with high stability. The oxides of the ternary Ti-Zr-Al-N thin films are also useful for the high-K, Si-gate oxides with a small leakage current.

Table 5.43. Sputtering Conditions for the Deposition of Ti-Al-N Thin Films

Sputter system	RF diode
Target	TiN-AlN mixed powder (2N), 100 mm diameter, stainless dish
Sputter gas	(1.5–5)×10^{-2} torr (Ar 5N)
Substrate	glass, alumina
Sputter power	300–400 W
Substrate temperature	150°–500°C
Growth rate	0.6–1.2 μm/hr

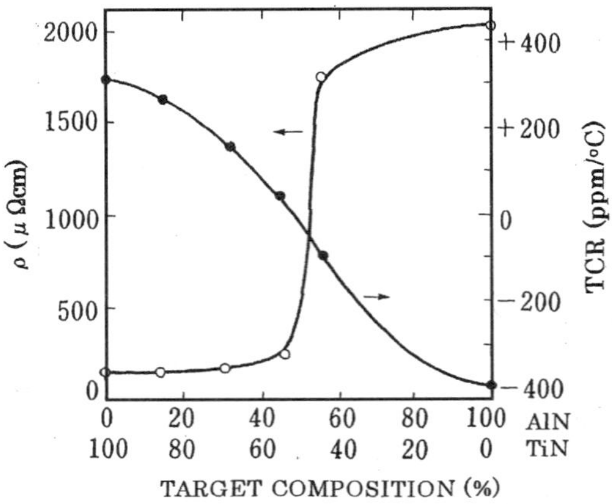

Figure 5.123. Electrical properties of sputtered Ti-Al-N films.

5.2.3 Si-N Thin Films

Amorphous films of plasma chemical-vapor-deposited (CVD) SiN are produced by the reaction of nitrogen and/or ammonia with silane. These films include hydrogen in the form of N-H and/or Si-H bonds.[255] Reduction of the hydrogen concentration in Si-N films is necessary for increasing their thermal and chemical stability.[256] It is noted that Si-N films with a low hydrogen concentration can be prepared by ion-beam sputter deposition.[257]

Figure 5.124 shows the construction of an ion-beam sputter system. A mixed gas of argon and nitrogen is introduced into a Kaufman-type ion source. A water-cooled Si target is reactively sputtered by the ion beam. The acceleration voltage and ion-beam current are 1200 V and 60 mA, respectively.

The temperature of the substrates, n-type Si (100) wafers, is controlled from room temperature to 200°C. The substrates are located almost parallel to the direction of the ion beam so that the incident ion beam grazes the surface during film growth. The vacuum chamber is maintained at about 10^{-4} torr during sputtering.

Under these conditions, transparent amorphous Si-N films are deposited at a deposition rate of 70 Å/min. Infrared absorption spectra suggest that the sputtered films exhibit a Si-N absorption at about 800 cm^{-1}, but absorption due to hydrogen bonds is barely detected. The

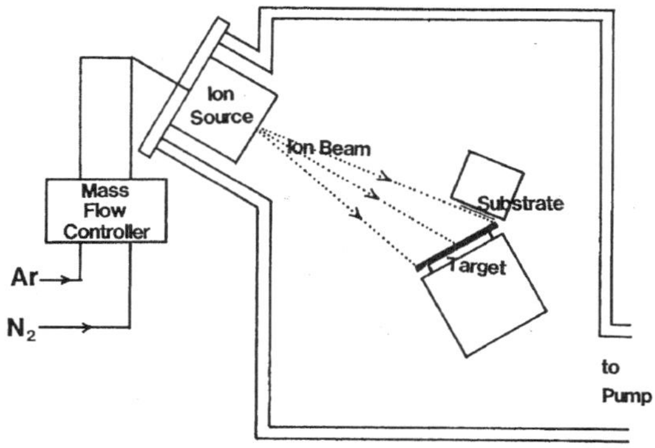

Figure 5.124. Ion beam sputtering system. Substrate surfaces are irradiated by the ion beam during the deposition. Substrates are located nearly parallel to the ion beam.

hydrogen concentration, measured using secondary ion mass spectrometry (SIMS), is found to be below 0.1%, which is much lower than that of plasma CVD Si-N films.[258]

Table 5.44 shows a summary of the physical properties of sputtered Si-N films and their sputtering conditions. It shows that hydrogen-free Si-N films are prepared by ion-beam sputtering at room temperature. Spectroscopy measurements of electron energy loss suggest that the chemical composition of the sputtered films is close to stoichiometric Si_3N_4 when the mixed-gas ratio $N_2/Ar > 4$. These sputtered films show high chemical/thermal stability similar to pyrolytic Si_3N_4 films.

5.3 CARBIDES AND SILICIDES

Carbides and silicides are known as high-temperature materials with strong mechanical strength similar to nitrides. For instance, silicon carbide (SiC) shows a high melting point of 2700°C with a Vickers hardness of 4000 kg/mm^2. The growth of single-crystal SiC films has been studied in relation to SiC thin-film devices including high-temperature SiC transistors and blue-laser diodes.[259] Diamond thin films are also of technological interest because of their potential applications in electronic devices capable of operating at high temperatures and under irradiation of cosmic rays.

Table 5.44. Sputtering Conditions and Physical Properties of Sputtered Si-N Films

Sputtering conditions:	
Target	Si (6N, 100 mm in diameter)
Sputter gas	N_2, mixed gas N_2/Ar (N_2/Ar = 2–6)
Acceleration voltage	1200 V
Ion beam	60 mA (25 mm in diameter)
Substrate	n-Si (100)
Substrate temperature, T_s	RT to ≈ 200°C
Deposition rate	70 Å/min
Film thickness	0.1–10 μm
Refractive index (at 6328 Å)	2.1–2.2 (N_2/Ar > 4)[a]
Etching rate (buffered HF at 20°C)	< 30 Å/min (N_2/Ar > 4)
Memory trap density: in MNOS structure	6×10^{11} cm^{-2} (T_s = RT, pure N_2) 1×10^{12} cm^{-2} (T_s = RT, N_2/Ar = 4) 6×10^{10} cm^{-2} (T_s = 200°C, postanneal at 400°C)
Permittivity	6–7
Dielectric strength	> 10^6 V/cm

[a] When N_2/Ar < 4, the sputtered films comprise a Si-rich Si_3N_4 showing high refractive index.

Several processes have been studied for preparing thin films of high-temperature materials. In general, these thin films include high amounts of lattice defects and also show poor adherence to the substrate due to their hardness.

5.3.1 SiC Thin Films

Various processes for making SiC films are available including vapor-phase reaction,[260] plasma reaction,[261] evaporation,[262] rf sputtering,[263] and ion plating.[264] Among these processes, one of the most convenient is rf sputtering from a SiC target.

The crystalline structure of rf-sputtered SiC films varies from the amorphous phase to the crystalline phase, depending mainly on the substrate temperature during deposition. Typical sputtering conditions are shown in Table 5.45.

Figure 5.125 shows a typical surface structure and electron diffraction patterns of sputtered films on Si(111) substrates.

Epitaxial β-SiC films on Si(111) substrates were obtained at a substrate temperature of 740°C as shown in Fig. 5.125a. The epitaxial relationship is SiC(111) || Si(111). Polycrystalline SiC films with the (220) plane parallel to the substrate surface were obtained for substrate temperatures higher than 550°C. A typical result is shown in Fig. 5.125b. Amorphous SiC films with a specular surface were obtained below 500°C as shown in Fig. 5.125c.

The crystalline films sometimes show the form with a hexagonal structure. A mixture of the α and β phases is also observed in these sputtered films. These sputtered films exhibit an infrared absorption band with a maximum at about 800 cm^{-1}, which corresponds to the lattice vibration of bulk SiC,[265] and also indicates the same mechanical hardness as the value of crystalline SiC.

The microhardness of sputtered SiC films is measured by pressing a diamond pyramidal indenter, such as is used in the Vickers test, and measuring the diagonals of the square indentation. Figure 5.126 shows a typical scanning electron micrograph of the indentation for amorphous SiC films sputtered onto sapphire substrates compared to those taken from the (001) sapphire substrate and (001) surface of a SiC single crystal. It shows that the diagonal of the indentation for the SiC film, and therefore

Table 5.45. Typical Sputtering Conditions for SiC Thin Films

Target	SiC ceramics (80 mm in diameter)
Sputtering gas	Argon (purity 99.9999%, 5 Pa)
Substrates	Fused quartz, silicon, alumina
Substrates temp.	200°–500°C
Target RF power (13.56 MHz)	1–3 W/cm^2
Target-substrate distance	30 mm
Deposition rate	0.1–1 μm/h
Film thickness	4–5 μm

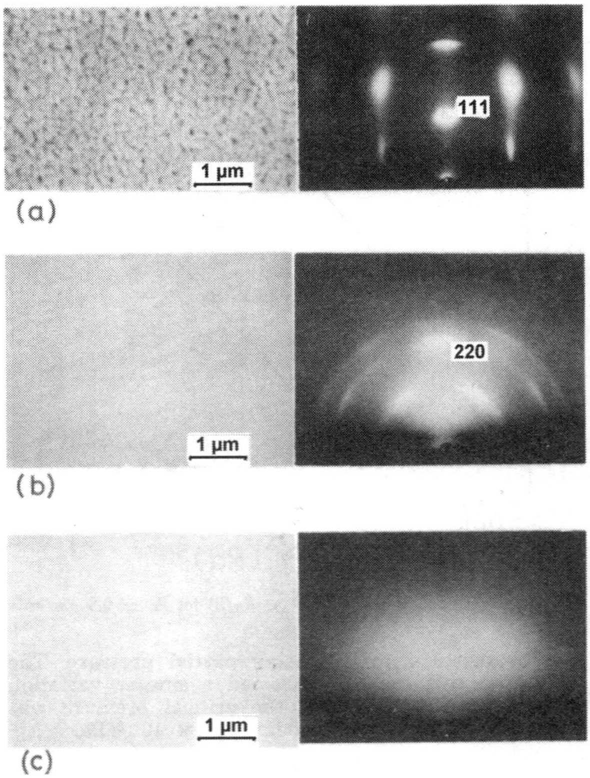

Figure 5.125. SEM images and RED patterns of SiC films of 0.3 μm thickness on Si (111) substrates. The films were sputtered from a SiC target at a deposition rate of 0.3 μm/hr for substrate temperatures of (*a*) 740°C, (*b*) 600°C, and (*c*) 200°C.

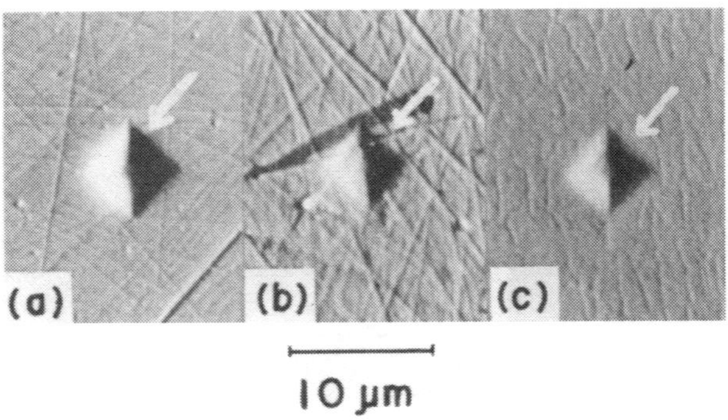

Figure 5.126. Typical scanning electron micrograph showing the indentation made by a diamond pyramidal indenter at an indenter load of 50 g: (*a*) on (001) sapphire; (*b*) on SiC films about 2.6 μm thick sputtered onto (001) sapphire substrates at 370°C with a deposition rate of 0.7 μm/hr; and (*c*) on an (001) SiC single crystal.

the hardness, is smaller than that for the sapphire substrate, and nearly equal to that for the single crystal. Similar results were also observed with polycrystalline α-SiC films. Figure 5.127 shows the Vickers hardness calculated from the diagonal of the indentation as a function of the indenter load for SiC films sputtered onto the substrate. Surface hardness decreases with an increase in indenter load. At heavy loads of more than 100 g, the hardness becomes equal to that of sapphire substrates (1900 kgmm^{-2}) since the diamond indenter completely penetrates the SiC film on the sapphire. With a light load of less than 25 g, the surface hardness tends to increase to 4000 kgmm^{-2}, corresponding to the hardness of the SiC layer. This value is nearly equal to that of bulk SiC.

The wear resistance of sputtered SiC films is evaluated by a cyclical wear test. Table 5.46 shows typical results of the wear test for sputtered amorphous SiC films compared with the wear of Pyrex glass and alumina plates (purity 97%). The wear of SiC films is much smaller than that of the Pyrex glass and alumina plates. Similar results were also found for polycrystalline SiC films. The table suggests that sputtered SiC films are useful for hard surface coatings.

Similar to SiC films, various kinds of rf-sputtered carbide films, such as B_4C, can be used for making hard surface coatings. However, as the hardness of the coating film increases, so does the internal stress contained in the film. This reduces the adherence of the film to the substrate, and an adhesion layer is necessary to make usable surface

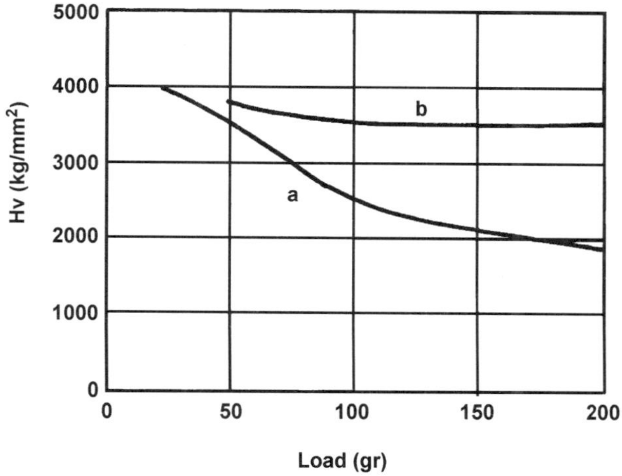

Figure 5.127. Vickers hardness as a function of indenter load: (*a*) for a SiC film, about 2.6 μm thick, sputtered onto (001) sapphire at 370°C with a deposition rate of 0.7 μm/hr; (*b*) for (001) SiC single crystal.

Table 5.46. Cyclical Wear Test Results for a Sputtered SiC Film Compared with Pyrex Glass and an Alumina Plate

Materials	Width of wear scar* (μm)	
	v_S = 1 mm/sec**	v_S = 4 mm/sec**
Pyrex glass	5.5	9
Alumina plate (purity 97%)	3.5	7
SiC film***	1	1

*Sliding distance 100 cm; load 4.0 g.
**v_S (sliding speeds) with a 0.7 mil diamond stylus.
***About 4.7 μm thick, sputtered onto pyrex glass at 370°C; deposition speed , 0.7 μm/hr.

coatings. Figure 5.128 shows a structure of the hard coating. The hard coating is composed of a multilayer SiC/SiC-O on a sapphire, Al_2O_3 substrate in which the SiC-O layer acts as the adhesion layer. A cross section of SEM images is shown in Fig. 5.128a. This multilayer is made by sputtering from a SiC target: the SiC target is first sputtered in a mixed gas of Ar and O_2, which results in the deposition of the adhesion layer, Si-C-O, and then the SiC hard coating is successively deposited by sputtering in pure Ar. The thickness distribution of Si, C, Al, and O atoms in the multilayer detected by XMA is shown in Fig. 5.128b. A mutual diffusion layer exists between the sapphire substrate and the adhesion layer. Table 5.47 lists the composition and mechanical properties of hard coatings made by the rf-sputtering process. The SiC glass systems are prepared by rf sputtering from a pressed target of mixed SiC and borosilicate glass powder. The rf-sputtered B_4C films show very poor adherence. To obtain surface coatings with high microhardness, mixed layer systems of SiC-B_4C are much more useful.

The electrical resistivity of sputtered SiC films is typically 2000 Ωcm at room temperature. The temperature variations of film resistance are reversible when the substrate temperature during deposition is higher than the test-temperature range. Figure 5.129 shows typical temperature variations of the resistance of SiC films deposited on an alumina substrate at a substrate temperature of 500°C measured at a temperature range between -100° and 450°C. It shows that the slope in the ln R versus $1/T$ plot varies with temperature. The slope increases with the increase of temperature and the value lies between 1600 and 3400 K.

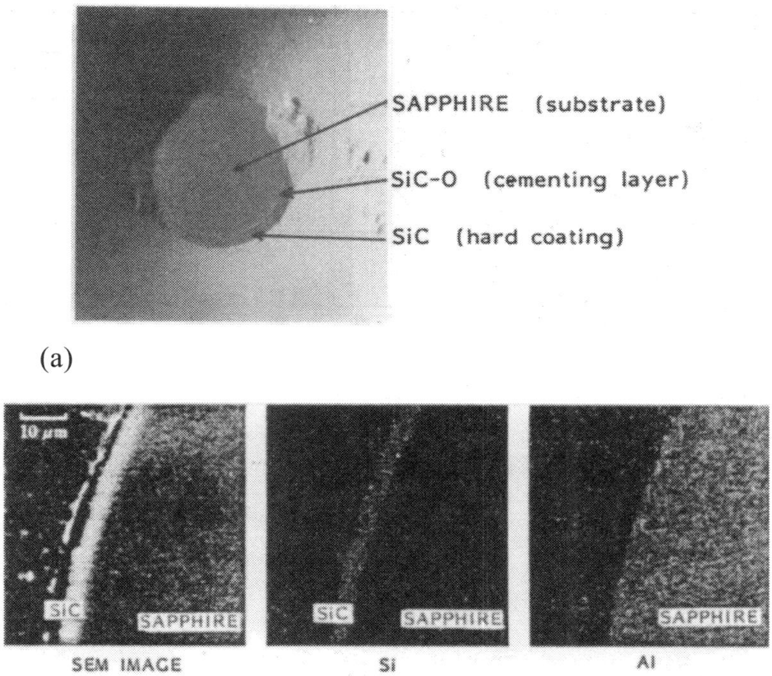

(a)

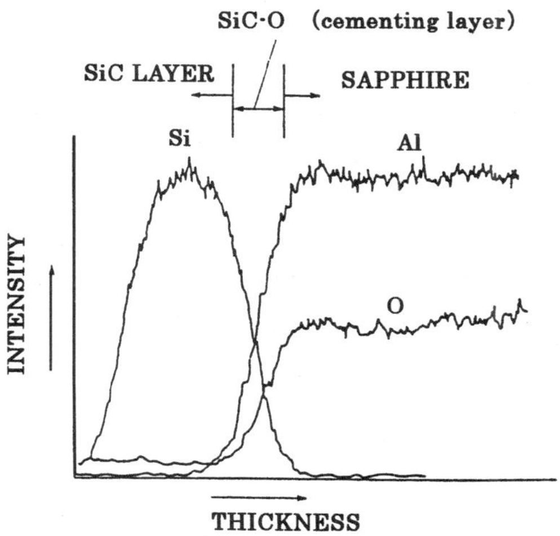

(b)

Figure 5.128. (*a*) A cross-sectional SEM image of a hard coating SiC layer on sapphire having cementing layer SiC-O, and (*b*) thickness distribution of Si, O and Al atoms detected by XMA.

Table 5.47. Summary of Properties in RF-Sputtered Hard Coating Films

Composition	Sputtering target*	H_V (kg mm^{-2})	Wear resistance**	Remarks
SiC	Pressed SiC	4000	1	
SiC + 1% glass	Mixture of SiC + borosilicate glass	3800	0.9	Good adherence
SiC + 5% glass		3300	0.1	
Si C + 10%glass		2400	0.03	
Si C + 25% B_4C	Mixture of SiC + B_4C	4500		
B_4C	Pressed B_4C	4800		Poor adherence

*Sputtering in argon at 300°C–900°C.
**Ratio of time required to a given wear-volume against an iron plate (S-15C).

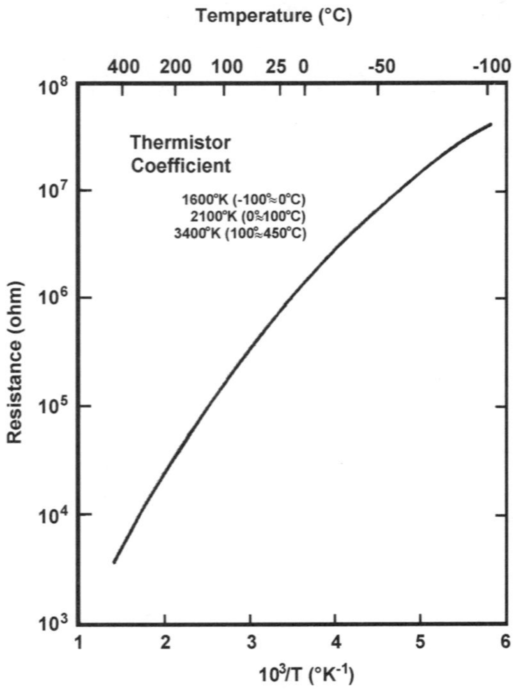

Figure 5.129. Typical temperature variation of the resistivity of sputtered SiC films, about 2 μm thick, deposited onto alumina substrate; zero power resistance vs reciprocal absolute temperature.

Sputtered silicon carbide thin films can be considered applicable for manufacturing silicon carbide thermistors as high-temperature sensors instead of silicon carbide single crystals.[266]

The construction of a silicon carbide thin-film thermistor is shown in Fig. 5.130. The thermistor is composed of silicon carbide thin-film layers of 2–5 μm thick, overlayed on substrate plates. The substrates are made of alumina ceramic of 0.6 mm thickness. One pair of comb electrodes is inserted between the silicon carbide layers and the substrate. Fired Pt layers about 10 μm thick are used as the comb electrode. The size of the substrate is 1 × 8 × 0.6 mm. The length and interval of the comb electrodes are 5 and 0.5 mm, respectively.

The thermistor is prepared by the following process: First, Pt comb electrodes are fired on an alumina wafer, 70 mm square. Then SiC thin films are deposited on the alumina wafer by rf sputtering. The temperature of the alumina wafer is kept at 500°–550°C during sputtering. The typical deposition rate is 0.5 μm/h. Finally, the alumina wafer is annealed in air at about 550°C for 20–100 hours and then broken into thermistor tips. Figure 5.131 shows the alumina wafer and the thermistor tip. The Pt lead wires (0.3 mm in diameter) are welded to the fired Pt electrodes of the thermistor tip. The thermistor tip is, if necessary, packed in an envelope.

SiC thin-film thermistors can be used for temperature sensing, temperature control, and flame detection with high reliability. They can operate between -100° and 450°C. The accuracy of temperature sensing or temperature control systems using these thermistors depends on the thermistor properties, i.e., thermistor resistance and thermistor coefficient. The accuracy of these properties are found to be 5% and 3%, respectively, when alumina substrates and fired Pt electrodes are used. When one uses silicon substrates and Cr/Au thin-film electrodes made by a photolithographic process, the accuracy of the thermistor resistance and thermistor coefficient are 1.5% and 0.5%, respectively. Figure 5.132 shows a typical SiC thin-film thermistor for high-precision use made by a photolithographic process. The tip dimension is 0.5 × 0.5 × 0.1 mm and the response time is found to be less than 0.1 second.

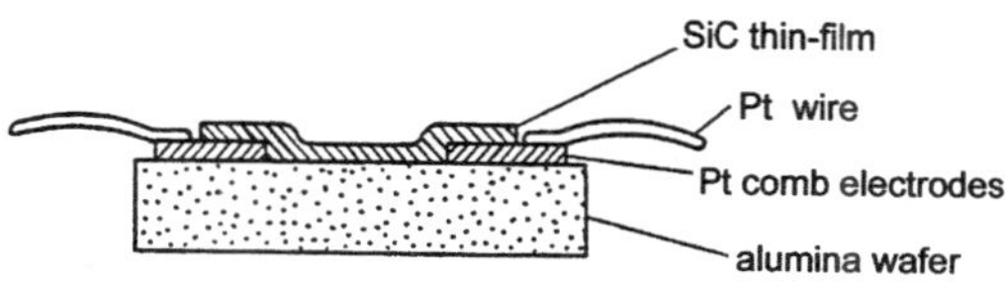

Figure 5.130. Construction of an SiC thin-film thermistor; the thermistor is composed of sputtered SiC thin-films on alumina substrate with fired Pt comb electrodes.

Typical thermistor properties of SiC thin-film thermistors are listed in Table 5.48, together with those of SiC single crystals. The SiC thin-film thermistor coefficient accuracy is much higher than that of the SiC single crystal.[267]

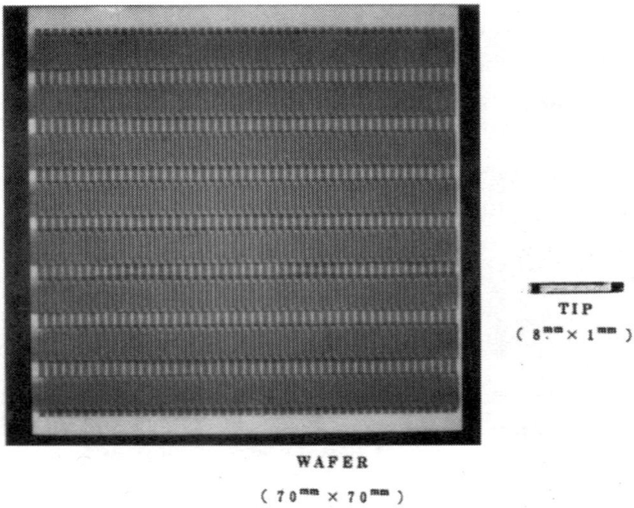

Figure 5.131. Photograph of the SiC thin-film thermistor wafer and thermistor tips.

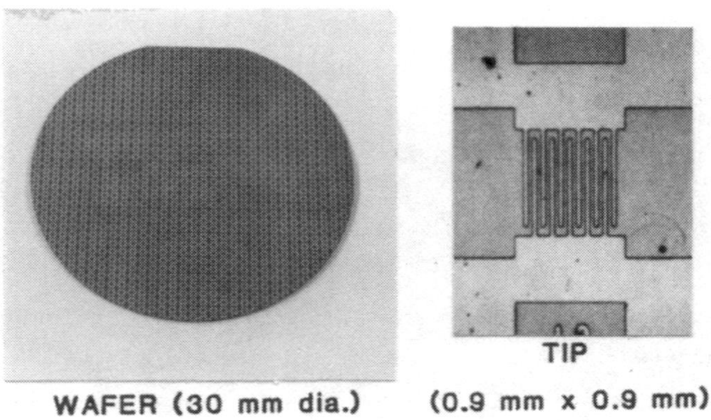

Figure 5.132. SiC thin-film thermistor tips formed on silicon wafer.

Table 5.48. Typical Characteristics of SiC Thin-Film Thermistors

	SiC thin film	SiC single crystal
Operating temperature range	-100°–450°C	-100°–450°C
Zero-power resistance	10 kΩ – 1 MΩ at 25°C	2600°C
Zero-power accuracy	< ±1.5% (thin-film electrode) < ±5% (fired electrode)	±2%
Thermistor coefficient, B (average value)	1600 K (-100°–0°C) 2100 K (0°–100°C) 3400 K (100°–450°C)	2000 K (25–125°C)
Accuracy	< ±0.5% (thin-film electrode) < ±3% (fired electrode)	±2.9%
Electrical stability	Resistance change < 3% (400°C, 2000 h)	

5.3.2 Tungsten Carbide Thin Films

Thin films of tungsten carbide have wide technological applications as wear-resistant and protective coatings on a variety of surfaces such as cemented carbide tools, steel, copper, and copper alloys. For normal steel, copper, and copper alloys, a coating of WC-Co has been found to be suitable. The presence of cobalt is essential in WC coatings to reduce both friction and wear. However, tungsten carbide films with fine grains are highly adherent to steel substrate and do not necessitate any cobalt addition.[42]

The thin films are prepared by direct sputtering from a tungsten carbide target, or by reactive sputtering of a tungsten target in a mixed gas of Ar and C_2H_2 as shown in Fig. 5.133. Typical sputtering conditions are shown in Table 5.49.[42]

Figure 5.134 shows x-ray diffraction patterns for tungsten carbide films sputtered on stainless steel at various substrate temperatures. It shows that a mixture of WC (cubic B_1), W_2C (hexagonal), and W_3C (A-15) cubic phase is formed at lower substrate temperatures (200°C). A single phase of WC is grown at higher substrate temperatures (400°–500°C). These sputtered films consist of a randomly shaped granular surface with a grain size of 400°–500 Å. A fractured cross section of the sputtered films shows a columnar structure consisting of fine columns with a width of 300 Å.

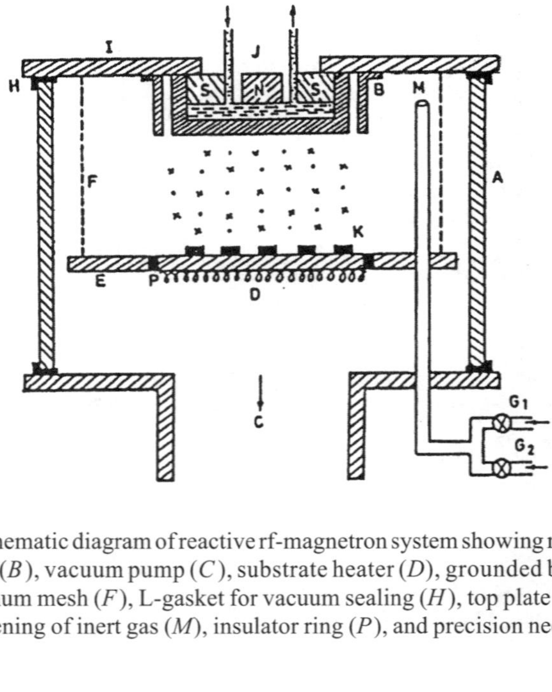

Figure 5.133. Schematic diagram of reactive rf-magnetron system showing metallicbelljar (*A*), target shield (*B*), vacuum pump (*C*), substrate heater (*D*), grounded baseplane (*E*), perforated aluminum mesh (*F*), L-gasket for vacuum sealing (*H*), top plate (*I*), target (*J*), substrate (*K*), opening of inert gas (*M*), insulator ring (*P*), and precision needle valves G_1 and G_2.[42]

Table 5.49. Sputtering Conditions for Deposition of WC Thin Films

Sputter system	planar magnetron
Target	tungsten
Sputter gas	Ar/C_2H_2
Gas pressure	2×10^{-2} torr
Substrate	304 stainless steel
Sputter power	4.5 W/cm^2
Substrate temperature	200°–500°C
Growth rate	0.36 µm/hr*

*4.9 µm/hr for the system shown in Fig. 5.135.

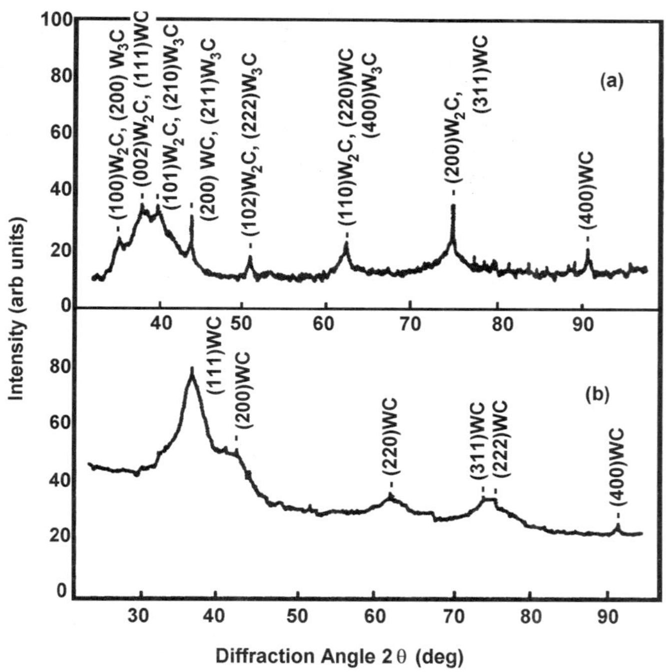

Figure 5.134. X-ray diffraction profiles of tungsten carbide films deposited on stainless steel at (*a*) 200°C, and (*b*) 400°C.[42]

Under the normal conditions of reactive sputtering shown in Fig. 5.133, carbides are known to form on the metal target surface and are subsequently sputtered off (normal mode deposition, NM). Due to the low sputtering yield of these compounds, the rate of deposition of the corresponding film is low. The rate is the same order as that of direct sputtering of the compound target.

Figure 5.135 shows a modified geometry of the magnetron sputtering system.[42] The reaction of the sputtered species from the target with the reactive gas occurs only in the vicinity of the substrate surface since separate zones of argon and acetylene are created by controlling the flow of the two gases such that the carbide formation on the tungsten target will be reduced. In this system, tungsten carbide films have been deposited on the stainless steel substrate at rates as high as that of pure tungsten, 4.9 μm/h, under sputtering conditions indicated in Table 5.49 (high rate mode of deposition, HRM). Table 5.50 summarizes the composition and physical properties of sputtered tungsten carbide thin films.[42]

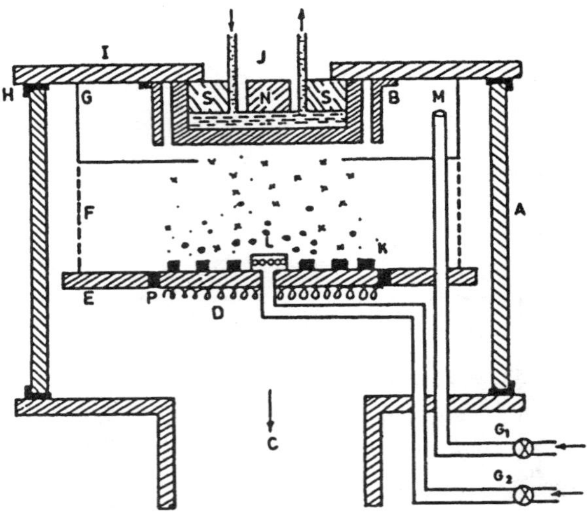

Figure 5.135. Schematic diagram of reactive rf-magnetron system in high rate mode (HRM) showing metallic belljar (*A*), target shield (*B*), vacuum pump (*C*), substrate heater (*D*), grounded base plate (*E*), perforated aluminum mesh (*F*), aluminum metallic partition (*G*), L-gasket for vacuum sealing (*H*), top plate (*I*), target (*J*), substrate (*K*), opening of reactive gas (*L*), opening of inert gas (*M*), insulator ring (*P*), and precision needle valves (G_1 and G_2).[42]

Table 5.50. Composition, Crystallographic Structure and Microhardness Variation of WC Thin Films[42]

Mode of deposition	Substrate Temp. (°C)	Composition			Crystallographic phase	Grain size (Å)	Micro-hardness (kg/mm^2)
		W	C	O			
NM	300	47	48	5	Single phase B1 f.c.c.	200	3200
NM	500	46	49	5	Single phase B1 f.c.c.	200	3200
HRM	300	58	37	5	Mixture of WC (hexagonal), W_3C(A-15) and carbon (graphitic and diamond phase). Graphitic phase in excess.	300	2365
HRM	500	38	59	3	Mixture of WC (hexagonal), W_3C(A-15) and carbon (graphitic and diamond phase). Diamond phase in excess.	300	2365

5.3.3 Mo-Si Thin Films

Thin films of silicides, such as Mo-Si and Cr-Si, have high electrical resistivity and are useful for making thin-film resistors. These silicide films are also used as Schottky gate materials for FETs (field-effect transistors) and interconnections for VLSI. Mo-Si films, for instance, are made by sputtering from a sintered Mo-Si target or a mixed powder of Mo and Si. Stable Mo-Si films are obtained at high substrate temperatures, 500° to 600°C. Figure 5.136 shows the electrical properties for different compositions of Mo-Si. The figure suggests that the Mo-Si films show zero temperature coefficients of resistance for up to 80% Si film composition. These Mo-Si films are of technological interest as heating elements for thermal heads used in thermal printers.

5.4 DIAMOND

Thin films of diamond are useful for making novel electronic devices. In the 1960s, this kind of film was first deposited by decomposition of CH_4 in a CVD system. In the 1970s, several methods were

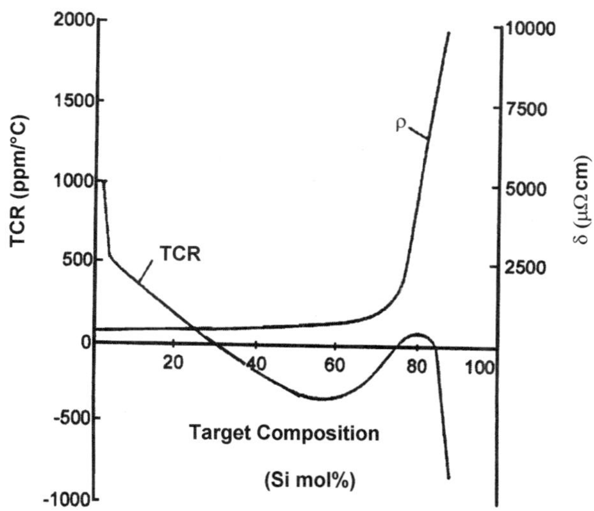

Figure 5.136. Resistive properties of sputtered Mo-Si thin films for various target compositions (sputtering gas: 20 mtorr; substrate temperature: 380°C; deposition rate: 300 Å/min).

considered for the deposition of diamond thin films, including plasma CVD, ion-beam deposition, and sputter deposition. In the CVD process, thin films of diamond were prepared at substrate temperatures of 800 to 1000°C. Ion-beam deposition and sputter deposition are attractive processes because it is possible to prepare thin films at room temperature due to their energetic adatoms.

Aisenberg and Chabot first tried to deposit thin films of diamond at room temperature by deposition of energetic carbon ions using ion-beam deposition. The carbon ions were accelerated at 40 eV by a biased field.[268] The resultant films were transparent with high electrical resistivity. Since their physical properties resemble diamond, these films were called diamond-like carbon (DLC) films. They may be composed of amorphous carbon with small diamond crystallites dispersed in the amorphous carbon network.

Sputter deposition of diamond films was first tried by Wasa and Hayakawa using rf-diode sputtering. They sputtered diamond powder in Ar and produced a transparent DLC film on a glass substrate at room temperature.[269] The film showed poor adherence to the substrate due to its hardness, as shown in Fig. 5.137.

In the 1980s, detailed studies were done on the sputter deposition of DLC films. Weissmantel, et al., deposited DLC films by sputtering from a graphite target using an ion-beam sputtering system.[270] These sputtered films can be used for optical hard coating in the infrared region.

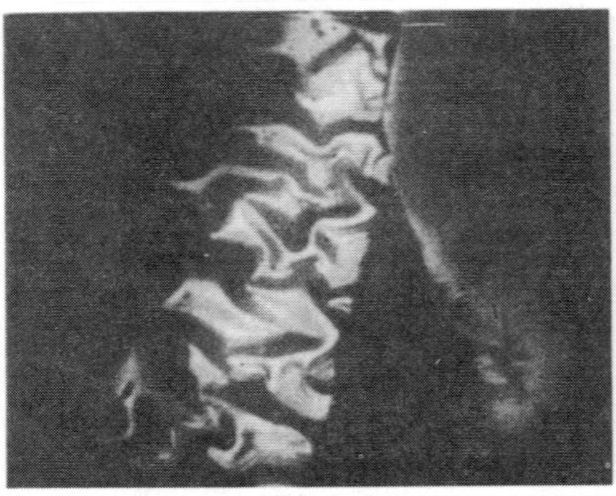

Figure 5.137. Micrograph of sputtered DLC thin film.

Bombardment of the substrate with hydrogen ions during ion-beam sputtering deposition enhanced the growth of the diamond structure and reduced the graphite composition in DLC films.[271]

The ion-beam sputter deposition system, used for the preparation of diamond films, is illustrated in Fig. 5.138. The graphite disk target (purity 5N, 100 mm in diameter) was bonded to the water-cooled holder. An electron-bombardment ion source was employed. The ion energy and ion current were 1200 eV and 60 mA, respectively. The ion-beam diameter was 25 mm. The incident angle of the ion beam was about 30 degrees to the target. The substrate was placed near the target as illustrated in Fig. 5.138. The ion beam sputtered the target and also bombarded the surface of the substrate at grazing incidence. The ion current densities were about 1 mA/cm^2 and 0.04 mA/cm^2 at the target and substrate, respectively. The ion beam that bombards the substrate can modify the deposited carbon film. Table 5.51 summarizes the sputtering conditions.

The optical transparency of the resultant films increased under irradiation of hydrogen ions, as shown in Fig. 5.139, and also increased their electrical resistivity. These results suggest that the graphite composition in the DLC films is reduced by the bombardment of hydrogen ions. The structural analysis of DLC films were studied in detail by Raman scattering spectra.

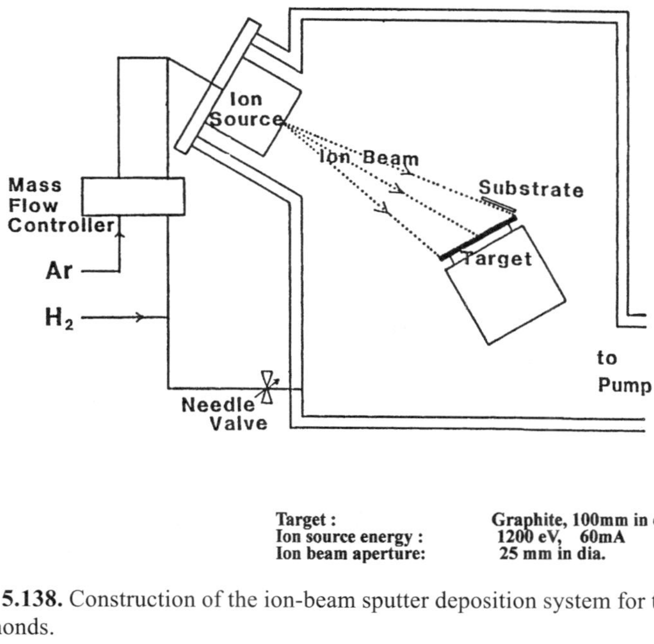

Target :	**Graphite, 100mm in dia.**
Ion source energy :	**1200 eV, 60mA**
Ion beam aperture:	**25 mm in dia.**

Figure 5.138. Construction of the ion-beam sputter deposition system for the deposition of diamonds.

Table 5.51. Sputtering Conditions for Depositing Diamond Thin Films

Sputtering system	ion-beam sputter
Target	graphite plate (100 mm in dia.)
Acceleration voltage	1200 V
Ion-beam current	60 mA
Beam aperture	25 mm in diameter
Gas pressure	$5\times10^{-5} - 2\times10^{-4}$ torr, Ar/H_2 mixed gas
Substrate	(111) Si
Substrate temperature	RT – 200°C
Target-ion source spacing	250 mm
Growth rate	0.3–0.4 μm/hr

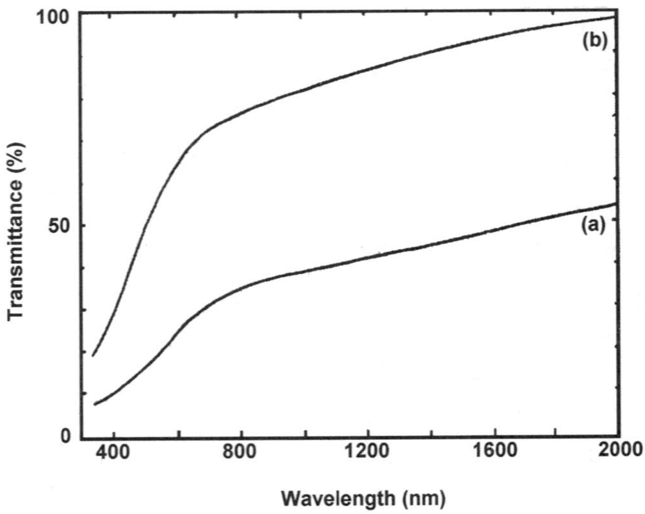

Figure 5.139. Optical absorption spectra of sputtered carbon films: (*a*) without hydrogen ion bombardment, (*b*) with hydrogen ion bombardment during deposition.

The most interesting phenomenon to come out of this study is that under bombardment of hydrogen ions, several diamond crystals are partially grown on DLC films, as shown in Fig. 5.140. The crystals exhibit the well defined morphology of cubic diamond and their lattice constant coincides with that of natural diamond. The effects of bombarding the substrate with hydrogen and argon ions are considered as follows: Energetic activation and rapid quenching occur at the surface of the film during deposition. Bombardment with argon ions supplies the thermal and displacement spikes to the carbon atoms on the substrate. With increased pressure, these spikes may create high-temperature regions. The microcrystals of diamond are partly formed around these spikes. These diamond regions barely grow in the film, because they are surrounded by nondiamond regions. The carbon atoms in these nondiamond regions are weakly bonded and are easily activated by hydrogen ion bombardment and change to the gas C_xH_y. The nondiamond regions may then be selectively removed.

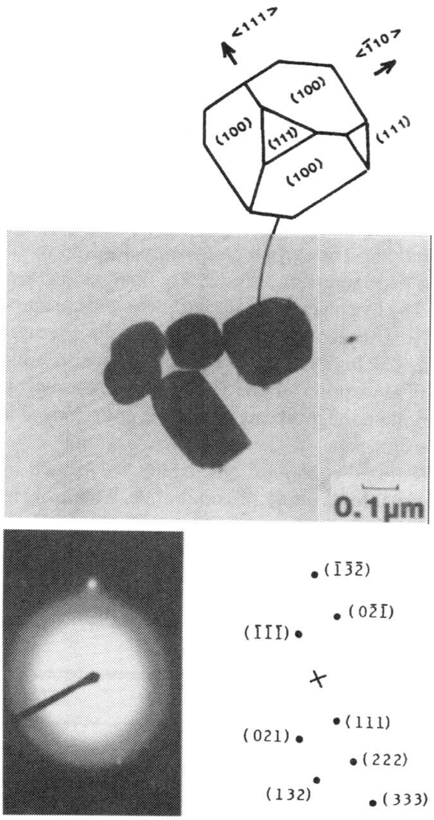

Figure 5.140. Sputtered cubic diamond crystals grown at room temperature.

The diamond regions formed by these mechanisms have many broken bonds, which make the structures very weak and places them in a high state of nonequilibrium. If the broken bonds are rapidly compensated by hydrogen, the diamond regions are quenched and may grow to a stable size. The bombarding hydrogen ions are so active that they compensate the broken bond and activate the growth of diamond.

Cubic diamond thin films are of interest for several applications, including heat sinks of Si-IC, diamond semiconducting devices, gigahertz diamond SAW devices, and electron emitters for flat displays.[272]–[274]

The effect of the incident angle of the Ar ion is also of interest. The grazing angle makes the cubic diamond, while increasing the incident angle makes the hexagonal diamond, as shown in Fig. 5.141. Hexagonal diamond is a metastable phase of bulk diamonds. Hexagonal diamonds are a promising material for future applications.

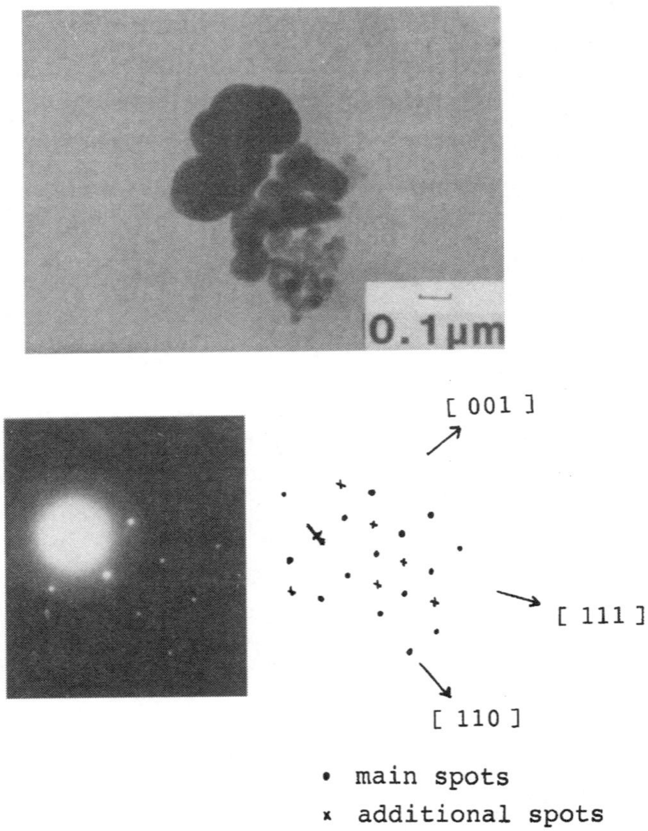

Figure 5.141. Sputtered hexagonal diamond crystals grown at room temperature.

5.5 SELENIDES

Thin films of selenides such as ZnSe are prepared by rf sputtering from a ZnSe cathode in an Ar atmosphere. Typical sputtering conditions are shown in Table 5.52.[3] In the rf-sputtering system, the anode is perforated and substrates are placed behind the anode at floating potential to reduce the effects of the high-energy sputtered atoms and secondary electrons from the cathode.

Transmission electron diffraction patterns from ZnSe layers sputtered onto the (100) plane of NaCl single crystals exhibit the polycrystalline or epitaxially grown monocrystalline phase shown in Fig. 5.142. The monocrystalline films consist of a single phase of cubic-ZnSe structure with the (100) plane parallel to the NaCl surface, but the polycrystalline films consist of two phases, a cubic ZnSe structure and a hexagonal ZnSe structure. The degrees of structural ordering depend on deposition rates and substrate temperatures as shown in Fig. 5.143. In this figure, a minimum epitaxial temperature T_c was found to exist for a given deposition rate R, and the exponential relationship $R = A \exp(-Q_d/kT_c)$ was obeyed for epitaxy. Here, A is a constant and Q_d is the energy of surface diffusion of adatoms. In the present case, we have $Q_d = 0.23$ eV. The satellite spots in Fig. 5.142 denote the existence of twins in the epitaxial layers. The crystalline size of the sputtered layer was in the order of 0.03 to 0.3 μm, as seen in the microstructure.

Table 5.52. Typical Sputtering Conditions for Depositing ZnSe Thin Films

Sputtering system	RF-diode (13.56 MHz)
Target	ZnSe disk (5N)
Sputter gas	5×10^{-3} torr (Ar 5N)
Spacing, target-anode	20 mm
Target-substrate	27 mm
RF-power	≈ 0.1–2 mA/cm^2, ≈ -900 – -1500 V
Substrate	(100)NaCl, (111), (100)GaAs, (111)Si
Growth rate	≈ 0.03–1 μm/hr

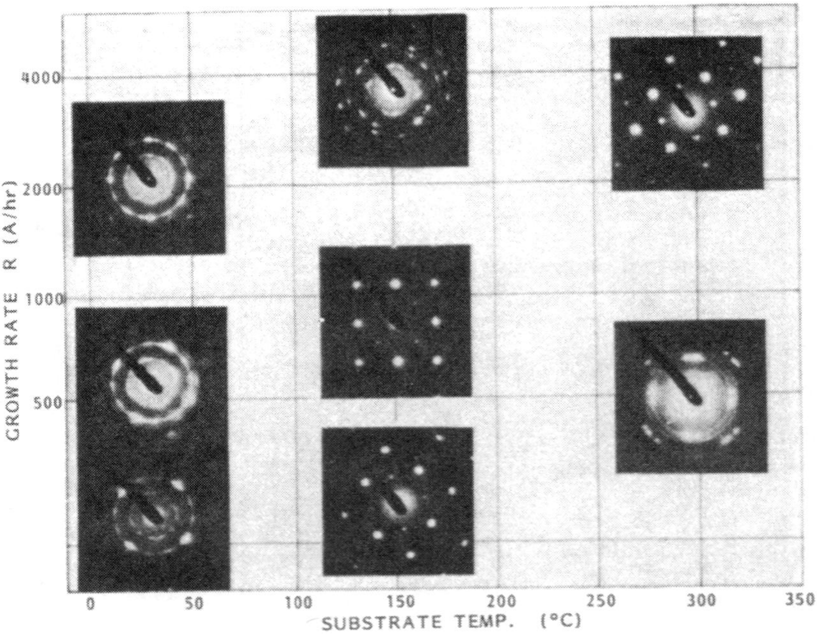

Figure 5.142. Transmission electron diffraction patterns for ZnSe films sputtered on (100) NaCl substrates.

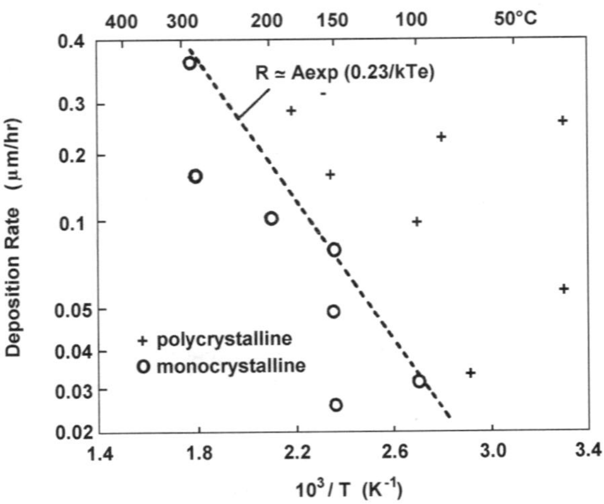

Figure 5.143. Deposition rate vs reciprocal substrate temperature showing the transition from the polycrystalline to monocrystalline structures for ZnSe films sputtered onto (100) NaCl substrates.

It is noteworthy, however, that single-crystal films are epitaxially grown at a low substrate temperature of 150°C, although bulk ZnSe single crystals are synthesized at about 1000°C. This suggests that the sputtering process lowers the synthesizing temperature of crystals. Thin films of ZnSe are promising materials for making heterojunction photodiodes, solar cells, and blue-light-emitting diodes.

Figure 5.144 shows a typical oscilloscope trace of a current-voltage characteristic for the ZnSe/Si diode measured in darkness at room temperature. The diode shows rectifying properties, and the direction of rectification suggests that the ZnSe layer is a n-type semiconductor. As shown in Fig. 5.145, the forward-current–voltage characteristics of the diodes exhibit ohmic currents at a low applied voltage, and space-charge-limited currents (SCLC) exhibit ohmic currents at a high applied voltage. The SCLC exhibited trap-controlled behavior similar to that occurring in vacuum-deposited n-type ZnSe, p-type Ge heterojunction diodes in which the SCLC flowed in a high-resistivity ZnSe layer.[275] The high resistivity may result from defects produced by the bombardment of energetic particles and/or inclusions of impure gases during deposition.

Several processes have been studied for making ZnSe single-crystal films with low defect densities; the MBE and MOCVD (metalorganic chemical vapor deposition) processes are considered the most promising. It is also reasonable to expect that lower defect densities may be achieved by controlling the impinging particle energy during sputter deposition.

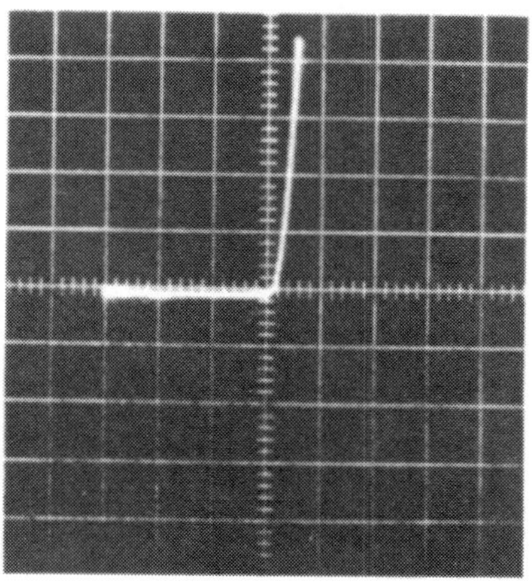

Figure 5.144. Forward-current–voltage characteristics for sputtered n-ZnSe/p-Si diode measured at room temperature in a dark space. Vertical scale: 10 μA/division. Horizontal scale: 1 V/division.

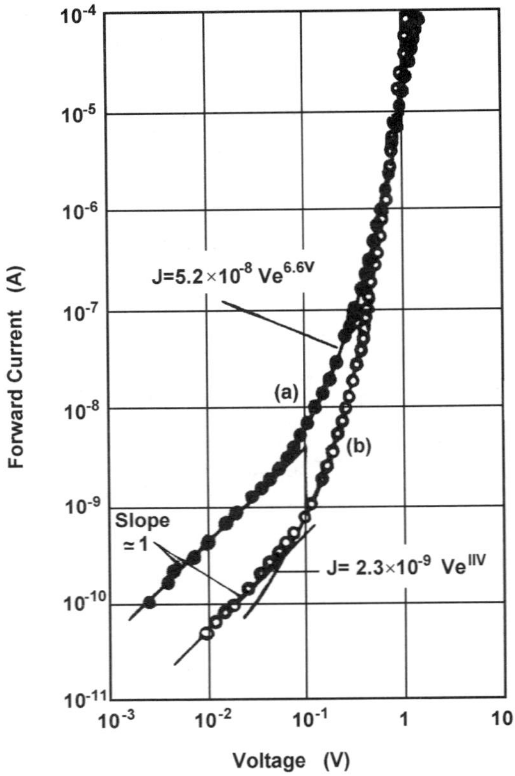

Figure 5.145. Dark-current–voltage characteristic for sputtered n-type ZnSe, p-Si heterojunction diode. Vertical scale: 10 μA/division. Horizontal scale: 1 V/division.

5.6 AMORPHOUS THIN FILMS

Amorphous materials are prepared by the quenching of melts or vapors.[276][277] Table 5.53 summarizes these quenching processes. Among these processes, vapor quenching is achieved by many thin-film deposition processes including vacuum deposition, sputter deposition, and CVD. Thin films of the amorphous phase are deposited at substrate temperatures below the crystallization temperatures of the thin films. Table 5.54 shows the crystallization temperature for various materials. The metals generally show low crystallization temperatures, while compounds such as oxides typically are much higher.[278][279]

Table 5.53. Summary of Quenching Process

Quenching process		Quenching speed (deg/sec)
from melt	annealing	$\approx 10^{-2}$–10^{-5}
	air-quenching	10
	liquid-quenching	10^3
	sputter-quenching	10^5
	roller-quenching	$\approx 10^6$–10^8
from vapor	vacuum deposition, CVD	$>10^9$
	sputtering	$>10^{16}$

Table 5.54. Crystallization Temperature

Materials	Crystallization temperature (K)	Materials	Crystallization temperature (K)
V	≈ 3–4	Cr_2O_3	718
Cr	≈220	MgO	598
Ga	≈ 10–60	NiO	558
Ge	743	Al_2O_3	1003
Si	993	Fe_2O_3	808
Bi	≈ 10–30	GaAs	~603
Se	≈300	SiO_2	948
Te	≈280	TiO_2	753
Sn-Cu	≈60	Ta_2O_5	1013

The crystallization temperature of metals will increase due to the inclusion of residual gas during deposition. For instance, the crystallization temperature of Fe thin films deposited at 10^{-12}–10^{-10} torr is 4 K. The temperature increases to 75 K when the O_2 partial pressure is 10^{-8} torr. A typical sputtering system for the deposition of amorphous thin films is shown in Fig. 5.146.

The temperature rise of the substrates during deposition is 200 to 500°C in a conventional diode-sputtering system when the substrates are not cooled by water. For making amorphous thin films, the substrates are cooled by water, liquid nitrogen, or He. Figure 5.147 shows a photograph of a sputtering system for depositing amorphous thin films. The magnetron target is useful in reducing the temperature rise of the substrates.

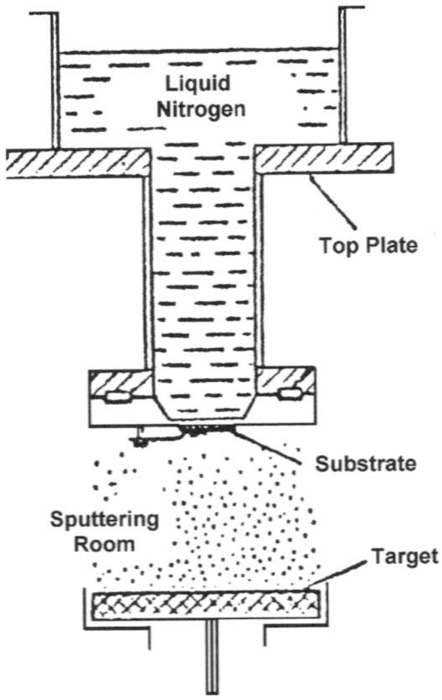

Figure 5.146. Construction of a sputtering system with cooled substrate for the deposition of amorphous films.

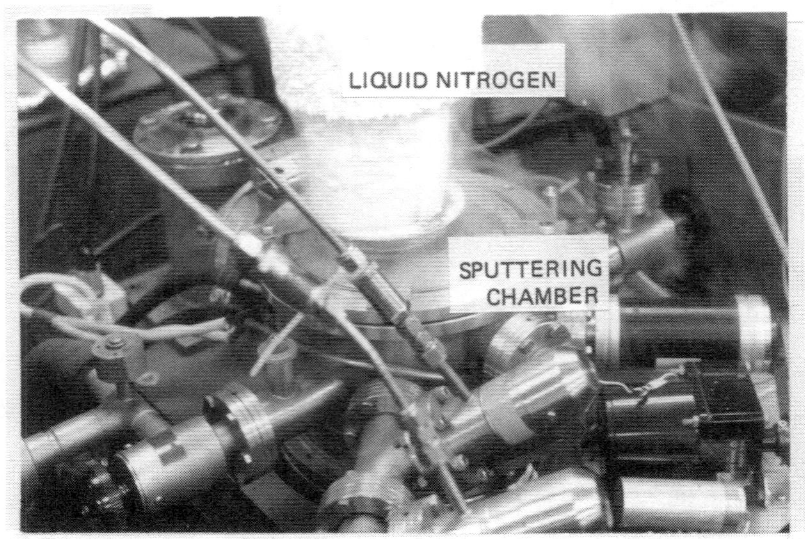

Figure 5.147. Photograph of sputtering system for depositing amorphous thin films.

In sputter deposition, the energy of adatoms is on the order of 1 to 10 eV; thus, the quenching time of these adatoms is estimated to be higher than 10^{16} K/sec since they will lose their energy in 10^{-12} sec, which corresponds to the frequency of the thermal lattice vibration of the substrate surface atoms. The quenching rate expected in sputter deposition is much higher than the value obtained in quenching from melts, as indicated in Table 5.53. This suggests that exotic structures may be prepared by the sputter deposition process.

5.6.1 Amorphous ABO_3

There has been a growing interest in amorphous states of ABO_3-type ferroelectric materials for both their physics and practical applications. The possibility of ferroelectricity in amorphous dielectrics was shown theoretically by Lines.[280] Experimental confirmation of this possibility is being explored. Glass, et al., prepared amorphous $LiNbO_3$ and $LiTaO_3$ by a roller-quenching method and reported dielectric anomalies with a peak of $\varepsilon_r > 10^5$ near the crystallization temperature.[281] These amorphous materials will exhibit other interesting properties such as ferroelectricity and high ionic conductivity.[282] Amorphous $PbTiO_3$ was prepared by Takashige, et al., using a modified roller-quenching method and their dielectric anomalies were reported.[283]

Sputter deposition on cooled substrates is an interesting method for the preparation of amorphous ABO_3 materials. Thin and uniform layers with a large area are easily obtained. Cooling the substrates during deposition suppresses crystallization of the deposited materials and makes it possible to produce amorphous materials.

Thin films of amorphous $LiNbO_3$ are made by rf sputtering from an $LiNbO_3$ compound target.[284] Typical sputtering conditions are shown in Table 5.55. An rf planar magnetron sputtering apparatus is employed for preparation of the films. A stainless steel dish, 10 cm in diameter and filled with $LiNbO_3$ powder, is used as a target. The $LiNbO_3$ powder is synthesized from Li_2CO_3 (purity 2N) and Nb_2O_5 (purity 3N). Substrates are polished fused quartz. Before the deposition of $LiNbO_3$, Au/Cr stripes are vacuum evaporated on the substrates as bottom electrodes for the measurement of dielectric properties. The substrate is attached to a water-cooled holder to avoid crystallization of the $LiNbO_3$ film during sputter deposition.

Table 5.55. Sputtering Conditions for the Preparation of Amorphous $LiNbO_3$ Thin Films

Sputtering system	RF diode
Target	$LiNbO_3$ powder
Sputter gas	Ar/O_2 (1:1)
Gas pressure	5×10^{-3} torr
Substrate temperature	water-cooling
Substrate	glass
RF-power	2.5 W/cm^2
Deposition rate	0.38 μm/hr
Film thickness	0.5–1.0 μm

Transparent and colorless films with smooth surfaces are obtained by sputter deposition. Figure 5.148a shows an x-ray diffraction pattern of a sputtered film, about 1 μm thick, without electrodes. The pattern exhibits halo diffraction which indicates that the film is in an amorphous state. In order to examine the crystallizing temperature, the films were annealed in air. From the x-ray diffraction pattern, it was found that films annealed at 350°C for 1 hour were still amorphous. Further annealing at 500°C for 1 hour caused crystallization of the films. Figure 5.148b shows an x-ray diffraction pattern of an annealed film which agrees well with that of $LiNbO_3$ powder (ASTM card # 20-631) shown in Fig. 5.148c. The film annealed at 500°C is considered to be polycrystalline $LiNbO_3$.

Figure 5.149 shows a typical temperature dependence of dielectric properties for as-sputtered amorphous $LiNbO_3$ films. In this figure, the ε_r of single-crystal $LiNbO_3$ along the a-axis and c-axis at 100 kHz reported by Nassau, et al., are plotted in dotted lines.[285] These dielectric properties differ from those of the roller-quenched samples: the dielectric anomalies are observed at about 350°C, which is 200°–300°C lower than that of the roller-quenched $LiNbO_3$.

Thin films of amorphous $PbTiO_3$ are also prepared by rf sputtering from a sintered $PbTiO_3$ target on cooled substrates.[286] In this case, there are Pb metal crystallites in as-sputtered $PbTiO_3$ films, and these films exhibit a high surface electrical conductivity of around 10 $(\Omega cm)^{-1}$ due to the hopping of electrons between the Pb metal crystallites. When the as-

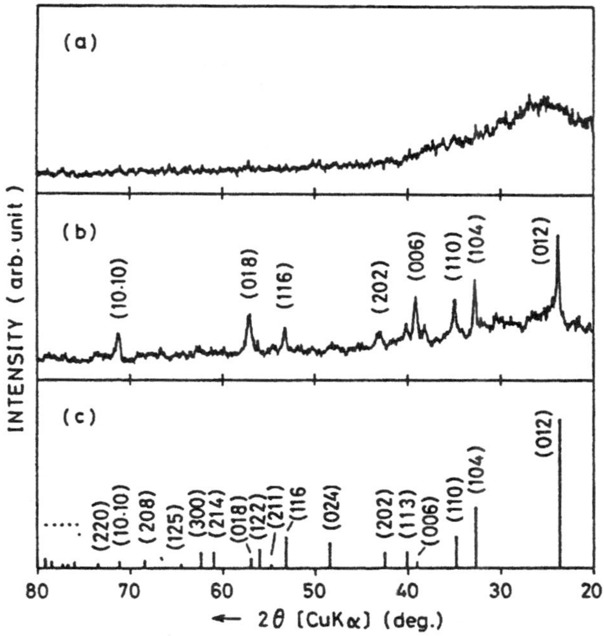

Figure 5.148. X-ray diffraction patterns of (*a*) as-grown film, (*b*) film annealed in air at 500°C for 1 hr, and (*c*) $LiNbO_3$ powder (ASTM card # 20-631).

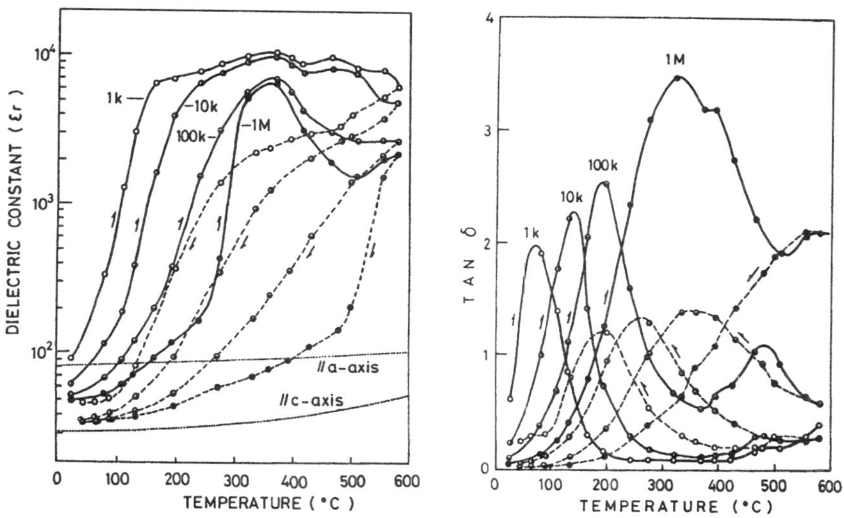

Figure 5.149. Temperature dependence of dielectric properties for the first cycle of heating and cooling of the as-grown $LiNbO_3$ films in amorphous state measured at 1 k, 10 k, and 1 MHz. Dotted lines indicate ε_r of single-crystal $LiNbO_3$ along a-axis and c-axis at 100 kHz.

sputtered films are annealed above 220°C, the surface electrical conductivity is reduced to 10^{-4} $(\Omega cm)^{-1}$. The annealing causes a decrease in Pb metal crystallites in the film. The dielectric constant of the annealed $PbTiO_3$ shows a broad maximum at 440°C. When the films are annealed above 500°C, they become polycrystalline $PbTiO_3$ films.

The maximum, sometimes observed in the roller-quenched $PbTiO_3$,[287] is similar to the temperature dependence of the dielectric constant of tetragonal red-PbO and are attributed to the presence of the red-PbO crystallites. A broad maximum is observed at 400°C in cooling. This peak may be blurred dielectric anomalies, which are attributed to nonstabilized ferroelectric ordering. This was also observed in the roller-quenched $PbTiO_3$.[283] These results suggest that the structure of the annealed film resembles that of roller-quenched amorphous $PbTiO_3$.

5.6.2 Amorphous SiC

The crystallization temperature of SiC thin films sputtered from an SiC target is about 500°C, below which the films are amorphous. However, the stable amorphous phase is obtained in SiC thin films reactively sputtered from a Si target in a C_2H_2 or CH_4 gas.[288] Typical sputtering conditions for the deposition of amorphous SiC films are shown in Table 5.56. Although the substrate temperature is higher than the crystallization temperature of conventional rf-sputtered SiC thin films, the

Table 5.56. Sputtering Conditions for the Preparation of Amorphous SiC Thin Films

Sputtering system	RF diode
Target	Si 6N,100 mm in dia.
Sputter gas	Ar/CH_4, C_2H_2
Gas pressure	2×10^{-3} torr
Substrate	Si (111)
Substrate temperature	200–740°C
RF power	3–4 W/cm^2
Deposition rate	0.2–0.6 µm/hr
Film thickness	0.8–2 µm

reactively sputtered SiC thin films show the amorphous phase. Their infrared transmission spectra are shown in Fig. 5.150. The large absorption band at 800 cm^{-1} is due to the fundamental lattice vibration of SiC.[289] The small absorptions due to Si-H and C-H are superposed on the spectra. The inclusion of hydrogen atoms in sputtered SiC films stabilize the amorphous phase.

Since the amorphous thin films often include microcrystallites of nanoscale, the amorphous thin films are promising materials for future applications.

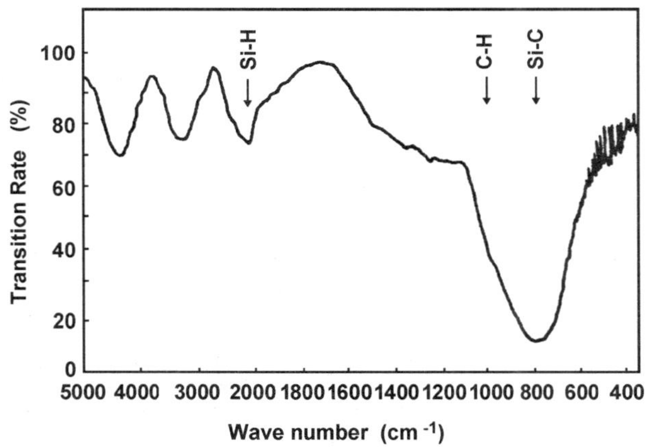

Figure 5.150. Infrared transmission spectra of sputtered a-SiC thin films.

5.7 SUPERLATTICE STRUCTURES

Intercalated structures consist of thin alternating layers grown by vacuum deposition such as the MBE process,[291] chemical vapor deposition including the ARE process,[292] sputtering deposition,[293] and chemical solution deposition.[294] Among these processes, the sputtering process allows the preparation of thin alternating layers of refractory metals and compounds such as oxides, nitrides, and III-V alloys. As described in Secs. 5.1.3.2 and 5.1.4.11, alternating layers of compound oxides, including perovskite dielectrics and superconductors, are made by sputtering from multitargets. Greene, et al., deposited the III-V compound alternating layers of InSb and GaSb (InSb/GaSb/InSb/...).[295] In multitarget sputtering, the substrate is continuously rotated through two or more electrically and physically isolated sputtering discharges. The structural properties depend on the layer thickness deposited per target pass and the rate of interlayer diffusion. Eltoukhy, et al., have deposited the InSb/GaSb superlattice structures with layer thicknesses ranging from 12.5 Å to 50 Å at substrate temperatures of 200° to 250°C.[296] They have suggested that the superlattice structures only a few monolayers thick may easily be grown by sputtering from the multitarget even at relatively high substrate temperatures.

Superlattice structures are also deposited by reactive sputtering. Figure 5.151 shows the construction of the sputtering system for making thin alternating layers of Nb/NbN. For the first layer, the Nb target is sputtered in Ar; later, in a mixed gas of Ar/N_2. Periodic change of the sputtering gas composition results in the formation of the superlattice Nb/NbN. Typical sputtering conditions are shown in Table 5.57. The Auger depth profile of the layered structure is shown in Fig. 5.152. The thickness of each layer is about 100 Å. The period of the layered structure corresponds to the variations of the sputtering gas composition. This sputtering process achieves a layer thickness that is thinner than 10 Å. Contamination by impurities can be reduced when the ion-beam sputtering system is used for deposition. Typical structural properties are shown in Fig. 5.153.[297]

The sputtering process is also a promising method for making the superlattice structure for refractory metals, oxides, and nitrides, which are useful for magnetic media, superconductors, dielectrics, and x-ray mirrors. However, energetic adatoms will produce lattice defects which can cause difficulty in making semiconducting devices.

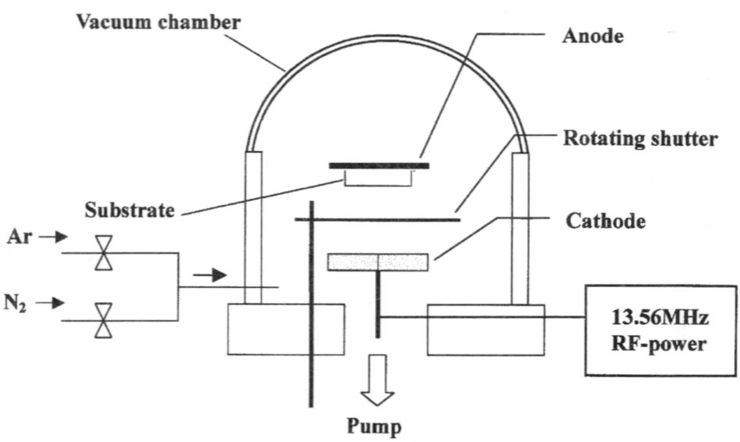

Figure 5.151. Construction of a reactive sputtering system for making thin alternating layers of Nb/NbN.

Table 5.57. Typical Sputtering Conditions for the Deposition of the Superlattice Structure Nb/NbN

	Nb Thin Layers	NbN Thin Layers
Sputtering gas	Ar (3×10^{-2} torr)	Ar (3×10^{-2} torr) N_2 (0.75×10^{-2} torr)
Substrate temperature	300°C	300°C
Deposition rate	50, 100 Å/min	100 Å/min

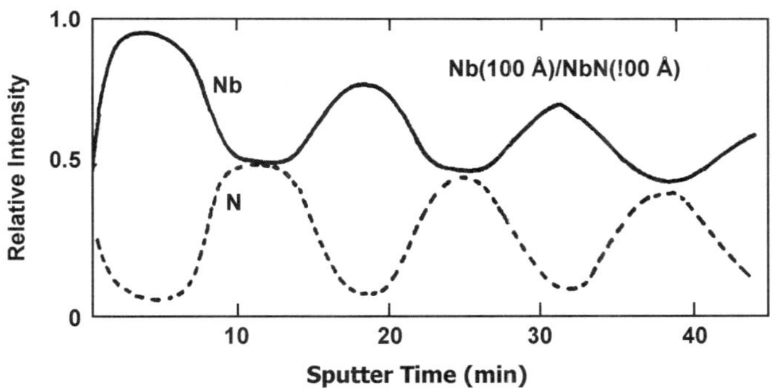

Figure 5.152. Depth profile of the Nb/NbN superlattice structure deposited by reactive sputtering.

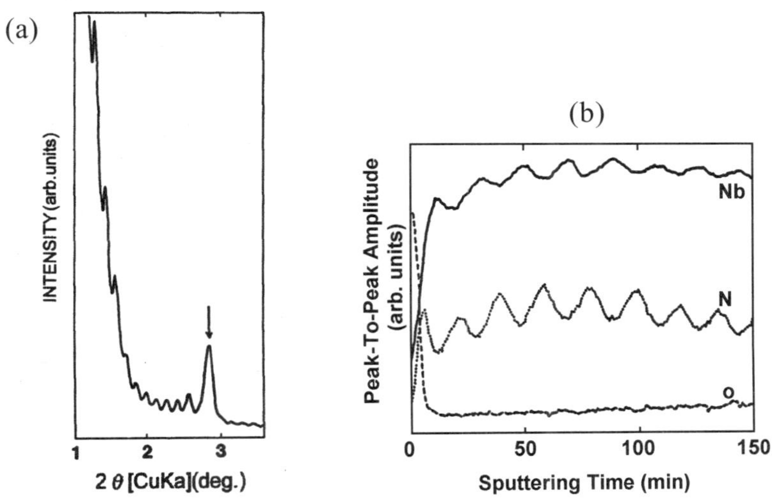

Figure 5.153. (*a*) The x-ray diffraction pattern and (*b*) the depth profile of the Nb/NbN superlattice structure prepared by ion bean sputtering.

5.8 ORGANIC THIN FILMS

Sputtered particles from an organic polymer target will be composed of atomic species and/or monomers of the target materials. Thin films of the organic polymer may be deposited since these sputtered particles may be polymerized in the sputtering gas discharge or on the substrate during film formation.

The organic polymer film is used as the target, which is fixed onto a cooled copper target holder with silicon grease to reduce the temperature rise of the target surface. Figures 5.154 and 5.155 show the infrared transmission spectra of organic thin films sputtered from a Teflon® and a polyimide target in Ar. The sputtered Teflon films exhibit the C-F absorption band corresponding to the bulk Teflon. The films are transparent and their physical properties resemble bulk Teflon; however, the sputtered polyimide films show different absorption from bulk polyimide, and thus may be significantly different from the bulk polyimide.

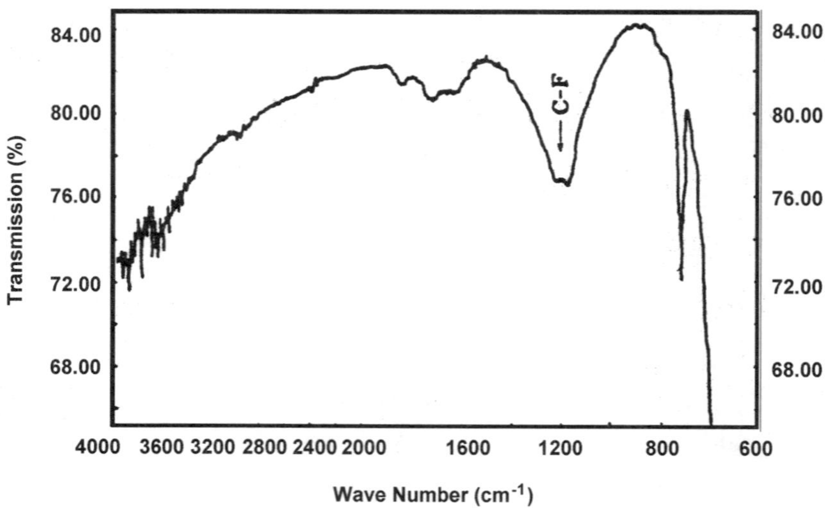

Figure 5.154. Infrared transmission spectra of sputtered Teflon® thin films.

Teflon® is a registered trademark of DuPont Chemical Co.

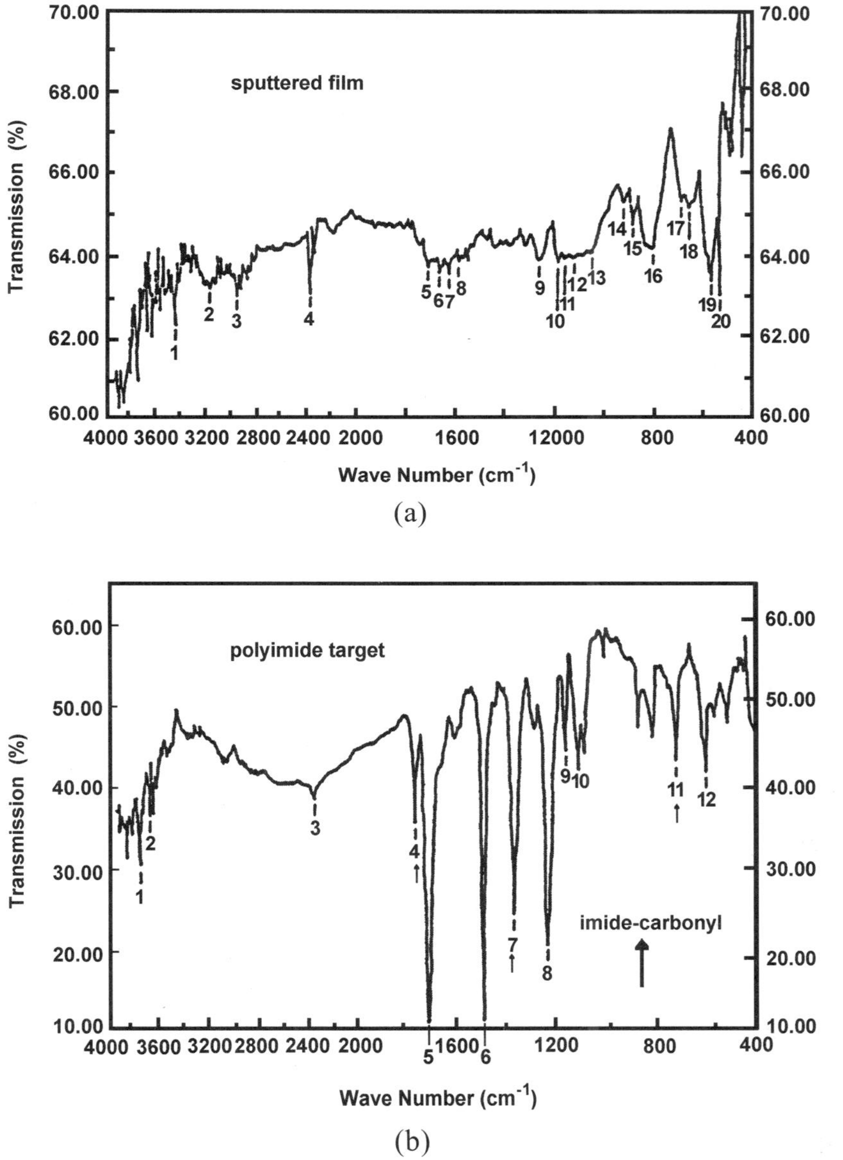

Figure 5.155. Infrared transmission spectra of (*a*) sputtered polyimide thin films and (*b*) polyimide target.

5.9 MAGNETRON SPUTTERING UNDER A STRONG MAGNETIC FIELD

Under a strong magnetic field (1000 G), a high-voltage discharge (1000 V) is kept in the magnetron sputtering system. The sputtering system operates at a low gas pressure of 10^{-4} to 10^{-5} torr. The high sputtering voltage and low working pressure will cause the impingement of high-speed sputtered atoms on the substrates. This may result in the lowering of the growth temperature as described in the synthesis of $PbTiO_3$ thin films. Unusual properties are also observed in these sputtered films. This section presents some of the interesting phenomena observed in sputtered films prepared by magnetron sputtering under a strong magnetic field.

5.9.1 Abnormal Crystal Growth

Polycrystalline ZnO films of hexagonal structure are prepared on a glass substrate by dc or rf sputtering from a zinc or ZnO target in an oxidizing atmosphere. Table 5.58 shows typical sputtering conditions and the crystallographic structure of ZnO films prepared in a conventional dc-sputtering system and in a dc-magnetron sputtering system. In the conventional sputtering system where the working pressure is 10 to 100 mtorr, it can be seen that the c-axis is preferentially oriented normal to the film surface, i.e., the (002) plane is parallel to the film surface. When ZnO films are prepared at a low working pressure of 1 mtorr or less in the magnetron sputtering system, the c-axis is predominantly parallel to the film surface, i.e., the (110) or (100) plane is parallel to the film surface.[298] Typical electron micrographs and electron diffraction patterns of these sputtered films are shown in Fig. 5.156. It shows that microstructures of ZnO films prepared by magnetron sputtering exhibit a pyramidal pattern, and are clearly different from those prepared by conventional sputtering. The microstructures of ZnO films prepared by conventional sputtering are composed of small hexagonal crystallites.

The c-axis orientation obtained in conventional sputtering is explained by the fact that surface mobility of adatoms is high during film growth, and sputtered films obey the empirical law of Bravais. The change in crystallographic orientation with these sputtering systems may be related to the difference in their working pressures. At low working pressure, the oxidation of the zinc cathode will not be completed during sputtering.

Moreover, the low working pressure causes the impingement of high-energy sputtered zinc atoms and/or negative oxygen ions on the substrates. This may lead to unusual nucleation and film growth processes in which Bravais' empirical law is inapplicable.

Table 5.58. Crystallographic Orientation of Polycrystalline ZnO Thin Films

Sputtering system[a]	Sputtering pressure (10^{-3} torr)	Substrate temperature (°C)	Deposition rate ($\mu m h^{-1}$)	Film thickness (μm)	Crystallographic orientation[d]
Conventional dc diode	35[b]	40	0.03	0.1	$C_{\perp}$
		100	0.15	0.3	$C_{\perp}$
		200	0.03	0.1	$C_{\perp}$
		200	0.15	0.3	$C_{\perp}$
		200	0.3	0.3	$C_{\perp}$
		300	0.03	0.1	$C_{\perp}$
		300	0.15	0.3	$C_{\perp}$
		300	0.3	0.3	$C_{\perp}$
dc magnetron	1[c]	40	0.03	0.1	$C_{\perp}$
		40	0.12	0.36	$C_{\perp}$.+ $C_{\parallel}$
		70	0.7	0.35	$C_{\parallel}$
		150	0.1	0.3	$C_{\parallel}$
		150	0.7	0.3	$C_{\parallel}$
		200	0.7	0.3	$C_{\parallel}$
		270	0.07	0.2	$C_{\parallel}$
		270	0.6	0.3	$C_{\parallel}$

[a]Pure Zn cathode, 7059 glass substrates.
[b]50% oxygen and 50% argon.
[c]30% oxygen and 70% argon.
[d]$C_{\perp}$, c axis normal to the film surface; $C_{\parallel}$, c axis in the film surface.

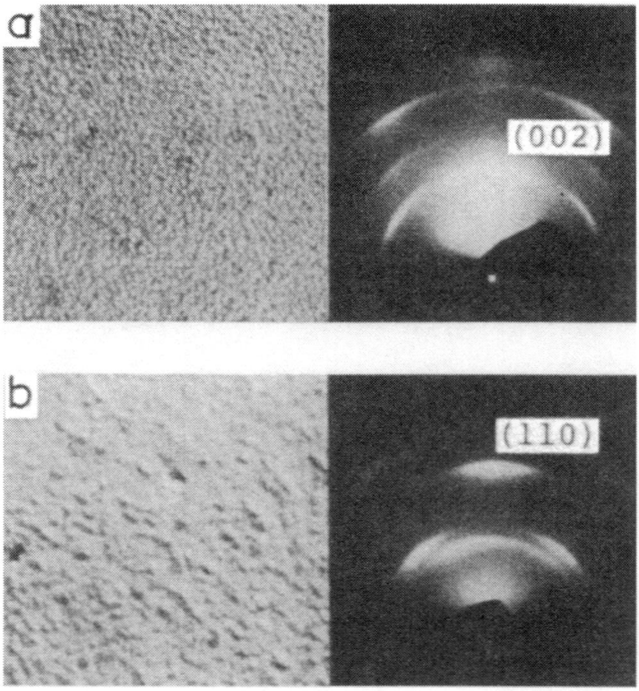

Figure 5.156. Electron micrographs and electron diffraction patterns for ZnO thin films 5000 Å thick on 7059 glass: (*a*) sputtered in a conventional diode system; (*b*) sputtered in a magnetron system.

5.9.2 Low-Temperature Doping of Foreign Atoms into Semiconducting Films

Cosputtering of foreign atoms seems to be useful for controlling the electrical conductivity of semiconducting films during sputtering deposition. Table 5.59 shows typical experiments for polycrystalline ZnO thin films in various sputtering systems. In the experiments an aluminum or copper auxiliary cathode was cosputtered with a zinc main cathode in an oxidizing atmosphere.[298]

It shows that in the conventional sputtering system, cosputtering of aluminum or copper hardly affects the conductivity of the resultant films, while in the magnetron sputtering system the cosputtering of aluminum or copper strongly affects the conductivity: aluminum increases the conductivity

Table 5.59. Electrical Conductivity of Polycrystalline ZnO Thin Films

Sputtering system[a]	Sputtering pressure (10^{-3} torr)	Content of foreign metals (at.%)	Substrate temperature (°C)	Deposition rate ($\mu m h^{-1}$)	Film thickness (μm)	Dark conductivity (Ω^{-1} cm^{-1})
Conventional dc diode	60[b]	0	300	0.25	0.5	1.6×10^{-6}
		0.2 (Al)	300	0.075	0.15	5×10^{-6}
		0.2 (Cu)	300	0.1	0.2	1.9×10^{-6}
dc magnetron	1[c]	0	40	0.03	0.1	1×10^{-6}
		0	200	0.7	0.3	1×10^{-4}
		1.3 (Al)	200	1.2	0.6	8×10^{-2}
		0.5 (Cu)	200	0.9	0.45	3×10^{-8}

[a]Pure zinc cathode, 7059 glass substrates.
[b]30% oxygen and 70% argon.
[c]50% oxygen, 50% argon.

by over three orders of magnitude and copper decreases it by approximately the same factor. This suggests that, in magnetron sputtering, aluminum is probably introduced as a donor and copper as an acceptor or a deep trap. In conventional sputtering, the cosputtered atoms stay mainlyat the crystal boundaries in the sputtered films and are probably not incorporated into the crystal lattice. Figure 5.157 shows the typical temperature variation of dark conductivity for ZnO films prepared by magnetron sputtering. The conductivity is controlled in the range of 10^{-1}–$10^{-8}\,\Omega^{-1}cm^{-1}$ by doping with foreign atoms in the cosputtering process.

Optical absorption measurements suggest that the forbidden gap width of these films is 3.3 eV, and that the acceptor or deep-trap level of copper is 2.5 eV below the conduction band at room temperature. The temperature dependence of the carrier concentration suggests that the donor level of aluminum is 0.08 eV below the conduction band. The doping of foreign atoms by the cosputtering process may possibly be the result of the impingement of high-energy sputtered atoms during film growth.

The highly conductive, aluminum-doped ZnO films can be utilized for making ZnO/Si heterojunction photodiodes and switching diodes. Figures 5.158 and 5.159 show a typical dark-current–voltage characteristic, and photo response of the photodiode, respectively. The diode was prepared by depositing an n-type aluminum-doped ZnO layer onto a p-type (111) silicon single-crystal wafer at about 200°C. The resistivity and thickness of the ZnO layer were 10 to 100 Ωcm and 0.6 μm, respectively. The resistivity of the silicon wafer was 20 Ωcm. The diode shows a photovoltage of about 100 mV for an open circuit and a photocurrent of 1 mA/cm^2 under irradiation from a tungsten lamp (2800°C) of 10^4 lux.

It is now believed that the magnetron sputtering system can be useful in industry for metallization of semiconductor devices because of its high deposition rate. However, the system offers even more attractive features when thin films of compounds such as oxides, nitrides, and carbides are deposited. Some experimental results obtained with the dc-magnetron sputtering system are summarized in Table 5.60. Further studies will undoubtedly bring success in the formation of unusual materials, which have yet to be obtained by vacuum-deposition processes.

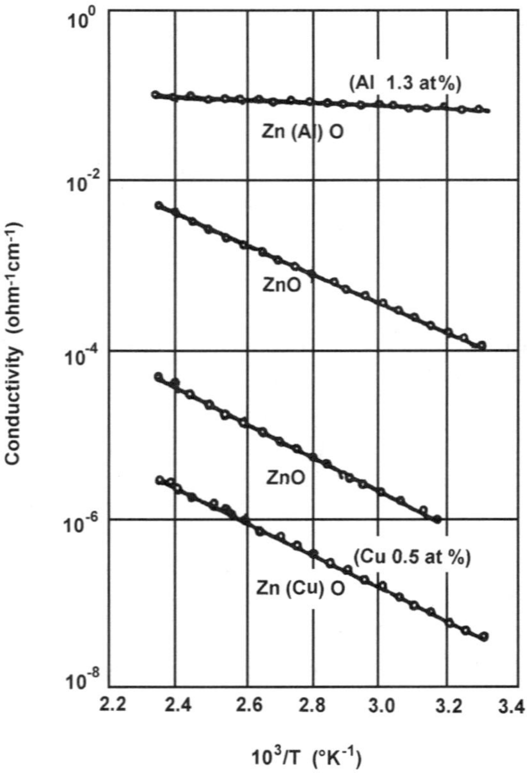

Figure 5.157. The temperature variation of the dark conductivity for ZnO films with and without cosputtered foreign metals prepared in a magnetron sputtering system.

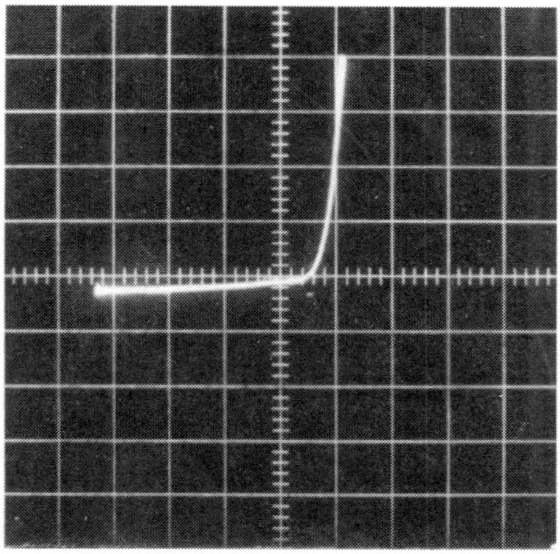

Figure 5.158. The dark-current–voltage characteristic for a sputtered n-p ZnO(Al)O/Si photodiode. Vertical scale: 100 μA/division. Horizontal scale: 2 V/ division.

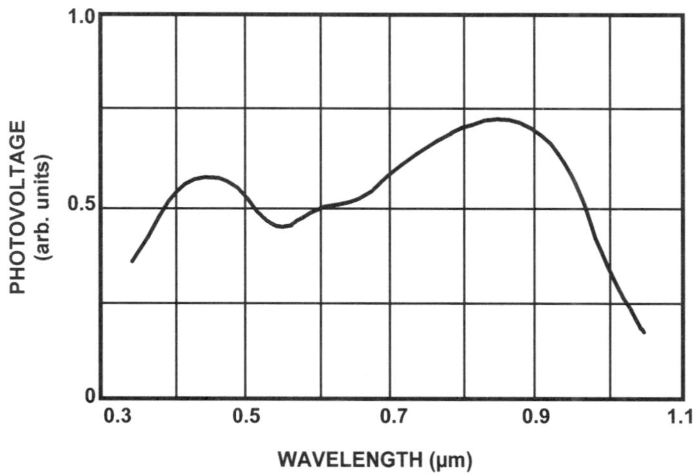

Figure 5.159. The spectral response of the photovoltage for an n-p Zn(Al)O/Si photodiode.

Table 5.60. Summary of Properties of Compound Films Prepared by dc-Magnetron Sputtering in a Strong Magnetic Field

Composition of Sputtered Films	Sputtering conditions					Film properties		Miscellaneous
	Target	Sputter Gas (torr)	Substrate[b] Temperature (°C)	Deposition Rate, R (μm h^{-1})	Rate Factor, K[c], (μm h^{-1} per V A cm^{-2})	Structure; Thickness (μm)	Electrical properties	
Al_2O_3	Al	Ar + O_2, $P_{O2} > 1\times10^{-4}$	≈ 200	0.5–1.0	0.24[d]	Amorphous; 0.1–0.3	$\varepsilon^* \approx 8$	
Ta_2O_5	Ta	O_2	≈ 200	0.1–0.3	0.11[e]	Amorphous; 0.1–0.3	$\varepsilon^* \approx 20$–30, tan $\delta < 0.01$, $V_b \approx 0.5$–1 MV cm^{-1},	
TiO_2	Ti	Ar + O_2, $P_{O2} > 3\times10^{-5}$	≈ 200	0.1–0.3	0.15[f]	Polycrystal, rutile, anatase; 0.1–0.5	$\varepsilon^* \approx 20$–60, tan $\delta \approx 0.01$–0.01, $V_b \approx 0.3$–0.7 MV cm^{-1}, TCC ≈ -300 to +300 ppm °C^{-1}	Optical absorption edge, 3.4 eV; at high P_{O2} strongly sensitive to water vapor
PbO	Pb	Ar + O_2 $P_{O2} \approx (2$–$3)\times10^{-4}$	40–200	2.0–6.0	3.0	Polycrystal, red; 0.1–1	$\varepsilon^* \approx 20$–25, tan $\delta \approx 0.04$–0.06, $V_b \approx 0.2$–0.5 MV cm^{-1} $\rho \approx 10^4$–10^5 Ω cm, TCC ≈ +700 to +900 ppm °C^{-1}	Optical absorption edge, 2.7 eV; photoconductive, max. at 2.7 eV; dielectric anomaly at 470°C
PbO_2	Pb	O_2	≈ 40	2.0	2.0	Polycrystal; 0.1–1	$\rho \approx 10^{-2}$–10^{-3} Ω cm	At high substrate temperature the composition becomes PbO

(cont'd.)

Table 5.60. *(cont'd.)* Summary of Properties of Compound Films Prepared by dc-Magnetron Sputtering in a Strong Magnetic Field

Composition of Sputtered Films	Sputtering conditions					Film properties		Miscellaneous
	Target	Sputter Gas (torr)	Substrate[b] Temperature (°C)	Deposition Rate, R ($\mu m\ h^{-1}$)	Rate Factor, K^c, ($\mu m\ h^{-1}$ per $V\ A\ cm^{-2}$)	Structure; Thickness (μm)	Electrical properties	
ZnO	Zn	Ar + O_2, $P_{O2} \approx (3\text{–}5) \times 10^{-4}$	40–270	0.03–1.0	0.3	Polycrystal, $c_{\parallel}(c_{\perp})$; 0.1–1	$\varepsilon^* \approx 8\text{–}10$, $\rho \approx 10^4\text{–}10^6\ \Omega$ cm	Optical absorption edge, 3.2 eV; photoconductive; piezoelectric
ZrO_2	Zr	Ar + O_2, $P_{O2} \approx 3\times10^{-4}$	≈ 200	0.1–0.3	0.3	Amorphous; 0.1–0.3	$\varepsilon^* \approx 13\text{–}20$, $\tan\delta \approx 0.003\text{–}0.007$, $V_b \approx 0.5\text{–}1$ MV cm^{-1}, TCC ≈ +100 to +300 ppm $°C^{-1}$	
Pb-Ti-O	Pb-Ti	Ar + O_2, $P_{O2} \approx 3\times10^{-4}$	100–300	0.3	0.3	Mixture of PbO, TiO_2, $PbTiO_3$ with Pb/Ti ≈ 1, 0.1–1.0	$\varepsilon^* \approx 120$, $\tan\delta \approx 0.005\text{–}0.015$, $V_b \approx 1\text{–}1.5$ MV cm^{-1}, TCC ≈ -100 to +100 ppm $°C^{-1}$	Dielectric anomaly at 490°C
AlN	Al	Ar + N_2, $P_{N2} \approx 3\times10^{-4}$	≈ 150	0.3	0.6	Amorphous; 0.1–0.5		
ZrN	Zr	Ar + N_2, $P_{N2} \approx 3\times10^{-4}$	≈ 150	0.7–2	0.42	Polycrystal; 0.1–0.5	$\rho \approx 1000\ \mu\Omega$cm, TCR ≈ -200 ppm $°C^{-1}$	
TiN	Ti	Ar + N_2, $P_{N2} \approx 3\times10^{-4}$	≈ 150	0.6–1.8	0.3	Polycrystal; 0.1–0.3	$\rho \approx 250\ \mu\Omega$ cm, TCR ≈ 150 ppm $°C^{-1}$	

(cont'd.)

Table 5.60. *(cont'd.)* Summary of Properties of Compound Films Prepared by dc-Magnetron Sputtering in a Strong Magnetic Field

Composition of Sputtered Films	Sputtering conditions					Film properties		Miscellaneous
	Target	Sputter Gas (torr)	Substrate[b] Temperature (°C)	Deposition Rate, R (μm h^{-1})	Rate Factor, K^c, (μm h^{-1} per V A cm^{-2})	Structure; Thickness (μm)	Electrical properties	
Ti-Zr-N	Ti-Zr	Ar + N_2, $P_{N2} \approx 3\times10^{-4}$	≈ 150	0.6–1.8	0.4	Polycrystal; solid solution of Ti-Z-N with Ti:Zr ≈ 3:7; 0.1–0.3	ρ ≈ 500 μΩ cm, TCR ≈ 0 ppm °C^{-1}	
Ti-Al-N	Ti-Al	Ar + N_2, $P_{N2} \approx 3\times10^{-4}$	≈ 150	0.6–1.8	0.4	Mixture of TiN, AlN with Ti:Al ≈ 54:46; 0.1–0.3	ρ ≈ 600 μΩ cm, TCR ≈ 0 ppm °C^{-1}	Cermet-like structure of conductive TiN and insulating AlN
Ti-Zr-Al-N	Ti-Zr-Al	Ar + N_2, $P_{N2} \approx 3\times10^{-4}$	≈ 150	0.6–1.8	0.4	Mixture of Ti-Zr-N, AlN with Ti:Zr:Al ≈ 1:2:1; 0.1–0.3	ρ ≈ 7800 μΩ cm, TCR ≈ -200 ppm °C^{-1}	Cermet-like structure of conductive Ti-Zr-N and insulating AlN

[a] Coaxial geometry; cathode current, 0.1–5 mAcm^{-2}; discharge voltage, 300–1000 V; cathode diameter, 20 mm; anode diameter, 60 mm; total sputtering gas pressure (0.6–1) × 10^{-3} torr; magnetic field, 1000–7000 G; substrates at anode.

[b] 7059 glass substrates.

[c] $K = R/V_s i_s$ where V_s is the discharge voltage and i_s is the cathode current density.

[d] For pure argon K ≈ 1.5.

[e] For pure argon K ≈ 0.7.

[f] For pure argon K ≈ 0.6.

REFERENCES

1. Foster, N. F., Coquin, G. A., Rozgonyi, G. A., and Vannatta, F. A., *IEEE Trans. Sonics Ultrason.*, SU-15:28 (1968)
2. Matsumoto, S., *Oyo Buturi,* 49:43 (1980)
3. Wasa, K., and Hayakawa, S., *Jpn. J. Appl.Phys.,* 12:408 (1973)
4. Park, R. M., Mar, H. A., and Salansky, N. M., *J. Vac. Sci. Technol.,* B3:373 (1985)
5. Stutius, W., *Appl. Phys. Lett.,* 33:656 (1978)
6. Nowicki, R. S., *J. Vac. Sci. Technol.,* 14:127 (1977)
7. Izama, T, Mori, H., Murakami, Y., and Shimizu, N., *Appl. Phys. Lett.,* 38:483 (1981)
8. Hartsough, L. D., and McLeod, P. S., *J. Vac. Sci. Technol.,* 14:123 (1977)
9. Shiosaki, T., *IEEE Ultrasonics Symp. Proc.,* p. 100 (1978)
10. Mitsuyu, T., Ono, S., and Wasa, K., *J. Appl. Phys.,* 51:2464 (1980)
11. Mitsuyu, T., Ono, S., and Wasa, K., *Proc. Symp. Fundamentals and Appl. Ultrasonic Electronics,* p. 55, Tokyo (1980)
12. Shiosaki, T., Yamamoto, T., Yagi, M., and Kawabata, A., *Appl. Phys. Lett.,* 39:399 (1981)
13. Mitsuyu, T., Wasa, K., and Hayakawa, S., *J. Electrochem. Soc.,* 123:94 (1976)
14. Takei, W. J., Formigoni, N. P., and Francombe, M. H., *J. Vac. Sci. Technol.,* 7:442 (1969)
15. Sugibuchi, K., Kurogi, Y., and Endo, N., *J. Appl. Phys.,* 46:2877 (1975)
16. Mitsuyu, T., Wasa, K., and Hayakawa, S., *J. Crystal Growth,* 41:151 (1977)
17. Payne, D. A., and Mukerjee, J. L., *Appl. Phys. Lett.,* 29:748 (1976)
18. Buchanan, M., Webb, J. B., and Williams, D. F., *Appl. Phys. Lett.,* 37:213 (1980)
19. Hori, M., Adachi, M., Shiosaki, T., and Kawabata, A., Fall Meeting of Jpn. Soc. Appl. Phys., paper 3p-F-10 (1978)
20. Mitsuyu, T., and Wasa, K., *Jpn. J. Appl. Phys.,* 20:L48 (1981)
21. Takada, S., Ohnishi, M., Hayakawa, H., and Mikoshiba, N., *Appl. Phys. Lett.,* 24:490 (1974)
22. Miyazawa, S., Fushimi, S., and Kondo, S., *Appl. Phys. Lett.,* 26:8 (1975)
23. Kwak, B. S., Erbil, A., Budai, J. D., Chisholm, M. F., Boatner, L. A., and Wilkens, B. J., *Phys. Rev.,* B49:14865 (1994)

24. Hamada, H., Morooka, H., and Hirai, H., Spring Meeting of Jpn. Soc. Appl. Phys., paper 27p-H-6 (1979)
25. Wasa, K., Hanada, Y., Satoh, T., Adachi, H., and Setsune, K., *J. Vac. Sci. Technol.,* A15:1185 (1997)
26. Okuyama, M., Usuki, T., Hamakawa, Y., and Nakagawa, T., *Appl. Phys.,* 21:339 (1980)
27. de Keijser, M., Cillessen, J. F. M., Janssen, R. B. F., de Veirman, A. E. M., and de Leeuw, D. M., *J. Appl. Phys.,* 79:393 (1996)
28. Okada, A., *J. Appl. Phys.,* 48:2905 (1977)
29. Kanno, I., Fujii, S., Kamada, T., and Takayama, R., *Appl. Phys. Lett.,* 70:1378 (1997)
30. Matsui, Y., Okuyama, M., and Hamakawa, Y., *1st Meeting Ferroelectric Mat. and Appl.,* paper 17-T-3, Kyoto (1977)
31. Ishida, M., Tsuji, S., Kimura, K., Matsunami, H., and Tanaka, T., *J. Cryst. Growth,* 45:393 (1978)
32. Sato, S., *Oyo Buturi,* 47:656 (1978)
33. Wasa, K., and Hayakawa, S., *Thin Solid Films,* 10:367 (1972)
34. Takeda, F., and Hata, T., *Jpn. J. Appl. Phys.,* 19:1001 (1980)
35. Shiosaki, T., Yamamoto, T., Oda, T., and Kawabata, A., *Appl. Phys. Lett.,* 36:643 (1980)
36. Shuskus, A. J., Reeder, T. M., and Paradis, E. L., *Appl. Phys. Lett.,* 24:155 (1974)
37. Liu, J. K., Lakin, K. M., and Wang, K. L., *J. Appl. Phys.,* 46:3703 (1975)
38. Agarwal, V., Vankar, V. D., and Chopra, K. L., *J. Vac. Sci. Technol.,* A6:2361 (1988)
39. Chopra, K. L., Agarwal, V., Vankar, V. D., Deshpandey, C. V., and Bunshah, R. F., *Thin Solid Films,* 126:307 (1985)
40. Mito, H., and Horiguchi, S., *Ionics,* 36:10 (1978)
41. Wasa, K., and Hayakawa, S., *Microelectron. Reliab.,* 6:213 (1967)
42. Srivastava, P. K., Rao, T. V., Vankar, V. D., and Chopra, K. L., *J. Vac. Sci. Technol.,* A2:1261 (1984); Srivastava, P. K., Vankar, V. D., and Chopra, K. L., *J. Vac. Sci. Technol,* A3:2129 (1985); ibid. A4:2819 (1986); Srivastava, P. K., Vankar, V. D., and Chopra, K. L., *Bull. Mater. Sci.,* 8:379 (1986); Srivastava, P. K. Vankar, V. D., and Chopra, K. L., *Thin Solid Films,* 166:107 (1988)
43. Wasa, K., Nagai. T., and Hayakawa, S., *Thin Solid Films,* 31:235 (1976)
44. Agarwal, V., Vankar, V. D., and Chopra, K. L., *J. Vac. Sci. Technol.,* A6:2341 (1988)

45. Wasa, K., Tohda, T., Kasahara, Y., and Hayakawa, S., *Rev. Sci. Instr.,* 50:1086 (1979)
46. Murayama, Y., and Takao, T., *Thin Solid Films,* 40:309 (1977)
47. Ono, H., Nishino, S., and Matsunami, H., *IECE (The Institution of Electronics, Information, and Communication Engineers) Japan, Tech. Reps.,* SSD 80–80: 125 (1981)
48. Shibahara, K., Nishino, S., and Matsunami, H., *Appl. Phys. Lett.,* 50:1888 (1987)
49. Kitabatake, M., and Greene, J. E., *Jpn. J. Appl. Phys.,* 35:5261 (1996)
50. Uchida, M., and Kitabatake, M., *Thin Solid Films,* 335:32 (1998)
51. Spitsyn, B. V., Bouilov, L. L., and Derjaguin, B. V., *J. Cryst. Growth,* 52:219 (1981)
52. Kitabatake, M., and Wasa, K., *J. Appl. Phys.,* 58:1693 (1985)
53. Deguchi, M., Kitabatake, M., Kitagawa, M., Hase, N., Kotera, H., and Shima, S., *Diamond and Related Mater.,* 6:367 (1997)
54. Matsumoto, S., Asakura, Y., and Hirakuri, K., *Appl. Phys. Lett.,* 71:2707 (1997)
55. Stringfellow, G. B., Stall, R., and Koschel, W., *Appl. Phys. Lett.,* 38:156 (1981)
56. Barnett, S. A., Bajor, G., and Greene, J. E., *Appl. Phys. Lett.,* 37:734 (1980)
57. Watkins, S. P., Brake, D. M., and Haacke, G., *J. Appl. Phys.,* 75:2952 (1994)
58. Amano, H., Kito, M., Hiramatsu, K., and Akasaki, I., *Jpn. J. Appl. Phys.,* 28:L2112 (1989)
59. Nakamura, S., Iwasa, N., Senoh, M., and Mukai, T., *Jpn. J. Appl. Phys.,* 31:1258 (1992)
60. Kim, W., Aktas, Ö., Botchkarev, A. E., Salvador, A., Mohammad, S. N., and Morkoç, H., *J. Appl. Phys.,* 79:7657 (1996)
61. Chang, C. A., Ludeke, R., Chang, L. L., and Esaki, L., *Appl. Phys. Lett.,* 31:759 (1977)
62. Greene, J. E., Wickersham, C. E., and Zilko, J. L., *Thin Solid Films,* 32:51 (1976)
63. Wu, C. T., Kampwirth, R. T., and Hafstrom, J. W., *J. Vac. Sci. Technol.,* 14:134 (1977)
64. Ihara, H., Kimura, Y., Okuyama, H., and Gonda, S., *IEEE Trans. Magn.,* MAG-19:938 (1983)
65. Kawamura, H., and Tachikawa, K., *Phys. Lett.,* 50A:29 (1974)

66. Gavaler, J. R., Greggi, J., Wilmer, R., and Ekin, J. W., *IEEE Trans. Magn.,* MAG-19:418 (1983)

67. Smith, W. D., Lin, R. Y., Coppola, J. A., and Economy, J., *IEEE Trans. Magn.,* MAG-11:182 (1975)

68. Dietrich, M., Dustmann, C. H., Scmaderer, F., and Whahl, G., *IEEE Trans. Magn.,* MAG-19:406 (1983)

69. Alterovitz, S. A., Woolam, J. A., Kammerdiner, L., and Luo, H. L., *Appl. Phys. Lett.,* 33:264 (1978)

70. Hamasaki, K., Inoue, T., Yamashita, T., and Komata,T., *Appl. Phys. Lett.,* 41:67 (1982)

71. Murakami, T., and Suzuki, M., *Jpn. J. Appl. Phys.,* 24 (Suppl. 24-2):323 (1985)

72. Terada, N., Ihara, H., Hirabayashi, M., Senzaki, K., Kimura, Y., Murata, K., and Tokumoto, M., *Jpn. J. Appl. Phys.,* 26:L508 (1987)

73. Suzuki, M., and Murakami, T., *Jpn. J. Appl. Phys.,* 26:L524 (1987)

74. Nagata, S., Kawasaki, M., Funabashi, M., Fueki, K., and Koinuma, H., *Jpn. J. Appl. Phys.,* 26:L410 (1987)

75. Adachi, H., Setsune, K., and Wasa, K., *Phys. Rev. B,* 35:8824 (1987)

76. Naito, M., Smith, D. P. E., Kirk, M. D., Oh, B., Hahn, M. R., Char, K., Mitzi, D. B., Sun, J. Z., Webb, D. J., Beasley, M. R., Fischer, O., Geballe, T. H., Hammond, R. H., Kapitulnik, A., and Quate, C. F., *Phys. Rev. B,* 35:7228 (1987)

77. Chaudhari, P., Koch, R. H., Laibowitz, R. B., McGuire, T. R., and Gambino, R. J., *Phys. Rev. Lett.,* 58:2684 (1987)

78. Oh, B., Naito, M., Arnason, S., Rosenthal, P., Barton, R., Beasley, M. R., Geballe, T. H., Hammond, R. H., and Kapitulnik, A., *Appl. Phys. Lett.,* 51:852 (1987)

79. Enomoto, Y., Murakami, T., Suzuki, M., and Moriwaki, K., *Jpn. J. Appl. Phys.,* 26:L1248 (1987)

80. Adachi, H., Setsune, K., Mitsuyu, T., Hirochi, K., Ichikama, Y., Kamada, T., and Wasa, K., *Jpn. J. Appl. Phys.,* 26:L709 (1987)

81. Adachi, H., Setsune, K., and Wasa, K., *Proc. 18th Int. Conf. Low Temp. Physics,* DM34, Kyoto (1987)

82. Dijkkamp, D., and Venkatesan, T., *Appl. Phys. Lett.,* 51:619 (1987)

83. Tsaur, B. Y., Dilorio, M. S., and Strauss, A. J., *Appl. Phys. Lett.,* 51:858 (1987)

84. Adachi, H., Setsune, K., Hirochi, K., Kamada, T., and Wasa, K., *Proc. Int. Conf. High-Temp. Superconductors and Mat. and Mech. Superconductivity,* Interlaken, Switzerland (1988); Ichikawa, Y., Adachi, H., Hirochi, K., Setsune, K., Hatta, S., and Wasa, K., *Phys. Rev. B,* 38:765 (1988)

85. Kang, J. H., Kampwirth, R. T., and Gray, K. E., *Appl. Phys. Lett.,* 52:2080 (1988); Kang, J. H., Kampwirth, R. T., Gray, K. E., Marsh, S., and Huff, E. A., *Phys. Lett. A,* 128:102 (1988)

86. Kuroda, K., Mukaida, M., Yamamoto, M., and Miyazawa, S., *Jpn. J. Appl. Phys.,* 27:L625 (1988)

87. Rice, C. E., Levi, A. F. J., Fleming, R. M., Marsh, P., Baldwin, K. W., Anzlower, M., White, A. E., Short, K. T., Nakahara, S., and Stormer, H. L., *Appl. Phys. Lett.,* 52:1828 (1988)

88. Kanai, M., Kawai, T., Kawai, M., and Kawai, S., *Jpn. J. Appl. Phys.,* 27:L1293 (1988)

89. Lolentz, R. D., and Sexton, J. H., *Appl. Phys. Lett.,* 53:1654 (1988)

90. Adachi, H., Kohiki, S., Setsune, K., Mitsuyu, T., and Wasa, K., *Jpn. J. Appl. Phys.,* 27:L1883 (1988)

91. Yamane, H., Kurosawa, H., Hirai, T., Iwasaki, H., Kobayashi, N., and Muto, Y., *Jpn. J. Appl. Phys.,* 27:L1495 (1988)

92. Ginley, D. S., Kwak, J. F., Hellmer, R. P., Baughman, R. J., Venturini, E. L., and Morosin, B., *Appl. Phys. Lett.,* 53:406 (1988)

93. Kang, J. H., Kampwirth, R. S., and Gray, K. E., *Phys. Lett. A,* 131:208 (1988)

94. Lee, W. Y., Lee, V. Y., Salem, J., Huang, T. C., Savoy, R., Bullock, D. C., and Parkin, S. S. P., *Appl. Phys. Lett.,* 53:329 (1988)

95. Hong, M., Liou, S. H., Bacon, D. D., Grader, G. S., Kwo, J., Kortan, A. R., and Davison, B. A., *Appl. Phys. Lett.,* 53:2102 (1988)

96. Adachi, H., Wasa, K., Ichikawa, Y., Hirochi, K., and Setsune, K., *J. Cryst. Growth,* 91:352 (1988)

97. Ichikawa, Y., Adachi, H., Setsune, K., Hatta, S., Hirochi, K., and Wasa, K., *Appl. Phys. Lett.,* 53:919 (1988)

98. Wasa, K., Adachi, H., Ichikawa, Y., Hirochi, K., and Setsune, K., *Proc. Conf. Sci. & Technol. Thin Film Superconductors,* Colorado Springs, CO (Nov. 1988)

99. Face, D. W., and Nestlerode, J. P., *Appl. Phys. Lett.,* 61:1838 (1992)

100. Adachi, H., Adachi, S., Ichikawa, Y., Setsune, K., and Wasa, K., *Jpn. J. Appl. Phys.,* 30:L39 (1991)

101. Hughes, R. A., Lu, Y., Timusk, T., and Preston, J. S., *Appl. Phys. Lett.,* 58:762 (1991)

102. Nonaka, T., Shimizu, K., and Kawai, T., *Physica C,* 185–189:1985 (1991)
103. Ikegawa, S., and Motoi, Y., *Appl. Phys. Lett.,* 68:2430 (1996)
104. Karimoto, S., and Naito, M., *Jpn. J. Appl. Phys.,* 38:L283 (1999)
105. Adachi, H., Mizuno, K., Satoh, T., and Setsune, K., *Jpn. J. Appl. Phys.,* 32:L1798 (1993)
106. Tsuei, C. C., Gupta, A., Trafas, G., and Mitzi, D., *Science,* 263:1259 (1994)
107. Moriwaki, Y., Sugano, T., Tsukamoto, A., Gasser, C., Nakanishi, K., Adachi, S., and Tanabe, K., *Physica C,* 303:65 (1998)
108. Krusin-Elbaum, L., Tsuei, C. C., and Gupta, A., *Nature,* 373:679 (1995)
109. Mizuno, K., Adachi, H., and Setsune, K., *J. Low Temp. Phys.,* 105:1571 (1996)
110. Adachi, H., Satoh, T., Ichikawa, Y., Setsune, K., and Wasa, K., *Physica C,* 196:14 (1992)
111. Norton, D. P., Chakoumakos, B. C., Budai, J. D., Lowndes, D. H., Sales, B. C., Thompson, J. R., and Christen, D. K., *Science,* 265:2074 (1994)
112. Allen, J. L., Mercey, B., Prellier, W., Hamet, J. F., Hervieu, M., and Raveau, B., *Physica C,* 241:158 (1995)
113. Adachi, H., Sakai, M., Satoh, T., and Setsune, K., *Physica C,* 244:282 (1995)
114. Maeda, T., Yoshimoto, M., Shimozono, K., and Koinuma, H. J. P., *Physica C,* 247:142 (1995)
115. Feenstra, R., Budai, J. D., Christen, D. K., and Kawai, T., *Appl. Phys. Lett.,* 66:2283 (1995)
116. Yamamoto, H., Naito, M., and Sato, H., *Jpn. J. Appl. Phys.,* 36:L341 (1997)
117. Coey, J. M., Berkowitz, A. E., Balcells, L. I., Putris, F. F., and Parker, F. T., *Appl. Phys. Lett.,* 72:734 (1998)
118. Li, X. W., Gupta, A., and Xiao, G., *Appl. Phys. Lett.,* 75:713 (1999)
119. Xiong, G. C., Li, Q., Ju, H., Bhagat, S. M., Lofland, S. E., Greene, R. L., and Venkatesan, T., *Appl. Phys. Lett.,* 67:3031 (1995)
120. Canedy, C. L., Ibsen, K. B., Xiao, G., Sun, J. Z., Gupta, A., and Gallagher, W. J., *J. Appl. Phys.,* 79:4546 (1996)
121. Asano, H., Hayakawa, J., and Matsui, M., *Jpn. J. Appl. Phys.,* 36:L104 (1997)
122. Konishi, Y., Kimura, T., Izumi, M., Kawasaki, M., and Tokura, Y., *Appl. Phys. Lett.,* 73:3004 (1998)
123. Tsukada, I., Nose, M., and Uchinokura, K., *J. Appl. Phys.,* 80:5691 (1996)

124. Liu, J. M., and Ong, C. K., *Appl. Phys. Lett.,* 73:1047 (1998)
125. Manako, T., Izumi, M., Konishi, Y., Kobayashi, K. I., Kawasaki, M., and Tokura, Y., *Appl. Phys. Lett.,* 74:2215 (1999)
126. Ueda, K., Tabata, H., and Kawai, T., *Science,* 280:1064 (1998)
127. Hatta, S., Shiono, T., Adachi, H., and Wasa, K., *Jpn. J. Appl. Phys.,* 26:L2082 (1987)
128. Schlomka, J. P., Tolan, M., and Press, W., *Appl. Phys. Lett.,* 76:2005 (2000)
129. Tanaka, C. T., Nowak, J., and Moodera, J. S., *J. Appl. Phys.,* 86:6239 (1999)
130. Kitabatake, M., and Wasa, K., *J. Appl. Phys.,* 58:1693 (1985)
131. Muller, K. H., *J. Appl. Phys.,* 59:2803 (1986)
132. Muller, K. H., *J. Vac. Sci. Technol.,* A5:2161 (1987)
133. Chubachi, N., *Oyo Buturi,* 46:663 (1973)
134. Galli, C., and Coker, J. E., *Appl. Phys. Lett.,* 16:439 (1970)
135. Machida, K., Shibutani, M., Murayama, Y., and Matsumoto, M., *Trans. IECE,* 62:358 (1979)
136. Ohji, K., Yamazaki, O., Wasa, K., and Hayakawa, S., *J. Vac. Sci. Technol.,* 15:1601 (1978)
137. Shiozaki, T., Ohnishi, M., Ohnishi, S., and Kawabata, A., *Proc. 1st Meeting Ferroelectric Mater. and Their Applications,* p. 7, Kyoto, Japan (1977)
138. Westwood, W. D., and Ingrey, S. J., *Wave Electronics,* 1:139 (1974/1975)
139. Hata, T., Wasa, K., and Hayakawa, S., *Thin Solid Films,* 7:135 (1971)
140. Foster, N. F., *J. Vac. Sci. Technol.,* 6:111 (1969)
141. Chubachi, N., *Proc. IEEE,* 64:772 (1976)
142. Tominaga, K., Iwamura, S., and Shintani, Y., *Jpn. J. Appl. Phys.,* 21:688 (1982)
143. Manabe, Y., Mitsuyu, T., Yamazaki, O., and Wasa, K., *Proc. 1988 Spring Meeting of Jpn. Appl. Phys.,* 28pS-8/II.
144. Mitsuyu, T., Ono, S., and Wasa, K., *J. Appl. Phys.,* 51:2464 (1980)
145. Mitsuyu, T., Yamazaki, O., Ohji, K., and Wasa, K., *J. Cryst. Growth,* 42:233 (1982)
146. Tiku, S. K., Lau, C. K., and Lakin, K. M., *Appl. Phys. Lett.,* 36:318 (1980)
147. Shiozaki, T., Ohnishi, S., Hirokawa, Y., and Kawabata, A., *Appl. Phys. Lett.,* 33:406 (1978)
148. Foster, N. F., Coquin, G. A., Rozgonyii, S. A., and Vannatta, F. A., *IEEE Trans. Sonics Ultrason.,* SU-15:28 (Jan. 1968)

149. Zhang,Y., Wang, Z., Cheeke, J. D. N., and Hickernell, F. S., *1999 IEEE Ultrason. Symp. Proc.,* p. 991, Nevada (Oct. 1999)

150. Wasa, K., Hayakawa, S., and Hada, T., *IEEE Trans. Sonics Ultrason.,* SU-21:298 (1974)

151. Lakin, K. M., *1999 IEEE Ultrasonic Symp. Proc.,* p. 895, Nevada (Oct. 1999)

152. Ishida, Y., Kobayashi, H., Nakamura, K., and Doi, A., *Proc. Int. Conf. New Piezoelectric Mater. and High Performance Acoustic Wave Devices,* p. 59, Tokyo (Jan. 2002)

153. Lemons, R. A., and Quate, C. F., *1974 IEEE Ultrasonic Symp. Proc.,* p. 4, Milwaukee (1974)

154. Kushibiki, J., and Chubachi, N., *IEEE Trans. Sonics & Ultrason.,* SU-32:189 (1985)

155., Hickernell, F. S., *J. Solid State Chemistry,* 12:225 (1975)

156. Kino, G. S., and Wagers, R. S., *J. Appl. Phys.,* 44:1480 (1973)

157. Ono, S., Wasa, K., and Hayakawa, S., *Wave Electronics,* 3:35 (1977)

158. Yamazaki, O., Mitsuyu, T., and Wasa, K., *IEEE Trans. Sonics and Ultrason.,* SU-27:369 (1980)

159. White, R. M., *Proc. IEEE,* 58:1238 (1970)

160. Hays, R. M., and Hartmann, C. S., *Proc. IEEE,* 64:652 (1976)

161. Setsune, K., Mitsuyu, T., Yamazaki, O., and Wasa, K., *1983 Ultrasonic Symp. Proc.* Atlanta, p. 467; Suhara, T., Shiono, T., Nishihara, H., and Koyama, J., *IEEE J. Lightwave Technol.,* LT-1:624 (1983)

162. Wasa, K., and Hayakawa, S., *Jpn. J. Appl. Phys.,* 10:1732 (1971)

163. Lou, L. F., *J. Appl. Phys.,* 50:555 (1979)

164. Joseph, M., Tabata, H., and Kawai, T., *Jpn. J. Appl. Phys.,* 38:L1205 (1999)

165. Onodera, A, Tamaki, N, Kawamura, Y, Sawada, T, and Yamashita, H., *Jpn. J. Appl. Phys.,* 35:5160 (1996)

166. Lenzo, P. V., et al., *J. Appl. Opt.,* 5:1688 (1966); Venturini, E. V., et al., *J. Appl. Phys.,* 40:1622 (1969)

167. Feldman, C., *Rev. Sci. Instrum.,* 26:463 (1955)

168. Wasa, K., and Hayakawa, S., *Thin Solid Films,* 52:31 (1978)

169. Kusao, K., Wasa, K., and Hayakawa, S., *Jpn. J. Appl. Phys.,* 7:437 (1968)

170. Kitabatake, M., and Wasa, K., *Jpn. J. Appl. Phys.,* Suppl. 24:33 (1985)

171. Adachi, H., Mitsuyu, T., Yamazaki, O., and Wasa, K., *Jpn. J. Appl. Phys.,* Suppl. 24:13 (1985)

172. Adachi, H., Mitsuyu, T., Yamazaki, O., and Wasa, K., *J. Appl. Phys.*, 60:736 (1986)

173. Kushida, K., and Takeuchi, H., *Appl. Phys. Lett.*, 50:1800 (1987)

174. Iijima, K., Kawabata, S., and Ueda, I., *Proc. 3rd Sensor Symp.*, p. 133, Tsukuba (1983)

175. Haertling, G. H., and Land, C. E., *J. Amer. Ceram. Soc.*, 54:1 (1974)

176. Adachi, H., Mitsuyu, T., Yamazaki, O., and Wasa, K., *Jpn. J. Appl. Phys.*, Suppl., 26:15 (1987); Adachi, H., Kawaguchi, T., Kitabatake, M., and Wasa, K., *Jpn. J. Appl. Phys.*, Suppl. 22-2:11 (1983)

177. Farnell, G. W., Cermak, I. A., Silvester, P., and Wong, S. K., *IEEE Trans. Sonics Ultrason.*, SU-17:188 (1970)

178. Keizer, K., and Burggraaf, A. J., *Ferroelectrics*, 14:671 (1976)

179. Smith, W. R., Gerard, H. M., Collins, J. H., Reeder, T. M., and Shaw, H. J., *IEEE Trans. MTT*, MTT-17:856 (1969)

180. Bechmann, R., *J. Acoust. Soc. Am.*, 28:347 (1956)

181. Gavrilyachenko, V. G., and Fesenko, E. G., *Sov. Phys. Crystallogr.*, 16:549 (1971)

182. Wasa, K., Yamazaki, O., Adachi, H., Kawaguchi, T., and Setsune, K., *J. Lightwave Technol.*, LT-2:710 (1984)

183. Tsai, C. S., Kim, B., and El-akkari, F. R., *IEEE J. Quantum Electron.*, QE-14:513 (1978)

184. Adachi, H., and Wasa, K., *ISIAT'83 & IPAT'83*, p. 951, Kyoto (1983); Adachi, H., Mitsuyu, T., and Wasa, K., *Jpn. J. Appl. Phys.*, Suppl. 24-1:121 (1985)

185. Higashino, H., Kawaguchi, T., Adachi, H., Makino, T., and Yamazaki, O., *Jpn. J. Appl. Phys.*, Suppl. 24-2:284 (1985)

186. Adachi, H., Mitsuyu, T., Yamazaki, O., and Wasa, K., *Jpn. J. Appl. Phys.*, Suppl. 24-2:287 (1985)

187. Zang, D. Y., and Tsai, C. S., *Applied Optics*, 25:2264 (1986)

188. Bednorz, J. G., and Müller, K. A., *Z. Phys.*, B64:189 (1986)

189. Caponell, D. W., Hinks, D. G., Jorgensen, J. D., and Zhang, K., *Appl. Phys. Lett.*, 50:543 (1987)

190. Wu, M. K., Ashburn, J. R., Torng, C. J., Hor, P. H., Meng, R. L., Gao, L., Huang, Z. H., Wang, Y. Q., and Chu, C. W., *Phys. Rev. Lett.*, 58:908 (1987)

191. Maeda, I., Tanaka, Y., Fukmori, M., and Asano, T., *Jpn. J. Appl. Phys.*, 27:L209 (1988)

192. Sheng, Z. Z., and Hermann, A. M., *Nature*, 332:138 (1988)

193. Sleight, A. W., Gillson, J. L., and Bierstedt, F. E., *Solid State Commun.,* 18:27 (1975)

194. Poole, C. P., Jr., Datta, T., and Farach, H. A., *Copper Oxide Superconductors,* John Wiley & Sons, New York (1988)

195. Fueki, F., Kitazawa, K., Kishio, K., and Hasegawa, T., *Proc. Srinagar Workshop on High Temperature Superconductivity* (A. K. Gupta, S. K. Joshi, C. N. R. Rao, eds.) World Scientific, p. 119, Singapore (1988)

196. Chaudhari, P., Collins, R. T., Freitas, P., Gambino, R. J., Kirtley, J. R., Koch, R. H., Laibowitz, R. B., LeGoues, F. K., McGuire, T. R., Penney, T., Schlesinger, Z., and Segmuller, A. M., *Phys. Rev. B,* 36:8903 (1987)

197. Adachi, H., Setsune, K., Hirochi, K., Kamada, T., Wasa, K., *Proc. Int. Conf. High-Temp. Superconductors and Mater. and Mechanisms of Superconductivity,* E12, Interlaken, Switzerland (1988)

198. Hammond, R. H., Naito, M., Oh, B., Hahn, M., Rosenthal, P., Marshall, A., Missert, N., Beasley, M. R., Kapitulnik, A., and Geballe, T. H., *Extended Abstracts for MRS Symp. on High Temperature Superconductors,* Anaheim, CA (1987)

199. Hirochi, K., Adachi, H., Setsune, K., Yamazaki, O., and Wasa, K., *Jpn. J. Appl. Phys.,* 26:L1837 (1987)

200. Nastasi, M., Arendt, P. N., Tesmer, J. R., Maggiore, C. J., Cordi, R. C., Bish, D. L., Thompson, J. D., Cheong, S. W., Bordes, N., Smith, J. F., and Raistrick, I. D., *J. Mater. Res.,* 2:726 (1987)

201. Hatta, S., Higashino, H., Hirochi, K., Adachi, H., and Wasa, K., *Appl. Phys. Lett.,* 53:148 (1988)

202. Tsaur, B. Y., Dilorio, M. S., and Strauss, A. J., *Appl. Phys. Lett.,* 51:858 (1987)

203. Narayan, J., Biunno, N., Singh, R., Holland, O. W., and Auciello, O., *Appl. Phys. Lett.,* 51:1845 (1987)

204. Berry, A. D., Gaskill, D. K., Holm, R. T., Cukauskas, E. J., Kaplan, R., and Henry, R. L., *Appl. Phys. Lett.,* 52:1743 (1988)

205. Adachi, H., Kohiki, S., Setsune, K., Mitsuyu, T., and Wasa, K., *Jpn. J. Appl. Phys.,* 27:L1883 (1988)

206. Higashino, H., Enokihara, A., Mizuno, K., Mitsuyu, T., Setsune, K., and Wasa, K., *5th Int. Workshop on Future Electron Devices High-Temperature Superconducting Electron Devices,* (FED HiTcSc-ED WORKSHOP), p. 267, Miyagi-Zao (1988)

207. Enokihara, A., Higashino, H., Setsune, K., Mitsuyu, T., and Wasa, K., *Jpn. J. Appl. Phys.,* 27:L1521 (1988)

208. Hirao, T., Kamada, T., Setsune, K., and Wasa, K., *Oyo Buturi,* 57:221 (1988)

209. Kitabatake, M., Mitsuyu, T., and Wasa, K., *J. Non-Cryst. Solids,* 53:1 (1982)
210. Oikawa, M., and Toda, K., *Appl. Phys. Lett.,* 29:491 (1976)
211. Dijkkamp, D., Venkatesan, T., Wu, X. D., Shaheen, S. A., Jisrawi, N., Min-Lee, Y. H., McLean, W. L., and Croft, M., *Appl. Phys. Lett.,* 51:619 (1987)
212. Fukami, T., and Sakuma, T., *Jpn. J. Appl. Phys.,* 20:1599 (1981)
213. Nakagama, T., Yamaguchi, J., Okuyama, M., and Hamakawa, Y., *Jpn. J. Appl. Phys.,* 21:L655 (1982)
214. Wehner, G. K., Kim, Y. H., Kim, D. H., and Goldman, A. M., *Appl. Phys. Lett.,* 52:1187 (1988)
215. Ihara, M., and Kimura, T., *5th Int. Workshop Future Electron Devices High-Temperature Superconducting Electron Devices,* (FED HiTcSc-ED WORKSHOP), p. 137, Miyagi-Zao (1988)
216. Wasa, K., Adachi, H., Ichikawa, Y., Setsune, K., and Hirochi, K., *Proc. Int. Workshop High* T_c *Superconductors,* Srinagar (1988)
217. Gurvitch, M., and Fiory, A. T., *Appl. Phys. Lett.,* 51:1027 (1987)
218. Ichikawa, Y., Adachi, H., Mitsuyu, T., and Wasa, K., *Jpn. J. Appl. Phys.,* 27:L381 (1988)
219. Takayama-Muromachi, E., Uchida, Y., Yukino, K., Tanaka, T., and Kato, K., *Jpn. J. Appl. Phys.*, 26:L665 (1987)
220. Jorgensen, J. D., Beno, M. A., Hinks, D. G., Solderholm, L., Volin, K. J., Hitterman, R. L., Grace, J. D., and Shuller, I. K., *Phys. Rev. B,* 36:3608 (1987)
221. Hayashi, S., Kamada, T., Setsune, K., Hirao, T., Wasa, K., and Matsuda, A., *Jpn. J. Appl. Phys.,* 27:L1257 (1988)
222. Ichikawa, Y., Kitabatake, M., Kohiki, S., Hatta, S., Adachi, H., Setsune, K., and Wasa, K., *Physica C,* 171:121 (1990)
223. Kambe, S., and Kawai, M., *Jpn. J. Appl. Phys.,* 27:L2342 (1988)
224. Li, H. C., Linker, G., Ratzel, F., Smithey, R., and Geek, J., *Appl. Phys. Lett.,* 52:1098 (1988)
225. Wu, X. D., Inam, A., Venkatesan, T., Chang, C. C., Chase, F. W., Barboux, P., Tarascon, J. M., and Wilkens, B., *Appl. Phys. Lett.,* 52:754 (1988)
226. Terashima, T., Iijima, K., Yamamoto, K., Bando, Y., and Mazaki, H., *Jpn. J. Appl. Phys.,* 27:L91 (1988)
227. Hatta, S., and Wasa, K., *Proc. 55th Magnetics Symp.,* p. 7, Tokyo (1988)
228. Setsune, K., Adachi, H., Kamada, T., Hirochi, K., Ichikawa, Y., and Wasa, K., *Extended Abstracts, 8th Int. Symp. Plasma Chem.,* p. 2335, Tokyo (1987)

229. Adachi, H., Setsune, K., and Wasa, K., *Jpn. J. Appl. Phys.,* Suppl. 26-3:1139 (1987)

230. Wasa, K., Kitabatake, M., Adachi, H., Setsune, K., and Hirochi, K., *AIP Conf. Proc.,* 165:37 (1988)

231. Kamada, T., Setsune, K., Hirao, T., and Wasa, K., *Appl. Phys. Lett.,* 52:1726 (1988)

232. Takayama-Muromachi, E., Uchida, Y., Ono, A., Izmi, F., Onoda, M., Matsui, Y., Kosuda, K., Takekawa, S., and Kato, K., *Jpn. J. Appl. Phys.,* 27:L365 (1988)

233. Adachi, H., Wasa, K., Ichikawa, Y., Hirochi, K., and Setsune, K., *J. Cryst. Growth,* 91:352 (1988)

234. Ichikawa, Y., Adachi, H., Setsune, K., Hirochi. K., Hatta, S., and Wasa, K., *Appl. Phys. Lett.,* 53:919 (1988)

235. Satoh, T., Yoshitake, T., Kubo, Y., and Igarashi, H., *Appl. Phys. Lett.,* 53:1213 (1988)

236. Abulfotuh, F., and Kazmerski, L. L., *Thin Film Technology Handbook* (A. Elshabini-Riad, and F. D. Barlow, III, eds.) p. 6-6, McGraw-Hill, New York (1997)

237. Hatta, S., Ichikawa, Y., Hirochi, K., Adachi, H., and Wasa, K., *Jpn. J. Appl. Phys.,* 27:L855 (1988)

238. Setsune, K., Hirochi, K., Adachi, H., Ichikawa, Y., and Wasa, K., *Appl. Phys. Lett.,* 15:600 (1988)

239. Ogale, S. B., Dijkkamp, D., Venkatesan, T., Wu, X. D., and Inam, A., *Phys. Rev. B,* 36:7210 (1987)

240. Wasa, K., Adachi, H., Ichikawa, Y., Hirochi, K., and Setsune, K., *Proc. ISS'88,* Nagoya (1988)

241. Kohiki, S., Hirochi, K., Adachi, H., Setsune, K., and Wasa, K., *Phys. Rev. B,* 38:9201 (1986); ibid, 39:4695 (1988)

242. Adachi, H., Kohiki, S., Setsune, K., Mitsuyu, T., and Wasa, K., *Jpn. J. Appl. Phys.*, 27:L1883 (1988)

243. Hatta, S., Hirochi, K., Adachi, H., Kamada, T., Ichikawa, Y., Setsune, K., and Wasa, K., *Jpn. J. Appl. Phys.,* 27:1646 (1988)

244. Bean, C. P., *Rev. Mod. Phys.,* 36:31 (1964)

245. Dinger, T. R., Worthington, T. K., Gallagher, W. J., and Sandstrom, R. L., *Phys. Rev. Lett.*, 58:2687 (1987); Van Dover, R. B., Schneemeyer, L. F., Gyorgy, E. M., and Waszczak, J. V., *Appl. Phys. Lett.,* 52:1910 (1988)

246. Matsushima, T., Ph.D Thesis, Osaka Univ. (Feb. 1991)

247. Enokihara, A. and Setsune, K., *J. Superconductivity,* 10:49 (1997)

248. Katsube, T., and Katsube, Y., *Oyo Buturi,* 49:2 (1980)

249. Ohata, Y., and Yoshida, S., *Oyo Buturi*, 45:43 (1977)

250. Buchanan, M., Webb, J. B., and Williams, D. F., *Appl. Phys. Lett.*, 37:213 (1980)

251. McLean, D. A., Schwartz, N., and Tidd, E. D., *IEEE Int. Conv. Record,* 12:pt-9, 128 (1964)

252. Wasa, K., and Hayakawa, S., *Thin Solid Films,* 52:31 (1978)

253. Wasa, K., and Hayakawa, S., *Thin Solid Films,* 10:367 (1972)

253. Wasa, K., and Hayakama, S., US Patent # 3803057 (1974)

255. Stein, H. J., Picraux, S. T., and Holloway, P. H., *IEEE Trans. Electron. Devices,* 25:587 (1978)

256. Fujita, S., Toyoshima, H., Ohishi, T., and Sasaki, A., *Jpn. J. Appl. Phys.*, 23:L268 (1984)

257. Chow, R., Lanford, W. A., Kem-Ming, W., and Rosler, R. S., *J. Appl. Phys.*, 53:5630 (1982)

258. Kitabatake, M., and Wasa, K., *Appl. Phys. Lett.*, 49:927 (1986)

259. Vossen, J. L., and Kern, W. (eds.), *Thin Film Processes,* p. 421, Academic Press, NY (1978)

260. Khan, I. H., and Summergrad, R. N., *Appl. Phys. Lett.,* 11:12 (1967)

261. Yoshihara, H., Mori, H., Kiuchi, M., and Kadota, T., *Jpn. J. Appl. Phys.*, 17:1693 (1978)

262. Learn, A. J., and Haq, K. E., *Appl. Phys. Lett.*, 17:26 (1970)

263. Wasa, K., Nagai, T., and Hayakawa, S., *Thin Solid Films,* 31:235 (1976)

264. Murayama, Y., and Takao, T., *Thin Solid Films,* 40:309 (1977)

265. Lipson, H. G., *Silicon Carbide* (J. R. O'Connor and J. Smiltens, eds.), p. 371, Pergamon, London (1960)

266. Wasa, K., Tohda, T., Kasahara, Y., and Hayakawa, S., *Rev. Sci. Instr.*, 50:1084 (1979)

267. Carborundum Co., Cat. No. 32602.

268. Aisenberg, S., and Chabot, R., *J. Appl. Phys.,* 42:2953 (1971)

269. Wasa, K., and Hayakawa, S., Japanese Patent # S51-84840 (1976)

270. Weissmantel, C., Bewilogua, K., Dietrich, D., Hinneberg, H. J., Klose, S., Nowick, W., and Reisse, G., *Thin Solid Films,* 72:19 (1980)

271. Kitabatake, M., and Wasa, K., *J. Appl. Phys.*, 58:1693 (1985)

272. Davis, R. F. (ed.), *Diamond Films and Coatings,* Noyes, NJ (1993)

273. Liu, H., and Dandy D. S., *Diamond Chemical Vapor Deposition*, Noyes, NJ (1995)

274. Pierson, H. O., *Handbook of Carbon, Graphite, Diamond and Fullerenes*, Noyes, NJ (1993)

275. Calow, J. T., Kirk, D. L., Owen, S. T. J., and Webb, P. W., *Radio and Elect. Eng.*, 41:243 (1971)

276. Tsuya, N., and Arai, K. I., *Jpn. J. Appl. Phys.*, 18:461 (1979)

277. Kitabatake, M., Mitsuyu, T., and Wasa, K., *J. Appl. Phys.*, 56:1780 (1984)

278. Kinbara, A., and Uosumi, K., *Oyo Buturi*, 45:1165 (1976)

279. Stroud, P. T., *Thin Solid Films*, 11:1 (1972)

280. Lines, M. E., *Phys. Rev. B*, 15:388 (1977)

281. Glass, A. M., Lines, M. E., Nassau, K., and Shiever, J. W., *Appl. Phys. Lett.*, 31:249 (1977)

282. Glass, A. M., Nassau, K., and Negran, T. J., *J. Appl. Phys.*, 49:4808 (1978)

283. Takashige, M., Nakamura, T., Tsuya, N., Arai, K., Ozawa, H., and Uno, R., *Jpn. J. Appl. Phys.*, 19:L555 (1980)

284. Mitsuyu, T., and Wasa, K., *Jpn. J. Appl. Phys.*, 20:L48 (1981)

285. Nassau, K., Levinstein, H. J., and Loiacono, G. M., *J. Phys. Chem. Solids*, 27:989 (1966)

286. Kitabatake, M., and Wasa, K., *J. Non-Cryst. Solids*, 58:1 (1982); Spring Meeting Phys. Soc. Japan, paper 2a-J-12.

287. Takashige, M., Cho, T., Nakamura, T., and Aikawa, Y., Spring Meeting Phys. Soc. Japan, paper 2a-J-12 (1982)

288. Tohda, T., Wasa, K., and Hayakawa, S., *J. Electrochem. Soc.*, 127:44 (1980)

289. Lipson, H. G., *Silicon Carbide* (J. R. O'Connor and J. Smiltens, eds.), p. 371, Pergamon Press, London (1960)

290. Moustakas, T. D., and Friedman, R., *Appl. Phys. Lett.*, 40:515 (1982)

291. Esaki, L., *Proc. 6th Int. Vac. Congr.*, Kyoto (1974); *Jpn. J. Appl. Phys.*, 13 (Suppl. 2, Pt. 1):821 (1974)

292. Pessa, M. V., Huttunen, P., and Herman, M. A., *J. Appl., Phys.*, 54:6047 (1983)

293. Wasa, K., Kitabatake, M., Adachi, H., Ichikawa, Y., Hirochi, K., and Setsune, K., *Thin Solid Films*, 181:199 (1989)

294. Chopra, K. L., private communication, Tokyo (Dec. 2000)

295. Greene, J. E., Wickersham, C. E., and Zilko, J. L., *J. Appl. Phys.*, 47:2289 (1976)

296. Eltoukhy, A. H., Zilko, J. L., Wickersham, C. E., and Greene, J. E., *Appl. Phys. Lett.*, 31:156 (1977)

297. Kitabatake, M., Ph.D. Thesis, Tohoku Univ., Sendai (1988)

298. Hata, T., Hayakawa, S., and Wasa, K., *Jpn. J. Appl Phys.,* 9:1078 (1970)

299. a. Denburg, D. L., *IEEE Trans. Sonics and Ultrason.,* Su-18:31 (1971)

b. Ieki, H., and Kadota, M., *IEEE Ultrason. Symp. Proc.,* p. 281 (1999)

c. Evans, D. R., Lewis, M. F., and Patterson, E., *Electron Lett.,* 7:557 (1971)

d. Larson, L. D., Winslow, D. K., and Zitelli, L. T., *IEEE Trans. Sonics and Ultrason.,* Su-19:18 (1972)

e. Hickernell, F. S., *J. Appl.Phys.,* 44:1061 (1973)

f. Hickernell, F. S., *IEEE Ultrason. Symp. Proc.,* p. 273 (1998)

g. Ohji, K., Tohda, T., Wasa, K., and Hayakawa, S., *J. Appl. Phys.,* 47:1726 (1976)

h. Wasa, K., Hayakawa, S., and Hada, T., *IEEE Trans. on Sonics and Ultrason.,* Su-21:298 (1974)

i. Chubachi, N., *Proc. IEEE,* 64:772 (1976)

j. Tiku, S. K., Lau, C. K., and Lakin, K. M., *Appl. Phys. Lett.,* 33:406 (1980)

k. Shiozaki, T., Ohnishi, S., Hirokawa, Y., and Kawabata, A., *Appl. Phys. Lett.,* 33:318 (1978)

l. Mitsuyu, T., Ono, S., and Wasa, K., *J. Appl. Phys.,* 51:2464 (1980)

m. Mitsuyu, T., Yamazaki, O., Ohji, K, and Wasa, K., *J. Crystal Growth,* 42:233 (1982)

300. a. Oikawa, M., and Toda, K., *Appl. Phys. Lett.,* 29:491 (1976)

b. Okada, A., *J. Appl. Phys.,* 48:2905 (1977)

c. Okuyama, M., Usuki, T., Hamakawa, Y., and Nakagawa, T.,*Appl. Phys.,* 21:339 (1980)

d. Adachi, H., Mitsuyu, T., Yamazaki, O., and Wasa, K., *J. Appl. Phys,,* 60:736 (1986)

e. Nakagawa, T., Yamaguchi, J., Usuki, T., Matsui, Y., Okuyama, M., and Hamakawa, Y., *Jpn. J. Appl. Phys.,* 18:897 (1979)

f. Ishida, M., Tsuji, S., Kimura, K., Matsunami, H., and Tanaka, T., *J. Crystal Growth,* 45:393 (1978)

6

Structural Control of Compound Thin Films: Perovskite and Nanometer Oxide Thin Films

The basic deposition process of compound thin films is described in Ch. 5. Among these compound thin films, the perovskite materials show a variety of functional properties including ferroelectricity, pyroelectricity, piezoelectricity, nonlinear optical properties, and superconductivity. The ferroelectric thin films of nanometer structure, in particular, are promising materials for future applications.

Perovskite thin films comprise complex chemical compositions. The ferroelectric property is changed by the chemical composition and/or the microstructure of the thin films. Structural control of ferroelectric thin films is important not only for the fabrication of layered perovskite electronic and photonic devices, but also for understanding their physical and chemical properties.

In this chapter, sputtered ferroelectric thin films of perovskite, including those with nanometer structures, are used as examples of compound thin films. The influences of sputtering parameters on the structure and dielectric properties of thin films are also discussed.

6.1 FERROELECTRIC MATERIALS AND STRUCTURES

6.1.1 Ferroelectric Materials

Ferroelectric materials were discovered in 1920 in the form of bulk single crystals of Rochell salt. Since then, a number of ferroelectric materials, including $NH_4H_2PO_4$ (ADP), KH_2PO_4 (KDP), $LiNbO_3$ (LN), $LiTaO_3$ (LT), $BaTiO_3$ (BT), $PbTiO_3$ (PT), Pb(Zr,Ti)O_3 (PZT), (Pb,La)TiO_3 (PLT), and (Pb,La)(Zr,Ti)O_3 (PLZT), were developed in the form of bulk single-crystal and bulk polycrystalline ceramics. Among these ferroelectric materials, perovskite (ABO_3) with oxygen octahedral structures is now widely used for the production of electronic components and microtransducers.

Ferroelectric thin films of perovskite structures were first deposited by the thermal evaporation of BT in the 1950s after the discovery of the BT single crystal.[1] In the 1960s, the sputtering process was applied to the deposition of perovskite materials. PT thin films were deposited by sputtering;[2] however, their structures and ferroelectric properties were not well-characterized. Detailed studies of ferroelectric thin films started in the 1980s.

Ferroelectric materials and deposition processes are shown in Tables 6.1 and 6.2, respectively.[3] Most of the early experiments were done using a range of film thicknesses from 100 nm to several micrometers. However, these ferroelectric thin films mostly exhibited polycrystal structure with an interfacial layer of low dielectric constant between the thin films and substrates.[4] Their dielectric constant decreases with the decrease of the film thickness when the film thickness is thinner than 100 nm. A controlled deposition of perovskite thin films with film thickness less than 100 nm is essential not only for understanding the size effect in ferroelectric thin films[5] but also for the application of ferroelectric thin films to electronic devices such as gigabite FEDRAMs (ferroelectric dynamic random access memories) as shown in Table 6.1. For better understanding of perovskite thin films, single-crystal thin films without the interfacial layer deposited by heteroepitaxial growth are useful. The typical substrates for the heteroepitaxial growth are listed in Table 6.3.[3]

Table 6.1. Ferroelectric Thin Films and Devices[3]

Function	Devices: Materials	Miscellaneous
Ferroelectricity	FEDRAM[1]: PZT, PLZT FESRAM[2]: BPZT, SBT FEMFET[3]: BMF	Nonvolatile High Ps, Pr PZT > 20 μC/cm^2
High permittivity	Capacitor for high count DRAM: SBT, ST, PZT, PLT	High permittivity PZT: 500–2000 Film thickness < 100 nm
Pyroelectricity	IR detector: PT, PLT	Sensitive/low noise PLT:$\gamma = 5.5 \times 10^{-4}$C/m^{2}K
Piezoelectricity	BAW/SAW: ZnO, AlN Filter: PZT, PLT Resonator Oscillator Delay line	High coupling for SAW ZnO/sapphire: $k^2 = 5\%$ High temperature stability ZnO/glass: TCD (temperature coefficient of delay) = 0
Electrostriction	Actuator: PLT, PZT MEMS[4]: ZnO	High sensitive
Acousto-optics	Integrated optics: ZnO, LN Channel switch: PLT Modulator: PLZT	Low working voltage High speed operation
Electro-optics	Integrated optics: LT, LN Coupler: BTO Channel switch: PLT Modulator: PLZT Optical shutter EO disk memory	Pockels effect (linear EO) LN, LT, BTO, PLZT Kerr effect (quadratic EO) PLT, PLZT: $R = 1 \times 10^{-16}$ m^2/V^2 (6328 Å)
BMF: $BaMgF_4$ BPZT: $BaTiO_3$-$PbZrO_3$ BST: $(Ba,Sr)TiO_3$ BTO: $Bi_4Ti_3O_{12}$	LN: $LiNbO_3$ LT: $LiTaO_3$ PLT: $(Pb,La)TiO_3$	PLZT: $(Pb,La)(Zr,Ti)O_3$ PZT: $Pb(Zr,Ti)O_3$ SBT: $SrBi_2Ta_2O_9$

[1] Ferroelectric dynamic random access memory.
[2] Ferroelectric static random access memory.
[3] Ferroelectric memory field effect transistor.
[4] Micro-electromechanical system.

Table 6.2. Thin Film Deposition Process[3]

Classification	Deposition System	Source Materials	Film Structure
Vapor phase deposition			
PVD	Thermal evaporation EB (electron beam) Crucible MBE	Individual metals Individual oxides Multisource	Uniaxial crystal by epitaxial growth (poly/single) tailoring FE by layer-by-layer dep. in situ poling
	Laser ablation	FE compounds Individual oxides Multitarget	
	Sputtering	FE compounds Individual metals Individual oxides Multitarget	
CVD	Low pressure CVD MOCVD Plasma assisted MOCVD	Individual halide Metal organic gas	Uniaxial crystal by epitaxial growth (poly/single) in situ poling
Chemical solvent deposition			
	MOD (metallo-organic deposition	Individual metal	Multiaxial (polycrystal)
		Organic gas	
	Sol-gel deposition		Ex situ poling (porous)

Table 6.3. Lattice Parameters of $PbTiO_3$ and Single-Crystal Substrates[3]

	Crystal System	Structure	Lattice Constant (Å)	Coefficient of Expansion (10^{-6}/K)
$PbTiO_3$	tetragonal	perovskite	a = 3.889	12
			c = 4.1532	
	cubic		a = 3.961	
Sapphire	trigonal	corundum	a = 4.763	$7.5 \approx 8$
MgO	cubic	NaCl	a = 4.203	13.8
$SrTiO_3$	cubic	perovskite	a = 3.905	10.8
$LaAlO_3$	pseudo cubic	perovskite	a = 3.792	10
YSZ	cubic	fluoride	a = 5.16	10
$KTaO_3$	cubic	perovskite	a = 3.989	6.7
Epitaxial relations:				
(111)PT//(0001) sapphire			(001)PT//(001)$LaAlO_3$	
(001)PT//(001)ST			(001)PT//(001)$KTaO_3$	
(100)PT//(100)MgO			(111)PT/(111)YSZ/(100)Si	

The microstructure of perovskite thin films is governed by the initial stage of the film's growth. The cooling conditions after deposition are also important when depositing ferroelectric thin films that have a controlled crystal structure.[6] Recent technological progress has enabled the deposition of perovskite materials that have such a controlled crystal structure.[7] The sputtering process also produces man-made ferroelectric thin films, including the superlattice of layered perovskite.[8] These man-made ferroelectric materials are essential for the design of novel nanometer materials.

6.1.2 Microstructure of Heteroepitaxial Thin Films

A typical structure of the heteroepitaxial thin film represented by schematic cross section is shown in Fig. 6.1. The heteroepitaxial thin film is either unstrained or strained. The unstrained structure includes misfit dislocations at the interface between the substrates and the thin films. The strained system is called *pseudomorphic* or *commensurate*. Due to the constraint on the in-plane lattice constant, the unit cell distorts as allowed by Poisson's ratio for an epitaxial layer. If the lattice constant of the thin film is longer than that of the substrate, the film is tetragonally distorted due to the compressive stress in the plane. If the lattice constant of the film is smaller than that of the substrate, the film is stretched in the plane and the height decreases.

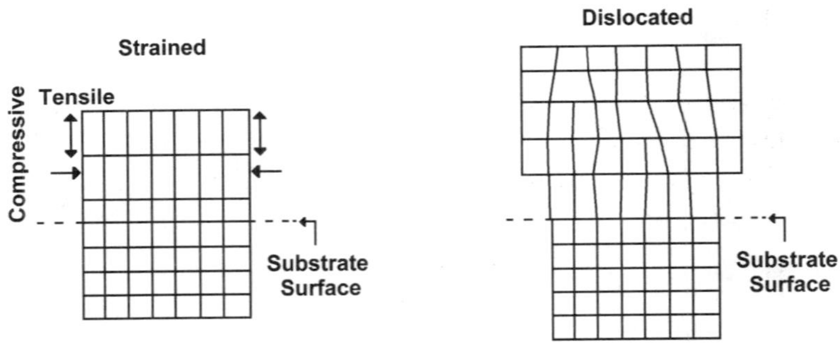

Figure 6.1. Lattice deformation of heteroepitaxial films.

The film growth process is considered to be as follows:[9]

- The initial layers in the growing film are strained elastically until the film's lattice constant equals that of the substrates.
- After a critical thickness (h_c) has been exceeded, it is energetically favorable for part of the strain to be relaxed by introducing misfit dislocations.
- For thicknesses much greater than h_c, all the strain is relaxed because of misfit dislocations.

The critical thickness, h_c, for a pseudomorphic epitaxy is derived by considering the film-thickness dependence of the strain energy and the dislocation energy, then minimizing the total energy.[10] If the dislocation spacing is greater than the film thickness, h, the extent of the strain field, r, is nearly equal to h, and the total energy, E_{tot}, is expressed by

Eq. (6.1) $$E_{tot} = \varepsilon_{\|}^2 Bh + \frac{\mu_f b(f - \varepsilon_{\|})[\ln(h/b)+1]}{2\pi(1-\nu)}$$

where μ_f is the shear modulus of the film, f is the lattice misfit, ν is Poisson's ratio for the film material, and $B = 2\mu_f(1 + \nu)/(1 - \nu)$. By setting $dE_{tot}/d\varepsilon_{\|} = 0$, the critical strain, $\varepsilon_{\|}^*$ becomes

Eq. (6.2) $$\varepsilon_{\|}^* = \frac{b[\ln(h/b)+1]}{8\pi(1+\nu)h}$$

The largest possible value of $\varepsilon_{\|}^*$ is the misfit f. The f is expressed by

Eq. (6.3) $$f = \frac{(a_f - a_s)}{a_s}$$

where a_f and a_s denote the lattice constants of the thin films and the substrates, respectively, and b is Burgers vector of the substrates.

Taking $\varepsilon_{\|}^* = f$, the critical thickness, h_c, is obtained:

Eq. (6.4) $$h_c = \frac{b[\ln(h_c/b)+1]}{8\pi(1+\nu)f}$$

This assumes that the epitaxial films include edge-type misfit dislocations that are perpendicular to the film plane, and that the Burgers vector, b, is parallel to the interface.

When the Burgers vector, b, is not parallel to the interface and the angle between the direction of the slip dislocation and the interface and/or the film plane is λ, the critical thickness is expressed by

Eq. (6.5) $$h_c = \frac{b[\ln(h_c/b)+1]}{8\pi(1+\nu)f\cos\lambda}$$

For perovskite materials, Burgers vectors <010> and <110>/2 are often observed. In the case of (001)PT/(100)ST, $f = (3.91\text{-}3.905)/3.905 = 0.0012$, and $\nu = 0.2$; taking $\lambda = 0$, and $b = a$ for the <010> direction, h_c becomes 65 nm; for the <110>/2 direction, $\lambda = 0$, $b = a/2^{1/2}$, and h_c becomes 45 nm. In the case of (111)PT/(0001)sapphire, $f = (2.84\text{-}2.75)/2.75 = 0.03$, and $\nu = 0.16$; taking $\lambda = 0$, $b = a/2^{1/2}$ for the <110>/2 direction, h_c becomes 0.5 nm; for the <010> direction, taking $\lambda = 35°$, and $b = a$, h_c becomes 1 nm.

This estimation means that a large misfit will give a small critical thickness, h_c. The PT thin films epitaxially grown on (0001) sapphire substrates usually include dislocations. If we consider that the thickness of the epitaxial PLZT thin films is several hundred nanometers, the epitaxial films will reasonably include the misfit dislocations since $h_c \ll 100$ nm.

A typical example of heteroepitaxial thin films and their microstructures is described using a sample PLT (20) heteroepitaxial thin film on (0001) sapphire with a film thickness of 400 nm. Sputtering conditions are shown in Table 6.4. In epitaxial conditions, PLT (20) thin films are deposited from the PLT (28) target. Figure 6.2 shows the RHEED patterns of the sputtered PLT thin films. The RHEED patterns suggest that epitaxial (111) PLT thin films are grown on (0001) sapphire as expected.

A cross-sectional TEM image with a selected area diffraction (SAD) pattern is shown in Fig. 6.3. The film shows a grain structure with a high-angle grain-boundary, although the SAD pattern shows the three-dimensional epitaxy. Each grain shows the same crystal orientation, both in the c-axis and in the a-b plane.

Figure 6.4 shows the TEM images of the plan view at various depths with transmission electron diffraction. Microcrystallites of about 10 nm to 100 nm in diameter are observed. The crystal orientation of each microcrystallite is the same. The microcrystallites and/or grains develop with an increase in the film thickness in order to decrease the strain energy according to the mismatch in the lattice constant and the thermal expansion coefficient between the films and the substrates.

Table 6.4. Sputtering Conditions for Heteroepitaxial PLT Thin Films on Sapphire Substrates

Target	PLT (28) ceramics powder
Target diameter	100 mm
Substrate	Sapphire (0001)
Target-substrate spacing	35 mm
Sputtering gas	Ar (60%) + O_2 (40%)
Gas pressure	0.5 Pa
Substrate temperature	500–700°C
RF power	150–250 W
Deposition rate	60–100 Å/min

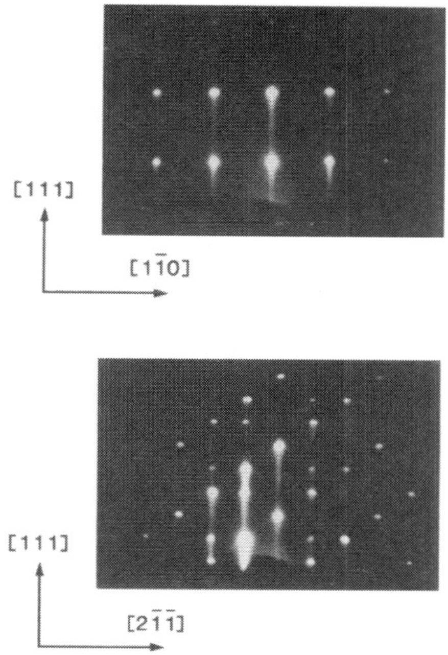

Figure 6.2. RHEED patterns of the sputtered PLT (20) thin films, 0.4 μm thick, deposited on (0001) sapphire substrate.

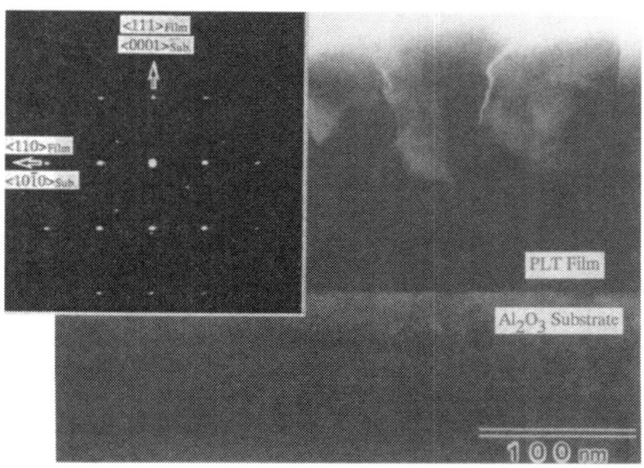

Figure 6.3. Cross-sectional TEM images of highly transparent PLT (20) thin films on sapphire substrates with the SAD pattern obtained at the thin film-substrate interface. The SAD pattern suggests that the PLT thin films are well-oriented in the c-direction.

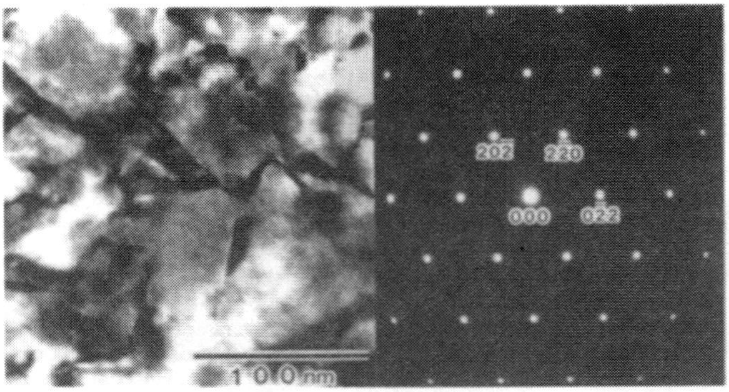

Figure 6.4. TEM image of the plan view with a SAD pattern for a PLT (20) thin film.

Ferroelectric thin films often display a multidomain structure such as the 90° domains shown in Fig. 6.5. The growth of 90° domains also reduces the strain in the films. A theoretical model of the domain-pattern formation uses the linear-elasticity theory and a Landau-Ginzburg-Demonshire-type phenomenological theory.[11] The mechanism of domain growth for epitaxial $PbTiO_3$ thin films is thought to be as follows: Starting with a PT thin film with a cubic crystal structure at an epitaxial temperature of 600°C, the crystal structure moves to a tetragonal structure due to the phase transition at the Curie temperature (about 490°C) during the cooling stage after deposition. The lattice of the PT thin film after cooling shows a strained structure due to the mismatch of the thermal expansion coefficients and lattice parameters between PT thin films and the substrates. When the film thickness is thicker than h_c, the domain structure grows in the thin film due to the minimizing of the strained energy. When we use $KTaO_3$ substrates with a small mismatch (a = 0.3899 nm at room temperature, and a = 0.4003 nm at 600°C), the thin film shows the layered structure of a-domain and c-domain, i.e., a/c/a/c/ structures, as shown in Fig.6.5. The PT thin film shows a single c-domain structure when the lattice mismatch is very small, as seen with a $SrTiO_3$ substrate. For the MgO substrate, which has poor lattice matches at both the a-axis and the c-axis (a = 0.4213 nm at room temperature, and a = 0.4239 nm at 600°C), the PT/MgO heterosystem is assumed to find the energy minimum by locking into the domain of the two-dimensional superlattice with the greatest atomic coincidences. The c- and a-domains will be mixed in the sputtered PT thin films, since the PT lattices of ($3a \times 3a$) are close to the MgO lattice of $2\sqrt{2}$ ($a \times a$), and the PT lattices of ($3b \times 3c$) are close to the MgO lattice of $2\sqrt{2}$ ($a \times a$).

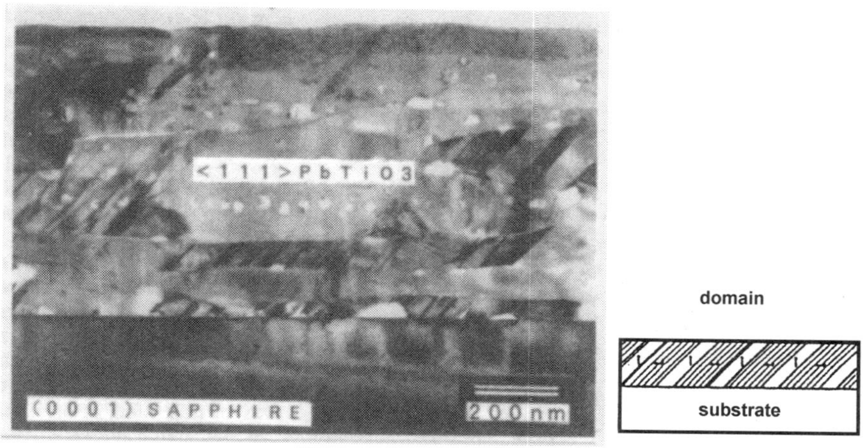

Figure 6.5. Cross-sectional TEM image of sputtered PT thin films on (0001) sapphire.

In heteroepitaxial films, the interface layer sometimes comprises different chemical compositions and/or different crystal phases. The formation of the interface layer is also caused by the interdiffusion between the thin films and the substrates. A typical SIMS depth profile of PLT thin films sputtered onto (0001) sapphire is shown in Fig. 6.6. The SIMS profile shows the interdiffusion of Al from the sapphire into the epitaxial PLT thin film. The interdiffusion makes the interfacial region. The structure of the interface layer is evaluated by the ellipsometric method.[12] Typical results are found in Fig. 6.7.[13] The interface comprises materials with a low optical index probably due to the change of the chemical composition and/or the crystal properties.

Both off-stoichiometry and nonuniformity of chemical composition also affect the microstructure of compound thin films. The secondary crystal phase is observed in the thin films. Typical TEM images with electron diffraction patterns for nonuniform PLT (20) thin films are shown in Fig. 6.8. The nonuniform films are not transparent. These opaque PLT thin films are grown due to the segregation of a secondary phase, like lanthanum-titanate.

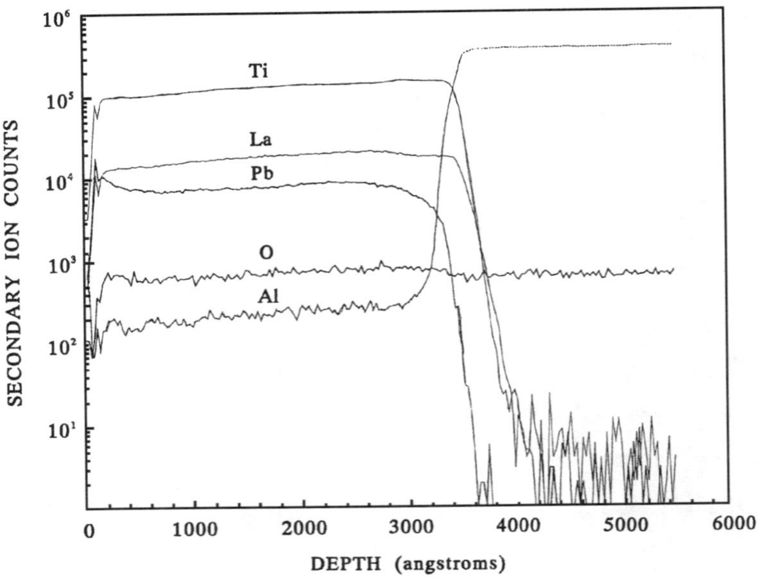

Figure 6.6. SIMS depth profile of the highly transparent PLT (20) thin films, 0.3 μm thick, deposited on (0001) sapphire substrate.

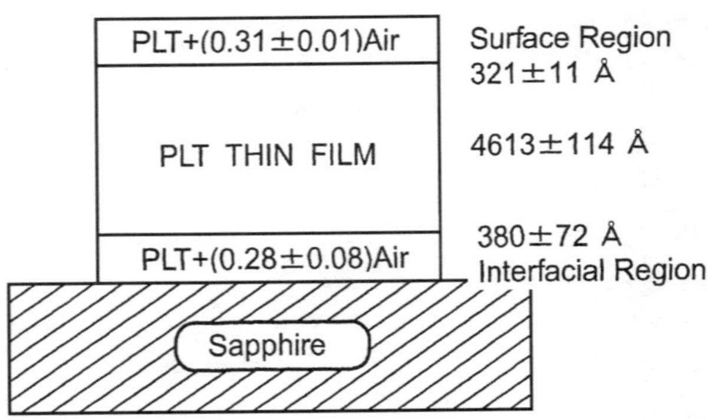

Figure 6.7. Evaluation of microstructure PLT (20) thin films on (0001) sapphire by spectroscopic ellipsometry.

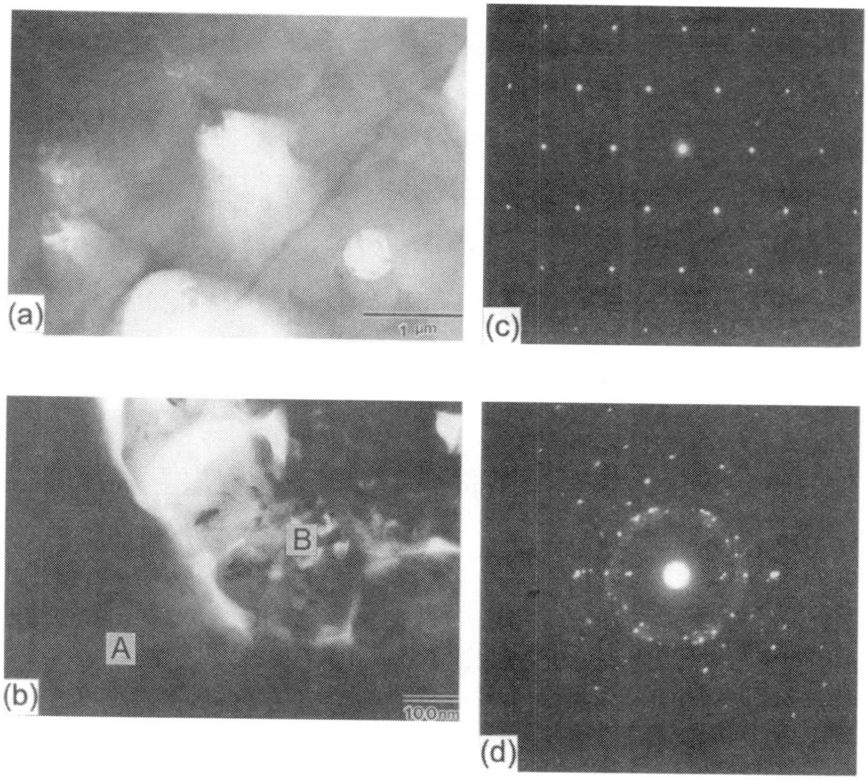

Figure 6.8. TEM images of the surface plan view with SAD patterns for an opaque PLT (20) thin film deposited onto a (0001) sapphire substrate; (*a*) surface images at low magnification, (*b*) segregated crystallites at high magnification, (*c*) SAD pattern at *A*, (*d*) SAD pattern at *B*. The SAD patterns show that zone A corresponds to PLT single crystal.

6.2 CONTROL OF STRUCTURE

The initial stage of the film growth comprises three modes as described in Ch. 2, Fig. 2.1: the island structure and the columnar structure growth modes of the Volmer-Weber and the Stranski-Krastanov types and the continuous thin film growth mode of the Frank-van der Merwe type. The careful selection of growth conditions provides perovskite thin films with controlled microstructures. Several examples of the deposition process for controlled microstructures are described in Sec. 6.2.1 for perovskite thin films.

6.2.1 Growth Temperature

The substrate temperature during film growth changes both the chemical composition and the crystal structure of the thin film. Figure 6.9 shows the variation of the crystal phase with the substrate temperature and the chemical composition of Pb-Ti-O thin films sputtered onto (100) MgO substrates. The films were deposited by sputtering from a mixed powder of PbO and TiO_2. The crystal phase of perovskite is observed in the narrow temperature range of 500° to 600°C for the stoichiometric composition for PbO in the -5% to +20% range. In a lower temperature range, the pyrochlore phase $Pb_2Ti_2O_6$ is dominant, and at a higher temperature range, the lead-deficient structure $PbTi_3O_7$ is dominant.

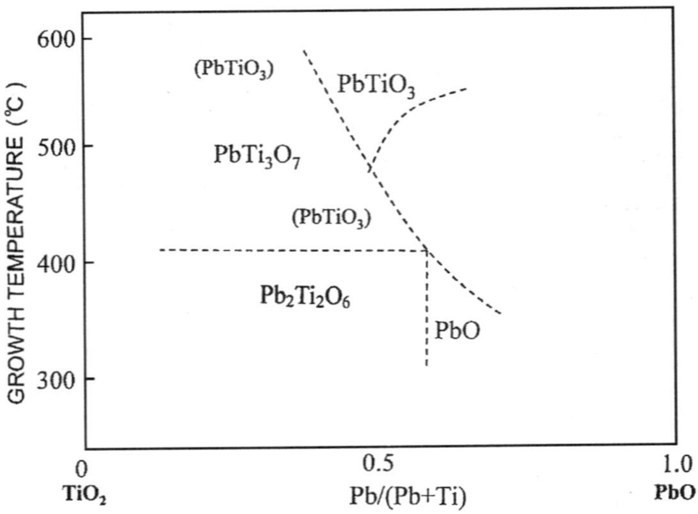

Figure 6.9. Crystal phase of sputtered Pb-Ti-O thin films onto (100) MgO substrates showing chemical composition of thin films vs growth temperature.

Figure 6.10 shows another example of perovskite thin films. The figure indicates XRD patterns for sputtered $Pb(Mg_{1/3}Nb_{2/3})O_3$ (PMN) thin films for different crystal phases. Bulk PMN is known as a relaxor ferroelectric material. It shows high piezoelectricity with an extremely large dielectric constant of over 10,000.[14] The Curie temperature of the PMN is changed by doping the PT. The PT-doped PMN is called PMNT.

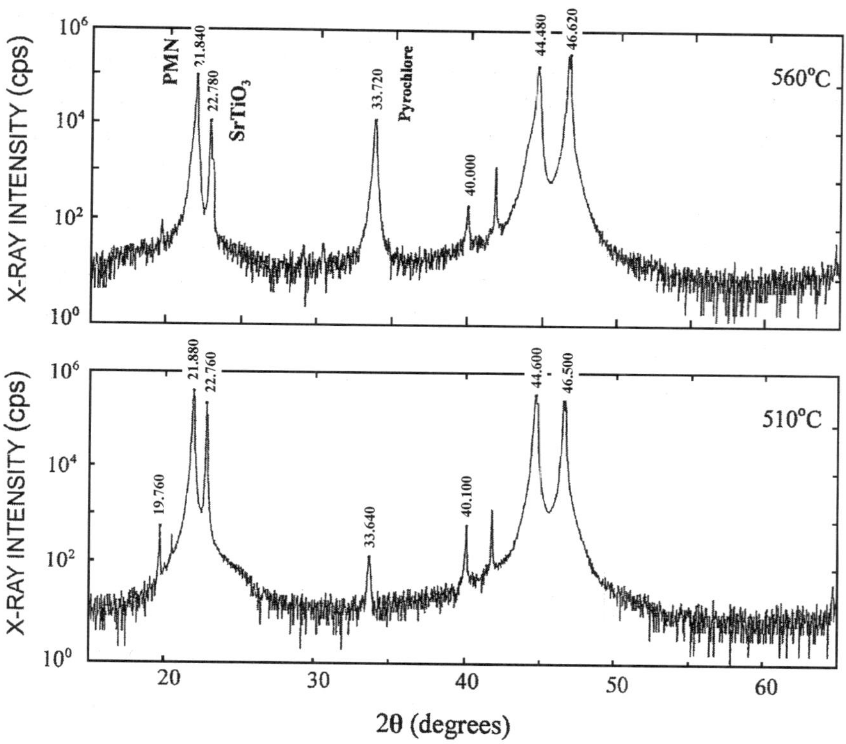

Figure 6.10. XRD patterns for sputtered PMN thin films grown on (100) $SrTiO_3$ at different growth temperatures (film thickness, 300 nm).

However, the deposition of PMN thin films with a single perovskite phase is difficult due to the appearance of a pyrochlore phase.[15] In the case of PMN thin films, the temperature dependency of the crystal phase is different from that of PT thin films. Figure 6.11 shows variations of the crystal phase of PMN thin films sputtered onto (100) MgO substrates. It shows that the perovskite phase is stable at a lower growth temperature zone, and the optimum growth condition for single-phase perovskite exists in a narrow temperature zone.[16] Figure 6.12 shows a typical cross-sectional TEM image of sputtered PMN thin films. The film shows a single-crystal-like continuous structure without grains.

The single-crystal-like structure is a uniform crystal growth. The uniform crystal growth is clearly observed for sputtered PMN thin films on

(100) $SrTiO_3$ substrates. The initial growth of PMN thin films on $SrTiO_3$ substrates shows an oscillation pattern in an XRD spectrum as shown in Fig. 6.13, due to the interference fringes of the x-rays. The interference fringes result from the extremely flat surface with uniform film thickness. Since the La-doped $SrTiO_3$ is conductive, the dielectric properties of PMN thin films are evaluated at the sandwich structure of Au/PMN/La-$SrTiO_3$. Figure 6.14 shows the dielectric properties of PT-doped PMN (PMNT) thin films. The PMN thin films show the typical dielectric properties of relaxor ferroelectrics. Frequency dispersion and broad peaks are observed at dielectric-constant /temperature curves, as shown in Fig. 6.14.

From these examples, it is evident that the growth temperature and chemical composition of the thin films need to be well-controlled in order to achieve precise control of the crystal phase of the thin films. If closer control of the structural properties of thin films is needed, i.e., microstructure and/or strain, the growth conditions need to be modified as described in Sec. 6.2.2.

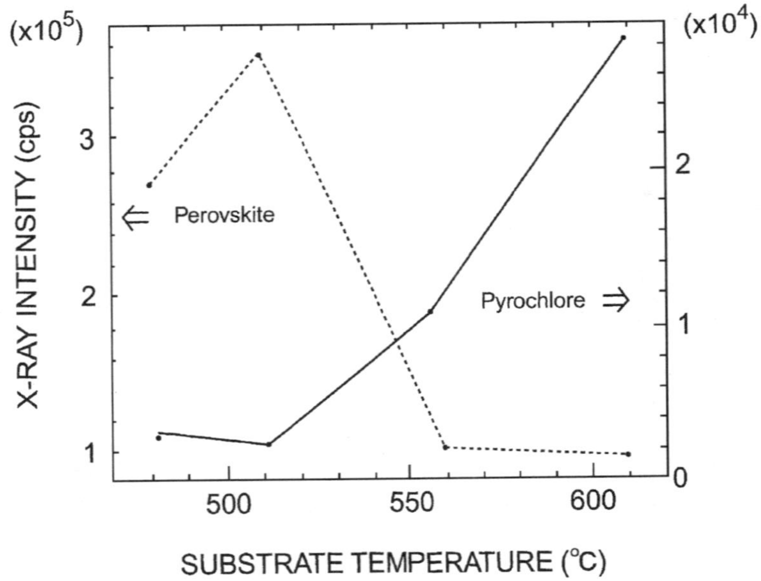

Figure 6.11. Variation of the crystal phase with growth temperature for sputtered PMN thin films on (100) $SrTiO_3$.

Figure 6.12. Cross-sectional TEM image of sputtered PMN thin films epitaxially grown onto (001) MgO.

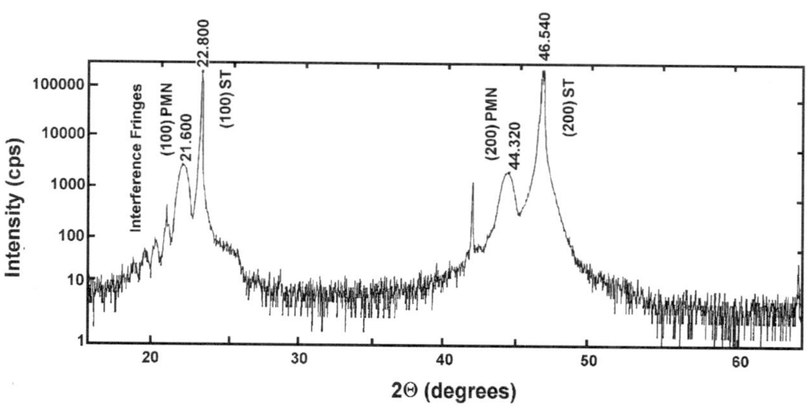

Figure 6.13. XRD patterns for a sputtered PMN ultrathin film (film thickness, 8 nm) epitaxially grown onto (100) $SrTiO_3$.

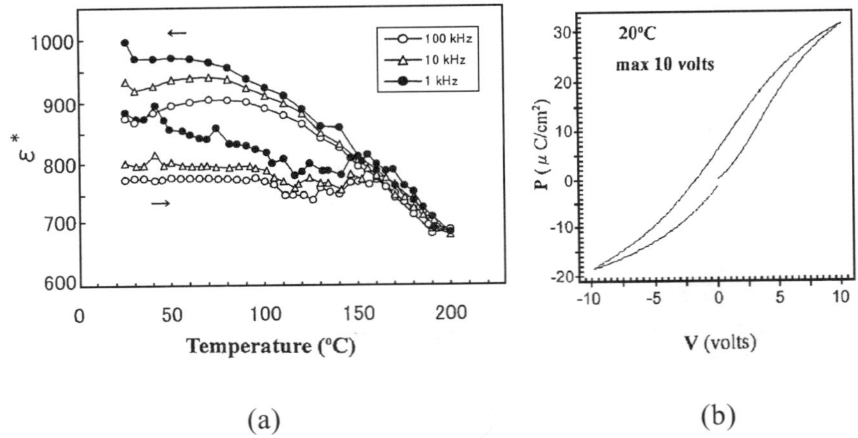

Figure 6.14. Dielectric properties of sputtered PT-doped PMN (0.8 PMN:0.2 PT) thin films grown on La-doped (100) $SrTiO_3$: (*a*) temperature variations of dielectric constant, (*b*) P-E hysteresis curves.

6.2.2 Buffer Layers and Graded Interfaces

The adatoms of compounds nonuniformly diffuse on the substrate surface at the initial stage of film growth. When PZT or PLZT thin films are epitaxially grown on (001) MgO or on (0001) sapphire substrates, a ZrO_2 layer will cover the substrate surface. The ZrO_2 layer suppresses the epitaxial growth of the perovskite structure. The deposition of the PT or PLT layer as a buffer layer before the deposition of PZT or PLZT stabilizes the growth of the perovskite structure.[17]

The other method of controlling the interface is the introduction of a graded interface. This reduces the dislocation density at the interface and improves the crystal structure of the heteroepitaxial thin films.

Figure 6.15 shows typical surface SEM images of PLZT (9/65/35) thin films epitaxially grown on (0001) sapphire. The PLZT thin films were directly deposited onto the (0001) sapphire at the epitaxial temperature of 600°C by a multitarget sputtering system. The film thickness is 400 nm and the sputtering conditions are shown in Table 6.5. Figure 6.16 shows the cross-sectional TEM image with SAD patterns. These SEM and TEM images show a polycrystalline structure with a rough surface. The interfacial region of amorphous and/or poor crystalline structure is observed in the TEM image. The SAD patterns show a streak pattern indicating the presence of a misalignment of the crystal orientation and/or the formation of the polycrystalline phase. The x-ray diffraction analysis suggests that the crystal orientation of the PLZT thin films does not simply obey the ideal epitaxial relation of (111) PLZT // (0001) sapphire. The PLZT thin films show a mixed orientation of (100), (111), and (001).

Figure 6.15. SEM images of the surface of sputtered PLZT (9/65/35) thin films, 400 nm thick, deposited onto (0001) sapphire substrates.

Table 6.5. Sputtering Conditions for Epitaxial PLZT Thin Films on Sapphire Substrates

Target diameter	60 mm
Target-substrate spacing	100 mm
Input power	Pb: 10–30 W La: 5–30 W Zr: 0–100 W Ti: 50–200 W
Sputtering gas	Ar/O_2=2/1
Gas pressure	3 Pa
Substrate temperature	700°C
Deposition rate	30–50 Å/min

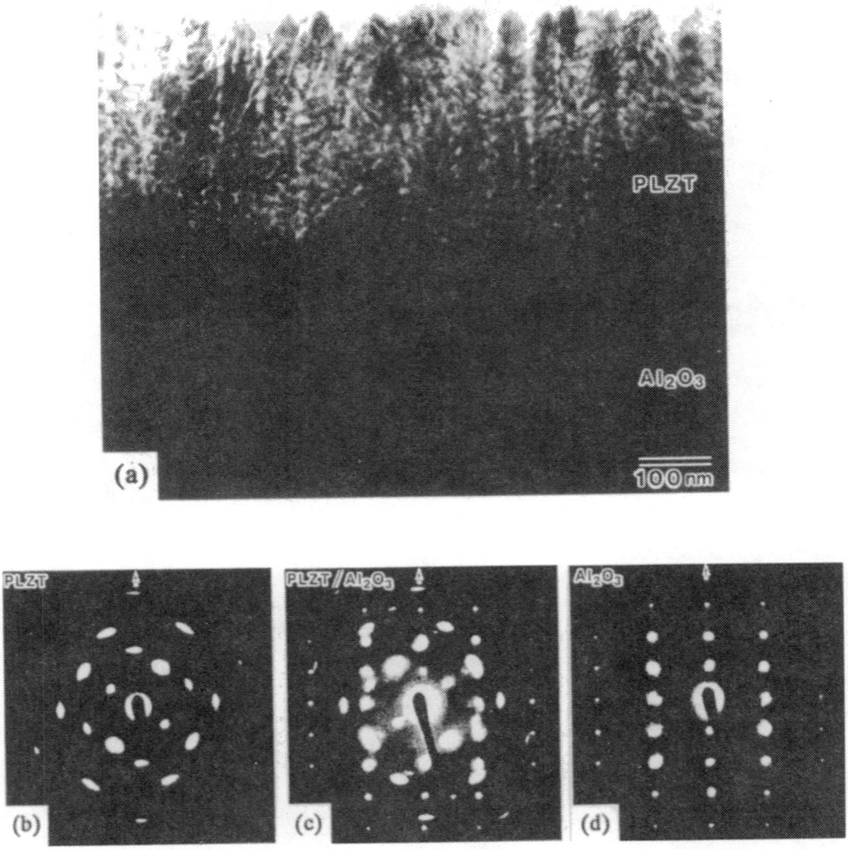

Figure 6.16. (*a*) Cross-sectional TEM images of PLZT (9/65/35) thin films, 400 nm thick, deposited onto a (0001) sapphire substrate, with SAD patterns obtained at (*b*) the film surface, (*c*) the interface, and (*d*) the sapphire substrate.

An atomic-resolution lattice image at the film-substrate interface is shown in Fig. 6.17. The interface layer disturbs the uniform growth of the (111) PLZT on (0001) sapphire. The interfacial region is composed of different microstructures: the region comprising ordered clusters of (111) PLZT crystallites, Fig. 6.17a; and the region comprising disordered clusters of (101) and/or (110) PLZT crystallites, Fig. 6.17b.

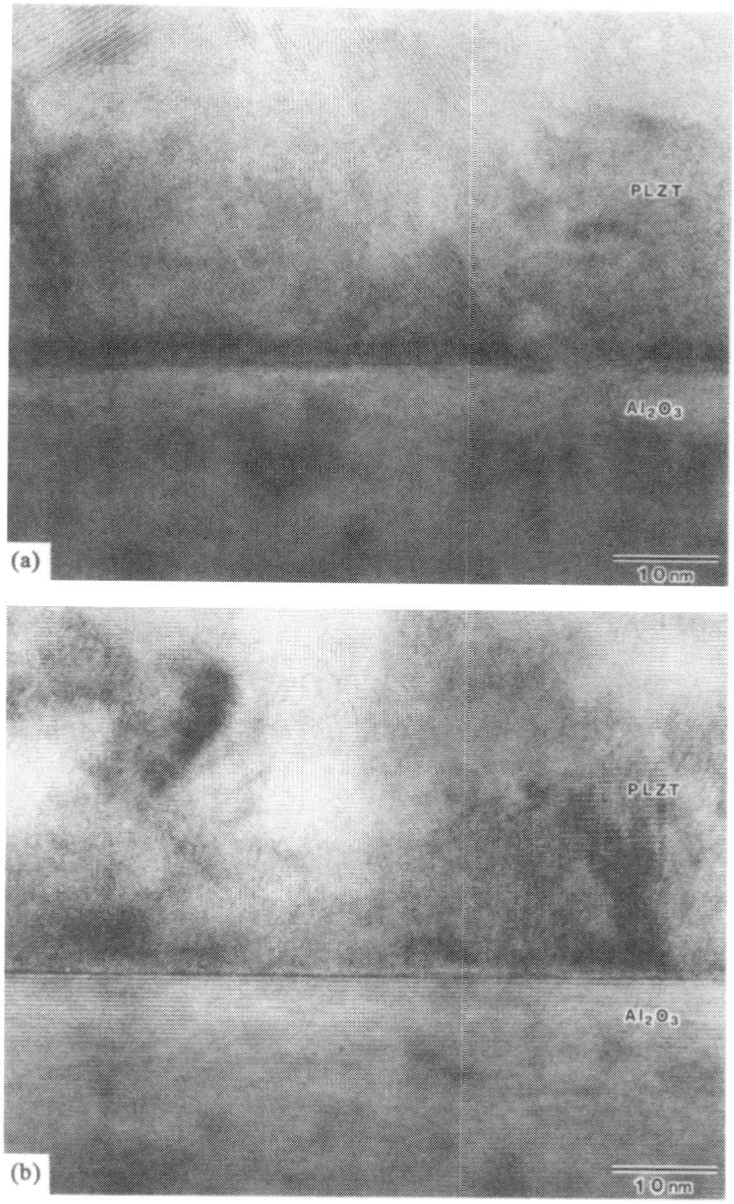

Figure 6.17. Atomic-resolution image of the PLZT/(0001)-sapphire interface showing (*a*) epitaxial growth of (111) PLZT on (0001) sapphire and (*b*) the growth of disordered clusters of (101) and/or (110) PLZT crystallites.

Figure 6.18 shows surface SEM images of PLZT thin films, 400nm thick, on (0001) sapphire with a PLT graded buffer layer. The buffer enhances the crystal growth of the PLZT thin films. The buffer layer was prepared by the following procedure:

- In the first stage, PLT thin films of about 20 nm thick were epitaxially grown on the sapphire.
- Then, input power levels of four targets, Pb, La, Zr, and Ti, were varied gradually to introduce Zr, and a graded composition layer of about 80 nm in thickness was fabricated.
- Finally, quaternary oxide of PLZT (9/65/35) of 400 nm in thickness was deposited.

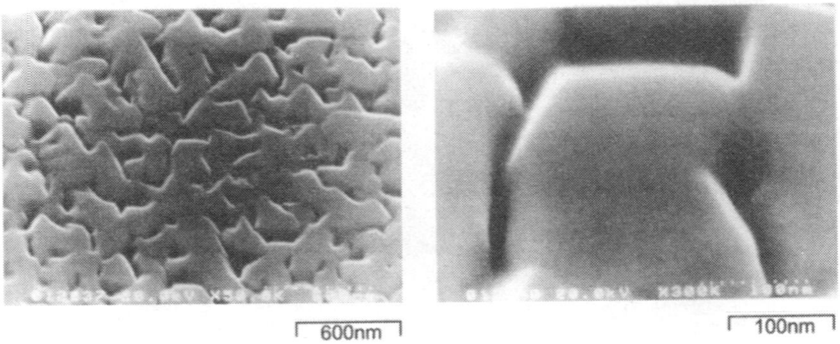

Figure 6.18. Surface SEM images and RHEED patterns of sputtered PLZT (9/65/35) thin films, 400 nm thick, on (0001) sapphire with the PLT (20) buffer layer, 100 nm thick, showing the well-oriented structure in the a- and b-directions.

The XRD analysis shows that the PLZT thin films with the graded buffer layer exhibit a single (111) orientation. The cross-sectional TEM images with SAD patterns at various surface positions are shown in Fig. 6.19. The SAD patterns show that the sputtered thin films have a single (111) orientation, although a twin structure is observed in the interface.

The atomic-resolution lattice images shown in Fig. 6.20 suggest that the PLZT thin films exhibit uniform epitaxial growth on the (0001) sapphire, corresponding to the ideal epitaxial relationship. The interface is coherent.[18]

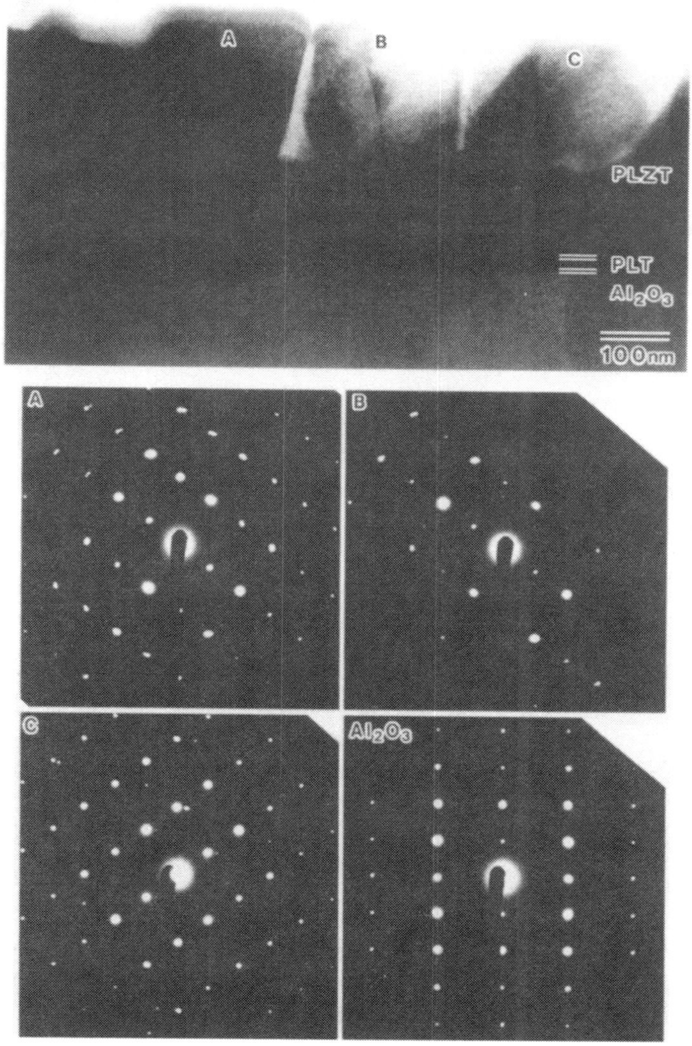

Figure 6.19. Cross-sectional TEM images and the SAD patterns of the PLZT (9/65/35) thin films, 400 nm thick, on a (0001) sapphire substrate with a PLT (20) buffer layer, 100 nm thick; SAD pattern at surface *A*, surface *B*, surface *C*, and the (0001) sapphire substrate.

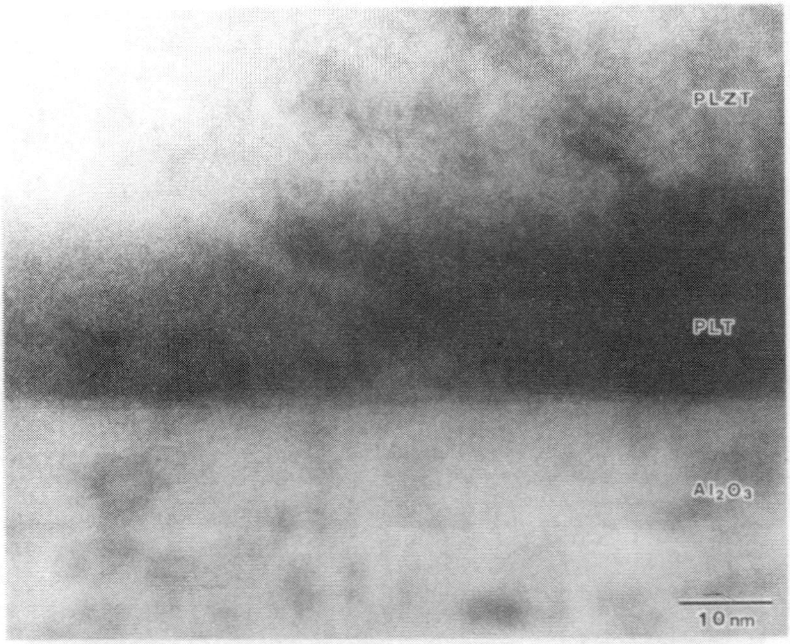

Figure 6.20. Atomic-resolution image of the PLZT/(0001)-sapphire interface showing the epitaxial growth of (111) PLT buffer layer on (0001) sapphire.

Table 6.6 shows the lattice parameters of sputtered PLZT thin films with and without the buffer layer. The lattice parameters of the sputtered film without the buffer layer are close to those of bulk materials. This indicates that films without buffer layers include the dislocated interfacial region and show the strain-relaxed structure. The buffer layer improves the coherency of the interface. The individual crystal grains are well-oriented in both the c- and the a-, b-directions, and they are connected in the plane. The graded buffer layer is useful for the improving the crystallinity of heteroepitaxial thin films.

The complex microstructure of the heteroepitaxial growth of perovskite materials is also observed in high-T_c superconducting thin films.[19] A suitable design of a graded buffer layer for the heteroepitaxial system results in continuous single-crystal thin films of perovskite materials.

Table 6.6. Lattice Spacing of PLZT Sputtered Thin Films of (111) Orientation on (0001) Sapphire

Samples	Position	Lattice spacing (Å) (001)	(110)
Without buffer layer	Surface	4.15	2.88
	Interface	4.15	2.89
With PLT buffer layer	Surface	4.10	2.87
	Interface	4.04	2.86

6.2.3 Cooling Rate

The substrate materials and the cooling rate also govern the structural properties after deposition. Kwak and Erbil,[20] and Speck and Pompe[21] studied the domain formation of epitaxial PT. The epitaxial temperature of PT thin films is 600°C, which is higher than the Curie temperature of bulk PT (490°C). They suggest that if the c-lattice parameter of PT is larger than that of the substrates, e.g., (001) $SrTiO_3$, the c-domain is predominant during deposition. If the c-lattice is smaller than substrate lattice, e.g., (001) MgO, the a-domain is predominant during deposition. The mixed domain structure of a/c-domains is developed during the cooling stage; the cooling rate affects the domain structure, and high cooling rates enhance the c-domain growth.

Figure 6.21 shows XRD patterns for different cooling rates of epitaxial PT thin films deposited by sputtering onto (001) MgO substrates.[22] Typical deposition conditions and the lattice constants for rapid and slow cooling rates are shown in Tables 6.7 and 6.8, respectively. A rapid cooling rate increases the population of the c-domain, while a slow cooling rate increases the population of the a-domain, as seen in Fig. 6.21a. The c-domain growth is caused by a preferential orientation. The ϕ scan of XRD analysis suggests that rapid cooling provides single c-domain (001) PT thin films with in-plane epitaxy, as shown in Fig. 6.21b. The film surface is smooth, and the a-domain growth is caused by epitaxial growth. The lattice parameters of PT thin films are shown in Table 6.8. For epitaxial

(100) PT thin films on (001) MgO, the c-axis is highly expanded due to the large lattice spacing of the MgO substrates. The lattice arrangement of PT thin films on MgO substrates is shown in Fig. 6.22. The lattice arrangement for the a-domain is not governed by the epitaxial relation PT (3b × 3c) // MgO $2\sqrt{2}$(a × a) for a two-dimensional superlattice. The a-domain PT thin films include twins due to the presence of two in-plane epitaxial directions for the c-axis.

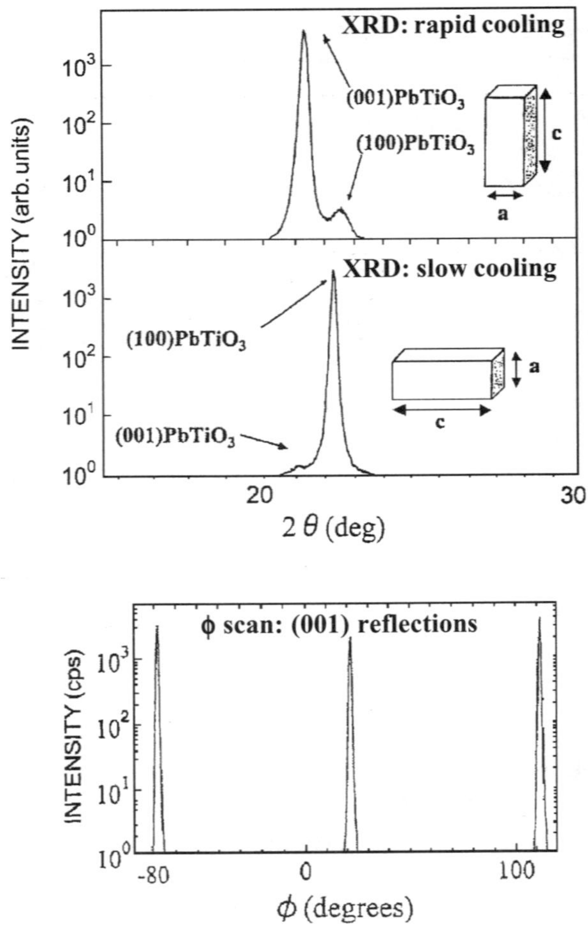

Figure 6.21. The XRD patterns of sputtered PT thin films, 120 nm thick, on (100) MgO with rapid cooling and slow cooling. The ϕ scan shows the (001) reflections for a PT film, 120 nm thick, grown on normal MgO, using in-plane XRD.

Table 6.7. Sputtering Conditions

Substrate	(100) MgO
Substrate temperature	600°C
Target composition	PT + PbO (10 mol%), PLT(21) + PbO (10 mol%)
Target diameter	80 mm
Input power density	1.6 W/cm^2
Sputtering gas	Ar/O_2 = 20/1
Gas pressure	1 Pa
Deposition rate	≈ 2–3 nm/min

Table 6.8. Lattice Constant for Sputtered $PbTiO_3$

System	Lattice Constant of $PbTiO_3$ (PT) Thin Films	
	Rapid Cooling	Slow Cooling
PT/MgO	c = 0.416 nm [+0.48%]* a = 0.394 nm [+1.02%]*	c = 0.419 nm [+1.02%]* a = 0.398 nm [+1.05%]*
	c_{bulk} = 0.4152 nm, a_{bulk} = 0.3903 nm	

* Lattice difference between thin films and bulk $PbTiO_3$.

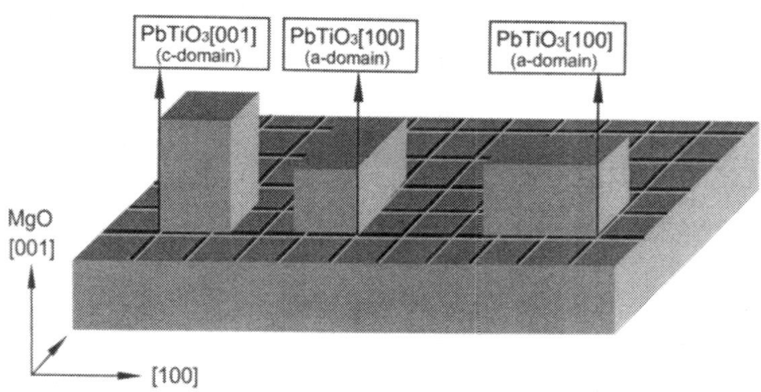

Figure 6.22. Lattice arrangements of sputtered $PbTiO_3$ on (100) MgO substrates.

When the a-lattice parameter of a thin film is close to the substrate lattice parameter, e.g., PT on (001) $SrTiO_3$, or the tetragonality of the perovskite thin films is close to unity, e.g., PLT, the cooling rate does not affect the crystal orientation.

6.2.4 Vicinal Substrates

When the growth of the epitaxial films is governed by island growth, the resultant thin films show nonuniform crystal growth. The growth of the epitaxial thin films on well-oriented and nominal single-crystal substrates is usually governed by island growth, since the density of the step on the crystal substrates is too small for the step-mediated growth and the lateral growth.

The vicinal substrates are useful for achieving the lateral growth of crystallites. Figure 6.23 shows a typical surface of the vicinal (001) $SrTiO_3$ (ST) substrate. The surfaces of vicinal substrates comprise the atomic step lines and terrace.[23] The ideal step height is a crystal unit of the substrate. The step height for ST substrates is 0.39 nm. The terrace length is $d = h/\tan\theta$, where h denotes the step height, and θ denotes the miscut angle. On the vicinal substrates, the crystal steps act as nucleation centers. Lateral growth is observed on the vicinal substrates if the adatoms have sufficient surface mobility to diffuse across the terrace. The lateral growth is called "step-flow growth." Figure 6.24 shows a typical surface AFM image and a cross-sectional TEM image with SAD patterns for PT thin films on a vicinal (001) ST substrate grown laterally. The surface shows the step lines and terrace. The sputtering conditions are shown in Table 6.9. The film shows continuous structure without grain boundaries.[24] The surface is atomically flat.

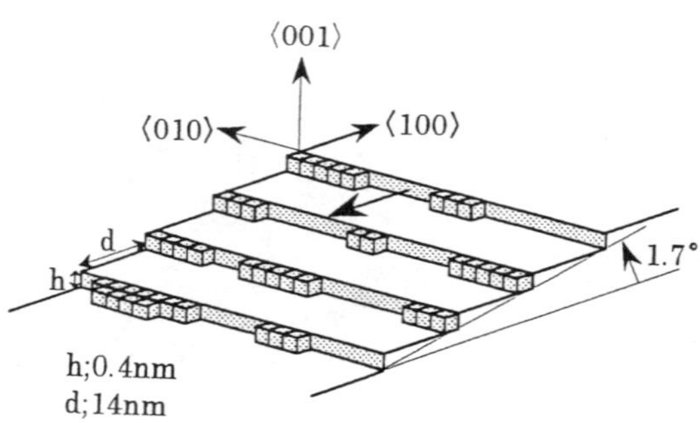

Figure 6.23. Typical surface model of miscut (001) ST substrate; miscut angle, 1.7°.

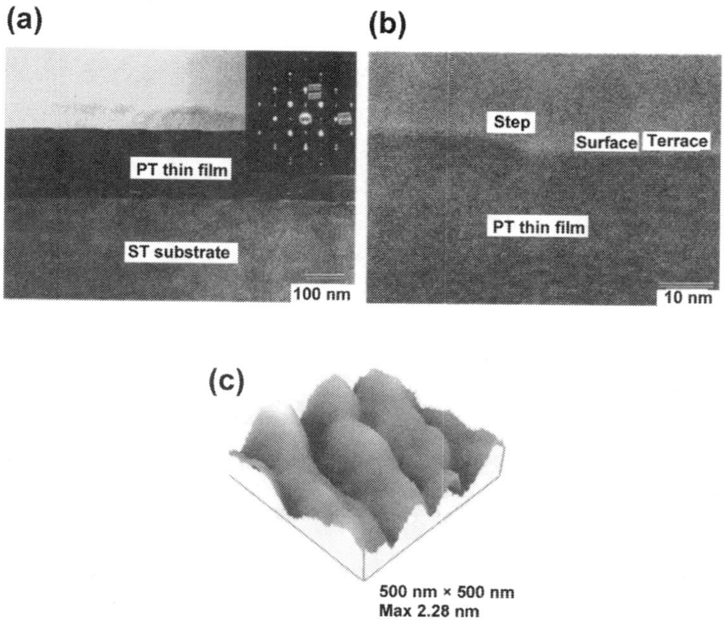

Figure 6.24. Cross-sectional TEM images of sputtered PT thin films on miscut (001) ST with SAD pattern at (*a*) film/substrate interface, and (*b*) lattice image of film surface; (*c*) surface AFM image.

Table 6.9. Sputtering Conditions

Sputter system	RF-planar magnetron
Target	$PbTiO_3$ powder
Target diameter	80 mm
Target-substrate spacing	30 mm
Substrate	(001) $SrTiO_3$, miscut (001) $SrTiO_3$
Substrate position	on axis
Substrate temperature	600°C
Sputtering gas	$Ar/O_2 = 20/1–2/1$ (0.5 Pa)
Growth rate	3 nm/min
Film thickness	5–300 nm

Figure 6.25 shows a typical XRD pattern. The interfering signals in the XRD pattern suggest that the films have an extremely uniform structure. The interface of the thin films and the substrates is coherent without an interfacial layer.

The surface mobility of adatoms increases with a rise in the substrate temperature during film growth. There is a critical temperature for the lateral growth for a given vicinal substrate. At a given substrate temperature, a critical miscut angle is present. Below the critical temperature or without the critical miscut angle, the island growth type governs film growth.[25] Typical conditions for lateral growth are shown in Fig. 6.26. The critical miscut angle is about one degree for PT thin films epitaxially grown on $SrTiO_3$ at 600°C.

Lateral growth is governed by the diffusion of adatoms on the terrace; the growth condition is affected by the diffusion-related deposition parameters. The oxygen partial pressure during deposition also affects the lateral growth process since an increase in the oxygen partial pressure decreases the surface mobility of adatoms. Figure 6.27 shows the results of experiments demonstrating that there is a critical oxygen partial pressure. Below the critical oxygen pressure, the film shows a continuous lateral-growth film structure. Above the critical oxygen pressure, the film comprises multigrains and/or a columnar structure as a result of island growth.[26]

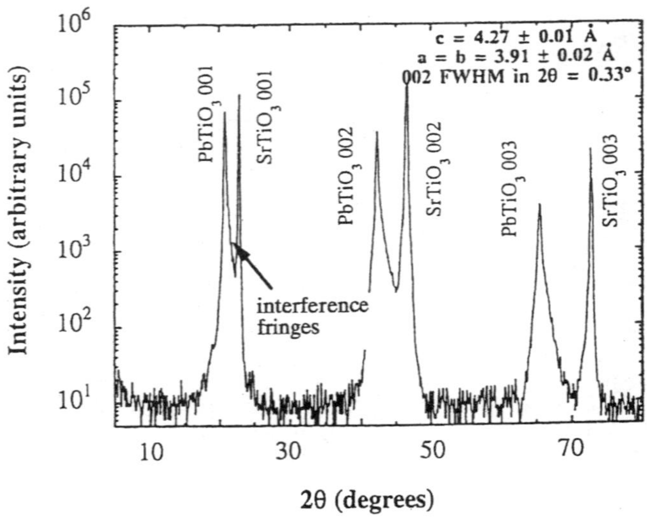

Figure 6.25. X-ray diffraction pattern of PT thin film, 40 nm thick, on miscut (1.7°) ST.

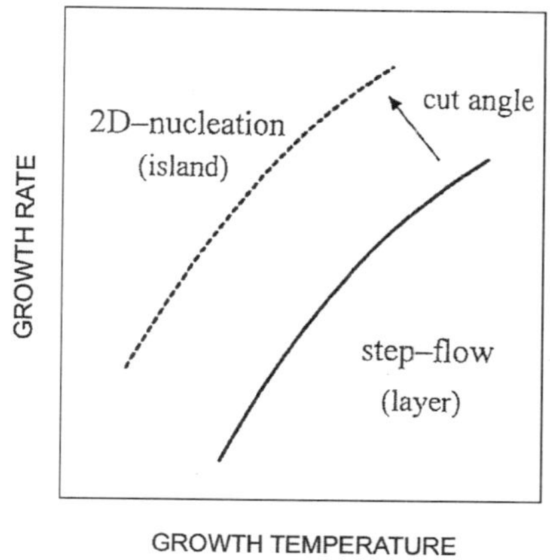

Figure 6.26. Thin film growth modes on a vicinal substrate.

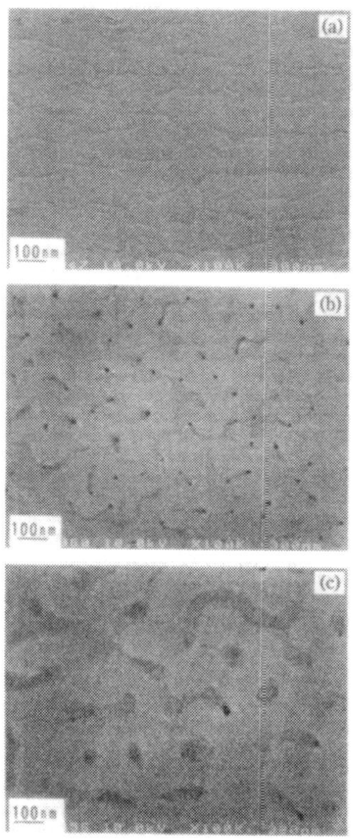

Figure 6.27. Surface SEM images of sputtered PT thin films, 120 nm thick, on miscut (001) ST for various oxygen partial pressures during sputtering deposition; (*a*) $Ar/O_2 = 20/5$, (*b*) $Ar/O_2 = 20/6$, (*c*) $Ar/O_2 = 20/7$.

The growth mode of thin films is closely related to the film growth conditions seen in Fig. 6.26. At a low substrate temperature, the films show an amorphous structure. There is little surface diffusivity and the adatoms get trapped in a conglomerate of noncrystal sites. At a high substrate temperature, the adatoms have sufficient surface mobility to arrange themselves epitaxially on the crystal substrates where epitaxial islands are formed. At higher substrate temperatures, the crystal growth occurs by the lateral growth of steps. The step-mediated growth is governed by nucleation at the steps and by diffusion of the adatoms on the terrace of the substrates. The surface diffusion coefficient (D_s), is expressed as

Eq. (6.6)
$$D_S = \frac{\tau_0^2}{t}$$

where τ_0 is an interstep distance (terrace width) and t is a characteristic time for the diffusion, which is a time interval for the growth of a monolayer or the average time for nucleation. If the growing atoms can reach the step before the average time for nucleation, t_{nuc}, they cannot form a nucleus and the growth is in the step-flow mode. If they cannot reach the step before t_{nuc}, they form nuclei resulting in 2-D nucleation. The long length of the terrace prohibits step-mediated growth under the conditions at a given growth temperature. The commercial, well-oriented substrates show large λ_0 (i.e., miscut angle of $\pm$ 0.3° with $\lambda_0 > 90$ nm). The vicinal substrates reduce the λ_0, which produces the step-flow growth under the conditions at a given growth temperature.[5] The epitaxial growth on the vicinal substrates is useful for the deposition of continuous single-crystal films of perovskite materials.

PT thin films are grown on ST substrates at 600°C. Since the growth temperature is higher than Curie temperature, T_c, of the PT (T_c = 490°C for bulk PT), the PT thin films show a cubic structure at the growth temperature (if we consider that the structure of the PT thin films is the same as the bulk PT). At the growth temperature, the lattice parameter of the bulk PT, $a_b = 0.397$ nm. The PT thin films are deformed into a tetragonal shape due to the small lattice parameter of the ST substrate ($a_{ST} = 0.393$ nm). The tetragonality of the PT thin films at the growth temperature, c_f/a_f, is 1.03, under the assumption $a_f = a_{ST}$ and $c_f/a_f = a_b^3$, where c_f and a_f denote the lattice parameter of the PT thin films. The tetragonal form is quenched during the cooling stage after deposition. During the cooling stage, the PT thin films are further deformed as a result of the phase transition from cubic to tetragonal below T_c of 490°C, resulting in the large tetragonality c_f/a_f = 1.06 to 1.1 at room temperature. Typical lattice parameters are shown in Table 6.10.[27]

The TEM cross-sectional atomic images of PT thin films on the vicinal (001) ST substrate are shown in Fig.6.28a at interface, and Fig. 6.28b at the surface where tetragonal lattice distortion is prevailing. The triclinic lattice distortion is not clearly observed. These lattice images show that the same a-lattice and c-axis parameters are obtained at both the interface and surface regions. The PT thin films show uniform lattice distortion in the whole area of the PT thin films. The a-lattice parameter and the c-lattice parameter are 0.39 nm and 0.43 nm, respectively, for a film thickness of 170 nm at the miscut angle of 1.7°.[28] This implies that the lattice distortion is not fixed during the film growth, but the lattice distortion will be fixed during the cooling stage after deposition.

The lattice distortion is reduced by the increase of miscut angles as indicated in Table 6.10. It was expected that the lattice distortion would disappear and the strain would be fully relaxed when the substrates were removed from the thin films. The strain relaxation was confirmed by the

Table 6.10. Typical Lattice Parameters for Sputtered $PbTiO_3$ Thin Films[1]

ST miscut angle (degrees)	PT film tilt angle (degrees)	a (nm)	c (nm)	c/a	
0.06	0.05[2]	0.3920	0.4275	1.09	island growth
0.87	0.94[3]	0.3964	0.4254	1.16	island growth
1.32	1.45[3]	0.3981	0.4229	1.06	step-flow growth
4.95	5.34[3]	0.3992	0.4208	1.05	step-flow growth
9.89	10.73[3]	0.3954	0.4219	1.07	step-flow growth

[1] On miscut (001) SrTi O_3 substrates at various miscut angles ($PbTiO_3$: 120 nm in film thickness, measured at room temperature).

[2] uphill

[3] downhill

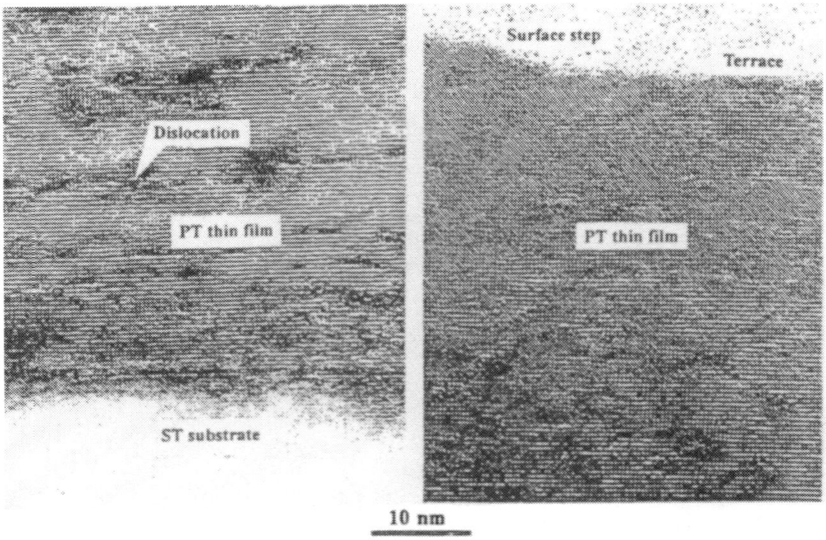

(a) (b)

Figure 6.28. Cross-sectional atomic lattice TEM images of epitaxial PT thin film (160 nm thick) on a miscut (001) ST substrate; (*a*) at the interface, (*b*) at the surface of the PT thin film.

XRD measurement of PT crystallites shown in Fig. 6.29. The PT crystallites were obtained by peeling PT thin films from the ST substrates. The lattice parameters were the same as those of bulk PT.

Figure 6.30 shows the cross-sectional TEM image of sputtered PT thin films on a vicinal (001) ST substrate at the miscut angle of ten degrees. The steps at the interface and/or the surface of the ST can be seen. The c-axis of the PT thin films is slightly tilted toward the direction of the substrate terrace normal. This indicates the presence of twins at the step region. The schematic structure of the twins is shown in Fig. 6.31.[29]

The tilt angle ϕ is expressed by the relation

Eq. (6.7) $$\phi = \tan^{-1}\left(\frac{a_s - c_f}{a_s}\right)\tan\theta \qquad \text{(uphill)}$$

where a_s and c_f denote the lattice parameters of the vicinal ST substrate and epitaxial PT thin films, respectively.[30] Typical experimental results are shown in Fig. 6.32.

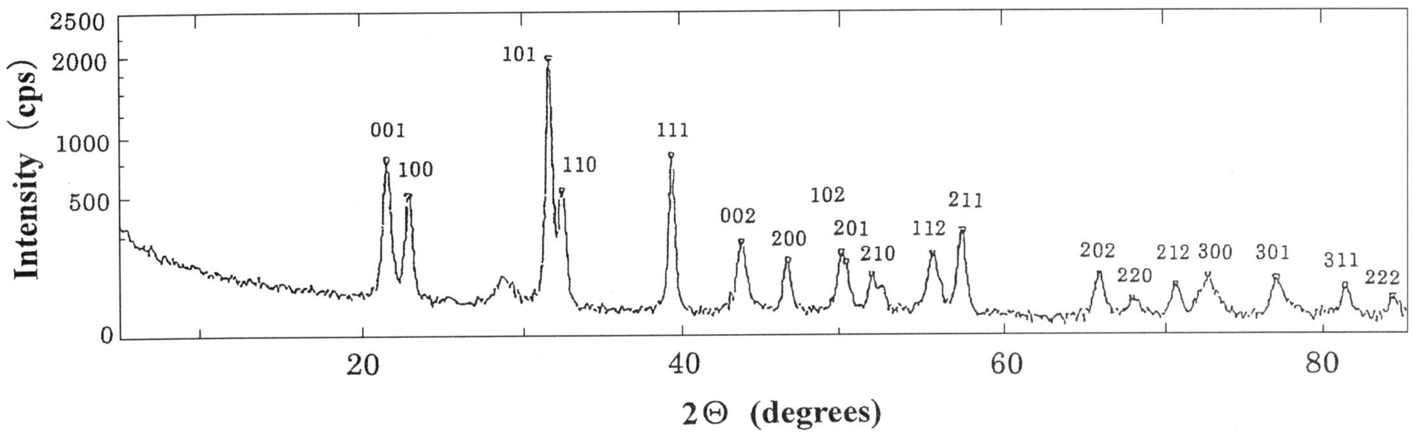

Figure 6.29. PT thin film crystallites with a θ–2θ diffraction pattern.

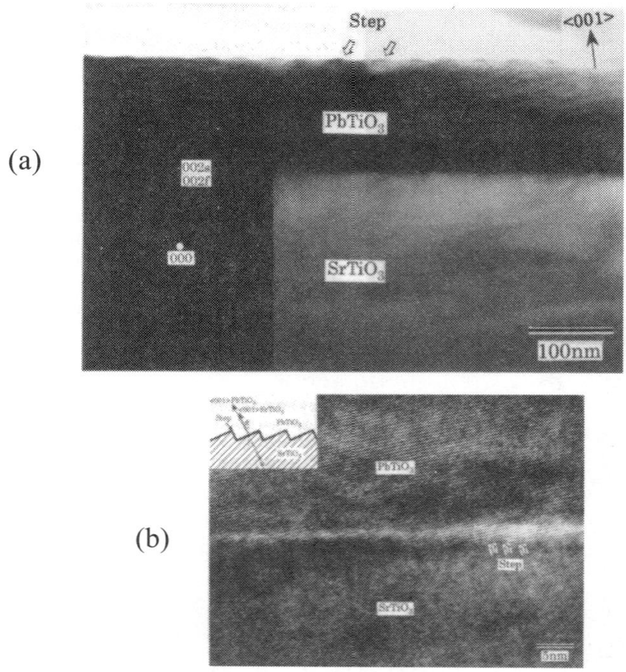

Figure 6.30. Cross-sectional TEM micrograph of PT thin films, 130 nm thick, on miscut (001) ST (miscut angle; 10°), (*a*) with SAD pattern, and (*b*) atomic image at the interface of PT thin film and ST substrate (f: film, s: surface).

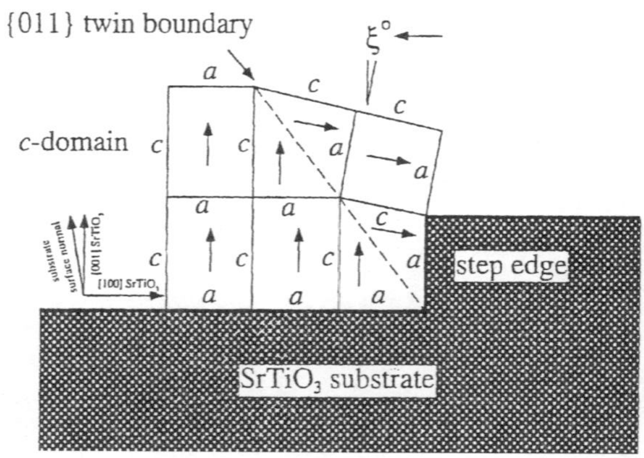

Figure 6.31. Schematic showing how the presence of step edges on a vicinal (001) ST substrate may lead to the preferred nucleation of a PT (011) twin boundary. The arrows indicate the direction in which the c-axis lies. The tetragonality of PT has been exaggerated for clarification.[29]

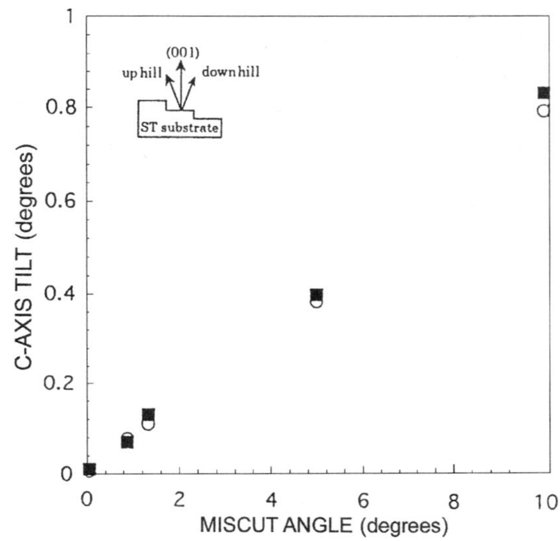

Figure 6.32. Graph of c-axis tilt angle for sputtered PT thin film, 130 nm thick, on miscut (001) ST as a function of substrate miscut angle, (■) measured, (O) calculated.

The terrace width and the step height are larger than the initial surface of the vicinal ST. This indicates that step bunching should be observed.[31] Step bunching was confirmed at the initial growth stage of the PT thin films as shown in Fig. 6.33. The step height shows saturation at 3 nm above a critical film thickness of 10 nm, as seen in Fig. 6.34. The step height of 3 nm is caused by three times step bunching.*

Continuous single-crystal thin films of PLT and PZT are also grown on vicinal substrates, as shown in Fig. 6.35. Conducting perovskite thin films of $SrRuO_3$ (SRO) are also grown on vicinal (001) ST substrates under step-flow growth. The lattice parameter of the SRO is shown in Table 6.11.[32]

The critical film thickness, h_c, for the PT thin films on the vicinal substrates is larger than the h_c of the PT thin films on nominal (well-oriented) substrates. The critical film thickness of the PT thin films on the vicinal substrates, determined experimentally, is 300 nm at the miscut angle of 1.7°. This value is two times larger than the PT thin films on the nominal

---

* Initial step height of film surface is a lattice spacing of the ST substrate, a_{ST}. The final step height after n times step bunching is calculated by $a_{ST}^{2^n}$ Taking $a_{ST} = 0.4$ nm and $n = 3$, the calculated value becomes 3.2 nm which is clost to the observed value.

substrates.[33] Heteroepitaxial perovskite thin films on the vicinal substrates are stable against mechanical shock, and highly adhesive to the substrates. The epitaxial thin films on well-oriented substrates often peel off one to two years after deposition, due to the accumulated stress. The vicinal substrates improve the long-term stability of the heteroepitaxial single-crystal thin films. The combination of the graded buffer layer and the vicinal substrates will provide high-quality heteroepitaxial perovskite thin films with long-term stability.

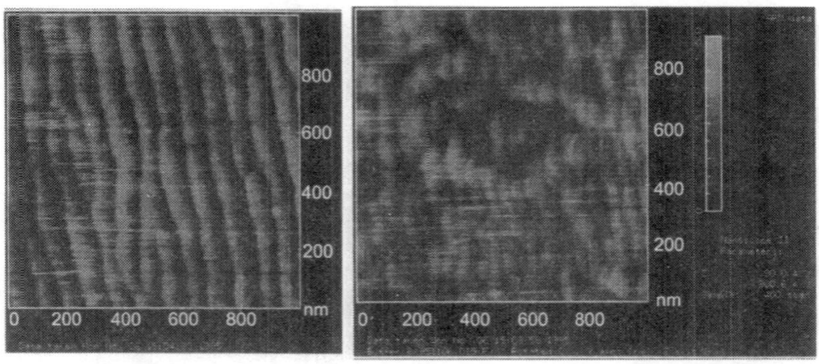

(a) (b)

Figure 6.33. AFM images of sputtered epitaxial PT thin films on miscut (001) ST for different film thicknesses: (*a*) 10 nm and (*b*) 5 nm.

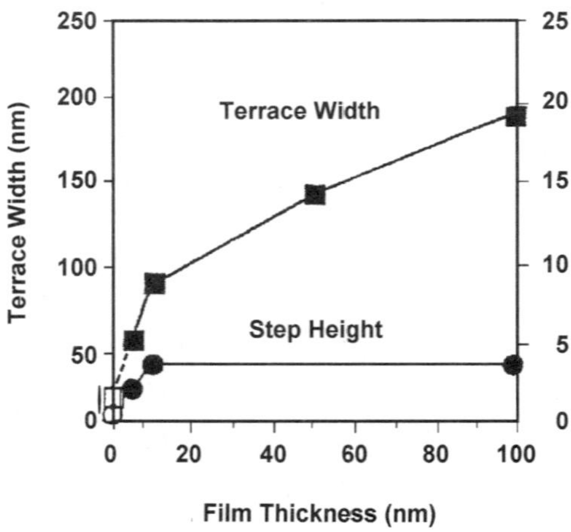

Figure 6.34. Variation of width of terrace and step height with film thickness.

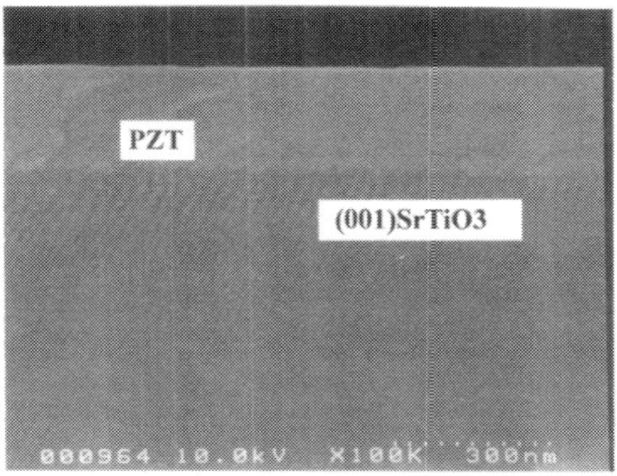

Figure 6.35. Cross-sectional SEM image for (001) PZT on vicinal (001) ST substrate. PZT film thickness is 220 nm.

Table 6.11. Space Groups, Lattice Parameters of $SrRuO_3$

Materials	Space Group	Lattice Parameters (Å) (ortho-rhombic)	Lattice Parameters (Å) (pseudocubic perovskite)	Lattice Distortion	Lattice Mismatch with $SrTiO_3$ (%)
$SrRuO_3$	Pbnm (no.62) $GdFeO_3$ type	$a_0 = 5.53$	$a_0' = 3.93 = d(110)$	89.6°	0.64
		$b_0 = 5.57$			
		$c_0 = 7.85$	$c_0' = 3.93 = d(002)$		

6.2.5 Dielectric Properties of Structure-Controlled Thin Films

The dielectric properties of ferroelectric thin films are affected by their microstructures and/or interface structure.

Ferroelectric thin films deposited by a sol-gel process show multidomains and/or grained structure, which is similar to bulk ceramics. Their ferroelectric properties are also similar to those of bulk ceramics.[34] The strain in sol-gel thin films is relaxed due to the presence of a grain and/or domain boundary. Sputtered ferroelectric thin films also show this behavior if the films include the domain and/or grain-structures.

If the thin films include strain, the strain modifies the ferroelectric properties. A typical example is the shift of Curie temperature. The two-dimensional compressive strain induced by the mismatch in the lattice parameter or the thermal expansion coefficient between the film and the substrate will elevate the Curie temperature.[35]

Continuous single-crystal thin films show a diffused temperature anomaly in their dielectric properties. The diffused temperature anomaly is passively observed in thin epitaxial ferroelectric films[36] and in epitaxial films on vicinal substrates.[37] Typical examples are shown in Fig. 6.36. Thin ferroelectric films and epitaxial films on vicinal substrates are coherent and tightly bonded to substrate surfaces. This results in a diffused temperature anomaly. The diffused temperature anomaly is also observed in the lattice parameter.

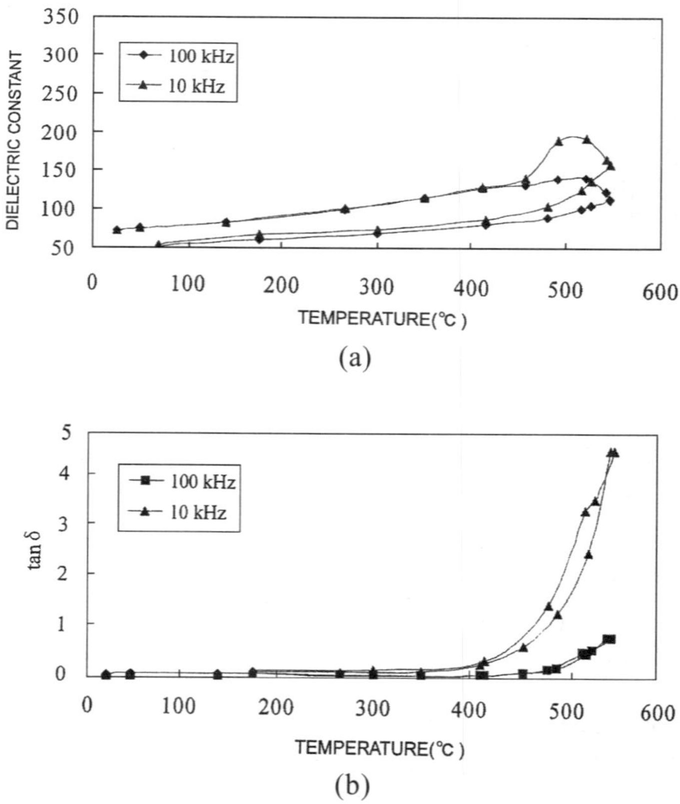

Figure 6.36. Typical temperature variations of (*a*) the dielectric constant and (*b*) loss factors for the single c-domain/single-crystal PT thin films on miscut (001) ST substrates (PT: 125 nm thick, ST: miscut angle, 1.7°).

Figure 6.37 shows the temperature variation of the lattice parameter for sputtered ferroelectric thin films. Epitaxial PT thin films with grain structure show the temperature anomaly at the Curie temperature. The shift of the Curie temperature from bulk PT is due to two-dimensional compressive stress. Compressive stress increases the Curie temperature and tensile stress reduces it. A phenomenological thermodynamic theory of ferroelectric thin films heteroepitaxially grown on foreign substrates can be applied to films with thicknesses much larger than the ferroelectric correlation length, i.e., film thicknesses >50 nm.[38][39] Typical results are shown in Fig. 6.38.[39] In this figure, the negative misfit strain corresponds to the in-plane compressive stress. The misfit compressive strain will reduce the dielectric constant, ε_{33}. Continuous single-domain PT thin films on vicinal substrates do not show the temperature anomaly. These results show the effects of the microstructure and/or the interfacial structure on the temperature anomaly of ferroelectric thin films.

The continuous single-domain structure modifies the P-E hysteresis curve as shown in Fig. 6.39.[40] The low-K interfacial region increases the coercive field. In Fig. 6.39, the coercive field is one order of magnitude higher than that of bulk materials, even if there is no interfacial region.

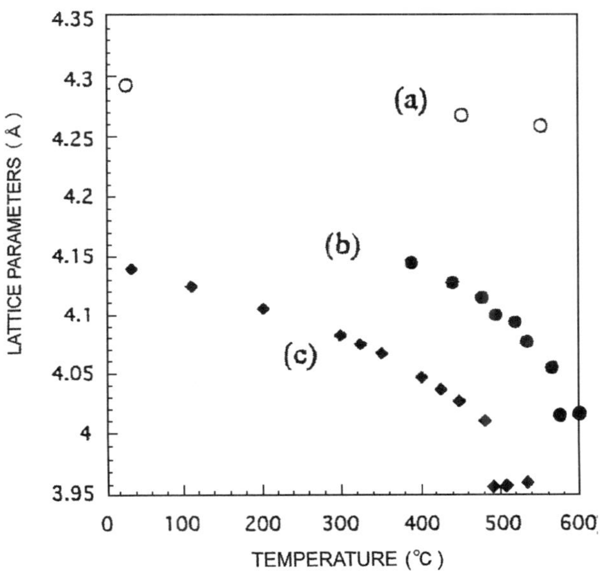

Figure 6.37. Temperature dependence of the c-lattice parameter (*a*) for PT thin films on miscut ST, (*b*) PT thin film on ST, and (*c*) PT powder.

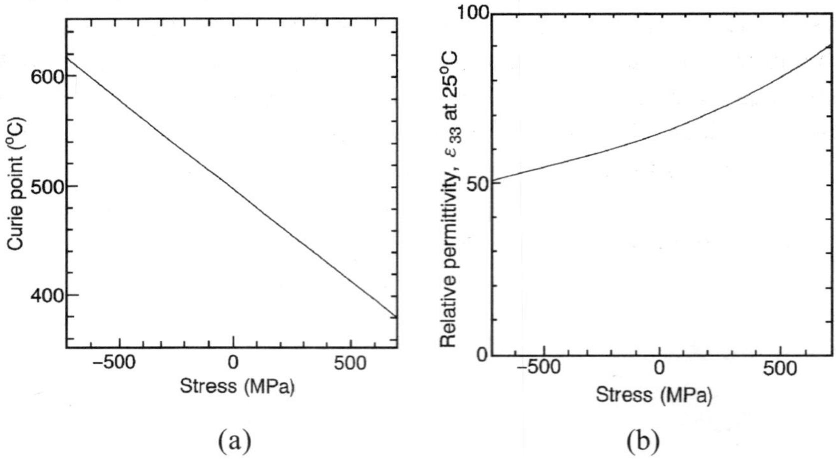

Figure 6.38. Stress effects on ferroelectric properties: (*a*) Curie temperature, (*b*) dielectric constant.[39]

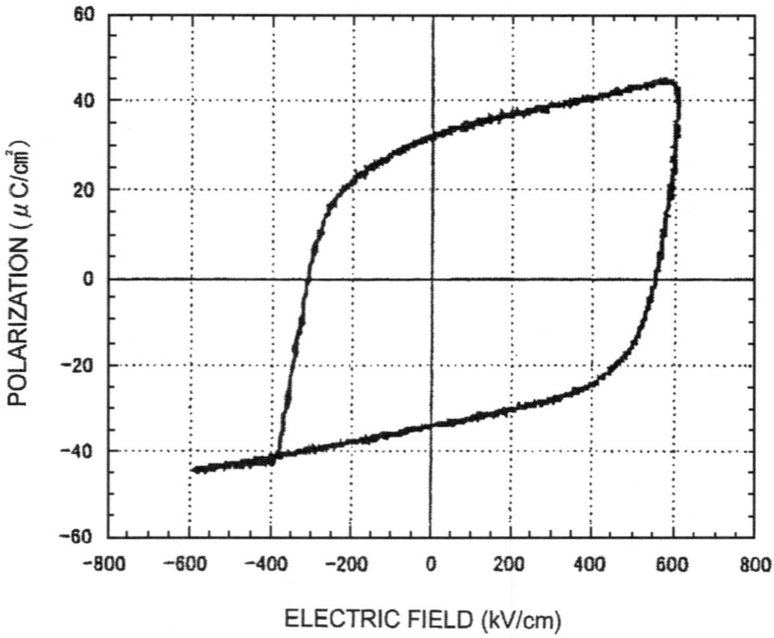

Figure 6.39. Typical P-E hysteresis curves for sputtered PT thin films on La-doped (100) ST measured at 1 kHz and at room temperature (PT thin film: 125 nm thick).

The sharp temperature anomaly is obtained in single-crystal thin films with relaxed structures. Heteroepitaxial films generally include stresses due to lattice mismatches between the thin films and the substrates. The interfaces of the thin films and the substrates are free for the relaxed films. The lattice parameters of the thin films are not governed by the substrate lattices. The lattice parameters of the relaxed films are almost the same as those of the bulk materials. A typical example is shown in Fig. 6.40.[6] The relaxed thin films are produced by sputtering PT onto (001) MgO with rapid cooling after deposition.

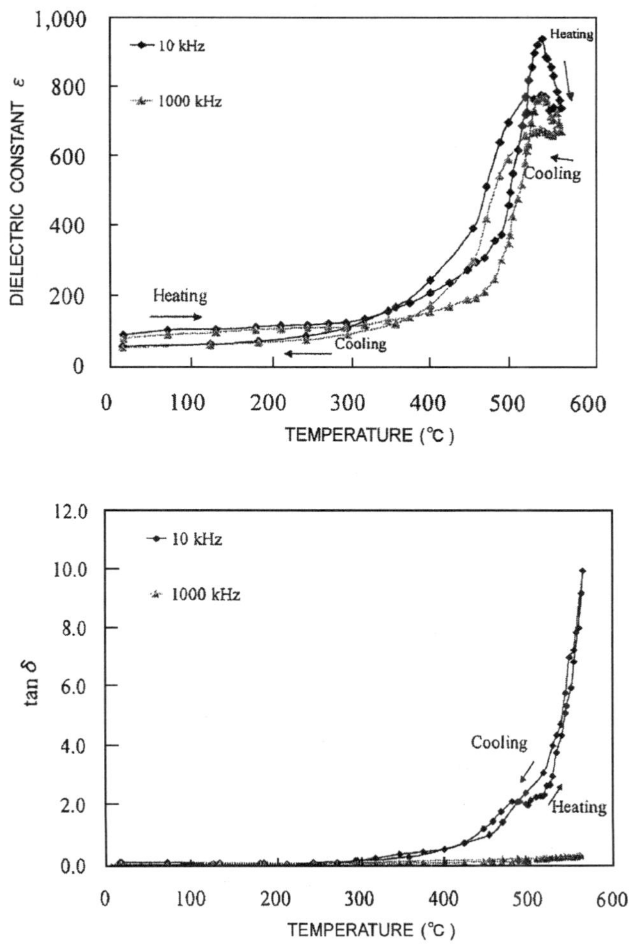

Figure 6.40. Dielectric properties for a sputtered (001) PT thin film (120 nm thick) on (001)Pt/(001)MgO substrates. PT thin films were quenched after deposition.

6.3 NANOMETER STRUCTURE

6.3.1 Nanometer Materials

Nanometer structures hold the promise of new materials for numerous functions. Table 6.12 shows the special features of nanometer structures.[41] Nanometer composites are among the most promising nanostructures. A schematic construction of a nanometer composite is shown in Fig. 6.41. Ceramic composites of nanometer structures are proposed by Newnham.[42] They comprise twelve types of structures, i.e., 0-0, 1-1, 2-2, 3-3, etc. The size of each element is in nanometers, and is on the order of the wavelength of an electron wave. The nanometer composites show the quantum effect. The 0-0 and 1-1 structures correspond to quantum dots and quantum wires, respectively. The traditional color glass comprising the metal elements for coloring is one example of a nanometer composite composed of quantum dots. The 2-2 composite is a multilayered structure that includes a man-made superlattice. Nanometer composites are important not only for their application in making electronic and photonic devices of the future,[43] but also for understanding the physical and chemical properties of thin films.

Nanometer composites are fabricated by a thin film deposition process and the microfabrication process (including electron-beam lithography for fabricating semiconducting ICs). For precise control of nanometer materials, a nanofabrication process that includes nanophotolithography is necessary. However, IC technology is limited to making nanoscale materials with a controlled size and/or population of quantum dots. For the fabrication of nanometer structures, the application of thin film deposition is important.[44]-[46] The codeposition process is useful for making the 0-0 structures. Figure 6.42 shows the photoluminescence spectrum of a CdS-dots/SiO_2-network provided by cosputtering CdS and SiO_2. The spectrum shows the shift of the luminescence peak toward a short wavelength.[47] Nanometer dots are also made by depositing a multilayer structure followed by the postannealing process. The stress observed on the surface of the heterostructure acts as the nucleation center for the fabrication of the nanometer dots.[48]

Precise control of the size and density of the quantum dots is achieved by depositing the island structure obtained at the initial growth of thin films. The control of the nucleation sites is essential for managing the

Table 6.12. Nanometer Materials and Applications[41]

Basic Structural Properties	New Materials & Devices
Nano Structure	
Nano-tailored geometry	Nanocomposites
Metastable structure	Nanotubes
Nanoscale inhomogenity	Nanowires/fibers
Thermodynamic effects	Nanobiomaterals
Nano Electronics	
Dimensionality effects	Single electron/molecule devices
Quantum effects	Carbon nanotube FET
Quantum dots/ wells	Magnetic nanowire for media
Spin tunneling effects	Giant magnetoresistance
Nanomagnetism	Magnetooptic devices
Ballistic transport	Ballistic devices
Biofunctinalized materials	Biofunctionalized sensors
Nano Chemicals	
Surface chemical reaction	Nanoparticle chemical catalysis
Nanoparticles/tubes	Nanotube-chemical reactors/catalysis
Extended solubility	Carbon nanotube chemical sensors
Radically modified phase diagram	Nanoscale metallurgy
Nano Mechanics	
Nanolayered structures	Superhardness
Mechanical strength	Protective, decorative
Surface modification	Lurication, tribological application
Nano Optics	
Quantum optics	PL/LED, quantum wells/dots lasers
Relaxed selected rules	Nanomicro-optics/ holographic devices
Plasmon effects	Nanooptical MEMS

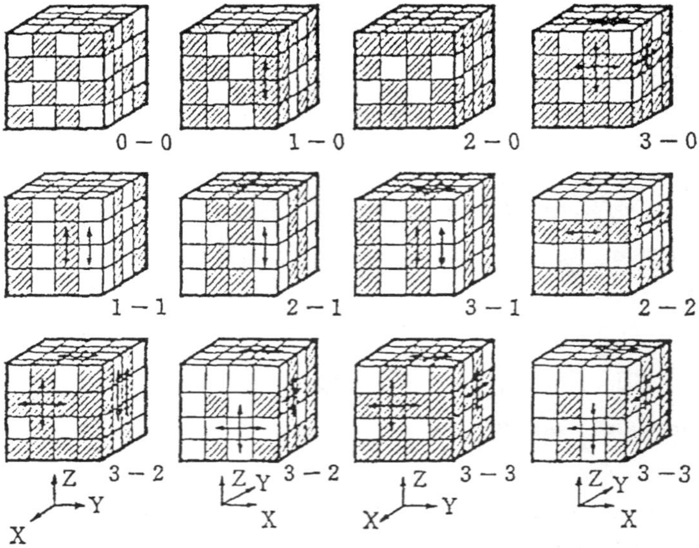

Figure 6.41. Construction of nanometer composites.[42]

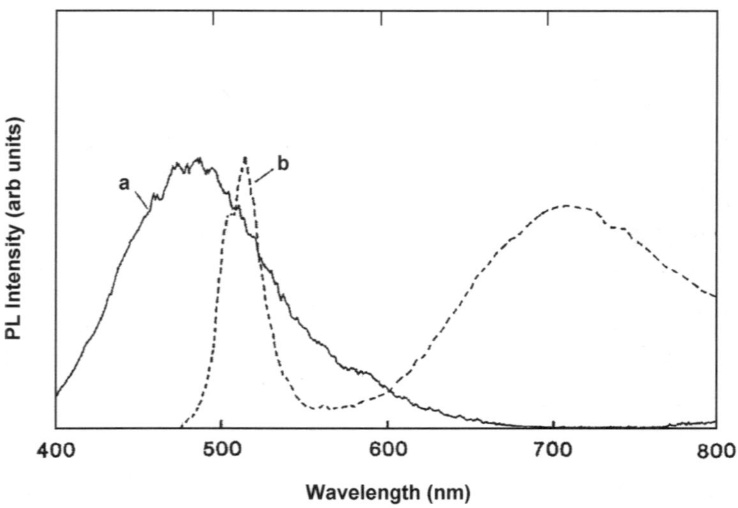

Figure 6.42. Photoluminescence spectrum of CdS-dots/SiO_2-network provided by co-sputtering of CdS and SiO_2 : (*a*) 18.5% CdS (*b*) CdS powder.

initial growth of the films. The nucleation sites are controlled by modifying the substrate surface by electron or ion-beam irradiation.

Typical examples of the control of initial film growth are shown in Figs. 6.43 and 6.44, and described below. Figure 6.43 shows typical surface SEM images of sputtered PT thin films on well-oriented (001) ST with a Pb-rich composition. A mixed sintered powder of PT and PbO was used for the sputtering target. Uniform growth of hexagonal crystallites is observed. The energy dispersive x-ray analysis (EDAX) measurement suggests the segregation of Pb. Increasing Pb enhances the crystal growth. Excess Pb acts as a nucleation center during film growth.

It was assumed that excess Pb would modify the step-flow growth. Figure 6.44 shows surface SEM and AFM images of sputtered PT thin films on a vicinal ST substrate with a Pb-rich composition. Note that the film growth process still follows the step-flow growth pattern. The step height and width of the terrace are of the same order as the stoichiometric PT thin films when the film thickness is less than 10 nm. The additional crystal growth, a screwlike structure, was superposed onto the step-flow growth surface when the film thickness further increased. The additional film growth is governed by the two-dimensional crystal growth on the surface of the PT thin films. The nucleation center is provided by the excess PbO and/or Pb. The stacked structure of the island growth makes the high-density quantum dots.

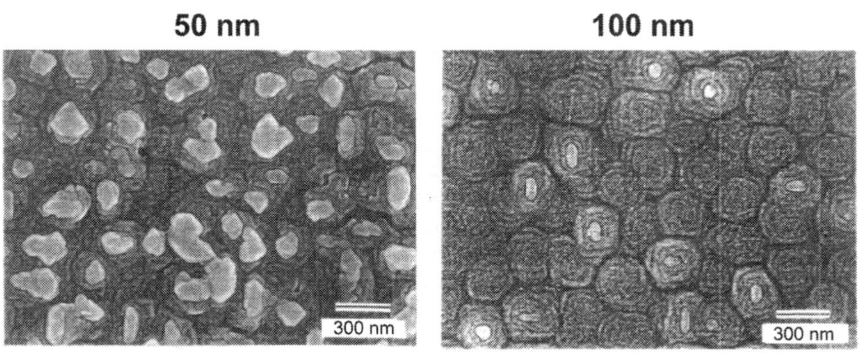

Figure 6.43. Surface SEM images of Pb-rich thin films of different thicknesses on a nominal ST substrate.

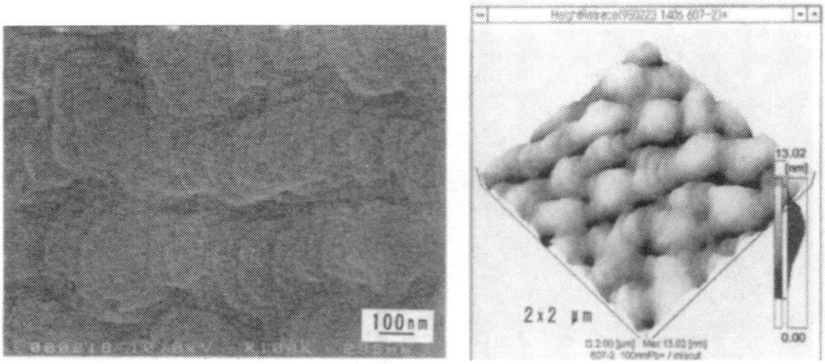

Figure 6.44. Surface SEM and AFM images of off-stoichiometric PT thin films epitaxially grown on miscut (001) ST substrate for the film thickness of 100 nm (see Fig. 6.43).

6.3.2 Nanometer Superlattice

Layered structures and superlattices provide thin films with a controlled nanometer structure. The multitarget sputtering mentioned in Ch. 4 is useful for the deposition of the layered structure of perovskite materials. Table 6.13 shows typical deposition conditions for a PT/PLT multilayered structure on (001) MgO and (001) Pt/(001) MgO substrate.[8] The layered structures are epitaxially grown on the MgO substrate. Typical XRD patterns are shown in Fig. 6.45. The satellite peaks show the formation of the superlattice.

Table 6.13. Sputtering Conditions

Substrate	(100) MgO, (100) Pt/(100) MgO
Substrate temperature	600°C
Target	PT + PbO (10 mol%) PLT (20) + PbO (10 mol%)
Sputtering gas	Ar/O_2 = 20/1 (0.5 Pa)
Cooling	Rapid (33°C/min)

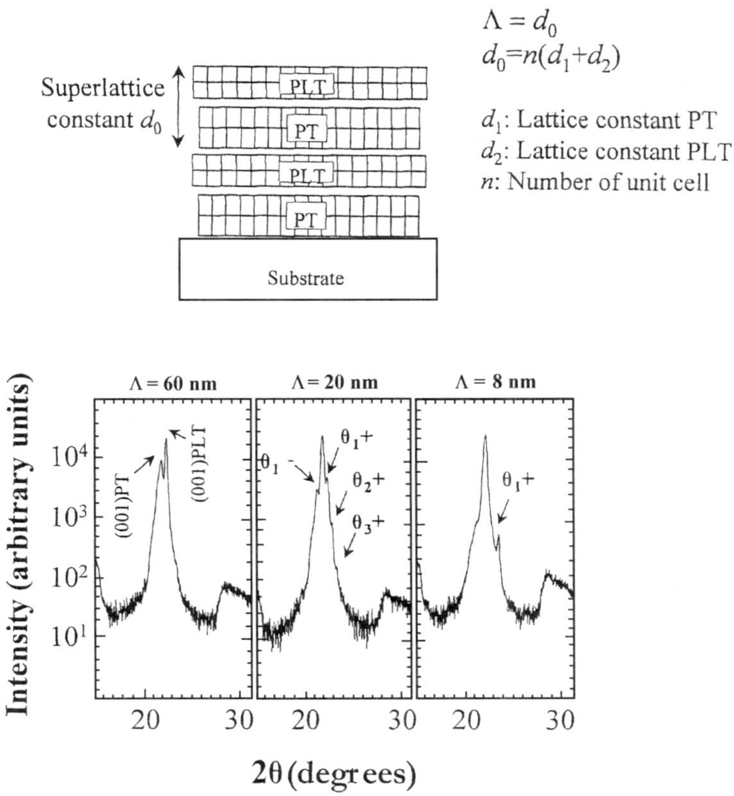

Figure 6.45. (*a*) Schematic illustration of PT/PLT superlattices and (*b*) typical XRD patterns of PT/PLT superlattices as a function of the modulated wavelength Λ, at Λ = 60 nm, 20 nm, and 8 nm: (100) MgO substrates.

A typical TEM image with a SAD pattern is shown in Fig. 6.46. The wavelength ranged from 16 nm to 600 nm. The SAD pattern shows that the superlattice was epitaxially grown on (001) MgO substrates. However, there is an interfacial region due to the lattice mismatch between the substrates and the multilayered PLT/PT. A layered system with improved crystallinity is obtained on $SrTiO_3$ substrates. A typical XRD pattern is shown in Fig. 6.47.

Layered structures are now considered to be the latest approach for creating new materials. For example, several layered structures have been proposed for ferroelectric materials. New man-made materials, including superlattices, may be easily produced by sputtering deposition.

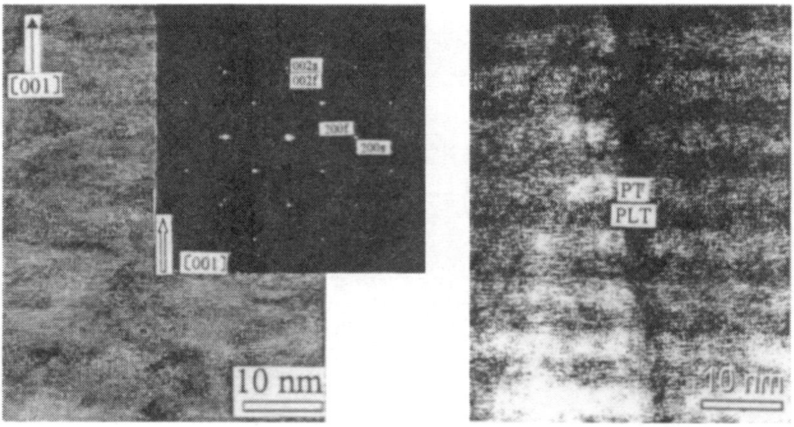

Figure 6.46. Cross-sectional TEM images with SAD pattern for a 240 nm PT/PLT superlattice on (001) MgO. The modulated wavelength, Λ, is 8 nm.

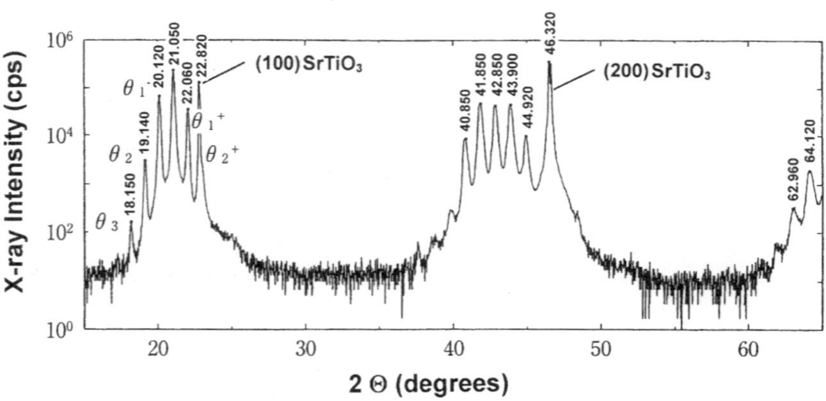

Figure 6.47. XRD patterns of PT/PLT superlattices grown on (100) $SrTiO_3$ substrates. The modulated wavelength, Λ, is 8 nm.

The superlattice of ferroelectric perovskite exhibits novel dielectric properties.[49] While the superlattice of perovskite high-T_c super conductors has been extensively studied as described in Ch. 5, the dielectric properties of ferroelectric superlattices are not well understood. Tabata and Kawai deposited a superlattice of BT/ST onto ST substrates by PLD and found an increase in the dielectric constant due to the strain.[50]

Erbil made superlattices of PLT/PT on ST substrates by MOCVD and found a giant dielectric constant due to the motion of the domain walls.[51]

Sputtering also provides the superlattice structures. The dielectric properties of superlattices at large wavelengths are treated as the series capacitance of the PT thin films, ε_{PT}, and PLT thin films, ε_{PLT}. The total dielectric constant, ε_T, is expressed by

Eq. (6.8) $$\frac{1}{\varepsilon_T} = \frac{1}{2}\left(\frac{1}{\varepsilon_{PT}} + \frac{1}{\varepsilon_{PLT}}\right)$$

Taking $\varepsilon_{PT} = 79$ and $\varepsilon_{PLT} = 413$, ε_T becomes 132.

Figure 6.48 shows the variation of the dielectric properties with the wavelength, Λ, measured at 100 kHz. The dielectric constant increases with a decrease of the wavelength. The relation Eq. (6.8) is valid for a wavelength of $\Lambda > 50$ nm.

The lattice parameters also change with the wavelength as seen in Fig. 6.49. The dielectric constant for a PT/PLT solid solution of the same overall composition is 200. When $\Lambda < 15$ nm, the dielectric constant of the superlattice exceeds that of the solid solution of the same overall composition. The increase in the dielectric constant is due to the strain and/or the additional electronic interaction between superlattice layers.

For the deposition of the nanometer superlattice, the usage of the sputtering process is not fully understood, since the irradiation of highly energetic particles possibly induces the interdiffusion between multilayers. However, experimental results suggest that the minimum thickness of each layer is as small as 1 to 2 nm. Sputtering is useful for the controlled deposition of nanometer superlattice of complex perovskite compounds.

6.4 INTERFACIAL CONTROL

The sputtering process was considered to be inapplicable for the Si MOS gate process since the highly energetic sputtered particles would induce irradiation damage to the semiconductive properties of the Si surface. However, the sputtering process, if it is carefully controlled to reduce the irradiation effects, shows a potential for depositing high-K Si gate oxides for an alternative to thermally grown SiO_2.

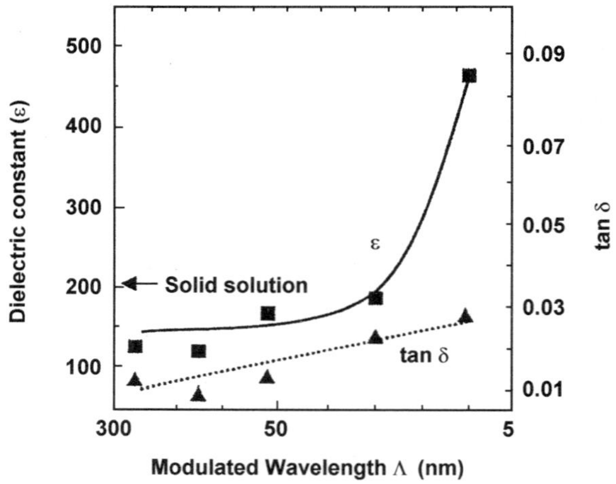

Figure 6.48. Variation of dielectric properties with wavelength Λ for a 240 nm PT/PLT superlattice on (001)Pt/(001)MgO measured at 100 kHz.

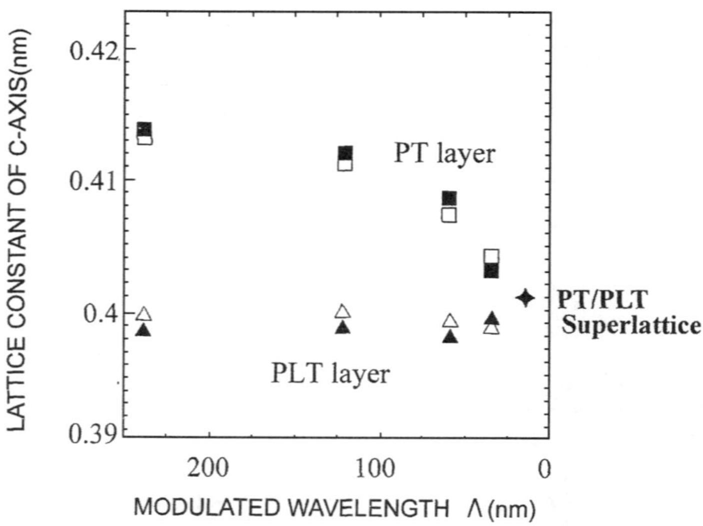

Figure 6.49. Variations of lattice parameters with wavelength for a 240 nm PT/PLT superlattice: ▲, ■, (PT/PLT)/(001)MgO; △, □, (PT/PLT)/(001)Pt/(001)MgO.

Several candidates for high-K materials are being considered, as shown in Table 6.14. The high-K materials should also be stable when in contact with Si, and show a small leakage current. High-K materials show a large mean atomic number per molecule. Some of the well-known high-K oxides are Ta_2O_5, Bi_2O_3, TiO_2, and/or HfO_2.[52] Oxides of the IIIB group, such as La_2O_3, are dielectrics with small leakage currents and high permittivities. The thermodynamic stability of the high-K-oxide/Si interface is being studied. Many high-K materials such as Ta_2O_5, TiO_2, $SrTiO_3$, and $BaTiO_3$ are thermally unstable when directly in contact with Si.[53] Among these high-K materials, ZrO_2 and HfO_2 are useful since they are thermally stable with crack-free structural properties.

The structure of gate oxides is usually amorphous, since polycrystalline oxide films include a microstructure and/or grains accompanied by a large leakage current. Uniform and grain-free structures are essential to reduce the leakage current. Single-crystal-like thin films with a continuous structure meet these requirements. Figure 6.50 shows examples of cross-sectional TEM images of YSZ (3% Y-doped ZrO_2) thin films sputtered onto (100) p-Si prepared at 700°C by using a planar magnetron. The thin film layer was deposited by direct sputtering of the YSZ target. The structure of YSZ layer shows a high density continuous structure, although the Si oxide layer of 3 nm is present at the interface.[54] It is interesting that the heterostructure shows the equivalent oxide thickness (EOT) of 4 nm with a small leakage current and a high breakdown voltage (5×10^{-8} A/cm^2 at 9.3 MV/cm), as shown in Fig. 6.51. These leakage currents are the same order of magnitude as those of thermally grown SiO_2 gate oxides. Deposition at a high temperature reduces the leakage current. When conventional magnetron sputtering is used, the oxide layer grown at a low temperature (i.e., 100° to 300°C) shows a large leakage current due to the low density of the resultant deposited films. The ion-beam sputtering described in Ch. 5 produces high-density oxide films at a low temperature.

The heterostructure YSZ/(100) Si shown in Fig. 6.51 includes the Si oxide layer at the YSZ-Si interface. A careful sputtering process reduces the thickness of the Si oxide layer and reduces the EOT. For the deposition of HfO_2,[55] the sputtering process includes two steps. In the first step, metal Hf is sputtered from an Hf metal target onto the Si surface to reduce the natural Si oxide at the initial surface of the Si; and in a second step, the HfO_2 layer is deposited by reactive sputtering the Hf target in the metal mode to reduce the development of the Si-oxide layer at the interface during the sputtering deposition. In a sample using this process, the resultant EOT values were reasonably reduced. The value was less than 1.5 nm.

The sputtering process still involves problems such as irradiation effects of the energetic particles from the plasma. The thermalization process to reduce the energy of the sputtered particles further improves the semiconductive properties.

Table 6.14. Basic Design of High-K Materials and Sputtering Conditions

Basic Material Selection	Basic Sputtering Conditions
Requirements	
Low leakage current continuous structure high density flat surface & interface	Suppressing island formation layer-by layer growth increasing surface mobility of adatoms optimum substrate temperature/growth rate
High-K at ultra thin films	
Thermodynamic stability ≈ 900°C	ion assisted deposition photo-assisted deposition
Low permeability to oxygen	Non oxidizing condition for Si surface
Examples	
Single metal oxides/oxide-nitrides metal elements[57] IIIa; Sc, Y, IVa; Ti, Zr,[58] Hf,[59] Va; Nb, Ta, Ln; La Binary/ternary metal oxides/nitrides high-K with low leakage current materials Ti with Ln (Ln <10%)[60] Ti with Zr, Al, Si (i.e., Ti-Zr-O, Ti-Zr-O-N)	Reactive sputtering at metal mode unbalanced magnetron high sputtering pressure for thermallization 1st step: Zr or Al metal sputtering at Ar, 2nd step: reactive sputtering of Zr or Hf substrate >200–300°C
Multi-layers for interface improvements high–K/(interface) low leakage materials Ti-Ln-O/Zr-O at interface Hf-O/Al-O at interface[61]	Ion-Beam sputtering ion beam sputtering of metal irradiation of oxygen ion or radical beam substrate : room temperature

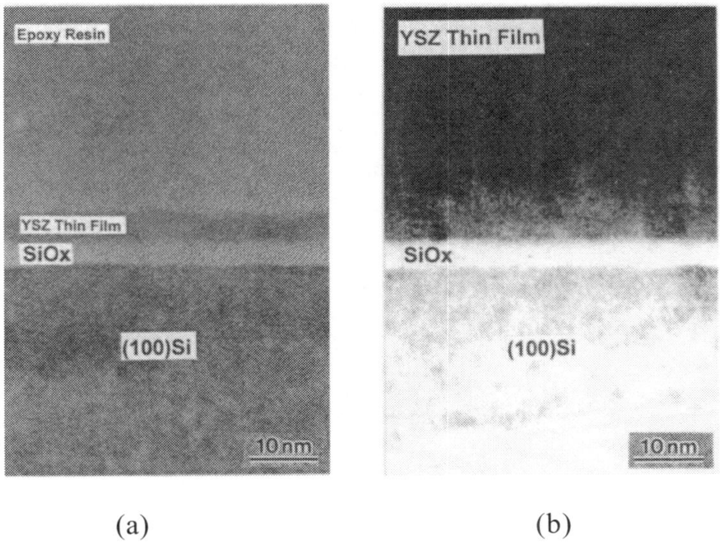

Figure 6.50. Cross-sectional TEM images of the heterostructure (002)YSZ/(100)Si for different deposition times of YSZ: (*a*) one min (3.5 nm film thickness), (*b*) 20 min (65 nm film thickness).

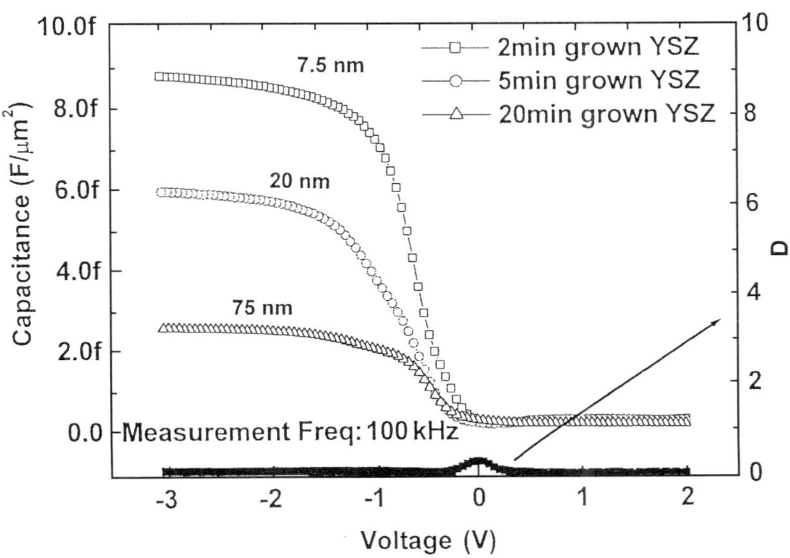

Figure 6.51. Capacitance-voltage curves and dissipation factor D for the heterostructure Au/YSZ/(100)Si for different film thicknesses of YSZ measured at 100 kHz.

The sputtering process was first used for the deposition of thin-film electronic devices (i.e., Pt thin-film resistors by F. Krüger in 1921).[56] Sputtering is old technology compared to MBE. However, the old technology is potentially an alternative to the newer MBE technology if we really understand and carefully design the sputtering process. The sputtering process is essential for producing new materials and devices in the present and in the future.

REFERENCES

1. Feldman, C., *Rev. Sci. Instr.*, 26:463 (1955)
2. Kusao, K., Wasa, K., and Hayakawa, S., *Jpn. J. Appl. Phys.*, 7:437 (1968)
3. Wasa, K., and Setsune, K., Ferroelectric Thin Films & Devices, *Wiley Encyclopedia of Electrical and Electronics Engineering*, John Wiley & Sons (1999)
4. Wasa, K., Adachi, H., and Kitabatake, M., *Ferroelectrics*, 151:1 (1994)
5. Tomashpolski, Y. Y., and Sevostianov, M. A., *Ferroelectrics,* 13:415 (1976)
6. Ai, R., Ito, H., Asayama, G., and Wasa, K., Superconducting and Related Oxides, Physics and Nanoengineering IV, (D. Pavuna and I. Bozovic, eds), *Proc. SPIE*, 4058:426, Orlando (Apr. 2000); Ai, R., MS thesis, Yokohama City Univ. (2001)
7. Sato, T., Wasa, K., Tabata, K., Adachi, H., Ichikawa, Y., and Setsune, K., *J. Vac. Sci. Technol.*, A13:1022 (1995)
8. Adachi, H., Mitsuyu, T., Yamazaki, O., and Wasa, K., *Jpn. J. Appl. Phys.*, Suppl., 26-2:15 (1987)
9. Tu, K. N., Mayer, J. W., and Feldman, L. C, *Electronic Thin Film Science*, Macmillan Publ. Co., New York (1992)
10. Langjahr, P. A., Lange, F. F., Wagner, T., and Rühle, M., *Acta Mater.*, 46:773 (1998)
11. Kwak, B. S., Erbil, A., Budai, J. D., Chisholm, M. F., Boatner, L. A., and Wilkens, B. J., *Phys. Rev. B*, 49:14865 (1994)
12. Trolier-McKinstry, S., *Ferroelectrics*, 152:169 (1994)
13. Wasa, K., Sato, T., Adachi, H., Setsune, K., and Trolier-McKinstry, S., Structural Control of Epitaxially Grown Perovskite Thin Films, *Proc. 3rd Int. Symp. on Sputter and Plasma Processes*, p. 83, Tokyo (1995)

14. Cross, L. E., *Ferroelectric Ceramics*, p. 1, Birkhδuser Verlag, Boston (1993)
15. Wakiya, N., Shiihara, J., Shinozaki, K., and Mizutani, N., *J. Solid State Chem.,* 142:344 (1999)
16. Wasa, K., Yamada, Y., Shimoda, M., Seo, S. H., Noh, D. Y., Okino, H., and Yamamoto, T., *Proc. Int. Conf. on New Piezoelectric Materials and High Performance Acoustic Wave Devices*, p. 35, Tokyo (2002)
17. Adachi, H., Mitsuyu, T., Yamazaki, O., and Wasa, K., *J. Appl. Phys.*, 60:736 (1986)
18. Wasa, K., Sato, T., Tabata, K., Adachi, H., Yabuuchi, Y., and Setsune, K., *J. Mater. Res.*, 9:2959 (1994)
19. Wasa, K., Adachi, H., Ichikawa, Y., Hirochi, K., Matsushima, T., Enokihara, A., Mizuno, K., Higashino, H., and Setsune, K., *Science and Technology of Thin Film Superconductors 2*, (R. D. McConnell and R. Noufi, eds.), p. 1, Plenum Press, New York (1990)
20. Kwak, B. S., Erbil, A., Wilkens, B. J., Budai, J. D., Chishlom, M. F., and Boatner, L. A., *Phys. Rev. Lett.*, 68:3733 (1992)
21. Speck, J. S. and Pompe, W., *J. Appl. Phys.*, 76:466 (1994); Speck, J. S., Daykin, A. C., Seifert, A., Romanov, A. E., and Pompe, W., *J. Appl. Phys.*, 78:1696 (1995)
22. Ai, R., Ichikawa, Y., and Wasa, K., *Hyoumen Kagaku*, 21:456 (2000)
23. Wasa, K., Haneda, Y., Sato, T., Adachi, H., Hayashi, S., and Setsune, K., *Jpn. J. Appl. Phys.*, 34:5132 (1995)
24. Wasa, K., Haneda, Y., Satoh, T., Adachi, H., and Setsune, K., *Appl. Surf. Sci.*, 121/122:152 (1997)
25. Kimoto, T., and Matsunami, H., *J. Appl. Phys.*, 75:850 (1994)
26. Wasa, K., Haneda, Y., Sato, T., Adachi, H., Kanno, I., Setsune, K., Schlom, D. G., Trolier-McKinstry, S., and Eom, C. B., *Korean J. Phys. Soc.*, 32:S1344 (1998)
27. Wasa, K., Ai, R., and Asayama, G., Modified Ferroelectric Properties of the Single Domain/Single Crystal $PbTiO_3$ Thin Films with Tightly Bonded Interface, *Superconducting and Related Oxides: Physics and Nanoengineering IV* (D. Pavuna and I. Bozovic, eds.), *Proc. SPIE,* 4058:295, Orlando (Apr. 2000)
28. Wasa, K., Haneda, Y., Sato, T., Adachi, H., Kanno, I., Schlom, D. G., Trolier-McKinstry, S., Gang, Q., and Eom, C. B., Step-Flow-Growth of Perovskite $PbTiO_3$ Thin Films Epitaxially Grown on a Miscut $SrTiO_3$ Substrate, *Superconducting and Related Oxides: Physics and Nanoengineering III* (D. Pavuna and I. Bozovic, eds.), *Proc. SPIE*, 3481:182, San Diego (Jul. 1998)

29. Theis, C. D., Investigation of Growth and Domain Structure of Epitaxial Lead Titanate Thin Films, MS Thesis, Penn State Univ. (1996)

30. Nagai, H., Structure of Vapour-Deposited Ga_xIn_{1-x} as Crystals, *J. Appl. Phys.*, 45:3789 (1974)

31. Wasa, K., Haneda, Y., Satoh, T., Adachi, H., and Setsune, K., *J. Vac. Sci. Technol.*, A15:1185 (1997)

32. Eom, C. B., Cava, R. J., Fleming, R. M., Phillips, J. M., van Dover, R. B., Marshall, J. H., Hsu, J. W. P., Krajewski, J. J., and Peck, W. F., Jr., *Science*, 258:1766 (1992)

33. Li, Z., Foster, C. M., Guo, D., Zang, H., Bai, G. R., Baldo, P. M., and Rehn, L. E., *Appl. Phys. Lett.*, 65:1106 (1994)

34. Li, J. F., Viehland, D., and Payne, D. A., *J. Korean Phys. Soc.*, 32:S1311 (1998)

35. Rossetti, G. A., Jr., Cross, L., and Kushida, K., *Appl. Phys. Lett.*, 59:2524 (1991)

36. Kanno, I., Ph. D Thesis, Osaka Univ. (1998)

37. Asayama, G., Wasa, K., and Ai, R., Temperature Variations of Dielectric Properties for Single Domain/Single Crystal Thin Films of Pb-Ti-O Families, in *Superconducting and Related Oxides: Physics and Nanoengineering IV* (D. Pavuna and I. Bozovic, eds.), *Proc. SPIE*, 4058:418, Orlando (Apr. 2000)

38. Pertsev, N. A., Zembilgotov, A. G., and Tagantsev, A. K., *Phys. Rev. Lett.*, 80:1988 (1998)

39. Yamamoto, T., and Matsuoka, H., *Jpn. J. Appl. Phys.*, 33:5317 (1994)

40. Wasa, K., Haneda, Y., Adachi, H., Kanno, I., Schlom, D. G., Trolier-McKinstry, S., Gang, Q., and Eom, C. B., *Integrated Ferroelectrics*, 21:451 (1998)

41. Chopra, K. L., private communication (Dec. 2001)

42. Newnham, R. E., Skinner, D. P., and Cross, L. E., *Mat. Res. Bull.*, 13:525 (1978)

43. Liz-Marzan, L. M., and Norris, D. J., *MRS Bull.*, 26:981 (2002)

44. Neshere, D., Hofmeiser, H., Levi, Z., Aneva, Z., *Vacuum*, 65:109 (2002)

45. Petroff, P. M., Medliros-Ribeiro, G., *MRS Bull.*, 21:51 (1996)

46. Kotov, N. A., *MRS Bull.*, 26:992 (2001)

47. Tanahashi, I., and Mitsuyu, T., private communication.

48. Cullis, A. G., *MRS Bull*, 21:21 (1996)

49. Li, S., Eastman, A., Vertrone, J. M., Newnham, R. E., and Cross, L. E., *Philos. Magazine*, 76:47 (1997)

50. Tabata, H., Tanaka, H., and Kawai, T., *Appl. Phys. Lett.*, 65:1970 (1994)
51. Erbil, A., Kim, Y., and Gerhardt, R. A., *Phys. Rev. Lett.*, 19:1628 (1996)
52. Harrop, P. J., and Champbell, D. S., *Handbook of Thin Film Technology* (L. I. Maissel and R. Glang, eds.), pp. 16–17, McGraw-Hill, NY (1970)
53. Hubbard, K. J., and Schlom, D. L., *J. Mater. Res.*, 11:2757 (1996)
54. Jeon, S., Matsuda, T., Unno, A., Wasa, K., Ichikawa, Y., and Hwang, H., *Vacuum*, 65:19 (2002)
55. Lee, B. H., Kang L., Qi, W. J., Nieh, R., Jeon, Y., Onishi, K., and Lee, J. C., *IEDM Tech. Digest* (1999)
56. Krüger, F., German Patent 336,469 (1921); Kisaka, S., *History of Science for Electronics*, 5th Ed., p. 148, Tokyo, Nikkan Kogyo (2003)
57. Billman, C. A., Tan, P. H., Hubbard, K. J., and Schlom, D. G., Alternate Gate Oxides for Silicon MOSFETs using High-K Dielectrics, *Mat. Res. Soc. Symp. Proc.*, 567: 409 (1999)
58. Qi, W. J., Nieh, R., Lee, B. H., Kang, L., Jeon, Y., Onishi, K., Nagai, T., Banerjee, S., and Lee, J. C., MOSCAP and MOSFET Characteristics using ZrO_2 Gate Dielectric Deposited Directly on Si, *IEDM Tech. Digest*, p. 145 (1999)
59. Lee, B. H., Kang, L., Qi, W. J., Neih, R., Jeon, Y., Onishi, K., and Lee, J. C., Ultrathin Hafnium Oxide with Low Leakage and Excellent Reliability for Alternative Gate Dielectric Application, *IEDM Tech. Digest*, p. 133 (1999)
60. Alers, G. B., van Dovar, R. B., Schneemeyer, L. F., Stirling, L., Sung, C. Y., Diodato, P. W., Liu, R., Wong, Y. H., Fleming, R. M., Lang, D. V., and Chang, J. P., Advanced Amorphous Dielectrics for Embedded Capacitors, *IEDM Tech. Digest*, p. 797 (1999)
61. Koo, J., Han, S., Choi, S., Kim, Y., paper TF-TuA9 Colorado, 49th AVS Int. Symp. (Nov. 2002)

7

Microfabrication by Sputtering

Atoms of a solid surface are removed under irradiation by energetic ions. This phenomenon is called "sputter etching." The sputter etching process is governed by collisions between the irradiated atoms and the surface atoms of the solid target. The interaction depth for sputter etching is around 100 Å beneath the surface. When the surface of a solid is bombarded by argon ions, the adsorbed gases are removed and a clean surface is produced. The thinning of specimens is achieved by argon ion bombardment for fabricating test pieces for electron microscope analysis. The sputter etching process is a key technology for the fabrication of Si ICs.

In this chapter, the sputter etching process for microfabrication is described in relation to the fabrication of thin-film electronic devices.

7.1 CLASSIFICATION OF SPUTTER ETCHING

A physical process governs the mechanism of sputter etching by the bombardment of energetic inert ions. Bombardments by chemically reactive ions deliver surface etching through chemical reactions. Since the sputter etching is processed in a vacuum using gas-discharge plasma, the etching process is called "dry etching." Dry etching is a key technology for making masking patterns for large scale integration (LSI) and detailed descriptions can be found in the literature.[1][2]

Figure 7.1 shows a classification of dry etching systems and Fig. 7.2 shows their basic construction.[3] Dry etching is classified into three types of etching processes according to the energy of incident species, i.e., ions and/or excited atoms.

At high-energies, ion-beam systems are used for sputter etching. The physical etching process is predominant and the process is called *ion-beam etching* (IBE). When a reactive gas is used in the IBE system, the process is called *reactive ion beam etching* (RIBE).

In medium-energy regions, diode glow-discharge systems are used for sputter etching. Diode systems use higher density plasmas than ion-beam systems. Since the reactive ion etching process is commonly used in the diode system, the diode etching system is simply called *reactive ion etching* (RIE). Modifications of RIE include narrow-gap discharge systems and magnetron discharge systems. Both chemical and physical processes are included in the medium-energy region.

In the low-energy region, radicals in the plasma govern the etching process. The mechanism for etching in the low-energy region is an enhanced chemical reaction due to the radicals generated in the plasma. Typical reactive gases used in the chemical etching process are shown in Table 7.1. The chemical etching process is isotropic and there is less radiation damage than with physical sputtering. The physical etching

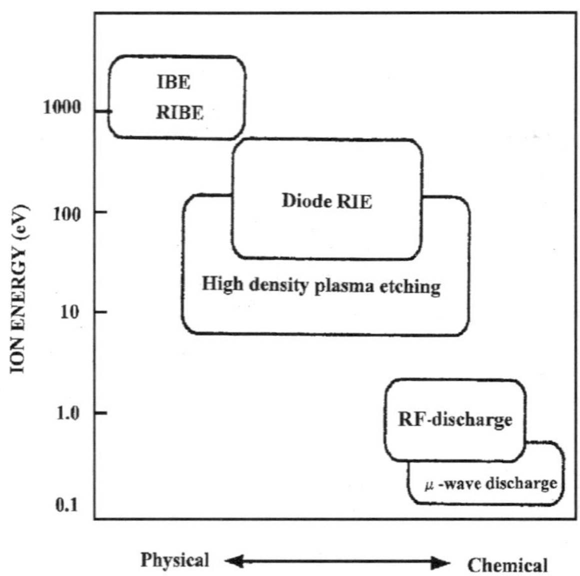

Figure 7.1. Classification of dry etching systems.[3]

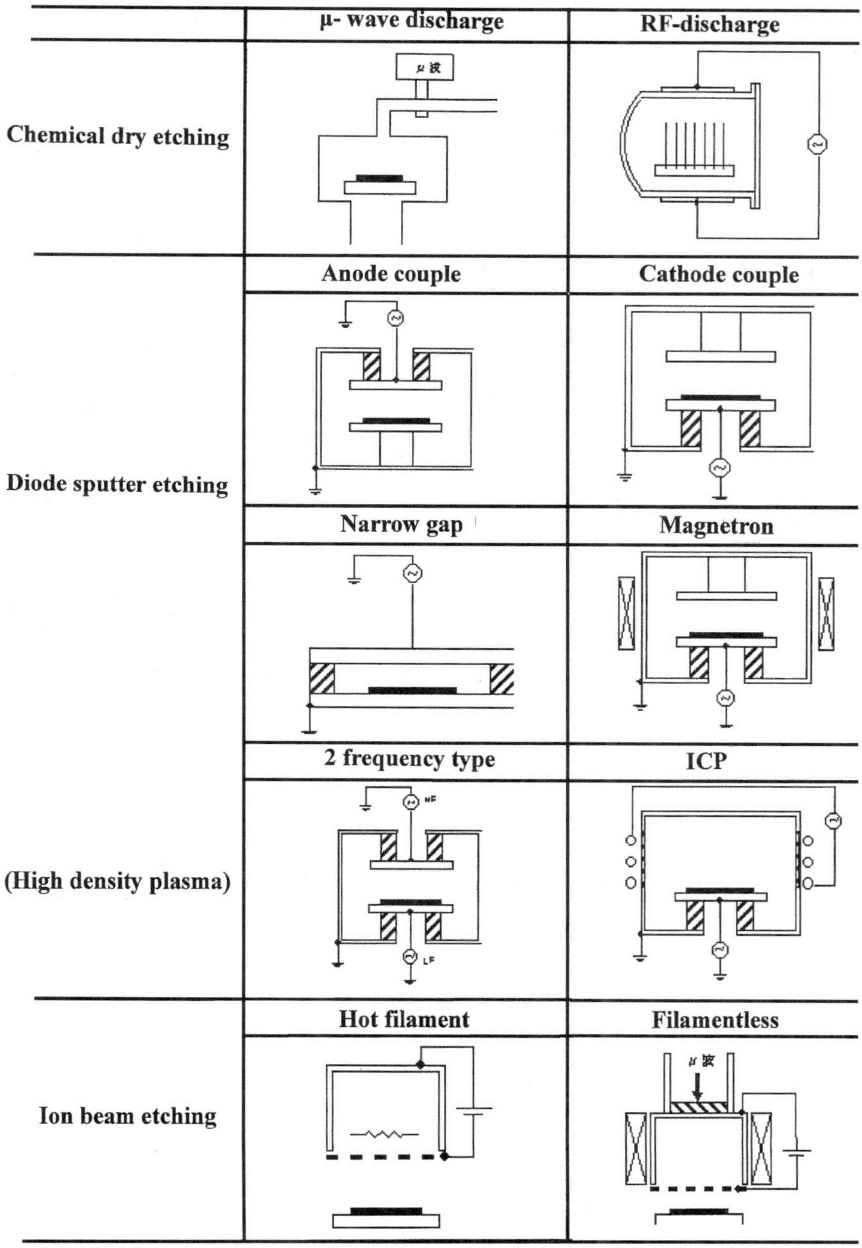

Figure 7.2. Diagrams of typical dry etching systems.[3]

Table 7.1. Reactive Gas for RIE *(Courtesy of Veeco)*

Materials	Gas
Poly-Si	Cl_2, Cl_2/HBr, Cl_2/O_2, CF_4/O_2, SF_6, Cl_2/N_2, Cl_2/HCl, $HBr/Cl_2/SF_6$
Si	SF_6, C_4F_8, CF_4/O_2, Cl_2, $SiCl_4/Cl_2/SF_6/N_2/Ar$, $BCl_2/Cl_2/Ar$, Cl_2/N_2Ar, $SF_6/O_2/HBr$
Si_3N_4	CF_4, CF_4/O_2, CF_4/H_2, CHF_3/O_2, C_2F_6, $CHF_3/O_2/CO_2$, CH_2F_2/CF_4, CHF_3/CF_4, $CF_4/O_2/N_2$, $CF_4/CHF_3/Ar$
SiO_2	$C_4F_8/O_2/Ar$, $C_5F_8/O_2/Ar$, $C_3F_6/O_2/Ar$, C_4F_8/CO, CHF_3/O_2, CF_4/H_2, $C_2F_6/He/CHF_3/O_2$
Al	BCl_3/Cl_2, $BCl_3/CHF_3/Cl_2$, $BCl_3/CH_2F_2/Cl_2$, BBr_3/Cl_2, $BCl_3/Cl_2/N_2$
Al_2O_3	BCl_3, $BCl_3/Cl_2/Ar$
Cr, CrO	Cl_2/O_2
GaAs	Cl_2, HCl/H_2, CH_4/H_2, $CH_4/H_2/Ar$, $Cl_2/BCl_3/Ar$, Cl_2/N_2
Mo, MoSi	CF_4/O_2, $CF_4/SF_6/O_2$
W	CF_4/O_2, SF_6, $SF_6/Ar/N_2$, SF_6/C_4F_8, SF_6/Ar
Ti	CF_4/O_2, Cl_2/N_2
Ta	$Cl_2/O_2/Ar$, $CF_4/CHF_3/O_2/SF_6$
Au	Cl_2, Cl_2/Ar
Pt	Cl_2/Ar, Cl_2/CF_4
ITO	HI, $C_3H_6O/O_2/Ar/HCl$, CH_4/H_2, CH_4/Cl_2, $HI/Ar/O_2$, HCl/CH_4
Polymide	O_2/SF_6, CF_4/O_2, O_2
Cu	Cl_2, $SiCl_4/Cl_2/N_2/NH_3$, $SiCl_4/Ar/N_2$, $BCl_3/SiCl_4/N_2/Ar$, $BCl_3/N_2/Ar$
Ta_2O_5	$CF_4/H_2/O_2$
PZT	$Cl_2/O_2/Ar$, $Cl_2/BCl_3/Ar$
STO	same as PZT
BST	SF_6/Cl_2

(cont'd.)

Table 7.1. *(cont'd.)*

Materials	Gas
TiN	$CF_4/O_2/H_2/NH_3$, C_2F_6/CO, CH_3F/CO_2, $BC_3/Cl_2/N_2$, CF_4
WN	F-compounds
TaN	same as Ta
Si-low-k	$CF_4/N_2/O_2$, C_2F_6, $C_4F_8/O_2/Ar$
Organic low-k	CH_4/N_2, H_2/N_2, NH_3/N_2

process is anisotropic. Desirable mixes of chemical and physical etching processes are delivered by ion-beam and reactive ion etching systems.

The selection of the discharge parameters including ion energy, partial and total gas pressure of the reactive gas, and its flow rate is essential to achieve the designed construction of electronic devices.

7.2 ION-BEAM SPUTTER ETCHING

The most useful system for sputter etching is the ion-beam system. The sputtering chamber is separated from the ion source, and the surface of the specimen is not revealed to the plasma generated in the ion source. The sputtering chamber is generally kept below 1×10^{-4} torr (1×10^{-2} Pa) during sputter etching.

Several ion sources are possible, including hot cathodes, cold cathodes, and plasmatrons.[4] Typical models are shown in Fig. 7.3.

The hot-filament cathode type is called a "Kaufman ion source" and is widely used for sputter-etching. In the Kaufman ion source, tungsten and/or oxide cathodes are used for the hot cathode. The reactive gas often damages the hot cathode so a hollow cathode ion source, where the hot cathode is mounted outside of the plasma, is used for generating reactive gas ions such as oxygen ions.

A typical Kaufman-type ion-beam etching system and a photograph of an ion-beam source are shown in Fig. 7.4. The ions are generated in the chamber by electron impact ionization. The ion source comprises a pair of parallel grids; one has floating potential and the other has negative potential to repel the electrons. The grid with floating potential has contact

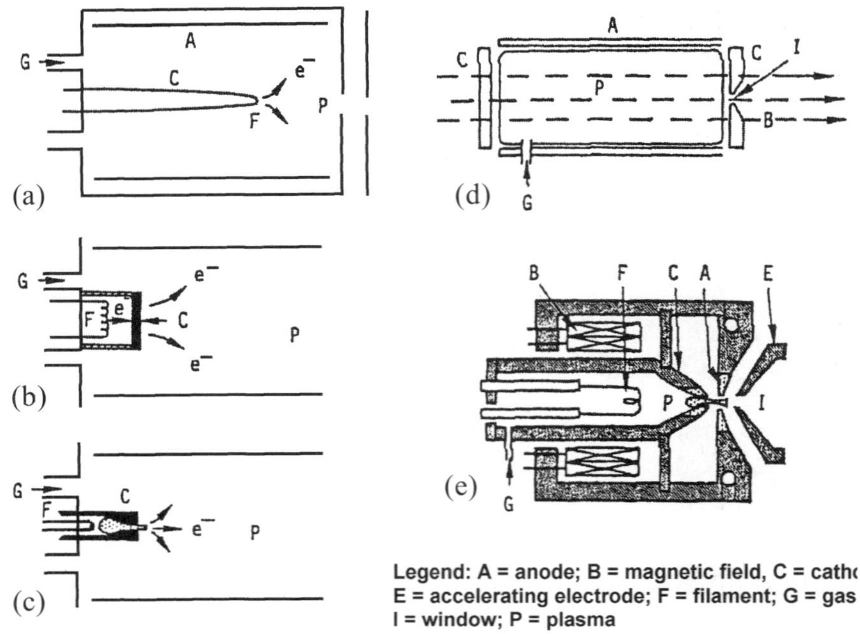

Figure 7.3. Typical models of ion sources; (*a*) filament cathode, (*b*) button cathode, (*c*) hollow cathode, (*d*) Penning type, (*e*) plasmatron.[4]

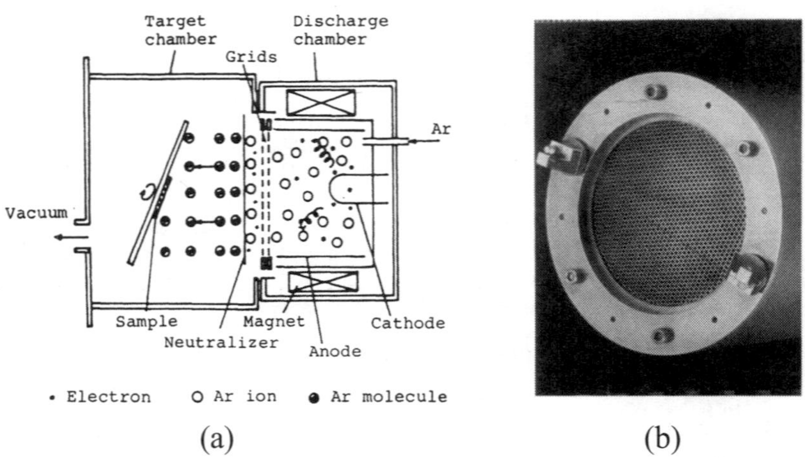

Figure 7.4. (*a*) Schematic diagram of an ion-beam etching apparatus with a neutralizer. Ions generated in a discharge chamber are collimated by grids and collide to etch the sample in the target chamber. (*b*) Photograph of an ion source. *(Courtesy of Veeco.)*

with the plasma. The negative grid extracts the ions from the chamber. In this configuration, the ion-current density, J_+, of the Kaufman-type ion source is limited by a Child-Langmuir equation. The Child-Langmuir equation is expressed by

Eq. (7.1)
$$J_+ = K\left(\frac{q}{m}\right)^{1/2}\left(\frac{V_t^{3/2}}{S^2}\right)$$

where q and m denote the electronic charge and mass of the ion, respectively; V_t denotes the voltage difference between the parallel grids; S is the spacing of the parallel grids; and K is a constant. For the parallel grids construction, K is expressed by

Eq. (7.2)
$$K = 4\left(\frac{e_0}{9}\right)$$

where e_0 denotes the transmission rate of the grids.[5] The acceleration voltage governs V_t. Small values of S give a high ion-current density under a given acceleration voltage. Since S shows a limitation in the reduction of spacing, the ion-current density shows an upper limit. The typical beam energy is 50 to 2000 eV, the ion beam current is 25 to 200 mA for a beam diameter of 5 cm, and the ion-current density is 1 to 10 mA/cm^2. The ion-beam etching system has several good points:

1. Samples are not in plasma
2. The incident angle of ions is controllable
3. Ion energy is controlled independent of working pressure
4. Ion-current density is controlled by either ion energy or gas pressure

The Kaufman-type ion source is essential for uniformly etching or directionally etching a variety of materials including non-reactive materials and multilayered stacks.

The ion-beam sputter process achieves an anisotropic etching pattern, while the chemical etching process produces isotropic etching. Typical etching rates of the ion-beam sputtering are listed in Table 7.2. The ion-beam sputtering process achieves directional etching, including vertical etching, in a trench structure. The etching rate strongly depends on the incident angle of the ion beam, as shown in Fig. 7.5. When photoresists are used for a mask pattern, suitable incident angles should be selected.

Table 7.2. Etching Rates by Ion Beam Sputtering at 500 eV, Ar^+, 1 mA/cm^2 *(Courtesy of Veeco)*

Target Material	Composition	Etch Rate (Å/min)
Silver	Ag	1400
Aluminum	Al	520
Alumina	Al_2O_3	100
Gold	Au	1400
Beryllium	Be	180
Bismuth	Bi	8500
Carbon	C	50
Cadmium	CdS (1010)	2100
Cobalt	Co	510
Chromium	Cr	530
Copper	Cu	820
Dysprosium	Dy	1050
Erbium	Er	880
Iron	Fe	380
Iron Oxide	FeO	470
Gallium Arsenide	GaAs (110)	1500
Gallium Gadolinium Garnet	GaGd	280
Gallium Phosphide	GaP (111)	1400
Gallium Antimonide	GaSb (111)	1700
Gadolinium	Gd	1000
Hafnium	Hf	590
Indium Antimonide	InSb	1300
Iridium	Ir	540
Lithium Niobate	$LiNbO_3$ (Y-cat.)	400
Manganese	Mn	870
Molybdenum	Mo	230
Niobium	Nb	390
Nickle	Ni	500

(cont'd.)

Table 7.2. *(cont'd.)*

Target Material	Composition	Etch Rate (Å/min)
Nickle Iron	NiFe	500
Osmium	Os	450
Lead	Pb	3100
Lead Telluride	PbTe(111)	3000
Palladium	Pd	1100
Platinum	Pt	780
Rubidium	Rb	4000
Rhenium	Re	470
Rhodium	Rh	650
Ruthenium	Ru	580
Antimony	Sb	3200
Silicon	Si	370
Silicon Carbide	SiC(0001)	320
Silicon Dioxide	SiO_2	400
Samarium	Sm	960
Tin	Sn	1200
Tantalum	Ta	380
Thorium	Th	740
Titanium	Ti	320
Uranium	U	660
Vanadium	V	340
Tungsten	W	340
Yttrium	Y	840
Zirconium	Zr	570
Resists	AZ 1350J	300
	COP	800
	PBS	900
	KTFR	290
	PMMA	560
	Riston 14	250
	Kodak 809	320
Glass(Na,Ca)		200
Stainless Steel		250

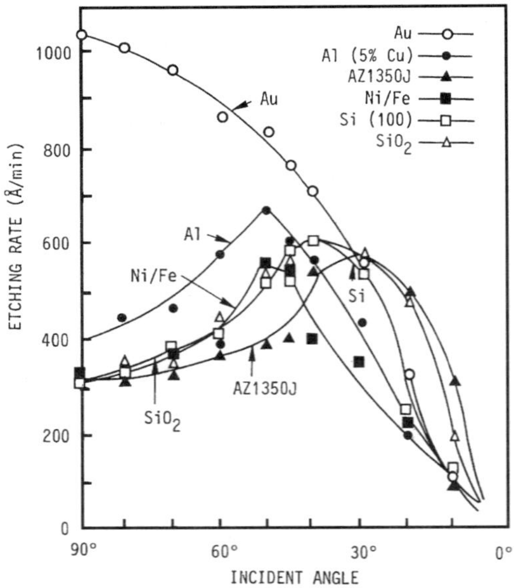

Figure 7.5. Etching rate vs incident angle of an argon ion beam. *(Courtesy of Veeco.)*

Photoresist mask patterns are often eroded, and an ideal vertical etching is not achieved. The sputter-etched structure may show tapered edges. In some cases, a trenching structure appears at the bottom of the etched pattern, as shown in Fig. 7.6.[6] The formation of the trenching structure results from the deposited materials being sputtered from the sidewall of the etched groove.

Ion-beam sputter etching is useful for microfabrication of thin films of alloys or compounds. Thin-film compounds of PLZT described in Ch. 5, for instance, are etched by the ion-beam sputtering system, and thin-film optical-channel waveguides are successfully fabricated. Typical constructions of optical-channel waveguides are shown in Fig. 7.7.[7]

The channel waveguides are classified into four types:

1. Raised-strip
2. Ridge
3. Embedded
4 Strip-loaded

The optical index at the inside of the optical channels is designed to have higher values than the optical index at the outside of the optical channels.

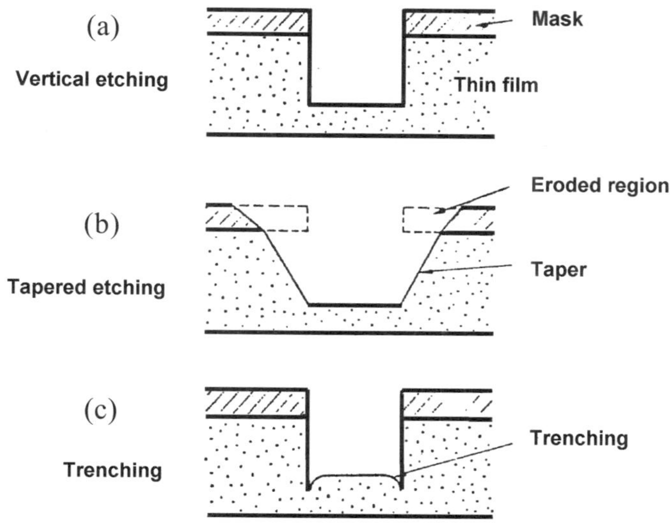

Figure 7.6. Typical cross-sections of etched grooves; (*a*) vertical etching, (*b*) tapered etching, (*c*) trenching.[6]

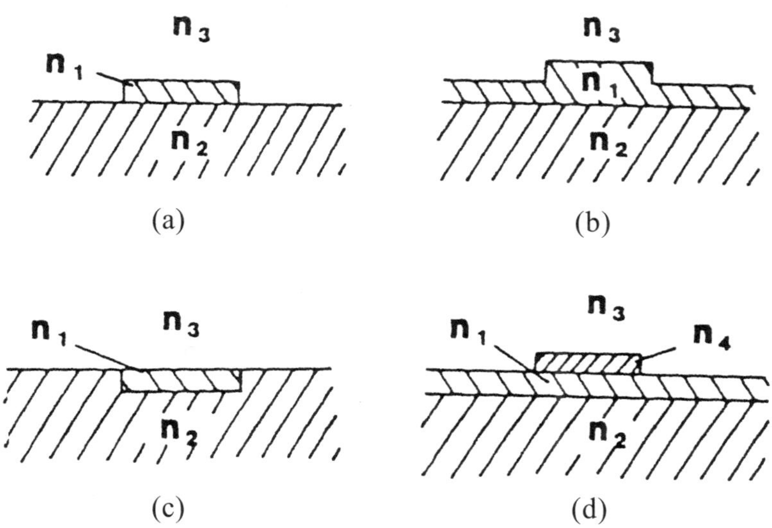

Figure 7.7. Cross-sectional view of four kinds of channel waveguides; (*a*) raised-strip type, (*b*) ridge type, (*c*) embedded type, and (*d*) strip-loaded type; n_1, n_2, n_3, n_4 are refractive indices of waveguide, substrate, environments, and loaded strip, respectively.[7]

Thin films of PLZT are epitaxially grown on sapphire substrates by sputtering deposition. These epitaxial PLZT thin films exhibit both excellent optical transparency and strong electrooptic effects.

The ridge-type channel waveguide pattern is drawn on PLZT thin films covered with a 0.4-μm-thick photoresist. The ion-beam etching conditions are shown in Table 7.3. The etching rate for PLZT thin films is 130 Å/min in an argon atmosphere. The ratio of the etching rate PLZT/photoresist (AZ1400) is around 1.2 at an ion-beam acceleration voltage of 550 V, whereas the photoresist pattern is deformed at an ion-beam acceleration voltage beyond 750 V. The waveguide patterns are etched at incident ion-beam angles of 30° to obtain tapered ridge walls. The waveguide fabrication process does not influence the atomic ratio of lead to titanium.

Figure 7.8a shows a cross-sectional view of a typical PLZT thin-film channel waveguide. The ion-beam sputtering process also successfully provides a curved optical channel waveguide. A typical curved waveguide is shown in Fig. 7.8b. These channel waveguides are useful for making thin-film integrated optical devices.

Ion-beam sputter etching is also useful for making the microstructures of the high-T_c superconducting thin films.[8][9] The sample is held on a sample table and tilted with respect to the direction of the incident ion beam. Typical etching conditions are shown in Table 7.4.

Table 7.3. Ion Beam Etching Conditions for PLZT Thin Films

Ar ion current	600 $\mu A/cm^2$
Acceleration voltage	550 V
Incident angle	30°
Argon pressure	1×10^{-4} torr
Etching rate of PLZT thin films	130 Å/min
Etching rate ratio PLZT thin films/AZ1400	1.2

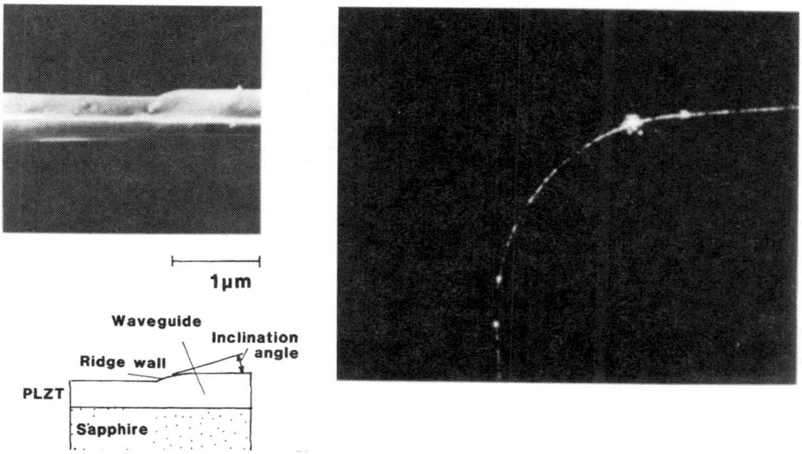

(a) (b)

Figure 7.8. (*a*) Cross-sectional view of thin-film optical waveguide. (*b*) Photograph of curved optical channel waveguide: line width is 10 μm, curvature $r = 1$ mm.

Table 7.4. Ion Beam Etching Conditions for High-T_c Superconducting Thin Films

Applied voltage/current	550 V/600 μA/cm^2
Argon pressure	1×10^{-4} Pa
Incident angle	45°– 90°
Sample table	Water cooling (10°C)
Sample	High-T_c superconducting films (Er-Ba-Cu-O, Ga-Ba-Cu-O)
Photoresists	OMR87 (negative) MP1400 (positive)
Etching rates	Superconductors: ≈ 250–300 Å/min Resists: ≈ 200–300 Å/min

High-T_c Gd-Ba-Cu-O thin films are prepared by rf magnetron sputtering on (100) MgO single-crystal substrates. The film thickness is around 5,000 Å. Figure 7.9 shows typical etching rates by argon ion bombardments for a high-T_c superconducting film and photoresist. The etching rates of high T_c superconducting films are higher than those of photoresists at incident angles over 45 degrees. This suggests that high T_c superconducting films can be patterned by argon ion-beam etching with a photoresist etching mask of about the same thickness as the high-T_c thin film.

The procedure for patterning the film is as follows: The OMR87 resist is spin-coated onto the high-T_c thin film and prebaked at 95°C for 30 minutes. The resist, about 1 μm thick, is irradiated with UV light by the contact printing method and then developed. After the resist is postbaked at 95°C for 1 hour, the high-T_c thin film is etched by argon ion beams for 25 minutes at a 60° incident angle with the patterned resist serving as the mask. The resist is finally removed by dissolving it in the organic liquid, 1,1,1-trichloroethane.

Figure 7.10 shows typical resistivity-temperature curves for the microstrip lines patterned by the ion-beam etching process. The resistance of the 10-μm strip shows approximately the same temperature dependence as that of the film before ion-beam etching, although the zero-resistance temperature slightly decreases for the 2-μm strip, high-T_c thin films. This suggests that the patterning of the high-T_c superconducting thin film is achieved by ion-beam etching without any post–heat-treatment. Generally, post–heat-treatments are necessary after microfabrication of high-T_c superconducting thin films when wet etching is conducted for pattern formation.

A Kaufman-type ion source is useful for ion-beam etching using an inert gas. For reactive ion-beam etching, a filamentless discharge is preferable. The electron cyclotron resonance-type cold cathode discharge is used for the high-current filamentless ion source. The electron cyclotron resonance (ECR) discharge is sustained in an rf electric field with a static magnetic field. The ECR conditions are given by

Eq. (7.3)
$$f = \frac{1}{2\pi}\frac{eB}{m}$$

where f denotes the frequency of the rf electric field, B the magnetic field strength, e and m, the electron charge and mass. For $f = 2.45$ GHz, B becomes 874 G.

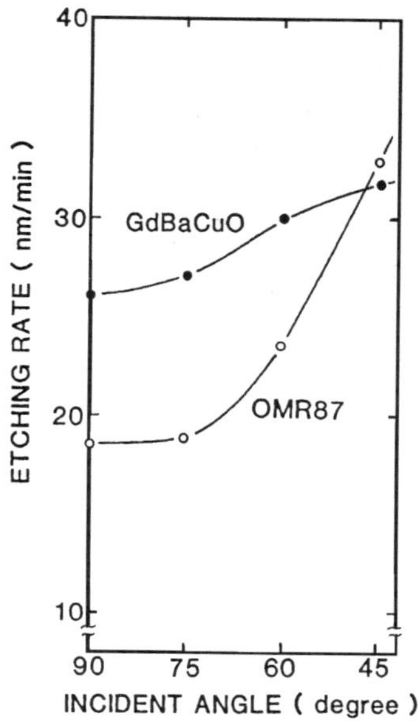

Figure 7.9. Etching rates as a function of incident angle of (•) Gd-Ba-Cu-O superconducting film and (o) negative-type photoresist, OMR87.

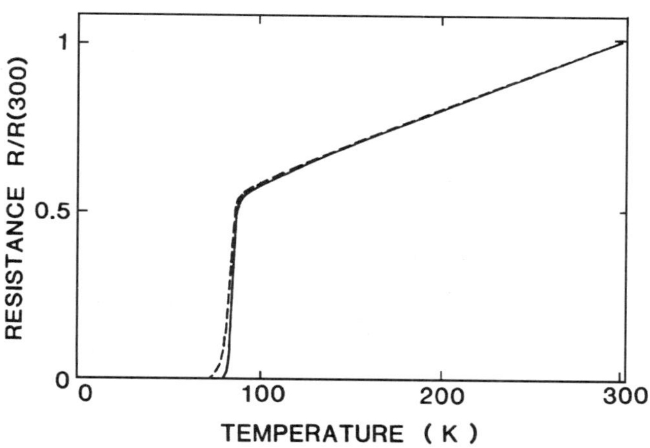

Figure 7.10. Temperature dependence of the resistance of Gd-Ba-Cu-O film before being patterned (*solid line*), and of the 2 μm wide strip line (*broken line*). The resistance is normalized to 300°K.

A typical ion-beam etching system is shown in Fig. 7.11. The system is not compact since it requires large water-cooled magnets. The operating pressure is as low as 10^{-5} torr, so the ions sputter the surface of the test pieces without any gas-phase collisions. This achieves the vertical etching and/or directional etching. A typical vertical etching structure is shown in Fig. 7.12.[10]

The inductively coupled rf ion-beam source is another type of filamentless ion-beam system. Figure 7.13 shows a diagram of and a photograph of an rf ion-beam source. This type of ion source is stable in a reactive gas. The diameter of the ion source ranges from 5 to 16 cm, ion-beam current is up to 1000 mA, and the beam energy ranges from 50 to 1500 V for commercially available ion sources. The linear ion source is also used for uniform processing of large-scale substrates. A photograph of a typical linear ion source is shown in Fig. 7.14.

The ion-beam sputtering system is also used for low-pressure sputter deposition, ion-assisted deposition, ion-beam deposition, and direct ion-beam deposition.[5]

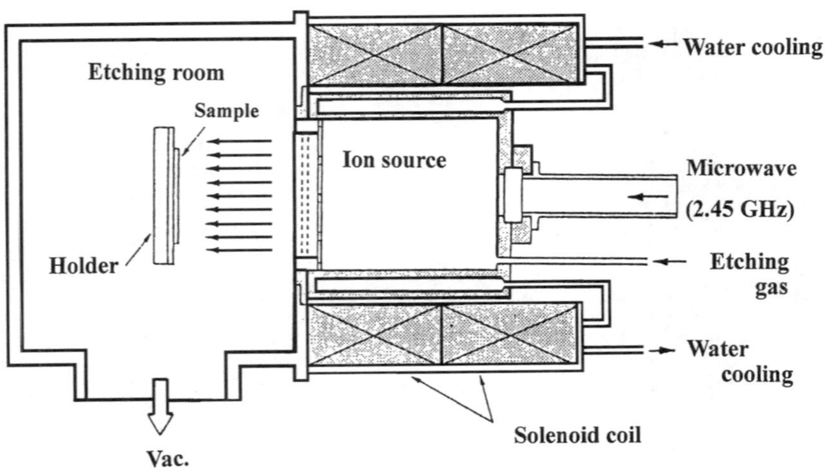

Figure 7.11. Diagram of a typical ECR ion-beam etching system.

Figure 7.12. Typical vertical structures of SiO_2 on Si, etched by an ECR ion-beam system. *(Courtesy of Anelva.)*

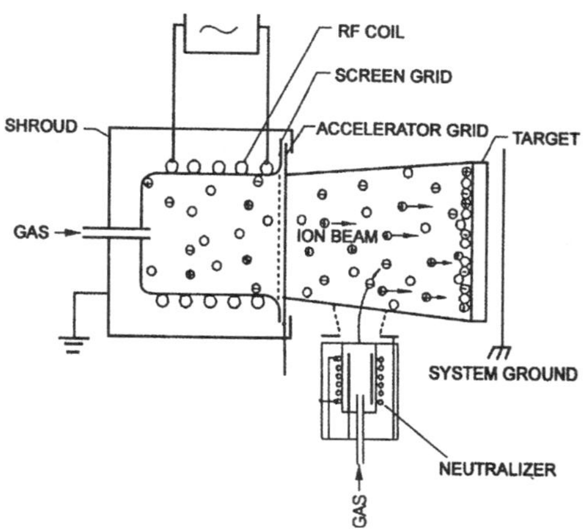

Figure 7.13. (*a*) Diagram of an rf ion source, and (*b*) a photograph of an ion source. *(Courtesy of Veeco.)*

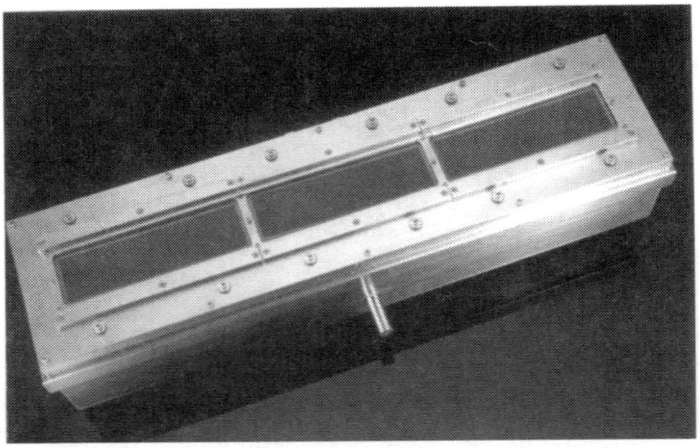

Figure 7.14. Photograph of a linear ion source. *(Courtesy of Veeco.)*

7.3 DIODE SPUTTER ETCHING

When the test samples are placed on the surface of the cathode in a diode sputtering system, the surfaces of the samples are etched by the incident ions. This kind of sputter etching system is very compact, although there are some drawbacks such as contamination of the samples due to residual gas and the temperature rise of the samples during sputter etching.

The diode sputter etching system is useful for RIE since a chemically reactive gas, such as oxygen or a halogen gas, can be used as the sputter gas. The current density of the Kaufman-type ion source is governed by the Child-Langmuir equation (space charge limited current, SCLC) as described in sec.7.2. In the diode sputter system, the ion-current density is more than 10 mA/cm^2.

Table 7.5 shows typical etching rates for various materials in diode RIE systems.[11] These etching systems are widely used for LSI microfabrication. Figure 7.15 shows a cross-sectional view of the submicron holes patterned by the diode etching system.

The RIE process produces microetching of chemically stable materials such as diamonds. Diamond is stable at temperatures as high as 1700°C in vacuum, and 600°C in air. The surface is conventionally etched by a molten potassium nitrate around 700°C–800°C. Reactive ion etching with oxygen results in surface etching even at room temperature. Figure 7.16 shows a typical etched surface of diamond (110) crystals.[12][13]

In the dry-etching process, the reacted products on the surface of the sample are vaporized. However, the products may remain on the etched surface when the reacted product is not volatile at the temperature of the sample. A critical example of this effect is the etching of Cu by CCl_2F_2.[14]

Low-temperature interdiffusion is often observed in a multilayered system during the plasma-etching process. Figure 7.17 shows a typical thickness distribution of compositions in a layered thin-film structure of Cu-thin-film/Cr-thin-film/glass-substrate after plasma etching by $CC1_2F_2$. Cr and/or Si out-diffuse onto the surface of the Cu even at a temperature of 120°C. Si may out-diffuse from the glass substrates.

Table 7.5. Diode RIE Etching Rates for Various Materials[11]

Materials	Etching rate (Å/min)	Etching ratio		Etching gas
		Materials/ Posiphotoresists	Materials/SiO_2	
Al	1,000	6	13	$BCl_3 + CCl_4$
Poly-Si	1,700	4	18	$CCl_4 + He + O_2$
Cr	200	1		$CCl_4 + O_2$
Mo	4,000	4	100	$CCl_4 + O_2$
GaAs	6,000	6		$CCl_4 + O_2$
SiO_2	600	3	9*	CHF_3
SiN_3	600	3	9*	CHF_3
PSG**	1,200	6	18*	CHF_3
TaN	120	0.5		$CF_4 + O_2$
TaSi	1,000	1		$CF_4 + O_2$
Ti	500	1	10	CF_4
WSi	2,000	2		$SF_6 + O_2$
InSb	300	1		Ar
Polyimide	2,000		100	O_2

* On Si (etching power 0.25 W/cm^2)
** 8 mol% P doped PSG

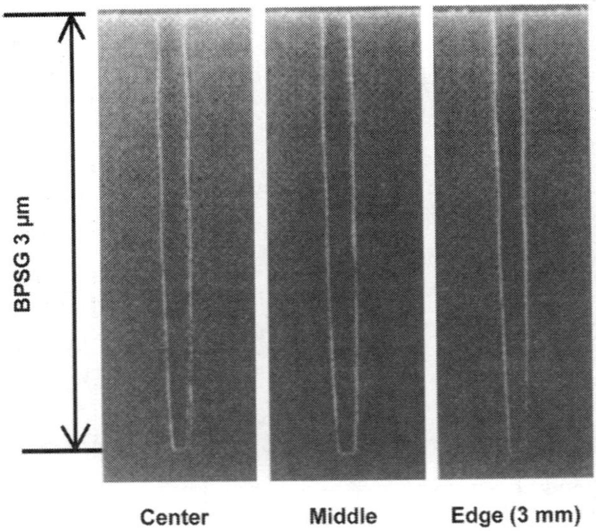

Figure 7.15. Narrow contact hole structures etched in SiO_2 on poly-Si using RIE in a 2-frequency diode sputter system with high density plasma.[22]

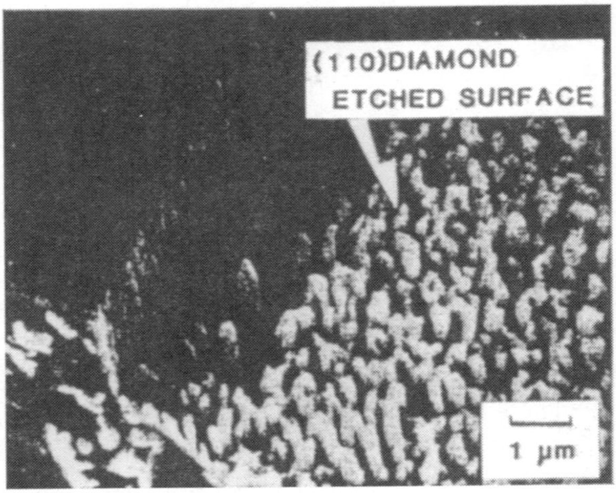

Figure 7.16. Surface of (110) diamond crystal etched by oxygen sputtering.

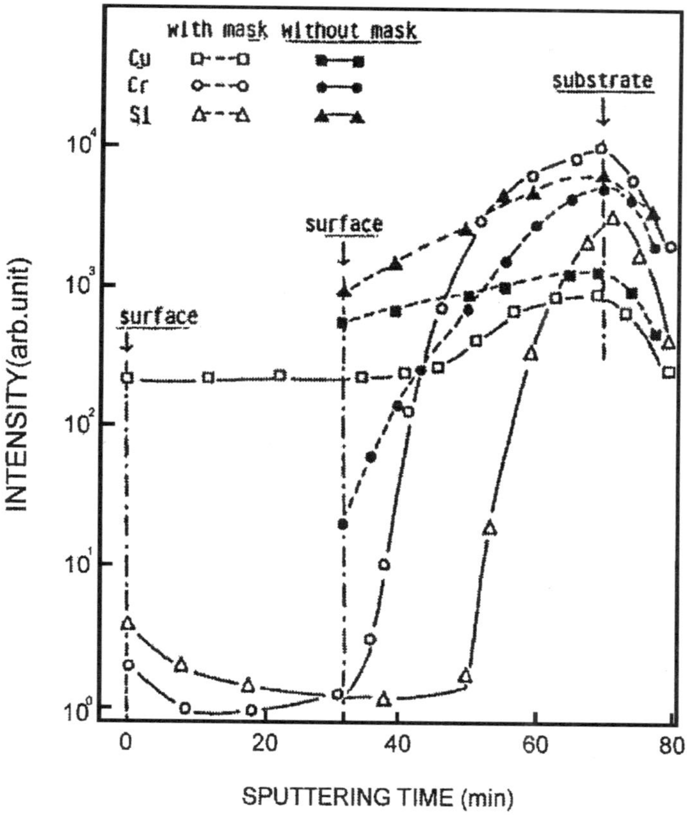

Figure 7.17. Depth profile of Cu/Cr/glass surface etched by CCl_2F_2 plasma.

ECR ion sources supply high-density plasmas. ECR discharge is also useful for high ion-current etching. However, the operating properties of the ECR system are strongly governed by the applied magnetic field. The stable operation of magnetically assisted plasma is guaranteed by the helicon plasma ion source. A helicon wave is present in the magnetically assisted plasma. The helicon wave propagates along the line of magnetic flux when electromagnetic energy is supplied to the magnetically assisted plasma. The helicon wave is observed in the plasma surrounding the ground. The electromagnetic energy of the helicon wave is transferred to electrons with high efficiency, and highly energetic electrons are generated. The highly energetic electrons generate a high-density plasma called helicon plasma. A diagram of the helicon plasma system is shown in Fig. 7.18. The antenna supplies 13.56 MHz of electromagnetic rf power. The magnetic field is needed to propagate the helicon wave in the plasma chamber. The cusp magnetic fields are superposed in order to confine the plasma. In the plasma chamber, the highly energetic electrons effectively ionize the reactive gas accompanied by radicals of the reactive gas. The highly ionized reactive gas and radicals enhance selectivity in the sputter-etching process. The typical ion-current density is 30 mA/cm^2 at a source power of 1 kW for an Ar plasma. The helicon plasma source also keeps the beam-like properties. The beam-like properties create sharp directional etching in the stack structure. The etching profile of the stack structure can be seen in Fig. 7.19.

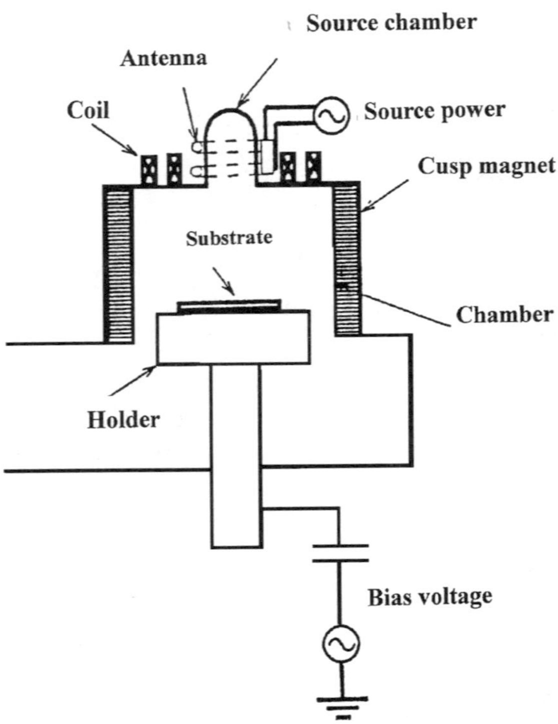

Figure 7.18. Construction of a helicon plasma system.[3]

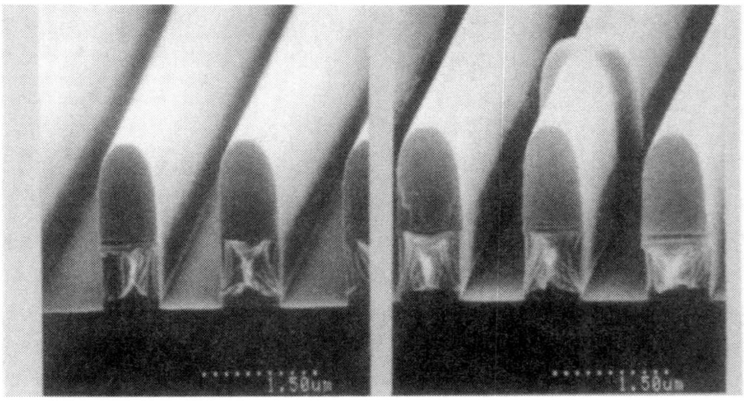

Figure 7.19. Structure of TiN/Al-Si-0.5%Cu/TiN/Ti stack etched by helicon plasma. *(Courtesy of Anelva.)*

7.4 DEPOSITION INTO DEEP-TRENCH STRUCTURES

Magnetron sputtering is widely used for the deposition of metal layers onto semiconductor wafers. The deposition process using a magnetron results in a very broad angular distribution of the depositing atoms, which is very useful for making planar films, but inappropriate for depositing films into deep structure. To overcome this fundamental problem, several directional sputtering technologies have been studied including collimated sputter deposition,[15] elevated temperature/reflow deposition,[16] long-throw sputter deposition,[17] and ionized magnetron sputter deposition.[18]

Among these sputtering technologies, long-throw sputter deposition is the leading candidate to overcome this fundamental problem. Typical step coverage deposited by the long-throw sputter process into the deep trench structure is shown in Fig. 7.20. Uniform step coverage can be seen at the center of the wafer in Fig. 7.20a. Nonuniform step coverage at the wafer edge is observed in Fig. 7.20b. Rossnagel proposed ionized long-throw magnetron sputtering to improve nonuniformity at the edge of the wafer.[19] If a small bias voltage is imposed on the wafer, the ionized metal ions from the plasma accelerate at normal incidence to the wafer surface resulting in the reduction of nonsymmetric deposition at the edge of the wafer. The results of a numerical model using the simulator SHADE are shown in Fig. 7.21.[20][21]

The nonuniformity at the wafer edge can also be improved by using a multitarget system. Figure 7.22 shows a model of one. Figure 7.23 shows the real structures of step coverage of Nb thin films deposited by a rotational target system and a single target system. The multitarget system shows potential for 0.1-mm Si technology with 300-mm Si wafers.

At present, the CVD process is widely used for deposition into the deep trench structure. However, a heavy environmental load is a by-product of the CVD process. The sputtering process, on the other hand, is essentially an environmentally benign production process.

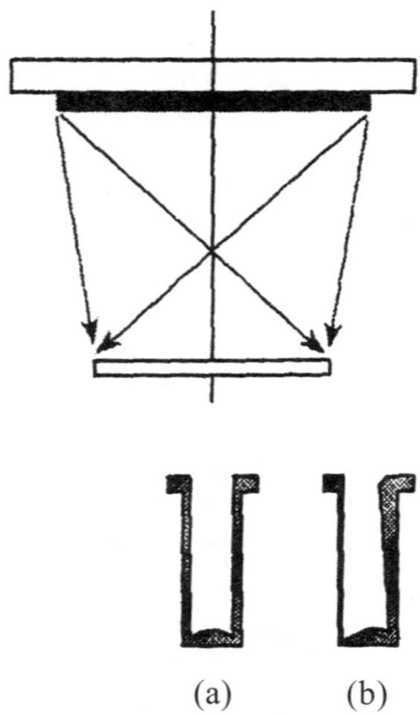

Figure 7.20. A schematic construction of the step coverage of a deep trench structure by a thin film deposited using a long-throw sputter system (*a*) at the center and (*b*) at the edge of the wafer.[23]

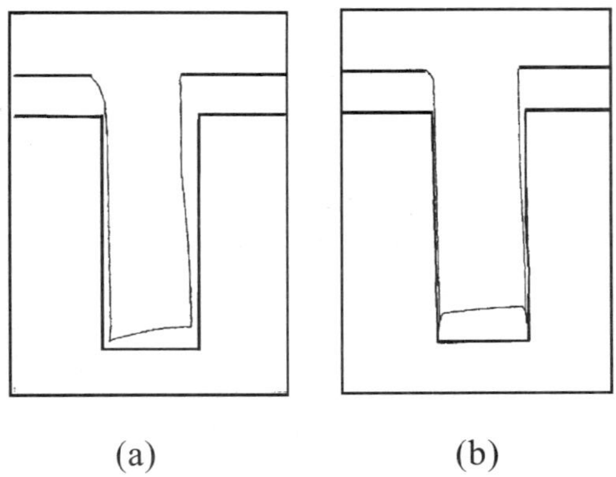

Figure 7.21. A schematic construction of the step coverage on a deep trench structure deposited by (*a*) conventional long-throw sputtering, and (*b*) an ionized long-throw sputter system.[20]

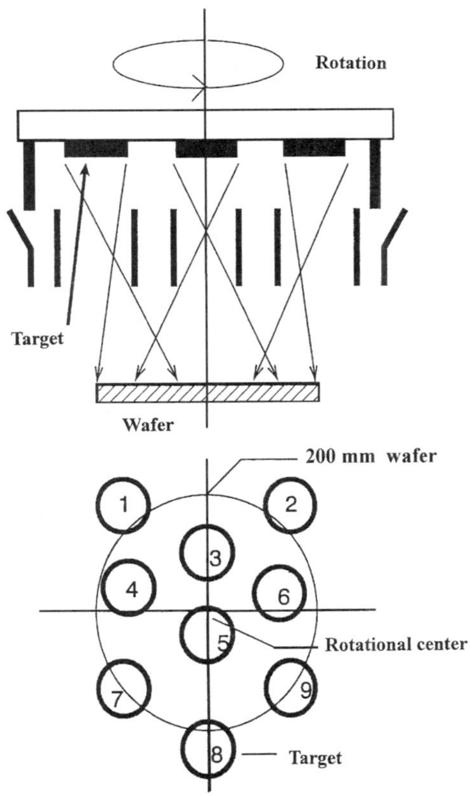

Figure 7.22. A schematic structure of a 9-multitarget long-throw sputtering system for 200 mm wafers.[23]

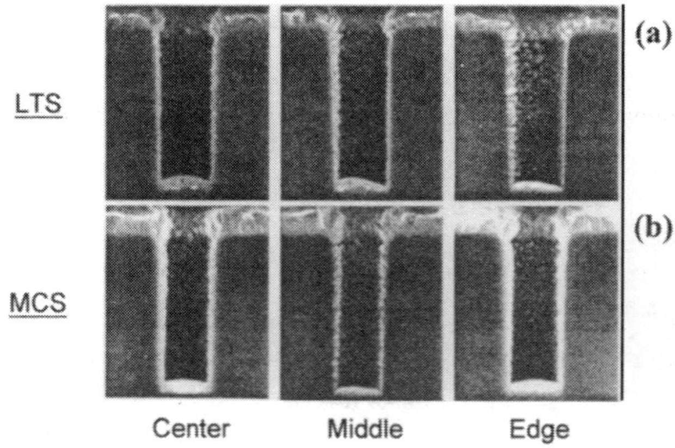

Figure 7.23. Cross-sectional images of the step coverage on deep trench structures by Nb thin films deposited using (*a*) conventional single long-throw sputtering cathode, and (*b*) a 9-multitarget long-throw sputter cathode.[23]

REFERENCES

1. Bunshah, R. F., (ed.), *Deposition Technologies for Films and Coatings*, Noyes Publ., NJ (1982)
2. Vossen, J. L., and Kern, W., (eds.), *Thin Film Processes*, Academic Press (1978)
3. Tsukada, T., private communication (Nov. 2002)
4. Wilson, R. G., and Brewer, G. R., *Ion Beams*, John Wiley & Sons, New York (1973)
5. Heem, J. E., *Vac. Technol. and Coating*, p. 46 (Sep. 2002)
6. Chapman, B., *Glow Discharge Processes*, John Wiley & Sons, New York (1980)
7. Kawaguchi, T., Adachi, H., Setsune, K., Yamazaki, O., and Wasa, K., *Applied Optics,* 23:2187 (1984)
8. Enokihara, A., Higashino, H., Setsune, K., Mitsuyu, T., and Wasa, K., *Jpn. J. Appl. Phys.*, 27:L1521 (1988)
9. Higashino, K., Enokihara, A., Mizuno, K., Mitsuyu, T., Setsune, K., and Wasa, K., *Proc. FED HiTcSc-ED Workshop,* p. 267, Miyagi-ZaO (Jun. 1988)

10. Anelva catalog, No. 154-14J.
11. Tsuzuki, K., Tabuchi, S., Banno, T., Kinbara, A., Nakagawa, Y., and Tsukada, T., *J. Vac. Sci. Technol.*, A11:672 (1993)
12. Wasa, K., *National Tech. Rept.*, 22:836 (1976)
13. Hayakawa, S., and Wasa, K., *Thin Film Technology*, p. 67, Kyoritsu, Tokyo (1982)
14. Kawaguchi, T., and Wasa, K., *Oyo Buturi,* 51:856 (1982)
15. Rosssnagel, S. M., Kinoshita, H., and Cuomo, J., *J. Vac. Sci. Technol.*, A9:261 (1991)
16. Ogawa, S., Nishimura, H., Kouzaki, T., and Sinclair, R., Advanced Metallization for ULSI Technologies (T. S. Cale and F. S. Pintchovski, eds.), *Mater. Res. Soc.*, p. 325, Pittsburgh (1992)
17. Cuomo, J., and Rossnagel, S. M., *J. Vac. Sci. Technol.*, A4:393 (1986)
18. Rossnagel, S. M., and Hopwood, J., *Appl. Phys. Lett.*, 63:3285 (1993)
19. Mayo, A. A., Hamaguchi, S., Joo, J. H., and Rossnagel, S. M., *J. Vac. Sci. Technol.*, B15:1788 (1996)
20. Hamaguchi, S., and Rossnagel, S. M., *J. Vac. Sci. Technol.*, B14:2603 (1996)
21. Rossnagel, S. M., *J. Vac. Sci. Technol.*, B16:2585 (1998)
22. Technical Review, Anelva, 7:21 (2000)
23. Technical Journal, ULVAC, No. 51 (1999)

Appendix

Table A.1. Electric Units, Their Symbols and Conversion Factors

Physical quantity	Name of electro-magnetic unit	Name of practical unit	Name of MKS unit	MKS unit (electro-magnetic unit*)	MKS unit (practical unit*)
Force	dyn	joule/cm	newton (joule/m)	10^5	$1/10^2$
Energy	erg	joule	joule	10^7	1
Power	erg/s	watt	watt	10^7	1
Potential (V)		volt	volt		1
Electromotive force (E)		volt/cm	volt/m		$1/10^2$
Current (I)		ampere	ampere		1
Resistance (R)		ohm	ohm		1
Resistivity (ρ)		ohm-cm	ohm-m		10^2
Conductance (G)		mho	mho (Siemens)		1
Conductivity (k)		mho/cm	mho/m		$1/10^2$
Electric charge, (q, Q)		coulomb	coulomb		1
Electric displacement (D)			coulomb/m^2		$4\pi/10^4$
Electric polarization (P)			coulomb/m^2		$1/10^4$
Capacitance (C)		farad	farad		1
Permitivity of free space (ε_0)			farad/m		$4\pi/10^9$

(cont'd.)

Table A.1. *(cont'd.)*

Physical quantity	Name of electro-magnetic unit	Name of practical unit	Name of MKS unit	MKS unit (electro-magnetic unit*)	MKS unit (practical unit*)
Magnetomotive force	gilbert	ampere-turn	ampere-turn	$4\pi/10$	1
Magnetic field strength (H)	oersted	ampere-turn/cm	ampere-turn/m	$4\pi/10^3$	$1/10^2$
Magnetic flux (Φ)	maxwell	maxwell	weber	10^8	10^8
Magnetic flux density (B)	gauss	gauss	weber/m² (tesla)	10^4	10^4
Intensity of magnetization			weber/m²	$10^4/4\pi$	$10^4/4\pi$
Magnetic charge (M)			weber	$10^3/4\pi$	$10^8/4\pi$
Magnetic moment (I)			weber-m	$10^{10}/4\pi$	$10^{10}/4\pi$
Inductance (L)		henry	henry	10^9	1
Magnetoresistance			ampere-turn/weber	$4\pi/10^9$	$4\pi/10^9$
Permeability of free space (μ_0)			henry/m	$10^7/4\pi$	$10^7/4\pi$
Magnetic susceptibility (x)			henry/m	$10^7/(4\pi)^2$	$10^7/(4\pi)^2$

* Example (force): 1 dyn = 10^{-5} N = 10^2 joule/cm

Table A.2. Fundamental Physical Constants

Velocity of light	$c = 2.998 \times 10^{8}$ m· sec $^{-1}$, – × 10^{10} cm ·sec^{-1}
Electron rest mass	$m_c = 9.11 \times 10^{-31}$ kg, – × 10^{-28} g
Proton rest mass	$m_p = 1.67 \times 10^{-27}$ kg, – × 10^{-24} g
Charge of electron	$e = 1.602 \times 10^{-19}$ C, 4.8×10^{-10} esu, 1.602×10^{-20} emu
Charge-to-mass ratio of electron	$e/m_c = 1.76 \times 10^{11}$ C·kg^{-1}, 5.27×10^{17} esu·g^{-1}, 1.76×10^{7} emu·g^{-1}
Planck's constant	$h = 6.626 \times 10^{-34}$ J·sec, – × 10^{-27} erg·sec, $h/2\pi = \hbar = 1.055 \times 10^{-34}$ J·sec, – × 10^{-27} erg·sec
Bohr radius	$= 5.29 \times 10^{-11}$ m, – × 10^{-9} cm
Bohr magneton	$= 9.27 \times 10^{-24}$ J·tesla^{-1}, – × 10^{-21} erg·gauss^{-1}
Magnetic moment	
Electron	$= 9.28 \times 10^{-24}$ J·tesla^{-1}, – × 10^{-21} erg·gauss^{-1}
Proton	$= 1.41 \times 10^{-26}$ J·tesla^{-1}, – × 10^{-23} erg·gauss^{-1}
Boltzmann's constant	$k = 1.38 \times 10^{-23}$ J·K^{-1}, – × 10^{-16} erg·K^{-1}, = 8.615×10^{-5} eV·K^{-1}
Avogadro's number	$= 6.02 \times 10^{23}$ mole^{-1}
Volume of 1 mole of an ideal gas	$= 2.24 \times 10^{-2}$ m^{3}·mole^{-1}, – × 10^{4} cm^{3}·mole^{-1}
Loschmidt's constant	$= 2.69 \times 10^{25}$ m^{-3}, – × 10^{19} cm^{-3}
Gas constant	$R = 8.31$ J·mole^{-1}·K^{-1}, – × 10^{7} erg·mole^{-1}·K^{-1}, 1.99 cal·mole^{-1}·K^{-1}
Faraday's constant	$= 9.65 \times 10^{4}$ C·mole^{-1}, 2.89×10^{14} esu·mole $^{-1}$, 9.65×10^{3} emu·mole^{-1}
Permittivity of free space	$\varepsilon_0 = 8.854 \times 10^{-12}$ C·V^{-1}·m^{-1}
Permeability of free space	$\mu_0 = 1.26 \times 10^{-6}$ H·m^{-1}

Table A.2. *(cont'd.)*

Standard value of the acceleration of gravity	$g = 6.67 \times 10^{11}$ N·m²·kg⁻², – × 10^{-8} dyn·cm²·g⁻²
Length	1 μ = 10^{-6} m, 10^{-4} cm
	1 Å = 10^{-10} m, 10^{-3} cm
	1 mil = 2.54×10^{-5} m, – × 10^{-3} cm
Work	1 dyn = 10^{-5} N
	1 g·wt = 9.81×10^{-3} N, – × 10^{2} dyn
Atmospheric pressure	1 torr = 1 mm Hg, 1.33 mbar, 1.33×10^{2} Pa
	1 atm = 760 torr, 1013 mbar
Temperature	0°C = 273.15 K
Energy	1 erg = 10^{-7} J
	1 cal = 4.18 J
1 eV	Wavelength = 1.24×10^{-6} m, – × 10^{-4} cm
	Frequency = 2.42×10^{14} sec⁻¹
	Wave number = 8.07×10^{5} m⁻¹, – × 10^{3} cm⁻¹
	Energy = 1.60×10^{-19} J, – × 10^{-12} erg
	Temperature = 1.16×10^{4} K

List of Acronyms

ADP	$NH_4H_2PO_4$
AES	Auger electron spectroscopy
AF	antiferro
AFM	atomic force microscope
ALOS	artificially made layered-oxide superconductor
ARE	activated reactive evaporation
a-Si	amorphous silicon
BAW	bulk acoustic wave
bcc	body-centered cubic
BCS	Bardeen-Cooper-Schrieffer
BLO	$Bi_{12}PbO_{19}$
BGO	$Bi_{12}GeO_{20}$
BT	$BaTiO_3$
BTO	$Bi_{12}TiO_{20}$
CVD	chemical vapor deposition
DLC	diamond-like carbon
DP	diffusion pump
DRAM	dynamic random access memory
ECR	electron cyclotron resonance
EDX (EDAX)	energy dispersive x-ray analysis

EELS	electron energy loss spectroscopy
EL	electroluminescence
EO	electro-optical
EOT	equivalent oxide thickness
EPMA	electron probe microanalysis
ESCA	electron spectroscopy for chemical analysis
EXFAS	extended x-ray absorption fine structure
fcc	face-centered cubic
FEDRAM	ferroelectric dynamic random access memory
FEM	finite element method
FEMFET	ferroelectric memory field effect transistor
FESRAM	ferroelectric static random access memory
FET	field effect transistor
FM	ferromagnetic
FWHM	full-width, half maximum (full width at one-half maximum height)
GMR	giant magnetoresistance
hcp	hexagonal close packed
HSP	hemispherical sputtering
IBE	ion beam etching
IBS	ion beam sputter
ICBD	ionized cluster beam deposition
ICP	inductively coupled plasma spectroscopy
IDE	interdigital electrodes
IDT	interdigital transducer
IP	ion plating
KDP	KH_2PO_4
LA	laser ablation
LAN	local area network
LCD	liquid crystal display
LD	laser diode
LED	light-emitting diode
LFS	laser-induced fluorescence spectroscopy
LN	$LiNbO_3$

LSI	large-scale integration
LT	$LiTaO_3$
MBE	molecular beam epitaxy
MEMS	microelectromechanical system
MF	medium frequency
MOCVD	metal-organic chemical vapor deposition
MOD	metal-organic deposition (chemical solvent deposition)
MOS	metal oxide semiconductor
MR	magnetoresistance
MPRI	multiphoton resonance ionization
MRDRAM	magnetoresistance dynamic random access memory
MSP	magnetron sputtering
NSC	negative space-charge
NTSC	National Television System Committee
PACVD	plasma-assisted chemical vapor deposition
PC	polycrystal; polycrystalline
P-CVD	plasma CVD
PDP	plasma display
PIXE	proton-induced x-ray emission
PL	photoluminescence
PLT	(Pb, La) TiO_3
PLD	pulsed laser deposition
PLZT	(Pb, La)(Zr, Ti)O_3
PMN	Pb($Mg_{1/3}$, $Nb_{2/3}$) O_3
P_s	spontaneous polarization
PSC	positive space-charge
PT	$PbTiO_3$
PVD	physical vapor deposition
PZT	Pb (Zr, Ti) O_3
QCOM	quartz crystal oscillator microbalance
RBS	Rutherford backscattering spectroscopy
RED	reflection electron diffraction

REM	reactive evaporation method
RHEED	reflection high-energy electron diffraction
RIBE	reactive ion-beam etching
RIE	reactive ion etching
SAD	selected area diffraction
SAW	surface acoustic wave
SBT	$SrBi_2Ta_2O_9$
SC	single crystal
SCLC	space-charge-limited current
SEM	scanning electron microscope
SIMS	secondary ion mass spectroscopy
SP	sputter; diode sputtering
SQUID	superconducting quantum interference device
SRO	$SrRuO_3$
ST	$SrTiO_3$
STM	scanning tunneling microscope
TCD	temperature coefficient of delay time
TCR	temperature coeffiecient of resistance
TEM	transmission electron microscope
TFT	thin film transistor
TIR	total internal reflection
TMR	tunneling magnetoresistance
TSP	triode sputtering
ULSI	ultra large scale integration
VE	vacuum evaporation
VIF	video intermediate frequency
VLSI	very large scale integration
XMA	x-ray microanalysis
XPS	x-ray photoelectron spectroscopy
XRD	x-ray diffraction

Index

A

B

C

D

F

G

H

I

J

K

L

M

N

O

P

T

Z